A First Course in
QUALITY ENGINEERING

Integrating Statistical and Management Methods of Quality

Second Edition

A First Course in QUALITY ENGINEERING

Integrating Statistical and Management Methods of Quality

Second Edition

K.S. Krishnamoorthi

with V. Ram Krishnamoorthi

CRC Press is an imprint of the
Taylor & Francis Group, an **informa** business

CRC Press
Taylor & Francis Group
6000 Broken Sound Parkway NW, Suite 300
Boca Raton, FL 33487-2742

Printed on acid-free paper
Version Date: 20150807

International Standard Book Number-13: 978-1-4398-4034-4 (Hardback)

Library of Congress Cataloging-in-Publication Data

Krishnamoorthi, K. S.
A first course in quality engineering : integrating statistical and management methods of quality. -- 2nd ed. / K.S. Krishnamoorthi, V. Ram Krishnamoorthi.
p. cm.
"A CRC title."
Includes bibliographical references and index.
ISBN 978-1-4398-4034-4 (hardback)
1. Quality control. 2. Process control--Statistical methods. I. Krishnamoorthi, V. Ram. II. Title.

TS156.8.K75 2011
658.5'62--dc23 2011032462

Visit the Taylor & Francis Web site at
http://www.taylorandfrancis.com

and the CRC Press Web site at
http://www.crcpress.com

Contents

Preface to the Second Edition xv

Preface to the First Edition xvii

Authors xxi

1. Introduction to Quality 1
1.1 An Historical Overview 1
1.1.1 A Note about "Quality Engineering" 7
1.2 Defining Quality 9
1.2.1 Product Quality vs. Service Quality 11
1.3 The Total Quality System 11
1.4 Total Quality Management 13
1.5 Economics of Quality 15
1.6 Quality, Productivity, and Competitive Position 16
1.7 Quality Costs 17
1.7.1 Categories of Quality Costs 18
1.7.1.1 Prevention Cost 18
1.7.1.2 Appraisal Cost 20
1.7.1.3 Internal Failure Cost 20
1.7.1.4 External Failure Cost 21
1.7.2 Steps in Making a Quality Cost Study 21
1.7.3 Projects Arising from a Quality Cost Study 26
1.7.4 Quality Cost Scoreboard 27
1.7.5 Quality Costs Not Included in the TQC 29
1.7.6 Relationship among Quality Cost Categories 31
1.7.7 Summary of Quality Costs 32
1.7.8 A Case Study in Quality Costs 32
1.8 Success Stories 39
1.9 Exercise 39
1.9.1 Practice Problems 39
1.9.2 Mini-Projects 43
References 46

2. Statistics for Quality 49
2.1 Variability in Populations 49
2.2 Some Definitions 50
2.2.1 The Population and a Sample 50
2.2.2 Two Types of Data 51
2.3 Quality vs. Variability 52

2.4 Empirical Methods for Describing Populations 53
2.4.1 The Frequency Distribution 53
2.4.1.1 The Histogram 53
2.4.1.2 The Cumulative Frequency Distribution 54
2.4.2 Numerical Methods for Describing Populations 59
2.4.2.1 Calculating the Average and Standard Deviation 61
2.4.3 Other Graphical Methods 61
2.4.3.1 Stem-and-Leaf Diagram 61
2.4.3.2 Box-and-Whisker Plot 63
2.4.4 Other Numerical Measures 64
2.4.4.1 Measures of Location 64
2.4.4.2 Measures of Dispersion 65
2.4.5 Exercise in Empirical Methods 65
2.5 Mathematical Models for Describing Populations 67
2.5.1 Probability 67
2.5.1.1 Definition of Probability 68
2.5.1.2 Computing the Probability of an Event 69
2.5.1.3 Theorems on Probability 72
2.5.1.4 Counting the Sample Points in a Sample Space 80
2.5.2 Exercise in Probability 85
2.5.3 Probability Distributions 87
2.5.3.1 Random Variable 87
2.5.3.2 Probability Mass Function 89
2.5.3.3 Probability Density Function 91
2.5.3.4 The Cumulative Distribution Function 92
2.5.3.5 The Mean and Variance of a Distribution 93
2.5.4 Some Important Probability Distributions 96
2.5.4.1 The Binomial Distribution 96
2.5.4.2 The Poisson Distribution 99
2.5.4.3 The Normal Distribution 101
2.5.4.4 Distribution of the Sample Average $\overline{X}$ 109
2.5.4.5 The Central Limit Theorem 110
2.5.5 Exercise in Probability Distributions 111
2.6 Inference of Population Quality from a Sample 113
2.6.1 Definitions 114
2.6.2 Confidence Intervals 115
2.6.2.1 CI for the μ of a Normal Population When σ Is Known 115
2.6.2.2 Interpretation of CI 116
2.6.2.3 CI for μ When σ Is Not Known 117
2.6.2.4 CI for σ^2 of a Normal Population 118
2.6.3 Hypothesis Testing 120
2.6.3.1 Test Concerning the Mean μ of a Normal Population When σ Is Known 121

2.6.3.2 Why Place the Claim Made about a Parameter in H_1? 123
2.6.3.3 The Three Possible Alternate Hypotheses 124
2.6.3.4 Test Concerning the Mean μ of a Normal Population When σ Is Not Known 125
2.6.3.5 Test for Difference of Two Means When σs Are Known 127
2.6.4 Tests for Normality 129
2.6.4.1 Use of the Normal Probability Plot 129
2.6.4.2 Normal Probability Plot on the Computer 130
2.6.4.3 A Goodness-of-Fit Test 133
2.6.5 The *P*-Value 134
2.6.6 Exercise in Inference Methods 135
2.6.6.1 Confidence Intervals 135
2.6.6.2 Hypothesis Testing 136
2.6.6.3 Goodness-of-Fit Test 137
2.7 Mini-Projects 137
References 142

3. Quality in Design 143
3.1 Planning for Quality 143
3.1.1 The Product Creation Cycle 143
3.2 Product Planning 144
3.2.1 Finding Customer Needs 145
3.2.1.1 Customer Survey 146
3.2.2 Quality Function Deployment 150
3.2.2.1 Customer Requirements and Design Features 152
3.2.2.2 Prioritizing Design Features 153
3.2.2.3 Choosing a Competitor as Benchmark 154
3.2.2.4 Targets 154
3.2.3 Reliability Fundamentals 155
3.2.3.1 Definition of Reliability 156
3.2.3.2 Hazard Function 156
3.2.3.3 The Bathtub Curve 158
3.2.3.4 Distribution of Product Life 161
3.2.3.5 The Exponential Distribution 161
3.2.3.6 Mean Time to Failure 162
3.2.3.7 Reliability Engineering 165
3.3 Product Design 166
3.3.1 Parameter Design 166
3.3.2 Design of Experiments 167
3.3.2.1 2^2 Factorial Design 168
3.3.2.2 Randomization 170
3.3.2.3 Experimental Results 171
3.3.2.4 Calculating the Factor Effects 172

3.3.2.5 Main Effects 173
3.3.2.6 Interaction Effects 174
3.3.2.7 A Shortcut for Calculating Effects 175
3.3.2.8 Determining the Significance of Effects 175
3.3.2.9 The 2^3 Design 178
3.3.2.10 Interpretation of the Results 182
3.3.2.11 Model Building 183
3.3.2.12 Taguchi Designs 184
3.3.3 Tolerance Design 185
3.3.3.1 Traditional Approaches 185
3.3.3.2 Tolerancing According to Dr. Taguchi 186
3.3.3.3 Assembly Tolerances 187
3.3.3.4 The RSS Formula 188
3.3.3.5 Natural Tolerance Limits 191
3.3.4 Failure Mode and Effects Analysis 191
3.3.5 Concurrent Engineering 195
3.3.5.1 Design for Manufacturability/Assembly 196
3.3.5.2 Design Reviews 197
3.4 Process Design 197
3.4.1 The Process Flow Chart 198
3.4.2 Process Parameter Selection: Experiments 200
3.4.3 Floor Plan Layout 204
3.4.4 Process FMEA 205
3.4.5 Process Control Plan 205
3.4.6 Other Process Plans 205
3.4.6.1 Process Instructions 205
3.4.6.2 Packaging Standards 207
3.4.6.3 Preliminary Process Capabilities 207
3.4.6.4 Product and Process Validation 207
3.4.6.5 Process Capability Results 208
3.4.6.6 Measurement System Analysis 208
3.4.6.7 Product/Process Approval 208
3.4.6.8 Feedback, Assessment, and Corrective Action ... 208
3.5 Exercise 209
3.5.1 Practice Problems 209
3.5.2 Mini-Projects 212
References 214

4. Quality in Production—Process Control I 217
4.1 Process Control 217
4.2 The Control Charts 218
4.2.1 A Typical Control Chart 219
4.2.2 Two Types of Data 221
4.3 Measurement Control Charts 221
4.3.1 $\overline{X}$- and R-Charts 222

4.3.2 A Few Notes about the $\overline{X}$- and R-Charts ... 228
4.3.2.1 The Many Uses of the Charts ... 228
4.3.2.2 Selecting the Variable for Charting ... 229
4.3.2.3 Preparing Instruments ... 230
4.3.2.4 Preparing Check Sheets ... 230
4.3.2.5 False Alarm in the $\overline{X}$-Chart ... 231
4.3.2.6 Determining Sample Size ... 231
4.3.2.7 Why 3-Sigma Limits? ... 232
4.3.2.8 Frequency of Sampling ... 233
4.3.2.9 Rational Subgrouping ... 233
4.3.2.10 When the Sample Size Changes for $\overline{X}$- and R-Charts ... 235
4.3.2.11 Improving the Sensitivity of the $\overline{X}$-Chart ... 236
4.3.2.12 Increasing the Sample Size ... 236
4.3.2.13 Use of Warning Limits ... 238
4.3.2.14 Use of Runs ... 238
4.3.2.15 Patterns in Control Charts ... 240
4.3.2.16 Control vs. Capability ... 240
4.3.3 $\overline{X}$ and S-Charts ... 242
4.4 Attribute Control Charts ... 245
4.4.1 The P-Chart ... 245
4.4.2 The C-Chart ... 248
4.4.3 Some Special Attribute Control Charts ... 251
4.4.3.1 The P-Chart with Varying Sample Sizes ... 251
4.4.3.2 The nP-Chart ... 254
4.4.3.3 The Percent Defectives Chart (100P-Chart) ... 255
4.4.3.4 The U-Chart ... 255
4.4.4 A Few Notes about the Attribute Control Charts ... 258
4.4.4.1 Meaning of the LCL on the P- or C-Chart ... 258
4.4.4.2 P-Chart for Many Characteristics ... 259
4.4.4.3 Use of Runs ... 259
4.4.4.4 Rational Subgrouping ... 259
4.5 Summary on Control Charts ... 262
4.5.1 Implementing SPC on Processes ... 263
4.6 Process Capability ... 268
4.6.1 Capability of a Process with Measurable Output ... 268
4.6.2 Capability Indices C_p and C_{pk} ... 269
4.6.3 Capability of a Process with Attribute Output ... 274
4.7 Measurement System Analysis ... 275
4.7.1 Properties of Instruments ... 275
4.7.2 Measurement Standards ... 277
4.7.3 Evaluating an Instrument ... 279
4.7.3.1 Properties of a Good Instrument ... 279
4.7.3.2 Evaluation Methods ... 279
4.7.3.3 Resolution ... 280

4.7.3.4 Bias 283
4.7.3.5 Variability (Precision) 284
4.7.3.6 A Quick Check of Instrument Adequacy 287
4.8 Exercise 289
4.8.1 Practice Problems 289
4.8.2 Mini-Projects 293
References 296

5. Quality in Production—Process Control II 297
5.1 Derivation of Limits 298
5.1.1 Limits for the $\overline{X}$-Chart 298
5.1.2 Limits for the *R*-Chart 301
5.1.3 Limits for the *P*-Chart 302
5.1.4 Limits for the *C*-Chart 303
5.2 Operating Characteristics of Control Charts 304
5.2.1 Operating Characteristics of an $\overline{X}$-Chart 304
5.2.1.1 Computing the OC Curve of an $\overline{X}$-Chart 305
5.2.2 OC Curve of an *R*-Chart 307
5.2.3 Average Run Length 309
5.2.4 OC Curve of a *P*-Chart 312
5.2.5 OC Curve of a *C*-Chart 313
5.3 Measurement Control Charts for Special Situations 315
5.3.1 $\overline{X}$- and *R*-Charts When Standards for μ and/or σ are Given 315
5.3.1.1 Case I: μ Given, σ Not Given 316
5.3.1.2 Case II: μ and σ Given 316
5.3.2 Control Charts for Slow Processes 319
5.3.2.1 Control Chart for Individuals (*X*-Chart) 320
5.3.2.2 Moving Average and MR Charts 322
5.3.2.3 Notes on Moving Average and Moving Range Charts 324
5.3.3 The Exponentially Weighted Moving Average Chart 326
5.3.3.1 Limits for the EWMA Chart 331
5.3.4 Control Charts for Short Runs 333
5.3.4.1 The DNOM Chart 333
5.3.4.2 The Standardized DNOM Chart 335
5.4 Topics in Process Capability 339
5.4.1 The C_{pm} Index 340
5.4.2 Comparison of C_p, C_{pk}, and C_{pm} 341
5.4.3 Confidence Interval for Capability Indices 342
5.4.4 Motorola's 6σ Capability 344
5.5 Topics in the Design of Experiments 349
5.5.1 Analysis of Variance 349
5.5.2 The General 2^k Design 355
5.5.3 The 2^4 Design 356

5.5.4 2^k Designs with Single Trial 357
5.5.5 Fractional Factorials: One-Half Fractions 359
5.5.5.1 Generating the One-Half Fraction 361
5.5.5.2 Calculating the Effects 361
5.5.6 Resolution of a Design 362
5.6 Exercise 369
5.6.1 Practice Problems 369
5.6.2 Mini-Projects 372
References 373

6. Managing for Quality 375
6.1 Managing Human Resources 375
6.1.1 Importance of Human Resources 375
6.1.2 Organizations 376
6.1.2.1 Organization Structures 376
6.1.2.2 Organizational Culture 378
6.1.3 Quality Leadership 380
6.1.3.1 Characteristics of a Good Leader 380
6.1.4 Customer Focus 381
6.1.5 Open Communications 383
6.1.6 Empowerment 385
6.1.7 Education and Training 387
6.1.7.1 Need for Training 387
6.1.7.2 Benefits from Training 388
6.1.7.3 Planning for Training 388
6.1.7.4 Training Methodology 390
6.1.7.5 Finding Resources 391
6.1.7.6 Evaluating Training Effectiveness 392
6.1.8 Teamwork 392
6.1.8.1 Team Building 393
6.1.8.2 Selecting Team Members 393
6.1.8.3 Defining the Team Mission 393
6.1.8.4 Taking Stock of the Team's Strength 394
6.1.8.5 Building the Team 394
6.1.8.6 Basic Training for Quality Teams 395
6.1.8.7 Desirable Characteristics among Team Members 396
6.1.8.8 Why a Team? 397
6.1.8.9 Ground Rules for Running a Team Meeting 397
6.1.8.10 Making the Teams Work 398
6.1.8.11 Different Types of Teams 399
6.1.8.12 Quality Circles 400
6.1.9 Motivation Methods 401
6.1.10 Principles of Management 402
6.2 Strategic Planning for Quality 402
6.2.1 History of Planning 402

6.2.2 Making the Strategic Plan 404
6.2.3 Strategic Plan Deployment 405
6.3 Exercise 407
6.3.1 Practice Problems 407
6.3.2 Mini-Project 408
References 408

7. Quality in Procurement 409
7.1 Importance of Quality in Supplies 409
7.2 Establishing a Good Supplier Relationship 410
7.2.1 Essentials of a Good Supplier Relationship 410
7.3 Choosing and Certifying Suppliers 411
7.3.1 Single vs. Multiple Suppliers 411
7.3.2 Choosing a Supplier 412
7.3.3 Certifying a Supplier 413
7.4 Specifying the Supplies Completely 414
7.5 Auditing the Supplier 415
7.6 Supply Chain Optimization 416
7.6.1 The Trilogy of Supplier Relationship 417
7.6.2 Planning 417
7.6.3 Control 418
7.6.4 Improvement 419
7.7 Using Statistical Sampling for Acceptance 420
7.7.1 The Need for Sampling Inspection 420
7.7.2 Single Sampling Plans for Attributes 422
7.7.2.1 The Operating Characteristic Curve 422
7.7.2.2 Calculating the OC Curve of a Single Sampling Plan 423
7.7.2.3 Designing an SSP 426
7.7.2.4 Choosing a Suitable OC Curve 426
7.7.2.5 Choosing a Single Sampling Plan 428
7.7.3 Double Sampling Plans for Attributes 431
7.7.3.1 Why Use a DSP? 432
7.7.3.2 The OC Curve of a DSP 432
7.7.4 The Average Sample Number of a Sampling Plan 434
7.7.5 MIL-STD-105E (ANSI Z1.5) 436
7.7.5.1 Selecting a Sampling Plan from MIL-STD-105E 438
7.7.6 Average Outgoing Quality Limit 447
7.7.7 Some Notes about Sampling Plans 451
7.7.7.1 What Is a Good AQL? 451
7.7.7.2 Available Choices for AQL Values in the MIL-STD-105E 451
7.7.7.3 A Common Misconception about Sampling Plans 452

7.7.7.4 Sampling Plans vs. Control Charts 452
7.7.7.5 Variable Sampling Plans 452
7.8 Exercise 453
References 454

8. Continuous Improvement of Quality 457
8.1 The Need for Continuous Improvement 457
8.2 The Problem-Solving Methodology 458
8.2.1 Deming's PDCA Cycle 458
8.2.2 Juran's Breakthrough Sequence 459
8.2.3 The Generic Problem-Solving Methodology 461
8.3 Quality Improvement Tools 464
8.3.1 Cause-and-Effect Diagram 465
8.3.2 Brainstorming 466
8.3.3 Benchmarking 467
8.3.4 Pareto Analysis 471
8.3.5 Histogram 472
8.3.6 Control Charts 474
8.3.7 Scatter Plots 476
8.3.8 Regression Analysis 479
8.3.8.1 Simple Linear Regression 479
8.3.8.2 Model Adequacy 481
8.3.8.3 Test of Significance 482
8.3.8.4 Multiple Linear Regression 487
8.3.8.5 Nonlinear Regression 489
8.3.9 Correlation Analysis 490
8.3.9.1 Significance in Correlation 491
8.4 Lean Manufacturing 493
8.4.1 Quality Control 496
8.4.2 Quantity Control 497
8.4.3 Waste and Cost Control 498
8.4.4 Total Productive Maintenance 499
8.4.5 Stable, Standardized Processes 500
8.4.6 Visual Management 500
8.4.7 Leveling and Balancing 502
8.4.8 The Lean Culture 504
8.5 Exercise 504
8.5.1 Practice Problems 504
8.5.2 Term Project 8.1 506
References 507

9. A System for Quality 509
9.1 The Systems Approach 509
9.2 Dr. Deming's System 510
9.2.1 Long-Term Planning 511
9.2.2 Cultural Change 512

9.2.3 Prevention Orientation 512
9.2.4 Quality in Procurement 513
9.2.5 Continuous Improvement 513
9.2.6 Training, Education, Empowerment, and Teamwork 514
9.3 Dr. Juran's System 518
9.3.1 Quality Planning 519
9.3.2 Quality Control 522
9.3.3 Quality Improvement 523
9.4 Dr. Feigenbaum's System 527
9.5 Baldrige Award Criteria 530
9.5.1 Criterion 1: Leadership 533
9.5.2 Criterion 2: Strategic Planning 534
9.5.3 Criterion 3: Customer Focus 535
9.5.4 Criterion 4: Measurement, Analysis, and Knowledge Management 537
9.5.5 Criterion 5: Workforce Focus 538
9.5.6 Criterion 6: Process Management 540
9.5.7 Criterion 7: Results 541
9.6 ISO 9000 Quality Management Systems 543
9.6.1 The ISO 9000 Standards 543
9.6.2 The Eight Quality Management Principles 544
9.6.3 Documentation in ISO 9000 545
9.7 ISO 9001:2008 Requirements 546
9.7.1 Quality Management System (4) 547
9.7.2 Management Responsibility (5) 548
9.7.3 Resource Management (6) 551
9.7.4 Product Realization (7) 552
9.7.5 Measurement, Analysis, and Improvement (8) 558
9.8 Six Sigma System 561
9.8.1 Six Themes of Six Sigma 562
9.8.2 The 6σ Measure 563
9.8.3 The Three Strategies 565
9.8.4 The Two Improvement Processes 566
9.8.5 The Five-Step Road Map 566
9.8.6 The Organization for the Six Sigma System 569
9.9 Summary of Quality Management Systems 569
9.10 Exercise 571
9.10.1 Practice Problem 571
9.10.2 Mini-Projects 573
References 573

Appendix 1 575

Appendix 2 587

Index 595

Preface to the Second Edition

"The average Japanese *worker* has a more in-depth knowledge of statistical methods than an average American *engineer*," explained a U.S. business executive returning from a visit to Japan, as a reason why the Asian rivals were able to produce better quality products than U.S. manufacturers. That statement, made almost 30 years ago, may be true even today as Japanese cars are continuously sought by customers who care for quality and reliability. Dr. Deming, recognized as the guru who taught the Japanese how to make quality products, said: "Industry in America needs thousands of statistically minded engineers, chemists, doctors of medicine, purchasing agents, managers" as a remedy to improve the quality of products and services produced in the U.S. He insisted that engineers, and other professionals, should have the capacity for statistical thinking, which comes from learning the statistical tools and the theory behind them. The engineering accreditation agency in the U.S., ABET, a body made up of academics and industry leaders, stipulates that every engineering graduate should have "an ability to design and conduct experiments, as well as to analyze data and interpret results" as part of the accreditation criteria.

Yet, we see that most of the engineering majors from a typical college of engineering in the U.S. (at least 85% of them by our estimate) have no knowledge of quality methods or ability for statistical thinking when they graduate. Although some improvements are visible in this regard, most engineering programs apart from industrial engineering do not require formal classes in statistics or quality methodology. The industry leaders have spoken; the engineering educators have not responded fully.

One of the objectives in writing this book was to make it available as a vehicle for educating all engineering majors in statistics and quality methods; it can be used to teach statistics and quality in one course.

The book can serve two different audiences: those who have prior education in statistics, and those who have no such prior education. For the former group, Chapter 2, which contains fundamentals of probability and statistics, needs to be only lightly reviewed; and for the latter, Chapter 2 must be strongly emphasized. Chapter 5, which has advanced material in theory of control charts, design of experiments, and process capability, should be included for the former group whereas it can be partly or completely omitted for the latter. Besides, a teacher has the option to pick and choose topics in other chapters as well, according to the needs of an audience.

The second edition has been completely revised to provide clearer explanation of concepts and to improve the readability throughout by adding a few figures and examples where necessary. An overview of lean manufacturing methods has been added to Chapter 8 as yet another set of tools for improving productivity and reducing waste so that quality products can be produced at minimum cost. More mini-projects, which expose the students to real-world quality problems, have been added to the end of each chapter. The coverage of ISO 9000 and the Baldrige Award Criteria has been updated to reflect the latest revisions of these documents.

The publishing staff at CRC Press have been extremely helpful in publishing this second edition. We are thankful to them. Some very special thanks are due to Ms. Cindy Carelli, our contact editor, for her prompt and expert response whenever help was needed.

Any suggestions to improve the content or its presentation are gratefully welcomed.

K.S. Krishnamoorthi
V. Ram Krishnamoorthi

Preface to the First Edition

In the 20-plus years that I have been teaching classes in quality methods for engineering majors, my objective has been to provide students with the knowledge and training that a typical quality manager would want of new recruits in his or her department. Most quality managers would agree that a quality engineer should have a good understanding of the important statistical tools for analyzing and resolving quality problems. They would also agree that the engineer should have a good grasp of management methods, such as those necessary for finding the needs of customers, organizing a quality system, and training and motivating people to participate in quality efforts.

Many good textbooks are available that address the topics needed in a course on quality methods for engineers. Most of these books, however, deal mainly with one or the other of the two areas in the quality discipline—statistical tools or management methods—but not both. Thus, we will find books on statistical methods with titles such as *Statistical Process Control* and *Introduction to Statistical Quality Assurance,* and we will find books on management methods with titles similar to *Introduction to Total Quality,* or *Total Quality Management.* The former group will devote very little coverage to management topics; the latter will contain only a basic treatment of statistical tools. A book with an adequate coverage of topics from both areas, directed toward engineering majors, is hard to find. This book is an attempt to fill this need. The term *quality engineering,* used in the title of this book, signifies the body of knowledge comprising the theory and application of both statistical and management methods employed in creating quality in goods and services.

When discussing the statistical methods in this book, one overarching goal has been to provide the information in such a manner that students can see how the methods are put to use in practice. They will then be able to recognize when and where the different methods are appropriate to use, and they will use them effectively to obtain quality results. For this reason, real-world examples are used to illustrate the methods wherever possible, and background information on how the methods have been derived is provided. The latter information on the theoretical background of the methods is necessary for an engineer to be able to tackle the vast majority of real-world quality problems that do not lend themselves to solution by simple, direct application of the methods. Modification and improvisation of the methods then become necessary to suit the situation at hand, and the ability to make such modifications comes from a good understanding of the fundamentals involved in the creation of those methods. It is also for this reason that a full chapter (Chapter 2) is devoted to the fundamentals of probability and statistics for those who have not had sufficient prior exposure to these topics. Chapter 2 can be skipped or quickly reviewed for those such as industrial engineering majors who have already taken formal classes in engineering statistics.

The use of computer software is indicated wherever such use facilitates problem solving through the use of statistical methods. The statistical software package Minitab has been used to solve many problems. The student, however, should understand the algorithm underlying any computer program before attempting to use it. Such understanding will help to avoid misuse of the programs or misinterpretation of the results. It will also help in explaining and defending solutions before a manager, or a process owner, when their approval is needed for implementing the solutions.

The management topics have been covered in a brief form, summarized from reliable references. Full-length books are available on topics such as supply chain management, customer surveys, and teamwork, and no attempt has been made here to provide an exhaustive review of these topics. The objective is simply to expose the student to the important topics, explain their relevance to quality efforts, indicate where and how they are used, and point to pertinent literature for further study. Through this exposure, a student should acquire a working knowledge of the methods and be able to participate productively in their use.

This book is organized into nine chapters, which are arranged broadly along the lines of the major segments of a quality system. The quality methods are discussed under the segment of the system where they are most often employed. This arrangement has been chosen in the hope that students will be better motivated when a method is introduced in the context that it is used. The book can be covered in its entirety in a typical one-semester course if students have had prior classes in engineering statistics. If time must be spent covering material in Chapter 2 in detail, then Chapter 5, which includes mostly advanced material on topics covered by other chapters, can be omitted entirely, or in part, and the rest of the book completed in one semester.

At the end of each chapter, one to three mini-projects are given. Many of these are real problems with real data and realistic constraints, and they do not have a unique solution. These projects can be used to expose students to real-world problems and help them to learn how to solve them.

In selecting the topics to include in this book, many judgment calls had to be made, first regarding the subjects to be covered and then concerning the level of detail to be included. Only experience will tell if the choice of topics and the level of detail are adequate for meeting the objectives. Any feedback in this regard or suggestions in general for improving the contents or their presentation will be appreciated.

I owe thanks to many colleagues and friends in the academic and industrial community for helping me to learn and teach the quality methods. Special thanks are due to Dr. Warren H. Thomas, my thesis advisor and chairman of the IE Department at SUNY at Buffalo when I did my graduate work there. He was mainly responsible for my choice of teaching as a career. When he gave me my first independent teaching assignment, he told me that "teaching could be fun" and that I could make it enjoyable for both the students and myself. Ever since, I have had a lot of fun teaching, and I have enjoyed every bit of it.

Baltasar Weiss and Spike Guidotti, both engineering managers at Caterpillar, Inc., Peoria, helped me learn how to work with people and protocols while trying to make quality methods work in an industrial setting. Their consistent support and encouragement helped me to take on many challenges and achieve several successes. I am deeply thankful to both.

The editorial and production staff at Prentice Hall have been extremely helpful in bringing this book to its final shape. The assistance I received from Dorothy Marrero in the early stages of developing the manuscript through the review/revision process was extraordinary, and beyond my expectations. In this connection, the assistance of numerous reviewers is appreciated: Thomas B. Barker, *Rochester Institute of Technology*; Joseph T. Emanuel, *Bradley University*; Jack Feng, *Bradley University*; Trevor S. Hale, *Ohio University*; Jionghua (Judy) Jin, *University of Arizona*; Viliam Makis, *University of Toronto*; Don T. Phillips, *Texas A&M University*; Phillip R. Rosenkrantz, *California State Polytechnic University*; and Ed Stephens, *McNeese State University*.

Finally, I owe thanks to all my students in the class IME 522 at Bradley, who helped me in testing earlier drafts of the manuscript. Many errors were found and corrected through their help.

K.S. Krishnamoorthi

Authors

K.S. Krishnamoorthi, PhD, is a professor of industrial engineering in the industrial and manufacturing engineering and technology department of Bradley University in Peoria, Illinois. He has a BE degree in mechanical engineering from the University of Madras, India, an MA in statistics and a PhD in industrial engineering from the University of Buffalo. He teaches statistics and quality engineering, among other industrial engineering subjects, and conducts research in the area of quality costs, process capability indices, and statistical thinking. He has provided consultation to several small and large corporations in process and product quality improvement using statistical tools.

V. Ram Krishnamoorthi, M.D., M.P.H. is an attending internist at the University of Chicago Medical Center and is an Assistant Professor of Medicine of the University of Chicago, Chicago, IL. He obtained an A.B. degree in economics from Princeton University and his M.D. and M.P.H. from Northwestern University Feinberg School of Medicine in Chicago. He is involved in health services research and quality improvement initiatives at the U of C Hospitals, and is interested in problems of quality and access to affordable health care at the national level.

1

Introduction to Quality

This chapter begins with an historical overview of how quality evolved to become a major strategic tool in managing businesses in the United States and in other parts of the world. It then goes on to explain the meaning of the term "quality." The need for a total quality system, accompanied by the total quality management philosophy, to achieve quality in products and services is then explained. The chapter concludes with a discussion on how quality impacts upon the economic performance of an enterprise, including a discussion on quality cost analysis. Detailed definitions of quality costs are given, along with an explanation of how a quality cost analysis is useful in evaluating the quality health of an organization, discovering opportunities to improve quality, and in reducing waste.

1.1 An Historical Overview

Striving for quality, in the sense of seeking excellence in any field, has always been a part of human endeavor. History provides numerous examples of people achieving the highest levels of excellence, or quality, in their individual or collective pursuits. The plays of Shakespeare, the music of Beethoven, the Great Pyramids of Egypt, and the temples of southern India are but a few examples. Quality appeals to the human mind and provides a certain sense of satisfaction, which is why most of us enjoy listening to a good concert, watching a good play, observing a beautiful picture, or even riding in a well-built car.

During the last 60 to 70 years, the term "quality" has come to be used in the marketplace to indicate how free from defects a purchased product is and how well it meets the requirements of its intended use. After the Industrial Revolution in the early 1900s, mass-production techniques were adopted for manufacturing large quantities of products in order to meet the increasing demand for goods. Special efforts had to be made to achieve quality in the mass-produced products because they were assembled from mass-produced parts. Variability or lack of uniformity in mass-produced parts created quality problems. For example, if a bearing and a shaft were chosen randomly from their lots and the bearing had one of the smallest bores among its lot and the shaft had one of the largest diameters among its lot, then they would be hard to assemble. If they were assembled at all, the assembly would fail early in its operation due to lack of sufficient clearance for lubrication. The larger this variability among the parts, the more severe this problem would be. Newer methods were therefore needed in order to guarantee the uniformity required in parts that were mass produced.

Dr. Walter A. Shewhart, working for the former Bell Laboratories in the early 1920s, employed the science of statistics to monitor and control the variability in manufactured products, and invented the control methods for this purpose. Drs. Harold Dodge and Harry Romig developed sampling plans that used statistical principles to ascertain the quality of a population of products from the quality observed in a sample drawn from it. These statistical methods, which were used mainly within the Bell telephone companies during the 1930s, were also used in the early 1940s in the production of goods and ammunition for the U.S. military in World War II. Some (Ishikawa 1985) even speculated that this focus on quality in military goods provided the United States and their allies in the war with an advantage that contributed to their eventual victory.

After World War II, the U.S. War Department was concerned about the lack of dependability of electronic parts and assemblies when they were deployed in their wartime missions. From this concern grew the science of reliability, which deals with the failure-free performance of products over time. This field matured during the early 1950s into a sophisticated discipline and contributed to improving the longevity—not only of defense products, but also of many commercial products such as consumer electronics and household appliances. The long life of refrigerators and washing machines, which we take for granted today, is in large measure the result of using reliability methods in their design and manufacture.

During the early 1950s, leaders in the quality field, such as Drs. W. Edwards Deming, Joseph M. Juran, and Armand V. Feigenbaum, redefined quality in several important ways. First, they established that quality in a product exists only when a customer finds that product satisfactory in its use. This was in contrast to an earlier understanding that a product had quality if it met the specifications selected by the manufacturer's designers, which might have been chosen with or without reference to the needs of the customer. Second, these gurus, as they were called, proposed that a quality product, in addition to meeting the needs of the customer, should also be produced at minimal cost. Finally, and most importantly, they claimed that, in order to create a product that will satisfy the customer both in terms of performance and cost, a system is necessary. This would be made up of all the units that contribute to the production of a quality item, every element of which will be focused on the common goal of satisfying the customer. With this premise, they proposed the concept of a "total quality system" whose elements would be defined along with their responsibilities and interrelationships. Guidelines would be laid down so that the components working together would optimize the system's goal of satisfying the customer's needs.

The management approach needed to complement such a quality system, the "total quality management philosophy" defined how the people working with the total quality system would be recruited, trained, motivated, and rewarded for achieving the system's goals. Although there was widespread acceptance of the need for a total quality system among the quality community in the United States and elsewhere, it was the Japanese who quickly embraced it, adapted it to their industrial and cultural environment, and implemented it to reap immense benefits. They called such a system "total quality control" or "company wide quality control." With the success of the Japanese, the rest of the world eventually saw the value in the systems approach to quality. The creation of the Standards for Quality Assurance Systems by the International Organization for Standardization (ISO 9000) in 1987 was the culmination of the worldwide acceptance of the systems approach to producing quality.

Another major milestone in the development of the quality discipline was the discovery, or rediscovery, of the value of designed experiments for establishing the relationship between process variables and product characteristics. The use of designed experiments to obtain information about a process was pioneered by the English statistician Sir Ronald Fisher in the early 1920s in the context of maximizing yield from agricultural fields. Although the methods had been used successfully in manufacturing applications, especially in the chemical industry during the 1950s, it was Dr. Genechi Taguchi, the Japanese engineer, who popularized the use of these experiments to improve product quality. He adapted the methods for use in product and process design and provided simplified (some would claim oversimplified) and efficient steps for conducting experiments to discover the combination of product (or process) parameters to obtain the desired product (or process) performance. His methods became popular among engineers, and experimentation became a frequently used method during quality improvement projects in industry.

The birth of the quality control (QC) circles in Japan in 1962 was another important landmark in the development of quality methodology. The Union of Japanese Scientists and Engineers (JUSE), which took the leadership role in spreading quality methods in Japan, had been offering classes in statistical quality control to engineers, managers, and executives starting in 1949. They realized that the line workers and line foremen had much to offer in improving the quality of their production and wanted to involve them in the quality efforts. As a first step, JUSE, under the leadership of Dr. Kaoru Ishikawa, began training workers and foremen in statistical methods through a new journal called *Quality Control for the Foreman*, which carried lessons in statistical quality control.

Foremen and workers assembled in groups to study from the journal and started using what they learned for solving problems in their own processes. These groups, which were generally made up of people from the same work area, were called "QC circles," and these circles had enormous success in solving quality problems and improving the quality of their products. Companies encouraged and supported such circles, and success stories were published in the journal for foremen. The QC circle movement, which successfully harnessed the knowledge of people working close to processes, gained momentum throughout Japan, and workers and foremen in large numbers became registered participants in QC circles. The success of QC circles is considered to be one of the most important factors in Japanese successes in quality (Ishikawa 1985). The details on how the QC circles are formed, trained, and operated are discussed in Chapter 6 under the section on teamwork.

The 1970s were important years in the history of the quality movement in the United States. Japanese industry had mastered the art and science of making quality products and won a large share of the U.S. market, especially in automobiles and consumer electronics. Domestic producers in the United States particularly automakers—lost a major share of their markets and were forced to close shops and lay off workers in large numbers. Those were very painful days for American workers in the automobile industry. It took a few years for the industry leaders to figure out that quality was the difference between their products and those of the Japanese. The NBC television network produced and aired a documentary in 1979 titled *If Japan Can ... Why Can't We?* The documentary highlighted the contributions of Dr. W. Edwards Deming in providing training for Japanese engineers and managers in statistical quality control and in a new management philosophy for achieving quality.

It is generally agreed that this documentary provided the rallying point for many U.S. industry leaders to start learning from the Japanese the secrets of quality methodologies. Many corporate leaders—such as Robert Galvin of Motorola, Harold Page of Polaroid, Jack Welch of General Electric, and James Houghton of Corning—led the way in spreading the message of quality among their ranks, and they had instant followers. The quality philosophy spread across U.S. industry and quality gurus such as Dr. Deming came home to provide training in the technological and managerial aspects of quality. Several U.S. corporations that had suffered losses to foreign competition started to recover using quality-focused management. Corporations such as Motorola, Xerox, IBM, Ford, General Motors, Chrysler, Corning, and Hewlett-Packard are examples of companies who began to regain market shares and economic strength with their new focus on quality. A new "quality revolution" had begun in the United States. The 1980s saw this recovery spreading across a wide industrial spectrum, from automobiles and electronics to steel and power to hotels and healthcare.

During the early 1980s, the electronics manufacturer Motorola launched a quality drive within its corporation using what they referred to as the "Six Sigma process." The major thrust of this process was to reduce the variability in every component characteristic to levels at which the nonconformity rates, or the proportion of characteristics falling outside the limits that were acceptable to the customer, would not be more than 3.4 parts per million. Such levels of uniformity, or quality, were needed at the component level, Motorola claimed, in order to attain acceptable quality levels in the assembled product. The Six Sigma process emphasized the use of statistical tools for process improvement and process redesign, along with a systematic problem-solving approach to improve customer satisfaction, reduce costs, and enhance financial performance. The statistical methods and the problem-solving methodology were taught to engineers, supervisors, and operators in a massive training program.

The success that Motorola achieved in quality and profitability through the implementation of the Six Sigma process attracted the attention of other corporations, such as General Electric, Allied Signal, and DuPont, who in turn gained enormously from its application. The list of followers grew, and the Six Sigma movement spread in the United States and abroad, helped by the formation of organizations such as the Six Sigma Academy in 1994, which provides training and advice on Six Sigma implementation. The Six Sigma process is discussed in more detail in Chapter 9 as part of the discussion about quality systems.

In 1988, the U.S. Congress established an award called the Malcolm Baldrige National Quality Award (MBNQA), named after President Reagan's secretary of commerce, who died in office. The award was established to reward those U.S. businesses that show the most progress in achieving business excellence through the use of modern methods for improving quality and customer satisfaction. This was an expression of recognition by the U.S. government of the need to focus on the quality of products and services for the U.S. economy to stay healthy and competitive.

The most recent development in the quality field has been the acceptance of quality as one of the strategic parameters for business planning along with the usual marketing and financial measures. Organizations that accept quality as a planning parameter will include quality goals in their strategic plans, along with marketing and financial goals. They will set goals—for example, for reducing the number of defectives that are shipped to customers or for improving customer satisfaction—just as they set goals for increasing sales, improving profit margins, or winning an additional market share.

Table 1.1 shows at a glance the major milestones in the progression of quality toward becoming an important parameter in economic activities in the United States. Although important activities relative to quality have been taking place in many parts of the world, especially in Japan, England, and other parts of Europe, nowhere else did quality come into such dramatic focus as it did in the United States.

The quality revolution, which has taken root in the United States and many other developed and developing countries throughout the world, will be an important phenomenon in the new global economy. Consumers are increasingly aware of the value of quality and will demand it.

TABLE 1.1

Major Events Related to the Quality Movement in the United States

1900	Post-Industrial Revolution era. Goods are mass produced to meet rising demands.
1924	Statistical control charts are introduced by Dr. Shewhart at Bell Labs.
1928	Acceptance sampling plans are developed by Drs. Dodge and Romig at Bell Labs.
1940	The U.S. War Department uses statistical methods and publishes a guide for using control charts.
1942	Several quality-control organizations are formed. Training classes are offered.
1946	American Society for Quality Control is organized.
1946	Dr. Deming is invited to Japan to help in their national census.
1949	JUSE begins offering classes in quality control to engineers and managers.
1950	Dr. Deming offers classes in quality methods to Japanese engineers, managers, and executives.
1950s	Designed experiments are used in manufacturing (chemical industry).
1950s	Study of reliability begins as a separate discipline.
1951	Dr. A.V. Feigenbaum proposes a systems approach to quality and publishes the book *Total Quality Control.*
1955	Beginning of quantity control concept (lean manufacturing) as part of the Toyota Production System (TPS) in Japan.
1960s	Academic programs in industrial engineering begin offering courses in statistical quality control.
1962	The quality control circle movement begins in Japan: workers and foremen become involved in statistical quality control.
1970s	Many segments of American industry lose to Japanese competition.
1979	NBC broadcasts the documentary *If Japan Can ... Why Can't We?*
1980s	American industry, led by the automobile companies, makes the recovery.
1981	Motorola introduces the Six Sigma process for quality improvement.
1987	The Malcolm Baldrige National Quality Award is established. The first edition of IS0 9000 is issued.
1990s	ISO 9000 standards gain acceptance in the United States.
2000s	Quality is becoming one of the strategic parameters in business planning.

Legislative bodies will require it because of its safety and health implications. Industrial managers will pursue it because of the reduction in waste, leading to a reduction in production costs and an improvement in profits. The reputation that a quality image brings is a marketing advantage. The focus on quality has extended to industries beyond manufacturing. Information technology, which drives the new economy, is expected to have its own share of quality problems of a similar kind to the old economy. According to one account: "If the construction of new homes in the United States were to have the same success rate as the development of new computer applications, the American urban landscape would resemble a war zone... Of every 100 new homes started, more than a third would never be completed at all and would be abandoned in various stages of disrepair. More than half the homes that did get finished would end up costing nearly twice as much as originally estimated and would also take more than twice as long to build. To make matters worse, these homes would be completed with only two-thirds of the originally planned floor space or rooms. On average, only around 16 homes would be completed on time and on budget" (Markus 2000). There seems to be plenty of opportunities to use quality methodology in the software industry to improve the quality of their products and reduce waste.

The pursuit of quality in the healthcare industry is another story. The Institute of Medicine (IOM), one of the United States National Academies charged with providing advice to the nation on medicine and health, said: "The U.S. healthcare delivery system does not provide consistent, high-quality medical care to all people.... Healthcare harms patients too frequently and routinely fails to deliver its potential benefits. Indeed between the healthcare that we now have and the healthcare that we could have lies not just a gap, but a chasm." This conclusion was contained in their report titled "Crossing the Quality Chasm: A New Health System for the 21st Century," published in March 2001. In an earlier report entitled "To Err Is Human," the IOM said: "at least 44,000 people, and perhaps as many as 98,000 die in hospitals each year as a result of medical errors that could have been prevented," making medical errors a worse cause for fatalities than car wrecks, breast cancer, and AIDS. The reports point to the lack of timely delivery of quality healthcare for people who need it.

There are several examples where healthcare organizations have successfully implemented quality methods to improve the quality of care. In one example, the Pittsburgh Regional Health Initiative (PRHI), whose mission is to improve healthcare in Southwestern Pennsylvania, achieved "numerous" successes in their effort to continuously improve operations, standardization, and elimination of errors. In one such success, they reduced catheter-related bloodstream infections at Allegheny General Hospital by 95% between 2003 and 2006, and reduced the number of deaths by such infections to zero. In another instance, they helped the Pittsburgh Health System Veteran Affairs reduce the rate of methicillin-resistant staphylococcus aureus infection from 0.97 per 1,000 bed-days of care in 2002 to 0.27 in 2004 (Krzykowski 2009). These instances, however, seem to be exceptions to the rule in the healthcare landscape. Opportunities seem to be abundant for improving healthcare delivery in the United States—both at the micro, process levels in hospitals and at the macro levels of policy-making—through the use of quality methodologies.

1.1.1 A Note about "Quality Engineering"

The experience of the past few decades has shown that quality is achievable through a well-defined set of methods used during the design, production, and delivery of products. The collection of these methods and the theoretical concepts behind them can be viewed as falling into an engineering discipline, which some have already called "quality engineering." The American Society for Quality, a premier organization of quality professionals, uses this name to signify the *body of knowledge* contributing to the creation of quality in products and services that leads to customer satisfaction. They even offer training programs in quality engineering and certification to those who pass a written examination and acquire a certain level of experience in the quality field.

The term "quality engineering" has been used in quality literature to denote many things. Some authors have used the term to refer to the process of improving product quality using improvement tools. Many have used it to signify the process of selecting targets and tolerances for process parameters through designed experiments. Some have used it to mean the selection of product characteristics that will satisfy the customer's needs. The term is used here, however, with a much broader meaning. In this book, quality engineering refers to the discipline that includes the technical methods, management approaches, costing procedures, statistical problem-solving tools, training and motivational methods, computer information systems, and all the sciences behind theses that are needed for designing, producing, and delivering products and services to satisfy customer needs.

The body of knowledge needed to make quality products has assumed different names at different times depending on the available set of tools at those times. It was called "quality control" when final inspection was the only tool being employed to achieve product quality before the product was shipped to the customer. When statistical principles were used to create control charts and sampling plans, it assumed the name "statistical quality control." It also took the name "quality assurance" at this time because the control charts were used to control the process upstream of final inspection and prevent defectives from being produced so as to assure defect-free shipments to customer. "Statistical process control" (SPC) was another term used at this time. Then came the addition of elements such as drawing control, procurement control, instrument control, and other components of a total quality system (as explained later) when people recognized that a system was necessary to achieve quality. When a new management philosophy became necessary to deal with the quality system, the body of knowledge was called "total quality management" (TQM). It was also known as "company wide quality control" (CWQC). Quality engineering became the name to indicate that the body of knowledge needed for quality includes the science, mathematics, systems thinking, psychology, human relations, organization theory, and the numerous methods created from them that are used during the design, production, and delivery of the product. It may be worth mentioning that the Six Sigma methodology that has become so popular in recent years, as a means of improving quality and reducing waste, encompasses almost the same set of knowledge that we refer to here as "quality engineering."

1.2 Defining Quality

We have, until now, used the term quality frequently without stopping to explain its meaning. It may seem that there is no need to do so, as the word is used liberally in both casual and professional contexts with tacit understanding of its definition. Yet, if we asked a sample of both the public and professionals for a definition, we may find as many definitions for the term as the number of people asked. However, a uniform definition is necessary when people of varying backgrounds in an organization are engaged in its pursuit. If a clear, consistent definition of what is meant by quality for a given organization and the products they create is not available, misunderstanding and confusion might result.

Some dictionary definitions of quality are "an inherent or distinguishing characteristic," "superiority of kind," and "degree or grade of excellence." These definitions are too general and nonspecific. More recent definitions proposed in the literature are more specific and practical: "fitness for use," "fitness for intended use," "ability to satisfy given needs," "conformance to specification," and "meeting the needs of the customer both in performance and price." According to the ISO 9000-2005 standards, quality is defined as the "degree to which a set of inherent characteristics fulfills requirements."

The most quoted definition for quality is that by Garvin (1984). According to him, quality has several dimensions:

1. Performance—the product's ability to do the work it is supposed to do.
2. Features—things that add to the convenience and comfort.
3. Reliability—the ability to perform without failure over time.
4. Conformance—the degree to which the product meets codes of a state or a community.
5. Durability—the length of time the product will last until it is discarded.
6. Serviceability—the ability for making repairs easily, quickly, and at a reasonable cost.
7. Aesthetics—sense appeal, such as color, sound, feel, and comfort.
8. Perceived quality—the impression the product creates in the customer's mind.

The above definition of quality by Garvin only reinforces the idea that quality has many aspects and cannot easily be defined in one simple phrase or sentence. It may have different meanings for different products, and even for the same product, it may have a different meaning for different users. However, for day-to-day communication, people have adopted some practical definitions, such as "fitness for use," originally proposed by Dr. Joseph Juran. This definition has been modified by others who recognize customer satisfaction as a necessary part of the definition. They therefore use an extended definition, such as "fitness for use and meeting or exceeding customer expectations."

Another definition of quality worth quoting is by Dr. A. Blanton Godfrey, a modern-day quality guru. In a 2002 article he says, "Unfortunately, defining quality abstractly is extremely difficult and not very useful. It is like defining the universe: Describing a portion of it seems to make sense, but extending this definition to cover all planets, stars and galaxies ultimately proves impossible." He goes on to say, "The most fundamental truth is that quality is relative: If a competitor can produce a similar part and sell it at a lower price, we lose ... The customer focuses simply on value, seeing it as a ratio of quality over price ... Only when we offer more value than our competitor do we really succeed." (Godfrey 2002)

This discussion on quality by Dr. Godfrey goes far beyond earlier definitions of quality as simply being fit for the customer's use. It includes, in addition, satisfaction of the customer, efficiency in production, and competitiveness in price. It conveys the idea that it is not enough for a business to make the product with the necessary characteristics to satisfy the customer's needs; the business should also be able to make profit by selling it and succeed as an enterprise.

Furthermore, these definitions do not provide enough clarity when we want to be able to *measure* quality so that it can be monitored and improved. For this purpose, a clearer, more decisive definition is necessary.

A complete definition of a product's *quality* begins with identifying the customers and determining their needs and expectations. A product designer takes these needs and expectations into account and selects features of the product to create a design that is responsive to the expressed needs. The designer also selects the targets and limits of variation (called "tolerances") for these features so that the product can be produced at a reasonable cost while meeting the customer's needs. These product features, which must be measurable, are called the "quality characteristics." The targets and tolerances together are called the "specifications" (or specs) for these characteristics. The collection of the quality characteristics and their specifications define the quality of the product.

When the product is created, the production personnel verify if the product's characteristics meet the targets and specifications chosen by the designer, which in turn will assure that the product, when delivered to the customer, will meet their needs. If the needs of the customer have been properly assessed, and if the quality characteristics and their specifications have been chosen by the designer to respond to those needs, and if the production team produces the product to conform to those specifications, then the product will meet the needs of the customer when delivered. The product is then considered to possess quality.

1.2.1 Product Quality vs. Service Quality

Most of the discussions in this book regarding quality methods are made in the context of producing a physical product, as it is easier to illustrate concepts of quality in this manner. However, we recognize that services, such as mail delivery, dry cleaning, maintenance, and security, must also be "produced" in order to meet the needs of the customers who use them. Even for these services, customer needs must be assessed; the service features must be chosen to meet those needs; and production and delivery must be performed in a manner to meet those needs. Only then will those services have quality.

Services can be divided into two categories: primary services, and secondary services. Services such as mail delivery by the U.S. Postal Service, financial services by a bank, and instructional services by a university, are examples of primary services, because the services mentioned are the major "products" of the respective organizations. On the other hand, services provided by manufacturers of garage doors and lawnmowers that help customers to install and use such products correctly are examples of secondary services. Similarly, the treatment received by a patient from doctors and nurses in a hospital for a health problem is a primary service while the pre-treatment reception and post-treatment counseling are secondary services. Often, secondary services (which are also called "customer services") are important and are needed for creating customer satisfaction in the primary product or service. The emergency road service provided by a car company is an example of this type of customer service.

Methodologies that have been developed to create quality in physical products are applicable, with some modifications, to quality in services as well. In the discussions of quality methods in this book, wherever the term "product" is used while describing a quality method, it can generally be taken to mean both a tangible product and an intangible service.

1.3 The Total Quality System

Many activities have to be performed by many people in an organization in the design and production of a product in order to meet the needs of the customer. The marketing department usually obtains the information on customer needs. Design engineers translate these needs into product characteristics and their specifications. Manufacturing engineers determine what materials to use and what processes to employ in order to make the product meet the required specifications. The production team follows the set of instructions generated by the manufacturing engineers and converts the raw materials into products with characteristics within the stated specifications. The packaging department design and produce packaging such that the product will reach the customer safely and without damage. All these are known as "line activities," because they are directly responsible for creating and delivering the product.

Several other supporting activities are associated with producing a quality product. Instruments must be properly chosen for the accurate and precise measuring of quality characteristics, and they must be maintained so that they retain their accuracy and precision. Training must be provided for all personnel who are involved in productive functions so that they know not only how to make the product but also how to make it a quality product. Equipment must be kept running and maintained so that they remain capable of meeting the required tolerances, and computer hardware and software have to be installed and maintained to gather data and generate information to facilitate decision making. These are referred to as "infrastructure activities." For all these quality-related activities, including the line and infrastructure activities, responsibilities need to be clearly assigned to various people in the organization. Rules are needed as to who has the primary and who has the supporting responsibility when multiple agencies are performing a function. The rules should also specify how differences, should they arise, be resolved. In other words, a system must be established in which several component agencies with assigned responsibilities and defined relationships will work together to meet the common goal of producing and delivering a product that will meet the customer's needs. Such a system is called a "total quality system" (TQS).

Figure 1.1 shows the components of a TQS enveloped by features of quality management philosophy. Notice that the system includes in addition to production processes, marketing, design engineering, process engineering, quality engineering, information systems, packaging and shipping, metrology, safety and environment protection, human resources, plant engineering, materials management, and customer service. Each of these functions has its main contribution to quality indicated. Each component of the system must perform its function well, and the functions must be controlled and coordinated so that they all work to optimize the system's common objective of satisfying the customer. Many of the functions are dependent on others performing their work satisfactorily. For example, the production function depends on procurement to bring parts and materials of the required quality. The procurement function depends on design activity for describing the specifications and quality requirements of the parts and materials to be procured. It also depends on stores to provide feedback as to how well suppliers are adhering to the agreed standards for supplies. Similarly, the inspection function depends on metrology for providing accurate and capable instruments for the correct verification of quality characteristics.

The control and coordination of the components in the TQS are accomplished by laying down the overall policies and procedures for the system—individual responsibilities, interrelationships, and control procedures—in written documents. These written policies and procedures define the TQS.

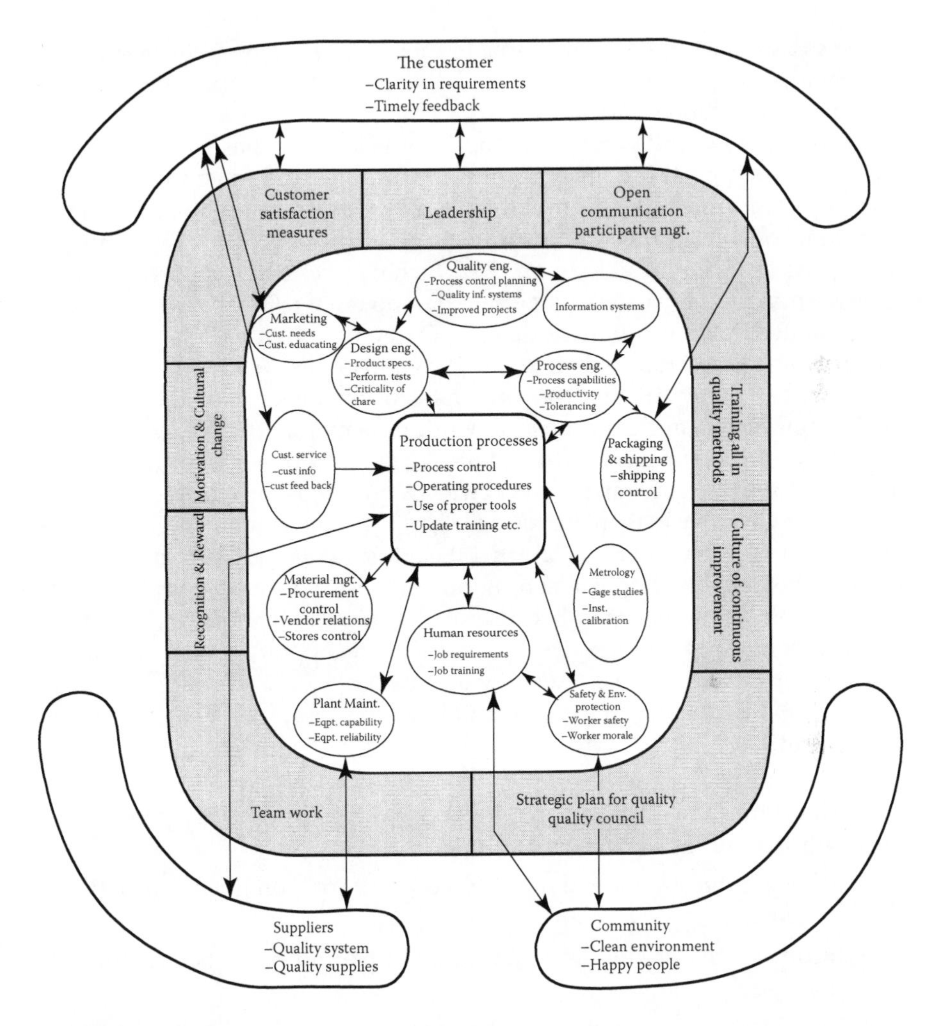

FIGURE 1.1
Model of a total quality management system.

1.4 Total Quality Management

A new management philosophy becomes necessary to manage the new quality system. In earlier days, when quality control was thought to be the sole responsibility of the inspection department, functions such as marketing, product development, and process development paid little attention to the quality of the finished product. Under the new management system, a cultural change becomes necessary in order to create awareness among *all* of the employees of an organization of the needs of the customers and to emphasize that each employee has a part to play in satisfying the customer.

The message that "quality is everyone's job" needs to be delivered across the organization—from the chief executive to those who put the products together on the shop floor.

Under the new philosophy, the top-level leadership has to first commit itself to producing and delivering quality products and must then demand a similar commitment from employees at all levels in the organization. They have to incorporate quality goals in their planning processes. The management process has to change to one in which everyone becomes involved in decision making rather than using an older model of ruling by fiat with decisions handed down through a hierarchy. Decisions have to be made based on factual information rather than on feelings and beliefs. All levels of employees must be trained in the new principles of business, with the customer as the main focus, and be trained to work in teams that strive for the common good of the business. There should be a free flow of information and ideas within the organization, and rewards must be set up to recognize any significant contributions made toward satisfying the customer. Finally, it is necessary to impress on everyone that the effort to improve products, and to make customers happier, is a continuous process. For one thing, the needs of customers keep changing because of changes in technology or changes in a competitor's strategies; then there are always opportunities to make a product better and more suited to the customer.

Thus, the total quality management (TQM) philosophy has the following components:

1. The commitment of top management to making quality products and satisfying the customer.
2. Focusing the attention of the entire organization on the needs of the customer.
3. Incorporating quality as a parameter in planning for the organization's goals.
4. Creating a system that will define the responsibilities and relationships of various components toward customer satisfaction.
5. Decision making based on factual information in a participative environment.
6. Training all employees in TQM, involving them, and empowering them as participants in the process.
7. Creating a culture in which continuous improvement will be a constant goal for everyone and contributions toward this goal will receive adequate reward.

Figure 1.1 shows the TQS complemented by the components of the TQM philosophy. The entire system, the combination of the technology-oriented TQS and the human-oriented TQM philosophy, is often referred to as the

"TQM System," or the "quality management system." Several models for the quality management system, such as the ISO 9000 standards and the MBNQA criteria, provide guidelines on how a quality management system should be designed, organized, and maintained. These are discussed in Chapter 9.

1.5 Economics of Quality

After Japan's defeat in World War II, its economy was in total disarray. The country was faced with the immense task of reconstructing its economic and civil systems. Since Japan is not endowed with many natural resources, the revival was based on importing raw materials and energy, making products using these, and then selling these products back to the foreign markets. Japan was, at the time, also handicapped by the prevailing perception in the world markets that the country's products were of notoriously poor quality. Hence, the revival of the economy was predicated on improving the quality of their products and gaining a new reputation for this improved quality. Japanese industrial leaders, with help from their government, sought and received assistance from a few well-known U.S. consultants, Dr. W. Edwards Deming being foremost among them. The Japanese learned, adapted, and implemented statistical methods to suit their needs. They learned how to organize TQM systems and produce quality products. They stormed the world markets with high-quality goods, including automobiles, electronics, steel, and chemicals. Japan became one of the economic powers in the world. Its economic success was achieved through a competitive advantage derived from the quality of its products.

Meanwhile, in the United States, the Detroit-based automobile industry lost a different kind of war during the late 1970s and early 1980s. Imports, especially those from Japan and Germany, were eating into the market share of U.S. automakers. American consumers were impressed by the quality of the imports, which had better features, ran longer without failures, and were cheaper to buy. U.S. automakers lost a sizeable portion of their market share, closed many plants, and laid off many workers. "All three major U.S. automakers were awash in red ink, GM's $762.5 million loss was its first since 1921. By virtue of its size, GM was not hit as hard as Ford and Chrysler, both of which hovered on the brink of ruin" (Gabor 1990). It took U.S. automakers almost 10 years before they learned their lessons on quality. However, they did learn how to listen to the needs of the customer, how to design the customer's needs into the cars, how to establish quality systems, and how to produce quality cars. They listened to the counsel of quality gurus such as Dr. Deming and embraced new approaches to managing people with trust and partnership. They then won back a major portion of their lost customers. Lack of quality was responsible for their downfall, and it was through embracing quality that they made an economic recovery.

1.6 Quality, Productivity, and Competitive Position

In the days when quality control was done by inspecting all of the units produced and then passing the acceptable products after rejecting those that were not, improvement in quality was accomplished by rejecting products, which resulted in a decrease in saleable products (see Old Model in Figure 1.2). This had given rise to the impression that quality improvement was possible only with a loss in output.

On the other hand, when product quality is achieved through a quality system, through use of methods that will prevent the production of unacceptable products, thus making the product "right the first time," then every product produced is a saleable unit (see New Model in Figure 1.2). In this new approach, improvement of quality is accompanied by improved productivity. That is, with the same amount of available resources—material, machinery, and manpower—more saleable units are produced. When overhead is spread over a larger number of units sold, the production cost per unit decreases. When a part of the cost reduction is passed on to the customer, it will result in better satisfaction for them, as they will be receiving a quality product at a reduced price. Improved customer satisfaction will create repeat as well as new customers and will lead to better market share.

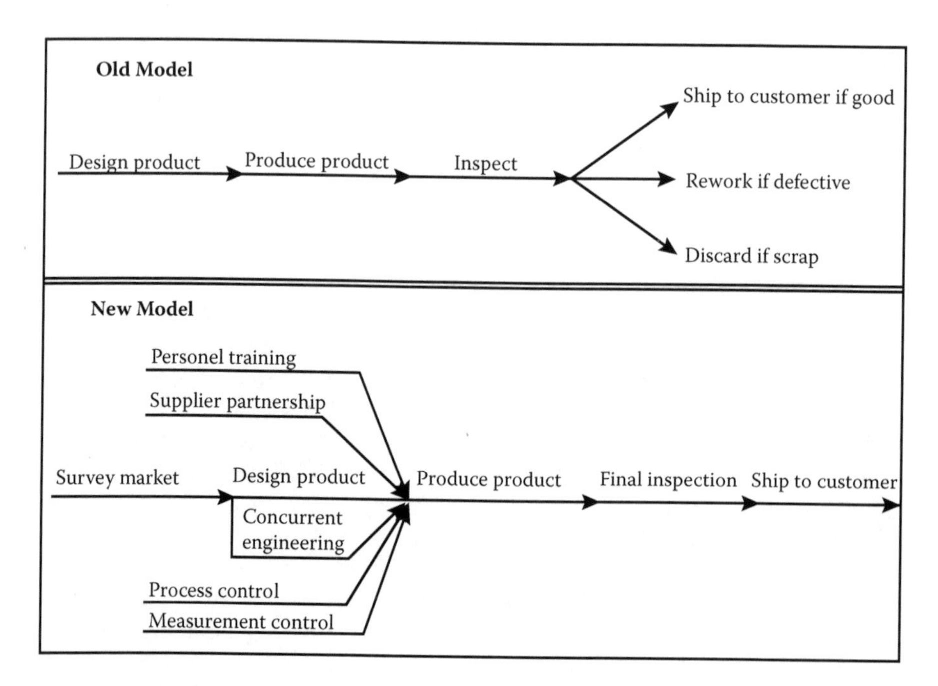

FIGURE 1.2
Old vs. new models of making quality products.

The improved market share improves profitability for the business, which in turn contributes to secure and lasting jobs for employees.

The above message that quality improvement through the prevention of defectives would result in increased productivity, improvement in profits, and an increase in market share—and hence more secure jobs for employees—was delivered by Dr. Deming to his Japanese audience as part of his new management philosophy. He delivered the same message to the American managers and executives through his book *Out of the Crisis* (Deming 1986).

The transformation of an organization from the old model to a new TQM organization has happened in many instances apart from the dramatic turnaround that occured within the U.S. automobile industry in the early 1980s. And it is happening even now in the 2000s, as seen in several examples that have been documented in the literature. For example, the website of the National Quality Program (http://www.baldrige.nist.gov/Contacts_Profiles.htm) contains many such examples where organizations have excelled in quality and achieved business success through the use of modern approaches to producing quality products and services.

1.7 Quality Costs

There were times, in the 1980s and before, when a company president or a business executive, when approached for a budget for a quality-improvement project, would respond with a question like "What will quality buy?" One has to refer to articles (e.g., McBride 1987) in the proceedings of the Annual Quality Congresses of the American Society for Quality during those years to understand the predicament of quality engineers and quality managers in trying to convince business executives of the need for investment in quality improvement. They were trying to "sell" quality to bosses who would understand its value only if expressed in terms of dollars and cents. Even today, there are business owners and company executives who have questions about the value of quality to a business. The concept of "quality costs," which translates the value of quality in dollars and cents, helps in providing answers.

A quality cost study summarizes the economic losses incurred by an organization when not producing quality products and highlights the potential for economic gains arising from producing quality products through the use of quality methods. The study involves gathering data as described below and computing a measure called the "total quality cost" (TQC).

The TQC is a summary measure that reflects (in economic terms) how well an organization performs in delivering products that satisfy its customers. It is, as defined above, the sum of the expenses required in producing quality products and the losses arising from producing defective products. The study undertaken to quantify the TQC may also reveal whether quality efforts are efficiently performed and if quality is produced at the right price. If quality is not produced at the right price, then the study will identify the areas in which efforts should be directed to improve the effectiveness of the system. The TQC can also be used to set strategic goals for quality, to prioritize improvement projects, and to monitor progress in quality efforts. The concept of quantifying the value of quality in terms of dollars and cents was first formally proposed by Dr. Armand V. Feigenbaum (1956) to communicate to business managers that quality has a positive influence on the profitability of a business.

A quality cost study consists of gathering cost data relative to the following four categories of quality-related expenses and then analyzing them in a manner to discover if the system produces quality economically. The four quality cost categories—prevention cost, appraisal cost, internal failure cost, and external failure cost—and their relationship to TQC are shown in Figure 1.3.

1.7.1 Categories of Quality Costs

The four categories of quality costs are further explained below and the possible sources of these costs are indicated. It should be noted that these quality cost categories only include operating costs and do not include any capital expenditure—capital expenditure being defined as expenditure on equipment that has a lifespan of more than one year. These quality costs are therefore referred to as "operating quality costs."

1.7.1.1 Prevention Cost

Expenses incurred for preventing the production of defective products are part of the prevention cost. Examples of expenditure under this category include:

a. Quality planning

This is the cost incurred in preproduction activities involving a review of drawings, specifications and test procedures, and the proper selection of

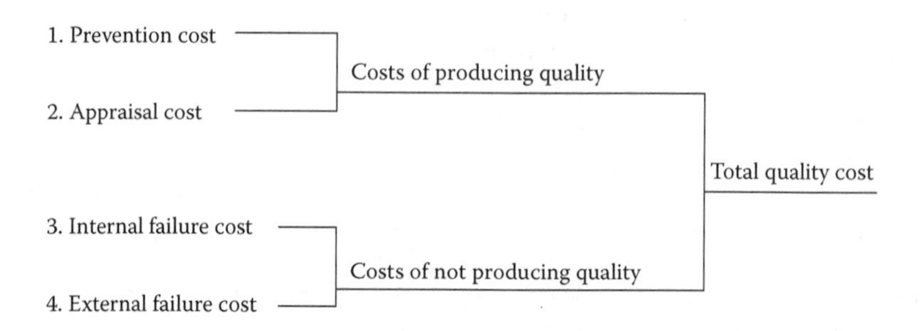

FIGURE 1.3
The four categories of quality costs.

process parameters and control procedures to assure the production of quality products. Evaluating the producibility of the product with available equipment and technology, selecting instruments and verifying their capabilities, and validating production processes are all part of the planning activity.

b. Training

This relates to training workers, supervisors, and managers in the fundamentals of quality methodology. All training costs do not fall in this category. The cost of training workers to make a product, for example, is not a quality cost. The additional cost involved in training workers to make a quality product, however, is a quality cost. For example, the cost of training given to a machinist in the use of a new CNC lathe does not belong in the quality cost category; the cost of training in the use of measuring tools, data collection techniques, and data analysis methods for studying quality problems does belong in this category.

c. Process control

The cost incurred in the use of procedures that help to identify when a process is in control and when it is not, so as to operate the process without producing defectives, would be part of this category. This would include the cost of samples taken, the cost of time for testing, and the cost of maintaining the control methods. If process control is part of the production operator's job, then the time that he or she spends for control purposes must be separated and the cost determined.

d. Quality information system

Gathering information on quality, such as data on quality characteristics, the number and types of defectives that have been produced, the quality performance of suppliers, complaints from customers, quality cost data, and so on, require computer time and the time of computer professionals. Additionally, cost is incurred in the analysis of the information and in the distribution of the results to the appropriate recipients. All these expenditures would be included in the prevention cost. It should be noted that the cost of a computer, which is a capital expenditure, would not be included in the quality cost. However, consumables, such as paper, storage disks, and minor software, or royalties paid, would be included in this category.

e. Improvement projects

The cost of special projects initiated to improve the quality of products falls under prevention. For example, the cost incurred in a project undertaken to make a quality cost study is a prevention quality cost.

f. System development

The amount of time and material expended in creating a quality system, including the documentation of policies, procedures, and work instructions, would be part of the prevention category. Preparation for system audit—for example, against ISO 9000—would also be part of the prevention category.

The items given above are some examples in which prevention quality costs are incurred. There may be many other places in which prevention costs may be incurred depending on the type of industry and the size of operations. The above list is a good starting point, and other sources that generate prevention costs should be included if and when they are discovered.

1.7.1.2 Appraisal Cost

Appraisal cost is the cost incurred in appraising the condition of a product or material with reference to specifications. The following are the major components of this cost:

a. Cost of inspection and testing of incoming material or parts at receiving, or at the supplier; material in stock; or finished product in the plant or at the customer.
b. Cost of maintaining the integrity of measuring instruments, gages, and fixtures.
c. Cost of materials and supplies used in inspection.

The inspection or verification of a product's condition at the end of a production process would be counted in the appraisal category. Inspection carried out on the product within the production process for purposes of process control, however, would be counted as prevention cost.

1.7.1.3 Internal Failure Cost

Internal failure cost is the cost arising from defective units produced that are detected within the plant. This is the cost arising from the lack, or failure, of process control methods, resulting in defectives being produced.

a. Scrap

This is the value of parts or finished products that do not meet specifications and cannot be improved by rework. This includes the cost of labor and material invested in the part or product up to the point at which it is rejected.

b. Rework or salvage

This is the expense incurred in labor and material to return a rejected product to an acceptable condition. The cost of storage, including heating and security, while the product is waiting for rework is included in this cost category.

c. Retest

If a reworked product has to be retested, the cost of retesting and re-inspection should be added to the internal failure cost.

d. Penalty for not meeting schedules

When schedules cannot be met because part of the production is rejected due to poor quality, any penalty arising from such a failure to meet schedules is an internal failure cost. Often, overproduction is scheduled to cover possible rejections. The cost of any overproduction that cannot be sold should be included in the internal failure category.

1.7.1.4 External Failure Cost

External failure cost is the cost arising from defective products reaching the customer. Making a bad product results in waste, and if it is allowed to reach the customer it results in a costlier waste. The external failure cost includes:

a. Complaint adjustment

This is the cost arising from price discounts, reimbursements, and repairs or replacements offered to a customer who has received a product of poor quality.

b. Product returns

When products are returned because they are not acceptable, then—in addition to the cost of replacement or reimbursement—the costs of handling, shipping, storage, and disposal would be incurred. Often, when the product includes hazardous material that needs special care for disposal, the cost of disposal of rejects can become very high. All these costs belong in the external failure costs.

c. Warranty charges

This quality cost category includes those failure costs that result from failures during the warranty period. Any loss incurred beyond the original warranty is not included in quality costs.

1.7.2 Steps in Making a Quality Cost Study

Step 1: Secure approval from upper management

Almost any organization can benefit from a quality cost study. It brings together useful information that has not been gathered before, and provides a perspective on the value of producing products that meet customers' needs. Many organizations do not know the losses they sustain in not making quality products. Based on this author's experience, a quality cost study followed by appropriate improvement projects could lead to substantial savings—a

30% to 50% reduction in quality costs—within a year or two. These savings usually go directly to improve the profit margin.

First, it is necessary to convince upper management of the necessity of a study and to obtain their approval. Because resources must be made available and help from people across the organization is needed to gather data, the study should have the backing of the people at the top.

The literature contains numerous examples (e.g., Naidish 1992; Ponte 1992; Shepherd 2000; Robinson 2000) that discuss successes achieved through quality cost analysis. Such examples can be used to convince upper management about the value of such a study. The case study that is described at the end of this chapter, which is a report on a real quality cost study, revealed that the estimated failure costs in a year amounted to about 94% of the company's profit from the previous year. In this case, a 50% reduction in failure costs would boost the company's profit by about 50%, an attractive proposition to any manager. In almost any situation, if no such studies have been previously made, the cost of making a quality cost study can easily be justified by the benefits to be derived from it.

Step 2: Organize for the study, and collect data

A quality cost study must be performed by a team with representation from various segments of the organization. The more these segments are represented on the team, the better the cooperation and information flow from the segments. Typically, the team will include members from manufacturing, sales, customer service, engineering, inspection, accounting, and quality control. A quality engineer will act as the secretary of the team. The accountant provides the authenticity needed when dealing with cost figures.

Before collecting the data, certain decisions have to be made on how the analysis will be performed. Specifically, a decision must be made on whether the study should be on an annual, semiannual, or quarterly basis. Similarly, a decision must be made on whether it is to be for the whole plant, a department, or a certain product line. For a large plant, it may be more practical to make the analysis on a quarterly basis. For small units, an annual study may be more appropriate. Similarly, if many products are produced in a plant, it may be more feasible to start the study with one major product and then expand it to other product lines. In smaller plants, analysis by individual products may not generate enough data for a study.

Collecting the data for a quality cost study is by no means an easy task, especially if it is being done for the first time. The data may not be readily available, because they are not normally flagged and captured in accounting information systems. They may be lying hidden in a manager's expense accounts, salaries and wage reports, scrap and salvage reports, customer service expenses, and several other such accounts. A large amount of data may have to be estimated by asking the concerned individuals to make educated guesses. However, most of the time, reasonable estimates are good enough for initial analysis. After a successful initial study, when the value of the quality cost study has been established and general agreement exists that the study should be repeated periodically, it may be possible to include additional codes in the accounting system so as to generate data that would be readily available for future studies.

The data collected can be very revealing, even in the raw form. Many managers will see for the first time just how much money can be lost by making bad products or how much is spent on inspections to maintain a quality image. Studies show (e.g., Naidish 1992) that many company executives are unaware of the magnitude of losses incurred by their organization due to the production and delivery of poor quality products. The study reported by Naidish indicated that when company executives were asked about the loss incurred by their organizations due to poor quality, about 66% either did not know anything or thought it was a small fraction—less than 5%—when the true state of affairs indicated a loss in the range of 20% to 40% of sales. In one instance, this author witnessed a scene in which the owner of a printing company was told that the annual cost of scrapped art prints in the plant were more than the annual profit. The president was shocked, jumped up from his seat, and announced that he wanted to fire every one of his employees in the room, which included his VP-operations, the plant superintendent, the chief inspector, and the chief accountant. The president had been totally unaware of the magnitude of the waste until then.

If the data are analyzed in the manner discussed below, then information can be generated in a more useful form.

Step 3: Analyze the data

There are two ways of analyzing quality cost data once they have been accumulated in the appropriate categories. Incidentally, in many situations, we may find that the category to which a particular quality-related expense belongs is unclear. The subcategories defined above as comprising of total quality cost are not exhaustive. Some industries will have their own unique expenditures belonging to quality costs but may not quite fit into the definition of the categories given above. In such situations, the basic definitions of prevention, appraisal, and failure costs must be kept in mind, and the quality-related expenditures should be allocated to the most appropriate category.

Analysis Method 1

Express the TQC as a percentage of some basis, such as net sales billed. Other bases, such as total manufacturing cost, direct labor, or profit, can also be used if they would highlight the magnitude of the quality costs in relation to the total operations. A basis that would normalize quality costs over changes occurring in the business volume would be preferred. Net sales billed is the commonly used basis. When the TQC is expressed as a percentage of sales, it can be used to monitor the quality performance of the organization over time and would enable a comparison with another organization of similar size or one with the best-in-class performance.

What is the optimal level for quality costs as a percentage of sales? The answer depends on the nature of the business, the maturity level of the quality system, how cost categories have been defined, and the management's attitude toward quality. The TQC cannot be reduced to zero because it includes prevention and appraisal costs, which always exist at some positive level. Some believe that failure costs must be reduced to zero or near zero no matter how much it costs in prevention and appraisal. Dr. Deming (1986), for example, believed that quality should not be judged by what it returns in dollars in the immediate future. Good quality earns customer satisfaction, which brings more customers, whereas poor quality spreads customer dissatisfaction and produces a loss of market share in the long term. According to Dr. Deming, continuously improving quality to reach excellence should be the goal, regardless of how much it costs. His message was: quality must be achieved at any cost.

On the other hand, others believe that expenditure in quality efforts must be made only to the extent that is justified by the economic returns they produce. Figure 1.4 depicts the relationship between product quality level and TQC. The graph, a popular reference among quality professionals, conveys the message that the TQC reaches a minimum at a particular quality level. Beyond this point, the cost of control becomes so expensive that it cannot be justified by the return it produces in terms of a reduction in failures. It conveys a commonly held view that quality has an optimum level based on the economics of a given situation, and that striving for quality beyond the optimum level may not be profitable. "Perfectionism" and "gold plating" are

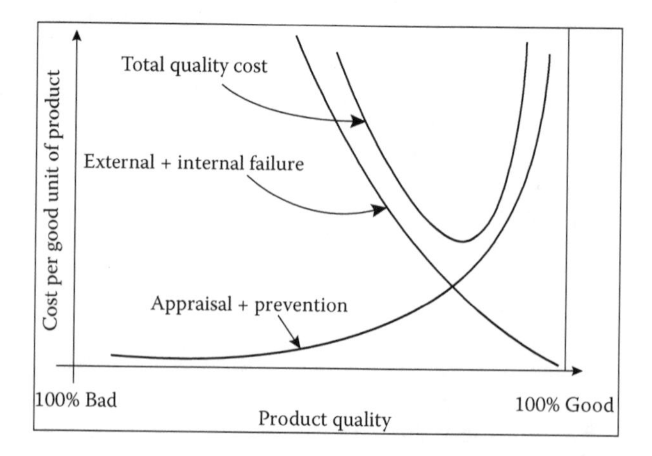

FIGURE 1.4
Relationship between quality and quality costs. (Reprinted from Juran, J. M. and F. M. Gryna, *Quality Planning and Analysis*, 3rd ed., McGraw-Hill, New York, 1993, with permission of McGraw-Hill Co.)

some of the terms used to refer to attempts to improve quality beyond this economic level.

Juran and Gryna (1993) state that this optimum has moved higher along the product quality scale in the U.S. industries over the years because of the availability of more efficient and automated methods of inspection. This means that it has become more economical to produce better quality, yet the message is still there that quality has an optimum limit.

Neither one of the above views can be the right policy for all situations. The policy applicable depends on the given situation. If, for example, the product is an automobile brake system, where a defective product could cause enormous property damage, injury to people, and loss of reputation for the company, then "quality at any cost" may be a proper goal. On the other hand, if the cost of marginal improvements in quality is high and the improvement will not produce commensurate customer satisfaction, then "quality at the most economic level" may be a justifiable policy. Thus the "optimal" level of the TQC as a percentage of sales depends on the particular philosophy that an organization adopts according to the requirements in its business environment.

We have experience with quality systems in which the failure quality costs were reduced to levels as low as 2% of sales with potential for even further reduction. This would indicate the possibility for reducing failure costs to very low levels. For low-precision industries, such as wire and nail manufacturers and furniture makers, the optimal percentage may be lower than that in high-precision industries, such as precision machine shops and instrument makers. However, in any given place, there will always be inexpensive improvement opportunities that can produce a sizeable return. Such opportunities should be continuously sought, and projects should be completed to improve quality and reduce waste.

Analysis Method 2

In this method, the costs under individual categories are expressed as percentages of the TQC, giving the distribution of the total cost among the four categories. Such a distribution usually reveals information on what is happening within the quality system. The following two examples illustrate the idea:

Example 1		**Example 2**	
Prevention	3%	Prevention	4%
Appraisal	20%	Appraisal	87%
Internal Failure	9%	Internal Failure	9%
External Failure	68%	External Failure	0%
Total	100%	Total	100%

In Example 1, it is obvious that the external failure is a major contributor to the quality costs and that reducing the external failure should be the first priority. External failures not only bring losses immediately in terms of customer complaints and customer returns, they also produce long-term losses due to loss of reputation leading to a loss in future sales. According to a rule of thumb that quality professionals use, if a defective unit caught within the plant costs $1 to repair, it will cost about $10 to repair it if it reaches the customer. Thus, stopping those defective units from leaving the door may save considerable external failure costs and therefore should be the first priority. Thus, a quick, short-term solution in this case would be to increase the final inspection to prevent the defective product from leaving the door. This will reduce the external failure costs, although it may result in increased internal failure costs. Appropriate prevention methods should then be initiated to reduce the internal failures to achieve further reduction in the TQCs.

In Example 2, the appraisal costs are very high, and the external failures are nonexistent, indicating a situation in which inspection is probably overdone, not done efficiently, or both. This may happen in organizations that depend on inspection to satisfy their customers and want to protect their reputation at any cost. It may be worthwhile for this system to review some of their inspection methods in order to make them more efficient or even eliminate them if they are found to be unnecessary.

1.7.3 Projects Arising from a Quality Cost Study

While reading the results from a quality cost study, it may sometimes be obvious what needs to be done in order to correct a quality problem as it is identified. For example, in one case in which the external failure costs were found to be high, it was discovered that the shipping department was sending products made for one customer to another customer! The problem was quickly corrected by installing a simple shipping control procedure. In another case, the inspection department was testing the final product under one set of conditions (110 V), whereas the customer was testing the product under a different set of conditions (240 V). This type of misinterpretation of a customer's needs is, unfortunately, a common cause of product rejection and customer dissatisfaction. Again here, the problem was corrected by a simple change to the testing procedure.

In other situations, however, the reasons for large failure costs, internal or external, may not be so obvious. Further investigation may be necessary to determine the root causes of the problems and take corrective action. In other words, projects must be undertaken, maybe by a team, to gather data, make analysis, discover solutions, and implement changes to effect quality improvement. Then, problem-solving tools, such as Pareto analysis, fishbone diagram, regression analysis, and designed experiments, may become necessary. These problem-solving tools, along with the systematic procedures for problem solving and project completion, are discussed in Chapter 8.

1.7.4 Quality Cost Scoreboard

After the initial study has been completed and a few quality-improvement projects have been successfully implemented, new codes can be created within an existing accounting system. The monograph by Morse et al. (1987), published by the Institute of Management Accountants, provides useful guidance on creating such a system. Quality cost data can then be accumulated on a routine basis so that future studies can be made without expending much labor. Periodic analysis and reporting would help in monitoring the progress that has been made in reducing these costs through identifying new opportunities for improvement. When quality cost studies are made on a continuing basis, a scoreboard can be generated to provide a pictorial perspective on how quality costs have been changing over time.

Figures 1.5 and 1.6 show the typical results of a quality cost study performed over a period of several years. Figure 1.5 shows the behavior of the TQC over time, and Figure 1.6 shows how the distribution of individual cost categories changes over time. Table 1.2 describes the conditions expected in organizations while on the quality improvement journey and it shows the appropriate responses to those conditions in order to achieve success in the improvement program.

Some of the best advice for people who want to use a quality cost study to pursue a quality improvement program comes from Ponte (1992). His advice

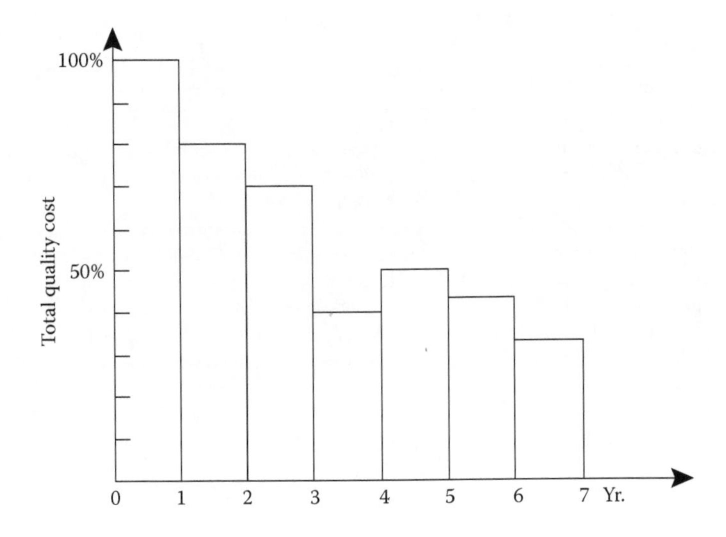

FIGURE 1.5
Typical progression of total quality costs over time. (Adapted and reproduced from Noz et al., *Transactions of the 37th Annual Quality Congress*, pp. 303. Milwaukee: American Society for Quality Control, 1983. With permission.)

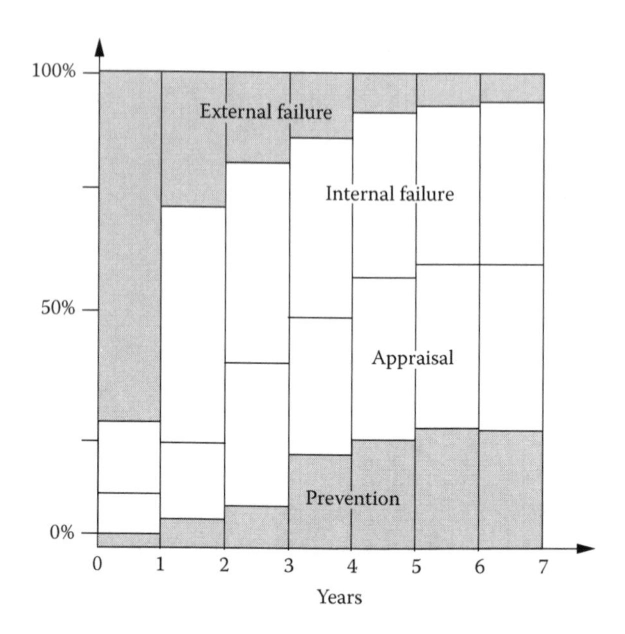

FIGURE 1.6
Typical changes in quality cost categories over time. (Adapted and reproduced from Noz et al., *Transactions of the 37th Annual Quality Congress*, pp. 303. Milwaukee: American Society for Quality Control, 1983. With permission.)

TABLE 1.2

Progression of Activities During a Quality Improvement Program Guided by Quality Costs

Year	Condition of the System	Appropriate Action
1	Large failure cost, most of it external. Low appraisal costs. Prevention almost nonexistent	Improve final inspection, and stop defectives from leaving the plant. Develop quality data collection. Identify projects, and begin problem solving
2	Failure costs still high. Prevention effort low	Establish process control. Improve data collection, data analysis, and problem solving
3	Process control installed. Some process capabilities known. External failures at low levels	Improve process control on critical variables based on capability data. Begin using experiments to select process parameter levels
4	Failure costs 50% of total quality costs	Continue using designed experiments for process improvements to reduce variability. Continue process control on critical variables
5	Improved processes	Continue process control. Look for project opportunities for further variability and cycle time reduction. Improve quality of service processes: maintenance, safety, scheduling, and customer service
6 and 7	Process in stable condition	Same as above

Source: Adapted and reproduced from Noz et al., *Transactions of the 37th Annual Quality Congress*, pp. 303. Milwaukee: American Society for Quality Control, 1983. With permission.

can be summarized in the following sequence of steps, which he calls the "cost of quality" (COQ) program:

1. Obtain a commitment from the highest level of management in the organization.
2. Choose a "focal person" to be responsible for promoting the COQ program.
3. Write a COQ procedure, which will include the definitions, responsibilities, training material, analyses to be made, reports to be generated, and the follow-up action needed.
4. Make the COQ procedure a dynamic document, requiring review and reassessment every 6 to 12 months.
5. Require a monthly COQ program review meeting, chaired by the CEO, as a mandatory requirement of the COQ procedure.
6. Train all involved employees in the principles and procedures of the COQ program.
7. Develop visibility for the program by creating a trend chart that shows the gains achieved. (Use manufacturing cost as the base rather than sales dollars, because the former will show the COQ as a larger percentage.)
8. Demand root-cause analysis for determining the causes of failures.
9. Demand closed-loop, positive corrective action response.

1.7.5 Quality Costs Not Included in the TQC

The TQC, as defined earlier in the chapter, is only an index to show the economic impact of achieving or not achieving product quality in a productive enterprise. It is a good barometer to indicate how well a system is performing in quality, but it is not an exhaustive measure that includes all economic outcomes pertaining to quality. Some organizations use only the failure costs (internal plus external) as an indicator of the progress made in quality, and they do not attempt to quantify the cost of prevention or appraisal. Although failure cost, or "poor quality cost" as some authors (e.g., Harrington 1987) call it, could serve the purpose of monitoring an organization's progress in quality, such an index will not reflect overspending in prevention or appraisal and may not give a complete picture. Therefore, TQC—as defined above, following the recommendation of the Quality Cost Committee of the American Society for Quality—is a more complete index.

We ought to realize at this point that several other costs arise when quality is not present in products, which are not included as part of the TQC index defined above. Some of them (Feigenbaum 1983; Juran 1988; Gryna 2000) are:

- Indirect quality costs
- Vendor quality costs

- Intangible quality costs
- Liability costs
- Equipment quality costs
- Life-cycle quality costs

Indirect quality costs come from redundant manufacturing capacity installed and production made to cover possible rejection due to poor quality. The excess inventory cost of material and parts, stored to accommodate possible rejections of the parts and material, are also sources of indirect costs.

Vendor quality costs are the excessive quality costs in appraisal and failures that are incurred by vendors who have not implemented a quality-assurance program themselves. In the absence of quality-assurance programs, the excess costs incurred in failures and other categories by the vendor are, indeed, passed on to the customer.

Intangible quality costs are losses resulting from customers forming a poor image of a company's product quality because of bad experiences. This includes loss of future sales, future customers, and general market share.

Liability costs are losses from liability actions by product users. It would include those costs arising from exposure to liability, even when no actual judgment is issued against the company. Such losses include expenses incurred for preparing to defend against liability actions (e.g., cost of legal fees, expert witnesses, data collection, etc.).

Equipment quality costs are the costs arising from capital equipment that have direct impact on quality, such as automated data collection systems, automatic test equipment (ATE), measuring equipment, and computer hardware and software meant specifically for processing quality information. The cost of space, heat, security, and other appurtenances needed for the equipment are part of this cost. The cost of equipment, as previously mentioned, is not normally included in quality costs. However, in some operations where ATEs—or the modern computerized measuring machines known as "coordinate measuring machines"—are used, this cost can be quite considerable, and so is included in the quality costs to fairly reflect the expenditure involved in creating quality.

Life-cycle quality costs are the costs arising from service and repair done to a product over a reasonable time after the warranty period. Normally, only service charges incurred within the warranty period are included as part of the external failure cost. It is possible to include the entire life-cycle cost of maintenance as part of quality costs because they reflect the original quality that was designed and built into the product. These data, however, may be a bit difficult to gather.

The quality costs discussed above exist in varying degrees in different organizations. Presumably, the reason that they are not included in the TQC index (defined by the American Society for Quality) is that they are very difficult to collect and quantify. Some of them have a long lag time between their occurrence and their actual showing up in company books.

As mentioned earlier, some companies use only the failure costs as the index to monitor the quality performance of a system. Others add some of the indirect costs to the TQC if they seem to be significant to their operations. It does not matter how the economic index is defined for a company as long as the same definition is used consistently over time so that comparisons are valid. However, if a comparison is sought between one company's TQC with another's—for example, a quality leader or a competitor—then the TQC needs to be made up of the same components in both systems for a valid comparison. That is when a standard definition for the four categories and the TQC is necessary.

1.7.6 Relationship among Quality Cost Categories

Often a quality professional has a question during his or her improvement journey: "How much more should I invest in prevention or appraisal in order to bring down the current failure costs by, for instance, 50%?" To answer such a question the interrelationship among the quality cost categories—the relationship that shows how a change in one category produces changes in others—is needed.

Several researchers have attempted to answer this question, yet no exact answers seem to be available. Statements, such as "an ounce of prevention is worth a pound of failure" (Campanella and Corcoran 1982), describe the relationship in a qualitative manner and indicate the leverage that prevention has in reducing failure costs. Krishnamoorthi (1989) attempted to obtain a relationship among the categories using regression analysis, treating the percentages in the four cost categories as random variables taking different values in different quality systems. Cost data were obtained from different quality systems, and the relationships among the categories were obtained using regression analysis, as shown in the following two equations:

$$E = \frac{5.9}{P} + \frac{298}{A} \tag{1.1}$$

$$I = \frac{121}{P} + 0.213A \tag{1.2}$$

where E, I, P, and A represent the percentages of external failure, internal failure, prevention, and appraisal costs of the TQC, respectively. These relationships can be interpreted as follows.

According to Equation 1.1, external failure decreases with an increase in prevention as well as appraisal, the decrease being much faster with an increase in appraisal than with prevention. According to Equation 1.2, internal failure decreases with an increase in prevention but increases with an increase in appraisal. These equations were developed from a mere 24 sets of sample data coming from vastly different types of industries, with a considerable amount of variability in them. As such, the equations are not suitable for making exact predictions about changes in the cost categories but can be used to follow the direction of changes in them. The usefulness of these results, however, lies in the fact that similar models can be developed from data from one company or one industry. When data are gathered from such similar sources, in which some homogeneity of data can be expected, models can be developed and used to make predictions about changes in one category due to changes in another, with some confidence.

1.7.7 Summary of Quality Costs

A quality cost study reveals quite a bit of information about the health of a quality system. It indicates whether quality is produced by the system at the right price. When the system health is not good, the study points to where deficiencies can be found so that action can be taken to address them.

The quality costs, which are the costs associated with control activities to produce quality products plus the losses arising from not producing quality products, are categorized into four categories: prevention, appraisal, internal failure, and external failure. The data collected from various sources that generate quality costs can be analyzed in two ways. The TQC, which is the sum of the costs in the four categories, expressed as a percentage of some chosen base representing the volume of business activity, can be used to follow the progress of quality performance of an organization over time. Alternatively, it can be used for comparing one company's performance with that of another, a leader or a competitor in the same industry. When the four categories are expressed as a percentage of the TQC, the places where deficiencies exist are revealed. The value of a quality cost study lies in pursuing such opportunities, discovering root causes, and implementing changes for improvement.

Many other quality costs, such as indirect costs, liability exposure costs, and so on, are not included in the TQC index because these are difficult to obtain and have a long lag time in coming. However, the TQC as described above, as recommended by the American Society of Quality's Quality Cost Committee, is a good index to reflect the economic benefits of implementing a quality assurance system.

Most of the models for quality systems, such as the ISO 9000, the MBNQA criteria, or the Six Sigma process (discussed in Chapter 9), recommend use of a performance measure to monitor the overall performance of an organization in both quality and customer satisfaction. The TQC is a good measure to use for this purpose.

1.7.8 A Case Study in Quality Costs

The case study provided below relates to ABCD Inc., a company that packages materials such as liquid soap, deodorant, and cooking oil for large customers who ship the material to ABCD in tank cars. The company then fills the liquids into bottles or cans in their filling lines, seals and labels them, packages them in cartons and skids, and ships them to the customers' warehouses. Sometimes, they do some mixing operations before or during filling the liquids into the containers. Customers' quality requirements include the appearance and strength characteristics of bottles and cans and a very demanding set of specifications for fill volume or fill weight of the contents of the bottles and cans.

The study made by D.H., their quality engineer, is called a "preliminary study," because information regarding several aspects of quality costs were either not available or were incomplete when this study was made. However, this study was the starting point of several improvement projects, which resulted in considerable improvements in operating profits of the company.

ABCD Inc.

Quality Improvement Program

Preliminary Quality Cost Study

Made by: D.H.

Important Notes

1. This study is a preliminary study in that several costs, which are unavailable or difficult to estimate, have not been included. For instance, this study does not attempt to estimate the proportion of management salaries that are part of quality costs.
2. Based on Note 1 above, one should realize that our total quality cost is actually much greater, probably 25–50% higher than the estimate presented here.
3. However, even considering Notes 1 and 2 above, this study still gives a good picture of the magnitude of our quality problems, and it provides an excellent measure of effectiveness for monitoring our quality improvement program.
4. Costs have been determined for the DDD plant only.
5. This study presents quality costs incurred in the period January–June XXXX, a six-month period, and extrapolates these costs to provide figures on a "per-year" basis.
6. Descriptions of all quality costs included in this preliminary study are given on the following pages. These quality costs have been put into three categories: failure costs (internal and external), appraisal costs, and prevention costs.

7. A summary and conclusion have been provided.
8. The assistance of L.M. and F.P. is greatly appreciated.

Failure Costs (Internal and External)

1. ABCD material loss

This category includes the costs of any material (chemical or component) that is lost because of overfill, changeovers, spills, or any other cause that results in using more material than should have used. This cost is calculated by taking the difference between the cost of the raw materials that theoretically should have been used to produce the finished product and the cost of the raw materials that were actually used to produce the finished product. No loss allowances are taken into account here. This category includes only those costs associated with ABCD-owned materials. (See Category 3 for the costs associated with customer-owned materials.)

Total material loss for six months: $507,000.

2. Rework labor

This category includes all labor associated with reworking any material that has been rejected either before or after leaving the plant. As a result, included in this cost are both the internal and external failure costs because of rework. These figures can be found in accounting documentation under Department 642.

Total rework labor for six months: $214,200.

3. Customer material loss

This category includes the cost of any customer-owned material (chemical or component) that is lost because of overfill, changeovers, spills, or any other cause that results in using more material than we should have used. Also included in this cost are customer-owned raw materials or filled stock that had to be destroyed. This category involves costs billed to us by our customers. We "pay back" our customers for overusing their material, which occurs during "reconciliation" meetings.

Total customer material loss for six months: $107,000.

4. Cost of producing off-quality products

This category includes a percentage of direct labor associated with the production of rejected material. For example, for the monthly reject percentage of 2.5% and a total direct production labor cost of $10,000 for that month, the cost of producing off-quality products would be $250 for that month. Because a large proportion of our rejected material is reworked and then sold, one could argue that the cost presented here in this category is not all quality cost. We recognize that there may be some overestimation of the failure cost here.

Total labor cost in rejected product for six months: $77,300.

5. Raw material destruct

This category includes the cost of any raw materials (chemical or component) that are not used for the conversion process but are destroyed because of off-quality rejection and the cost of raw material components that are scrapped during a rework operation. This figure is based only on the cost of ABCD-owned materials.

Total loss from raw material destruct for six months: $24,700.

6. Filled stock destruct

This category includes the cost of filled stock that is destroyed. The material destroyed here is scrap because the product has been produced and rejected, but rework is not possible. This figure is based only on the cost of ABCD-owned material.

Total loss from filled stock destruct for six months: $4700.

7. Disposal

This category includes only the cost of disposing of the crushed cans. The figure given here is based on a cost of $85 per trip and an average number of pickups of 1.5 per week.

Total loss from material disposal for six months: $3300.

8. Re-sampling

This category includes the cost of re-inspecting products after they have been reworked. The figure for this cost is calculated by multiplying the total number of re-sampling hours by an average hourly wage.

Total loss from re-sampling for six months: $3300.

9. Downtime

Although total downtime per month is available and somewhat categorized, considerable difficulty is encountered when attempting to estimate the proportion of downtime that results from quality problems. As a result, I have not included the downtime caused by poor quality in this study.

10. Obsolete material

I have not included loss caused by obsolescence in this study.

11. Customer service

The percentage of the salaries in the customer service department equivalent to the proportion of time dedicated to dealing with customer complaints, returned material, allowances, and so on, should be included in our total quality cost. However, since these figures are not readily available and would require considerable estimation, I have not included them here.

12. Managerial decision

One of the most costly quality costs of our organization is the cost associated with the time that managers have to spend on quality-related problems. Not only is this cost a big one, it also is a difficult one to estimate and monitor. Hence, as stated at the beginning of this study, I have not attempted to estimate these costs. We must realize, however, that this quality cost is very significant.

Total failure cost for six months: $941,500.

Appraisal Costs

1. Chemical QA

This category includes all labor and other expenses associated with the chemical quality assurance (QA) department. These figures can be found in accounting documentation under Department 632.

Total cost of chemical QA for six months: $164,600.

2. Physical QA

This category includes all labor and other expenses associated with the physical and component QA department. These figures can be found in accounting documentation under Department 633. Subtracted from these figures (and put in prevention costs) are those costs associated with quality reporting.

Total cost of physical QA for six months: $162,800.

3. Inspection and testing

This category includes all labor associated with inspecting and testing products for the first time. Fifty percent of the cost of the line process control coordinators (PCCs) was allocated to inspection and testing, and the other 50% was allocated to the prevention costs. This cost was estimated by D.R.

Total inspection and testing cost for six months: $119,000.

4. Materials and services

This category includes the cost of all products and materials consumed during testing. This cost was estimated by D.W.

Total material and service cost for six months: $30,000.

5. Test equipment

This category includes the cost of maintaining the accuracy of all test equipment. Any cost involved in keeping instruments in calibration is included in this cost. This cost was estimated by D.W.

Total cost due to test equipment for six months: $12,500.

Total appraisal cost for six months: $488,900.

Prevention Costs

1. Process control

This category includes all labor associated with assuring that our products are fit for use. Included in this cost is 50% of the cost of the line PCCs; the other 50% of the cost of the line PCCs was allocated to appraisal costs. This cost was estimated by D.R.

Total cost for six months: $119,000.

2. Improvement projects and new-products review

This category includes all labor associated with reviewing new products for preparing bid proposals, evaluating potential problems, making quality improvements, and so on. Included in this cost is the proportion of the salaries of technical service and product development personnel corresponding to the percentage of time spent on quality activities. After consultation with the personnel involved, the following figures were obtained. The cost of service to preparation of cost quotes, preparation of specifications, regulatory activities, preparation of batch packets, new-product review meetings, and so on is about $28,700. The cost of technical services quality-related activities is about $33,800.

Total cost of improvement projects for six months: $62,500.

3. Quality reporting

This category includes the cost of making quality reports to management. This cost has been estimated and taken out of the physical QA.

Total cost of quality reporting for six months: $8500.

4. Training

The cost of all quality-related training that was undertaken by anyone in the company during the six-month period should be included in this section. These costs are not readily available and, hence, are not included in this study.

Total prevention cost for six months: $190,000.

Summary	
Failure costs	$941,500
Appraisal costs	$488,900
Prevention costs	$190,000
Total quality cost for six months:	$1,620,400
Sales for January–June, XXXX	$24,287,000
Quality cost as percentage of sales	6.67%
Failure cost as percentage of sales	3.88%
Profit for January–June, XXXX	$1,005,000
Total quality cost for the period	$1,620,400
Quality cost as percentage of profit	161%
Failure cost as percentage of profit	94%
Interrelationship of quality cost categories:	
Failure cost	58%
Appraisal cost	30%
Prevention cost	12%
Total quality cost	100%

Conclusions

1. The estimated total quality cost for ABCD Inc., is $3,240,800 per year. The actual quality cost is probably 25–50% greater because of the several costs that were not readily available.
2. The total quality cost of 6.67% of total sales should be compared with other companies in our industry. We will attempt to find such comparable values.
3. The total quality cost is much greater than our profit. Any quality cost that is eliminated directly becomes profit. In other words, our profit over the first six months of this year would have been $2,625,400 (instead of $1,005,000) if no quality costs were incurred.
4. Internal failure costs account for 58% of all quality costs. This percentage is too high. More resources should be allocated to prevention.
5. The two highest failure cost categories—and hence the ones that should be tackled first—are:
   a. Material losses
   b. Rework labor
6. This cost study will be repeated periodically to monitor our progress in making quality improvements.

1.8 Success Stories

Many organizations have successfully made use of the methodologies available for creating quality products and satisfying customers, and have achieved business success. Many stories can be found in the literature where failing organizations were revived to reach profitability by using the systems approach to making and delivering quality products to customers. Each year the Baldrige National Quality Program publishes stories about winners of the Malcolm Baldrige National Quality Award on their website (http://www.baldrige.nist.gov/Contacts_Profiles.htm). Anyone who reads a few of these stories will become convinced about the value of the quality methodologies that have been developed by the leaders and professionals in the quality field. These methods, discussed in the following chapters of this book, can be put to use by anyone wishing to make quality products, satisfy customers, and succeed in business.

1.9 Exercise

1.9.1 Practice Problems

1.1 Pick out the appraisal quality cost from the following:

- **a.** Fees for an outside auditor to audit the quality management system
- **b.** Time spent to review customers' drawing before contract
- **c.** Time spent in concurrent engineering meetings by a supplier
- **d.** Salary of a metrology lab technician who calibrates instruments
- **e.** None of the above

1.2 The cost of writing the operating procedures for inspection and testing as part of preparing the quality management system documentation should be charged to:

- **a.** Appraisal costs
- **b.** Internal failure costs
- **c.** Prevention costs
- **d.** External failure costs
- **e.** None of the above

1.3 Which of the following will be considered a failure quality cost?

- **a.** Salaries of personnel testing repaired products
- **b.** Cost of test equipment

c. Cost of training workers to achieve production standards

d. Incoming inspection to prevent defective parts coming into stores

e. All of the above

1.4 In the initial stages of implementing a quality cost study, the data is likely to show:

a. Large prevention costs with small failure costs

b. Large appraisal costs with large prevention costs

c. Small prevention costs with large failure costs

d. Small total quality costs

e. None of the above.

1.5 One of the following is not a quality cost:

a. Cost of inspection and test

b. Cost of routine maintenance of machinery

c. Cost of routine maintenance of instruments

d. Salary of SPC analysts

e. None of the above

1.6 Operating quality costs can be related to different volume bases. A commonly used base is:

a. Indirect labor cost

b. Overhead maintenance cost

c. Processing costs

d. Sales dollars

e. None of the above

1.7 Analyze the following cost data:

- Employee training in problem solving—$25,000
- Scrap and rework—$250,000
- Re-inspection and retest—$180,000
- Disposition of rejects—$100,000
- Vendor system audits—$5000
- Incoming and final inspection—$100,000
- Repair of customer returns—$40,000

Considering only the quality costs in the above data, we might conclude:

a. Prevention is >10% of total quality costs (TQC)

b. Internal failure is >70% TQC

c. External failure is >10% of TQC
d. Appraisal costs >20% of TQC
e. All of the above

1.8 The percentages of total quality costs are distributed as follows:

- Prevention: 2%
- Appraisal: 23%
- Internal failure: 15%
- External failure: 60%

A recommendation for immediate improvement:

a. Increase external failure
b. Invest more money in prevention
c. Increase appraisal
d. Increase internal failure
e. Do nothing

1.9 A practical, simple definition for the quality of a product is:

a. It shines well
b. It lasts forever
c. It is the least expensive
d. It is fit for the intended use
e. It has a brand name

1.10 According to a more complete definition of quality, in addition to the product meeting the customer's needs, it should be:

a. Produced at the right price through efficient use of resources
b. Packaged in the most colorful boxes
c. Advertised among the widest audience possible
d. Produced by the most diverse group of people
e. Sold in the niche market

1.11 One of the following is not part of total quality management:

a. Commitment of top management to quality and customer satisfaction
b. The inspection department being in total control of quality
c. The entire organization being focused on satisfying the needs of the customer
d. Creating a culture where continuous improvement is a habit
e. Training all employees in quality methods

1.12 A batch of thermostats failed in the test for calibration at the end of the production line. After being individually adjusted, they were retested. The cost of this retest will be part of:

- **a.** Prevention quality cost
- **b.** Appraisal quality cost
- **c.** Retest quality cost
- **d.** Internal failure quality cost
- **e.** External failure quality cost

1.13 Indirect quality costs are the costs incurred:

- **a.** By vendors, which are then charged indirectly to the customer.
- **b.** In building additional capacity or inventory to make up for units lost due to poor quality.
- **c.** In lost future sales because of poor quality image.
- **d.** In making defective products because of the fault of indirect labor.
- **e.** Because of product failures after the warranty has expired.

1.14 Of the following statements, which is true in the context of relationships among quality cost categories?

- **a.** When prevention cost is increased, appraisal cost increases.
- **b.** When appraisal cost is increased, prevention cost increases.
- **c.** When internal failure cost is increased, external failure cost increases.
- **d.** When the appraisal cost is increased, internal failure cost increases.
- **e.** When external failure cost is increased, internal failure cost increases.

1.15 One of the following cannot be used as a basis for monitoring the annual total quality figure:

- **a.** Annual sales
- **b.** Annual labor cost
- **c.** Annual profit
- **d.** Annual production cost
- **e.** None of the above

1.16 A quality cost study in an organization revealed the following distribution of cost among the four categories:

- Prevention: 4%
- Appraisal: 80%

- Internal failure: 15%
- External failure: 1%

A reasonable conclusion from such an analysis is that:

a. Their external failure is excessive.
b. They are overspending in prevention.
c. They are spending an adequate amount in appraisal.
d. They are overdoing inspection.
e. Their internal failure cost is just right.

1.9.2 Mini-Projects

Mini-Project 1.1

This exercise is meant to provide an experience in defining the quality of a "product." Often, the first challenge for an engineer is to identify the "products" produced in an organization and the characteristics that define the quality of those products. The engineer must then come up with schemes for measuring the quality characteristics and select a set of specification limits—targets with allowable variation—that are used to measure the quality of the products. Here, the student is asked to select a service environment to perform the exercise, because it is more challenging to identify the above features in a service environment than in a physical product environment. If the exercise can be done successfully in the service environment, it will be easier to do the same in a tangible product environment.

Choose any two of the systems listed below, and identify for each the "products" produced (at least three), the quality characteristics of each product (at least five for each product), a measuring scheme for each quality characteristic, and a set of specifications for each characteristic. Provide the answers in the format shown below for an example system.

1. A university library
2. An academic department in a university
3. A city fire department
4. A post office
5. A laundromat
6. A motel
7. The maintenance department of a production facility
8. The marketing department of a company
9. Any other system that produces a service

Example System: A University Library

"Product"	Quality Characteristic	Measuring Scheme	Set of Specifications
Computers to use	Availability	Average time people wait	Less than 5 minutes
	Variety of applications	Number of applications available	At least 15 different application programs
	Reliability of computers	Average number of computer breakdowns/month	Not more than one computer/month
	Printer availability	Average waiting time for prints	Not more than 5 minutes
	Newness of technology	Years since most recent upgrade	Not more than 1 year
Books to borrow	Size of collection	Number of books in shelves	Minimum of 2 million
Research assistance	... ...	... ...	... ...

Mini-Project 1.2

This is an exercise in determining which expense belongs in which quality cost category and then analyzing the cost data to reveal any problems in the system. The data come from a chemical plant with annual sales of $100 million. Their operating budget (manufacturing cost) is $34 million. Analyze the data, and draw conclusions about the current health of the quality system. The remarks in parentheses are meant to explain the nature of some of the expenditures, which otherwise may not be readily apparent.

Quality Cost Study of XYZ Chemical Plant—19XX	$/Month
1. Quality planning	70
2. Vendor qualification (supplier certification)	7,150
3. Acceptance inspection and test planning (planning for inspection)	100
4. Process equipment design for quality	370
5. Quality task teams (special projects related to quality)	1,940
6. Turnaround planning—Equipment (quality-related scheduled maintenance)	330
7. Turnaround planning—Instruments (quality-related scheduled maintenance)	650
8. Quality profiling (determining product specs to meet customer needs)	1,000
9. Quality statistics (collection of data and analysis)	420
10. Planning quality motivational programs	170
11. Production engineers on quality	70
12. Process control engineering—QA department	27,050
13. Process control engineering—Engineering department	12,500
14. Systems conformance (audit to verify system's conformance)	70

15. Design/development of quality measurement/control equipment	
Engineering department	740
QA department	900
16. Quality training	1,050
17. Raw material acceptance testing	3,345
18. Process testing—Operators	13,300
19. Review production analysis data	530
20. Environmental testing (as it relates to product quality)	70
21. QA in no. 1, no. 2, lot analysis, instrument calibration, and aging checks	61,313
22. Off-grade burned	3,300
23. Off-grade blended off-labor	210
24. Off-grade storage with interest	3,200
25. Reprocessing/sold at reduced cost	5,300
26. Partial credit issued—Customer kept product	760
27. Bulk returns	2,750
28. Packaged returns	3,500
29. Returns to other locations	420

Mini-Project 1.3

A university ombudsman has the responsibility of resolving grievances brought to him/her by students relating to academic work, such as disputes on a grade, how the student was treated in the class, or any other issue related to academic work. The ombudsman mainly facilitates communication between the student and a teacher or any academic administrator who has jurisdiction in the matter. The ombudsman can also help in mediating the resolution of disputes by virtue of his/her position as an experienced, neutral faculty member. Furthermore, the ombudsman makes recommendations to the university administrators on matters of policy to help resolve student disputes expeditiously or to eliminate causes that give rise to grievances.

How would you measure the quality of service of an ombudsman to the university? Identify the services ("products") delivered, the quality characteristics of those services, and how they will be measured. Use the format suggested in Mini-Project 1.1.

Mini-Project 1.4

This project deals with a "large" system where the people involved are counted in millions and the costs and benefits are counted in billions. It takes a bit of experience to get used to thinking in millions and billions.

The U.S. Social Security system was created by an act of U.S. Congress in 1935 and signed into law by President Franklin Roosevelt. The main part of the program includes providing retirement benefits, survivor benefits, and disability insurance (RSDI). In 2004, the U.S. Social Security system paid out almost $500 billion in benefits. By dollars paid, the U.S. Social Security program is the largest government program in the world and the single greatest expenditure in the U.S. federal budget. More on the Social Security system can be obtained from their website: http://www.ssa.gov/.

How would you define the quality of service provided by the Social Security system? What are the "products," and what are the characteristics and performance measurements to be taken to monitor the service quality of the Social Security system?

References

Campanella, J., and F. J. Corcoran. 1982. "Principles of Quality Costs." *Transactions of the 36th Annual Quality Congress*. Milwaukee, WI: A.S.Q.C.

Deming, W. E. 1986. *Out of the Crisis*. Cambridge, MA: MIT Center for Advanced Engineering Study.

Feigenbaum, A. V. 1956. "Total Quality Control." *Harvard Business Review* 34 (6): 93–101.

Feigenbaum, A. V. 1983. *Total Quality Control*. 3rd ed. New York: McGraw-Hill.

Gabor, A. 1990. *The Man Who Discovered Quality*. New York: Time Books. Reprinted, New York: Penguin Books, 1992.

Garvin, D. A. 1984. "What does Product Quality Really Mean?" *Sloan Management Review* 26: 25–43.

Godfrey, A. B. 2002. "What is Quality?" *Quality Digest* 22 (1): 16.

Gryna, F. M. 2000. *Quality Planning and Analysis*. 4th ed. New York: McGraw-Hill.

Harrington, H. J. 1987. *Poor Quality Cost*. Milwaukee, WI: A.S.Q.C. Quality Press.

Institute of Medicine (IOM). 1999. *To Err is Human: Building a Safer Health System*. Washington, D.C.: National Academy Press. www.nap.edu/books/0309072808/html/.

Institute of Medicine (IOM). 2001. *Crossing the Quality Chasm: A New Health System for the 21st Century*. Washington, D.C.: National Academy Press. www.nap.edu/books/0309072808/html/.

Ishikawa, K. 1985. *What is Total Quality Control? The Japanese Way*. Translated by D. J. Lu. Englewood Cliffs, NJ: Prentice-Hall.

Juran, J. M. (editor-in-chief). 1988. *Quality Control Handbook*. 4th ed. New York: McGraw-Hill.

Juran, J. M., and F. M. Gryna. 1993. *Quality Planning and Analysis*. 3rd ed. New York: McGraw-Hill.

Krishnamoorthi, K. S. 1989. "Predict Quality Cost Changes Using Regression." *Quality Progress* 22 (12): 52–55.

Krzykowski, B. 2009. "In a Perfect World." *Quality Progress* 32–34.

Markus, M. 2000. "Failed Software Projects? Not Anymore." *Quality Progress* 33 (11): 116–117.

McBride, R. G. 1987. "The Selling of Quality." *Transactions of the 41st Annual Quality Congress*. Milwaukee, WI: A.S.Q.C.

Morse, W. J., H. P. Roth, and K. M. Poston. 1987. *Measuring, Planning, and Controlling Quality Costs*. Institute of Management Accountants, Montvale, NJ: Institute of Management Accountants.

Naidish, N. L. 1992. "Going for the Baldridge Gold." *Transactions of the 46th Annual Quality Congress*. Milwaukee, WI: A.S.Q.C.

Noz Jr., W. C., B. F. Redding, and P. A. Ware. 1983. "The Quality Manager's Job: Optimize Cost." *Transactions of the 37th Annual Quality Congress*, A.S.Q.C., Milwaukee, WI: A.S.Q.C.

Ponte, A. G. 1992. "You Have Cost of Quality Program, So What!" *Transactions of the 46th Annual Quality Congress*. Milwaukee, WI: A.S.Q.C.

Robinson, X. 2000. "Using Cost of Quality with Root Cause Analysis and Corrective Action Systems." *Transactions of the 54th Annual Quality Congress*.Milwaukee, WI: A.S.Q.C.

Shepherd, N. A. 200. "Driving Organizational Improvement Using Cost of Quality: Success Factors for Getting Started." *Transactions of the 54th Annual Quality Congress*. Milwaukee, WI: A.S.Q.C.

2

Statistics for Quality

2.1 Variability in Populations

The effort needed to design, produce, and deliver products to satisfy customer needs often requires the use of a few statistical tools. For example, when a product is produced in large quantities and we want to verify whether the entire population of the product being produced is indeed meeting the customer requirements, we take a sample of units from the population and verify if the sample units meet the requirements. Then we project the quality of the entire population from the quality observed in the sample.

If there is no variability in the population—that is, every unit in the population is identical—the task of projecting the quality of the population from sample is very simple. Inspection of just one sample unit is enough to project the quality of the entire population. If the units are not all identical—that is, if variability exists—then checking one unit will not be enough to verify the quality of the whole population. Questions arise as to how many more units should be checked, and what relationship exists between the quality of a sample and the quality of the population. If a sample contains 10% defectives, does it mean that the population also has 10% defectives? The answer is "no," as we will see later. How, then, can we estimate the amount of defectives in the population? Unfortunately, variability exists in every population and these questions arise in almost every situation where the quality of a population is to be measured and verified.

The science of statistics provides answers to these questions. It provides the means for describing and quantifying the variability in a population and offers methods for "estimating" population quality from sample quality in the presence of variability. In addition, several other methods have been created by statisticians using principles of statistics that are useful in creating, verifying, and ensuring quality in products. The major statistical methods of this kind are:

No.	Method	Used for
1	Control charts	Controlling processes to produce products of uniform quality
2	Sampling plans	Determining the acceptability of lots based on sample observations taken from them
3	Designed experiments	Determining the best combination of process parameter levels to obtain desired levels of product characteristics
4	Regression analysis	Determining the process variables that affect the quality in product characteristics and the extent of their effect
5	Reliability engineering	Understanding the factors that affect the life of parts and assemblies, and improving their longevity
6	Tolerancing	Determining allowable variability in product and process variables so that products can be produced economically while meeting customer needs

These methods are discussed in the appropriate context in different chapters of this book. To develop a good understanding of these methods, however, a good understanding of the fundamentals of probability and statistics is essential, since such understanding contributes to the correct and effective use of the methods. A basic discussion of the principles of probability and statistics is undertaken in this chapter as preparation for later discussion of the statistical tools of quality. The objective here is to provide the student with sufficient knowledge of the fundamentals so that he or she will be able to appreciate fully the statistical tools employed in quality engineering and use them effectively.

This chapter is broadly divided into three sections (2.4, 2.5, and 2.6). In Section 2.4, the discussion focuses on empirical methods for describing populations with variability. These methods, which are based on the analysis of data gathered from populations, include some graphical and some numerical methods. In Section 2.5, the discussion focuses on the mathematical approach to modeling such populations using probability distributions. The mathematical methods used for estimating population quality from sample quality are discussed in Section 2.6. First, a few terms need to be defined.

2.2 Some Definitions

2.2.1 The Population and a Sample

The term "population" refers to the collection of all items that are of interest in a given situation. The term "sample" refers to a subset chosen from the population. Whenever we use the term sample, we really mean a "random sample." A random sample is a sample taken in such a manner that each item in the population had an equal chance of being included in the sample.

A brief discussion of how a random sample is selected may be in order here. The best way to select a random sample is to first identify all members of the population with serial numbers—say, 1 to 1000 if the population has 1000 units. Then, random numbers, as many as the number of units required in the sample, are chosen from a random number table or through the use of a random number generator in a computer or a calculator. If a sample of 30 units is needed, then 30 random numbers are chosen in the range of 1 to 1000. Those items in the population that bear the 30 selected random numbers constitute the random sample.

This would be the ideal way to select a random sample, but there are situations where it may not be possible to use the ideal approach. In some instances, it may not be practical to identify all the items in a population using serial numbers, or it may not be possible to remove from the lot those identified to be in the sample. Compromises may be necessary to accommodate practical problems that would prevent selection of a truly random sample. In those situations, it is necessary to select the sample units making sure that the procedure does not follow a pattern, such as taking all units from the top row, from the corners, from the center, and so on. Such a procedure can still yield reasonably good random samples. Whenever possible, however, the rigorous approach that will not compromise the randomness of the sample should be used, as the randomness of the sample is critical to obtaining satisfactory results from most of the statistical tools.

2.2.2 Two Types of Data

When samples are inspected, observations are generated and the collection of these observations is called "data." Because data come from samples, they are referred to as sample data.

The data are classified into two main categories: measurement data, and attribute data. Measurement data, or variable data, as they are sometimes called, come from measurements of characteristics such as length, width, weight, diameter, amount of impurity, and so on. They are in units such as inches, pounds, and cubic centimeters. Examples of measurement data are 2.25 in., 5.22 lb., 256 cc, and so on. Attribute data come from inspection in which the units are classified based on attributes such as finish, color, taste, smell, or fit. The results from attribute inspection will be in counts, such as number too rough, number too dark, number too tight, and so on. In a majority of attribute inspections, the items inspected are classified into one of two categories, such as good/bad, tight/loose, or dark/light, and the resulting data are the proportion of one category in the total number of units in the sample. The reason for differentiating between the two types of data is that they require two different types of statistical methods to analyze them because they are mathematically different. The variable data that can assume values continuously in an interval needs a mathematical analysis different from that needed for the attribute data, which assume discrete values, usually in whole numbers.

2.3 Quality vs. Variability

As we said earlier, every population has some degree of variability in it. Whether it is the diameter of bolts, the hardness of piston rings, or the amount of cereal in a box, all populations have variability in them. It is this variability—the difference from unit to unit—that is the cause for poor quality, and therefore needs to be measured and managed in the effort to achieve quality in products. We will see several examples in this and later chapters that will illustrate the idea that excess variability is the cause of poor quality and waste. One simple example is given now to show how excessive variability causes poor quality and waste, while smaller variability avoids waste and results in good quality.

Example 2.1

Figures 2.1a and 2.1b show the plots of fill-weights in twenty 20-lb. boxes of nails filled in two automatic filling lines at a nail manufacturer. The graphs show that the average weight of nails in the boxes is about the same in both the lines, but some fill-weights from Line A fall below the lower specification, indicating that those boxes are not acceptable to the customer. Problems in Line A must be resolved.

Solution

In situations like in Line A, the usual solution attempted will be to raise the dial-setting to increase all the fill-weights so that those below the specification are all brought above it. This, however, will increase the fill-weights in boxes that are already full and over-full, resulting in the manufacturer giving away more nails than necessary. On the other hand, if someone can recognize the excess variability among the fill-weights, discover and eliminate the sources of the excess variability, and make the fill-weights more uniform, then the fill-weights below specification can be avoided without the need for overfilling those that are

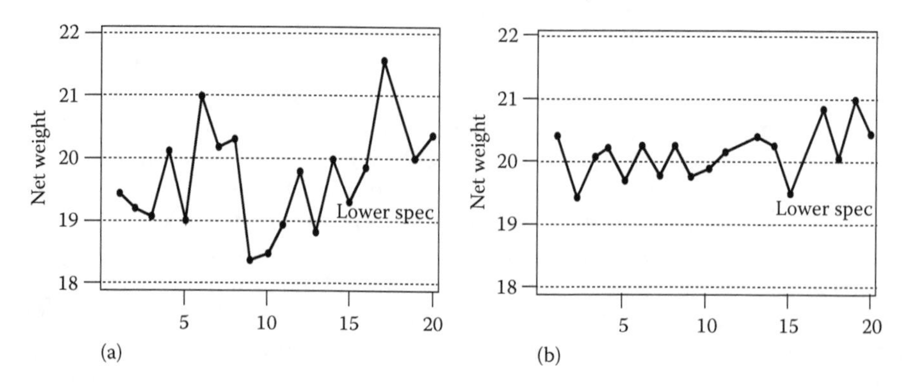

FIGURE 2.1
Net weight of nails in twenty 20-lb. boxes from two filling lines.

already full. There will be savings for the manufacturer, and the customer will not receive any box with a fill-weight less than the specification. Figure 2.1b is, in fact, the graph of fill-weights of the Line A after such steps were taken to reduce variability. A leaky shutter in one of the filling chutes was responsible for most of the variability in the filling in Line A.

2.4 Empirical Methods for Describing Populations

2.4.1 The Frequency Distribution

The frequency distribution is the tool used to describe the variability in a population. It shows how the units in a population are distributed within the possible range of the values. A frequency distribution can be drawn for an entire population by taking a measurement on each unit of the population, though this will take enormous time and resources. Thus, frequency distributions are usually made from data obtained from a sample of the population. Such a frequency distribution from sample data is called a "histogram." The histogram can be taken to represent the frequency distribution of the whole population if the sample had been taken randomly and the sample size had been large (≥50). The method of drawing a histogram is described next.

2.4.1.1 The Histogram

The histogram is a very useful tool for understanding the variability in a population. Many quality problems relating to the populations of products can be resolved by studying the histogram made from sample data taken from them. Many computer programs can prepare a histogram given a set of data; however, an understanding of how the histogram is made is necessary in order to interpret the computer-drawn histogram correctly. The method of making the histogram is illustrated in the following steps.

Step 1: Draw a (random) sample of 50 or more items from the population of interest, and record the observations.

Step 2: Find the largest and the smallest values in the data, and find the difference between the two. The difference is known as the "range" (R) of the data. Divide the range into a convenient number of cells (also called "class intervals," "bins," and so on). A few rules are available in determining the number of cells. One such rule is: the number of cells $k = 1 + 3.3$ ($\log n$), where n is the number of observations in the data. The result from the rule should be taken as an approximate guideline to be adjusted to suit the convenience of a given situation, as shown in the example below.

Step 3: Determine the limits for the cells, and tally the observations in each cell. The tally gives the frequency distribution of the data in the number of values falling in each cell. Calculate the percentage frequency distribution by calculating the percentage of data falling in each cell out of the total number of values in the sample.

Step 4: Draw a graph with the cell limits on the x-axis and the percentage frequency on the y-axis using convenient scales. The resulting graph is a histogram.

Example 2.2

The data below represent the amount of deodorant, in grams, filled in a filling line of a packaging firm. Draw the histogram of the data.

Solution

Step 1: The data are in Table 2.1. $n = 100$.

Step 2: Largest value = 346. Smallest value = 175. Range = 171.

Using the formula for number of cells: $k = 1 + 3.3(\log 100) = 7.6 \simeq 8$.

Next, cell width = 171/8 = 21.375. This is an inconvenient cell width, however, so we will make the cell width = 20 and then the number of cells = 10.

Steps 3 and 4: Table 2.2 shows the cell limits and the tally of the data. Figure 2.2 shows the histogram. In this histogram, the bars represent the percentage of data that lie in each of the cells. Histograms are also drawn with the bars representing the frequency of occurrence or number of values falling in each cell. Both histograms will look alike except for the scale on the y-axis.

2.4.1.2 The Cumulative Frequency Distribution

Another graph of the data that provides some very useful information is the cumulative frequency distribution. This is a graph of the cumulative percentage frequencies shown in Table 2.2. Figure 2.3 shows the percentage cumulative frequency distribution of the data in Table 2.1. With this graph, we can readily see the percentage of the data that lie below any value. For example, from Figure 2.3, we can read that approximately 40% of fill-weights are below 260 g and the value above which 10% of the fill-weights lie is 320. Note that although the graph is drawn from sample data, the conclusions that we draw can be stated as conclusions for the entire population, as the size of the sample was large.

Quite a bit can be learned about a population by studying the histogram drawn for it. The histogram gives an idea as to where on the x-axis, or around which central point, the population is distributed. It also gives an idea of the dispersion, or variability, in the population. By comparing the histogram with the applicable specifications, questions such as "Are all the measurements in the populations falling within specification?" or "Is the

TABLE 2.1

Data on Fill-Weight in Deodorant Cans in Grams

265	197	346	280	265	200	221	265	261	278	234	265	187	258	235	269	265	253	254	280
205	286	317	242	254	235	175	262	248	250	299	214	264	267	283	235	272	287	274	269
263	274	242	260	281	246	248	271	260	265	268	267	300	250	260	276	334	280	250	257
307	243	258	321	294	328	263	245	274	270	260	281	208	299	308	264	280	274	278	210
220	231	276	228	223	296	231	301	337	298	215	318	271	293	277	290	283	258	275	251

TABLE 2.2

Frequency Distribution Table

<table>
<tr><th>No. of Cells</th><th>Cell Limits</th><th>Tally</th><th>Freq. Distr.</th><th>% Freq. Distr.</th><th>% Cum. Freq. Dist.</th></tr>
<tr><td>1</td><td>160–180</td><td>|</td><td>1</td><td>1</td><td>1</td></tr>
<tr><td>2</td><td>181–200</td><td>|||</td><td>3</td><td>3</td><td>4</td></tr>
<tr><td>3</td><td>201–220</td><td>||||||</td><td>6</td><td>6</td><td>10</td></tr>
<tr><td>4</td><td>221–240</td><td>|||||||||</td><td>9</td><td>9</td><td>19</td></tr>
<tr><td>5</td><td>241–260</td><td>||||||||||||||||||||||</td><td>22</td><td>22</td><td>41</td></tr>
<tr><td>6</td><td>261–280</td><td>||||||||||||||||||||||||||||||||||</td><td>34</td><td>34</td><td>75</td></tr>
<tr><td>7</td><td>281–300</td><td>|||||||||||||||</td><td>15</td><td>15</td><td>90</td></tr>
<tr><td>8</td><td>301–320</td><td>|||||</td><td>5</td><td>5</td><td>95</td></tr>
<tr><td>9</td><td>321–340</td><td>||||</td><td>4</td><td>4</td><td>99</td></tr>
<tr><td>10</td><td>341–360</td><td>|</td><td>1</td><td>1</td><td>100</td></tr>
<tr><td></td><td></td><td>Total</td><td>100</td><td>100</td><td></td></tr>
</table>

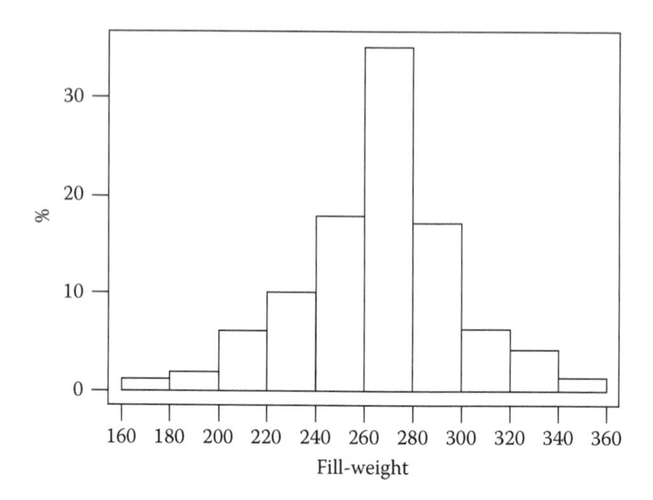

FIGURE 2.2
Histogram of fill-weights in Table 2.1.

variability within an acceptable level?" can be answered. Furthermore, the histogram can point to directions for solving any problems that may exist. Two examples are given below to illustrate how making a histogram can help in discovering what is wrong with a population and what should be done to correct it.

Example 2.3a

A manufacturer of thermostats used in toasters and ovens was experiencing considerable returns from customers because the thermostats failed in calibration

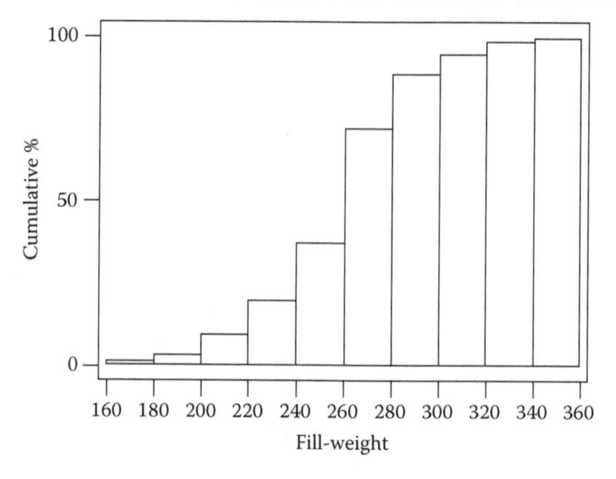

FIGURE 2.3
Cumulative distribution of fill-weights.

tests. The failures were traced to the width of a small key-way, called the "dog-width," on a tiny shaft 3/8 in. in diameter. These key-ways were typically cut on a milling machine that had two cutting heads and the cut shafts were collected in a common hopper below the two cutters. The uniformity of the dog-widths had to be ensured in order to prevent calibration failures.

Solution

A sample of 100 shafts was taken from the hopper of one of the milling machines and the widths were measured. The histogram made from these measurements is shown in Figure 2.4, where the lower and upper specification limits are indicated as LSL and USL respectively. The histogram typifies a *bimodal distribution*, which is a frequency distribution with two modes, or two peaks, indicating that two different distributions are mixed together in this population—one within the specification, and one almost outside the specification.

It was easy to guess that the difference in the two distributions might have been caused by the difference in the way that the two cutting heads were cutting the key-ways. Fifty shafts were taken separately from each of the two cutting heads, and separate histograms were drawn. It was easy to then see which cutter was producing key-ways distributed within specification and which was producing the distribution that was not within specification. Once this was explained and understood, the mechanic set out to discover the physical differences between the two cutter heads. The differences were found, and the faulty cutter was adjusted following the example of the good cutter. After adjustment, the key-way widths were all within specification, and the calibration problem was almost eliminated.

Incidentally, we note from the above example that the histogram, like other statistical methods, only points to the direction; knowledge of the technology

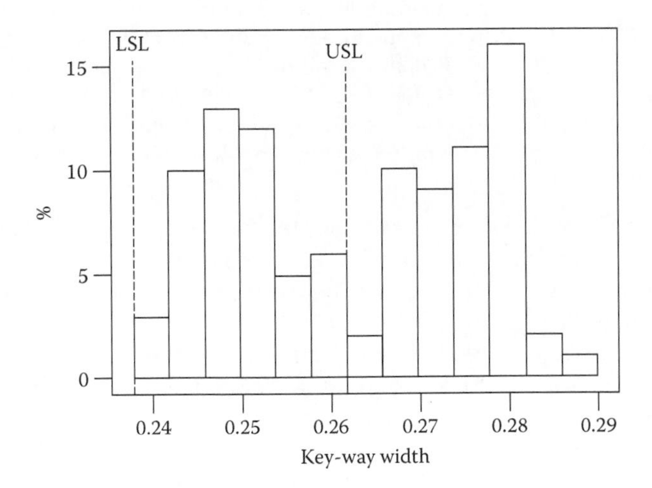

FIGURE 2.4
Histogram of key-way widths on a thermostat calibration shaft.

and the mechanics of the process is necessary in order to discover and implement a solution.

Example 2.3b

In an iron foundry that makes large castings, molten iron is carried from the furnace where iron is melted to where it is poured into molds over a distance of about 150 yards. This travel, as well as some time spent on checking iron chemistry and deslagging, causes cooling of the molten iron. The process engineer had estimated that the temperature drop due to this cooling to be about 20°F and, accordingly, had chosen the target temperature at which the iron is to be tapped out of the furnace as 2570°F so that the iron would be at the required temperature of 2550°F at pouring. The final castings, however, showed burn-in defects that were attributed to the iron being too hot at pouring. The castings with such defects were absolutely not acceptable to the customer.

Solution

The temperatures of iron at the tap-out at the furnace and the pouring at the moulds were taken over several castings, and histograms were made as shown in Figure 2.5a and 2.5b. Figure 2.5b shows that many pouring temperatures are out of specification on the high side, confirming the fact that many molds were poured too hot, thus causing the defects. The center of the pouring temperatures is at about 2560°F. The center of the tap-out temperatures in Figure 2.5a is about 2570°F, indicating that the average drop in temperature between tap-out and pouring is only about 10°F, whereas the process engineer had assumed it to be 20°F. Since the actual drop in temperature is smaller than expected, the iron remained hotter when poured. When this was brought to the attention of the process engineer, the tap-out target was adjusted down to 2560°F and the pouring temperatures started falling within the specification. The histogram of pouring temperatures also showed more variability in the actual pouring temperatures than was allowed by the specifications. This was addressed by taking steps to make the tap-out temperatures more consistent and by reducing the variability in waiting times, some of which were easily avoidable. As a consequence, the casting defects attributed to hot iron were almost eliminated.

This is an example where the histograms not only helped in discovering the causes of a problem, but also helped in communicating information to the people concerned in a clear manner so that everyone could understand the problem and agree on solutions.

[Disclaimer: This, as well as many other examples and exercises in this book, is taken from real processes, and the scenarios described represent true situations. Some of the names of product characteristics and process parameters, however, have been changed so as to protect the identification or the source of data. The numbers that represent targets and specifications have also been occasionally altered to protect proprietary information.]

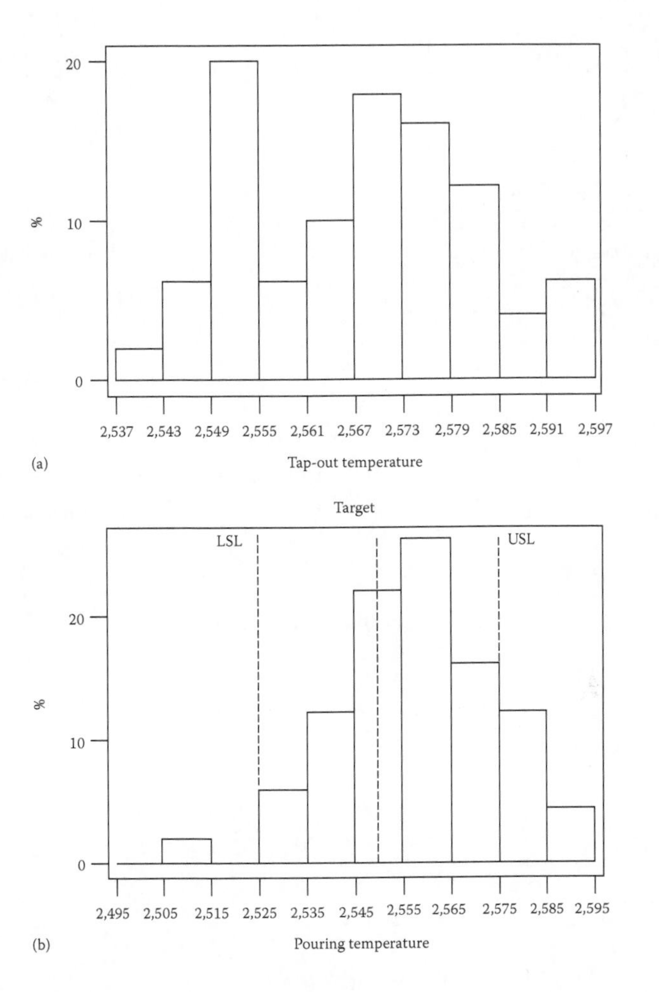

FIGURE 2.5
(a) Tap-out temperature of iron. (b) Pouring temperature of iron.

2.4.2 Numerical Methods for Describing Populations

When histograms are prepared to study the distributions of populations, several different types of distributions are encountered. A few of these types are shown in Figure 2.6. Of all the different types of distributions, the one that is most commonly encountered has a symmetrical, bell shape and is called the "normal distribution." Many measurements, such as length, weight, and strength, are known to follow this distribution. When the distribution of a population has this shape, just two measures—the average and the standard deviation of the distribution—are adequate to describe the entire distribution. (How it is so will be explained later while discussing the normal distribution.) The average and the standard deviation are estimated from sample data using the following formulas, where X_i is an observation in the sample and n is the sample size.

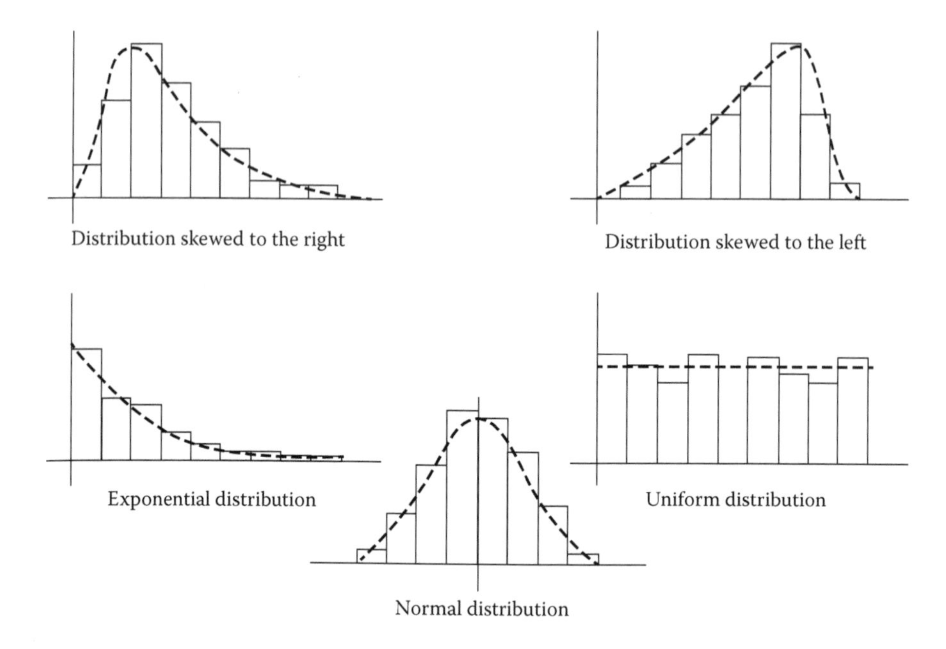

FIGURE 2.6
Different types of frequency distributions of populations.

$$\text{Average: } \overline{X} = \frac{\sum_i X_i}{n}$$

$$\text{Standard deviation: } S = \sqrt{\frac{\sum_i (X_i - \overline{X})^2}{n-1}}$$

The average represents the location of the distribution in the x-axis, or the center point around which the data are distributed. The standard deviation represents the amount of dispersion, or variability, in the data about the center. Because these measures are computed from sample data, $\overline{X}$ is called the "sample average" and S is called the "sample standard deviation."

When the sample is large, such as $n \geq 50$, the sample average and sample standard deviation can be considered the average and standard deviation of the entire population from which the sample was taken. (If the sample is not large, then the sample measures should not be equated to population measures, and the methods of inference, confidence interval, and hypothesis testing, discussed later in this chapter, must be used to estimate the population measures.)

2.4.2.1 Calculating the Average and Standard Deviation

The average is the simple arithmetic sum of all observations in the data divided by the number of observations. For the data shown in Table 2.1, $\overline{X} = 264.05$. The standard deviation is the square root of the "average" of the squared deviations of the individual values from the average of the data. Note that the formula uses $(n - 1)$ rather than n in the denominator for finding the "average" of the squared deviations. This is done for a good reason, which will be explained later when discussing the properties of a good estimator. The standard deviation can also be calculated by using another formula, which is mathematically equivalent to the one given above:

$$S = \sqrt{\frac{n\sum X_i^2 - \left(\sum X_i\right)^2}{n(n-1)}}.$$

This formula involves a bit less arithmetic work and is said to be computationally more efficient. The standard deviation of the fill-weights of the deodorant cans is calculated in the following steps:

$$\sum X_i = 26,405$$

$$\sum X_i^2 = 7,073,907$$

$$S = \sqrt{\frac{100(7,073,907) - (26,405)^2}{100(99)}} = 32.05$$

2.4.3 Other Graphical Methods

The histogram was introduced earlier as a way of describing the variability in a population. Besides the histogram, two other graphical methods are also useful in describing populations with variability. These are: the stem-and-leaf diagram, and the box-and-whisker plot.

2.4.3.1 Stem-and-Leaf Diagram

The stem-and-leaf (S&L) diagram is a graphical tool similar to the histogram, except it retains more information from the data. For a given set of data, the stems are somewhat equivalent to the cells of a histogram. The leaves are the values branching from the stems. For example, in Figure 2.7, the observations 231 and 234 are included in the branch with stem 23 and are denoted by the leaves 1 and 4, respectively. Figure 2.7 shows the S&L diagram drawn for the data in Table 2.1 using Minitab software.

Cum. Counts	Stems	Leaves
1	17	5
2	18	7
3	19	7
6	20	058
9	21	045
13	22	0138
19	23	114555
26	24	2235688
37	25	00013447888
(21)	26	000012334455555577899
42	27	01124444566788
28	28	0000113367
18	29	0346899
11	30	0178
7	31	78
5	32	18
3	33	47
1	34	6

FIGURE 2.7
Stem-and-leaf diagram of the amount of deodorant in cans.

Note that the data points within a stem are arranged in an ascending order and the diagram gives the cumulative counts of observations against each stem at the leftmost column. The cumulative counts below the central stem should be interpreted a bit differently from those above the central stem. For example, there are 6 values in the stems 20 and below; that is, there are 6 values ≤ 208. Similarly, there are 26 values in the stems 24 and below, which means that there are 26 values ≤ 248. On the other hand, there are 5 values in the stems 32 and above; that is, there are 5 values ≥ 321. Similarly, there are 18 values in the stems 29 and above, which means that there are 18 values ≥ 290. Finally, the count given against the central cell, in parenthesis, is not a cumulative count; it is the number of counts in the central stem.

The S&L diagram drawn as above helps in identifying the ordered rank of values in the data and makes it easy to compute the percentiles of the distribution. A percentile is the value below which a certain percentage of the data lies. We denote the *p*-th percentile by X_p to indicate that *p*% of the data lies below X_p. Thus, X_{25} and X_{75} represent the 25th and 75th percentiles, respectively, below which 25% and 75% of the data lie. The following names and notations are also used for these percentiles:

X_{25} is called the "first quartile" and is denoted as $Q1$.

X_{50} is called the "second quartile" or "median" and is denoted as $Q2$ or $\tilde{X}$.

X_{75} is called the "third quartile" and is denoted as $Q3$.

The percentiles are computed as follows. Suppose there are n observations in the data and the *p*-th percentile is required. The data are first arranged in an ascending order as in the S&L diagram, and the value at the $(n + 1)p/100$th location is identified. If $(n + 1)p/100$ is an integer, then the percentile is one of the values in the data. If $(n + 1)p/100$ is a fraction, then the percentile is determined by interpolating between the two values straddling the percentile, as shown in the following example.

Suppose we want the 30th percentile of the data in the above S&L diagram, where $n = 100$. Then, $(n + 1)30/100 = 30.3$. Thus, the 30th percentile lies between the 30th and the 31st ordered values in the data, one-third of the way above the 30th value. The 30th and 31st ordered values in the example are 251 and 253, respectively. Hence, the 30th percentile is 251.67. Suppose we want the 25th percentile of the data in the above example. Then, $(n + 1)25/100 = 25.25$. Therefore, the 25th percentile lies between the 25th and the 26th ordered values, one-quarter of the way above the 25th value. In this example, the 25th and the 26th ordered values are both 248, so $X_{25} = 248$. Similarly, to find the 75th percentile, $(n + 1)75/100 = 75.75$, so the 75th percentile lies between 75th and the 76th ordered values, three-quarters of the way above the 75th value. In this example, the 75th and the 76th values are both equal to 280; therefore, $X_{75} = 280$.

2.4.3.2 Box-and-Whisker Plot

The box-and-whisker (B&W) plot is another compact way of representing a population with variability, and it is especially useful when comparing several distributions with respect to their central value and dispersion. The B&W plot is a graph showing several percentiles of a given set of data. It shows the smallest value, X_0; the largest value, X_{100}; the median, X_{50}; and the first and the third quartiles, X_{25} and X_{75}, respectively. The box is made with the ends corresponding to the quartiles, and the whiskers extend up to the smallest value on one side of the box and the largest value on the other. The median is marked inside the box. Figure 2.8a shows the generic model of a B&W plot. Figure 2.8b shows the B&W plots of the pouring and tap-out temperatures of Example 2.3b arranged side by side so that a comparison can be made of the two distributions.

We can see from the B&W plots in Figure 2.8b that the median of the tap-out temperature is 2569°F and the median of the pouring temperature is 2559°F. The tap-out and pouring temperatures both seem to have about equal variability in that the quartile boxes are both of equal spread. The distribution of the pouring temperature seems to be symmetrical because the quartiles are approximately equidistant from the median. The distribution of tap-out temperature seems to be somewhat skewed, with the first quartile a bit farther from the median than the third quartile. We can also observe that the difference between the centers of the distributions is about 10°F, confirming a fact observed while studying their histograms. The plot made by Minitab shows the existence of an extreme value, an "outlier," indicated by a plot separated from the main B&W diagram in the pouring-plot. An outlier is a value that lies beyond three standard deviations from the median. The B&W plot facilitates making such comparative studies of populations.

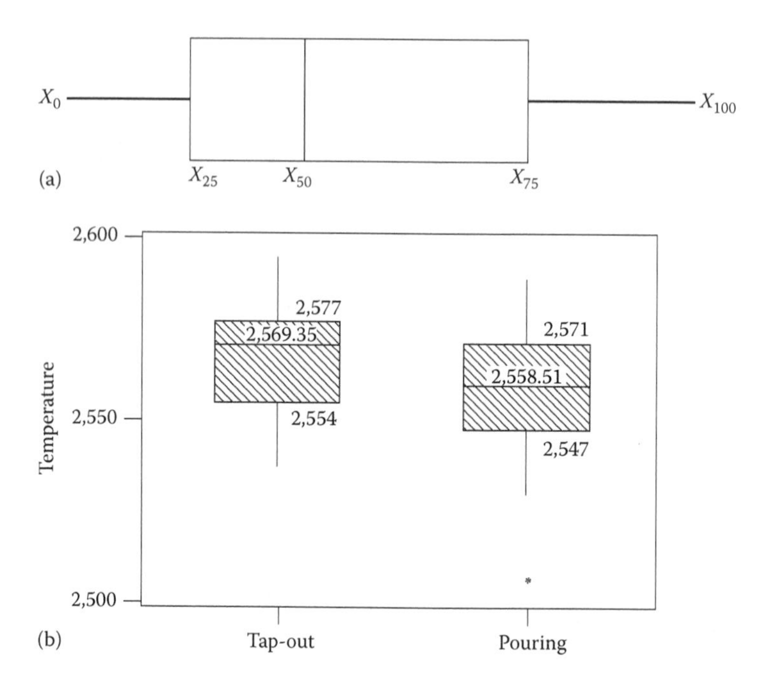

FIGURE 2.8
(a) A box-and-whisker plot. (b) Box-and-whisker plots of data from pouring and tap-out temperatures.

2.4.4 Other Numerical Measures

Besides the average $\overline{X}$ and standard deviation S, a few other measures obtained from sample data are also used to describe the location and dispersion of data.

2.4.4.1 Measures of Location

The median, denoted by $\tilde{X}$, is the middle value in the data. As stated before, it is the 50th percentile of the data. For data obtained from normally distributed populations, the median and the mean will be approximately the same.

The mode, denoted by M, is the value that occurs in the data with the most frequency. The mode has to be found from a large set of data. The mode has no meaning for a small set of data.

The midrange is the average of the maximum and minimum values:

$$\text{Midrange} = (X_{\text{max}} + X_{\text{min}})/2$$

Although the median and mode, and to a lesser extent the midrange, are used to indicate the location of a distribution, the average $\overline{X}$ is the most commonly used measure of location for distributions encountered in quality engineering.

2.4.4.2 Measures of Dispersion

The range, denoted by R, is the difference between the largest value and the smallest value in the data:

$$R = X_{max} - X_{min}$$

Another quantity that is sometimes used as a measure of variability is the inter-quartile range (IQR). It is the difference between the third quartile and the first quartile:

$$IQR = X_{75} - X_{25} (Q3 - Q1)$$

Although the standard deviation S is the most often used measure of variability in data, the range R is a popular measure of variability in quality engineering because of the simplicity of its calculation. It also has an appeal on the shop floor, because how it represents the variability in data can be easily explained and understood. The IQR represents variability in the data in the same way as the range does, but the IQR eliminates the influence of extreme values, or outliers, by trimming the data at the two ends.

For the example data in Table 2.1:

$$\text{Median} = 265$$

$$\text{Mode} = 265 \left(\text{see the S\&L diagram in Figure 2.7}\right)$$

$$\text{Midrange} = (346 + 175)/2 = 260.5$$

$$R = 346 - 175 = 171$$

$$IQR = 280 - 248 = 32$$

2.4.5 Exercise in Empirical Methods

2.1 The following represents the game time, in minutes, for the Major League Baseball games played in the United States between March 30 and April 9, 2003, sorted by league. Compare the times between the two leagues in terms of the average and the variability. A TV producer wants to know, for purposes of scheduling her crew, what the chances are that a game will end in 180 min. in each of the leagues. She also wants to know the time before which 95% of the games will end in each league? (Data based on report by Garrett 2003.)

American League

174	125	130	155	170	225	159	315	143	147	160	173	178	178	194	120	202
143	146	149	179	181	183	185	175	161	169	179	188	193	200	156	161	197
145	162	167	172	179	181	134	174	203	222	174	188	176	231	176	131	126

National League

164	171	174	180	183	197	209	149	150	153	214	146	142	155	136	150	154	170
155	162	164	173	179	208	156	164	166	169	161	169	172	177	201	278	135	172
171	173	188	133	162	163	163	190	195	197	169	172	178	209	182	237	151	178

2.2 A project for improving the whiteness of a PVC resin produced the following two sets of data, one taken before and the other taken after the improvement project. The whiteness numbers came from samples taken from bags of PVC at the end of the line. Compute the average, median, and other quantities used to describe the two data sets, and make the B&W plots of both data sets. If larger values for whiteness show improvement, has the whiteness really improved? (Data based on report by Guyer and Lalwani 1994.)

Before Improvement

149	147	148	155	163	154	152	151	155	166	153	153	151	152	174	154	152	150
154	150	149	168	153	153	152	151	153	152	149	153	155	151	168	165	151	153
154	153	149	148	155	154	150	149	150	152	152	153	155	162	151	152	149	153
151	154	163	152	151	154	155	149	170	155	154	155	152	155	150	152	154	151

After Improvement

155	166	168	167	169	167	168	170	166	158	165	157	166	168	165	158	169	170
170	167	164	166	171	156	169	167	167	158	170	167	167	169	170	168	168	165
165	165	166	169	165	170	166	167	165	167	170	168	164	167	165	164	169	165
166	169	170	170	158	167	166	166	168	169	168	155	151	169	165	165	168	168

2.3 Each observation in the following set of data represents the weight of 15 candy bars in grams recorded on a candy production line. The lower and upper specifications for the weights are 1872 and 1891 g, respectively. Draw, both manually and using computer software, a histogram, an S&L diagram, and a B&W plot for the data. Calculate, both manually and using computer software, the $\overline{X}$, S, median, mode, range, and *IQR*. Is the population of candy bar weights in specification? (Data based on report by Bilgin and Frey 1999.)

1894	1890	1893	1892	1889	1900	1889	1891	1901	1891	1889	1890	1881	1889
1895	1891	1891	1891	1889	1901	1890	1891	1881	1896	1890	1889	1893	1890
1883	1895	1890	1890	1889	1891	1886	1890	1900	1890	1889	1891	1880	1888
1893	1893	1898	1889	1890	1893	1889	1891	1889	1895	1891	1890	1891	1894
1892	1891	1890	1890	1891	1897	1892	1890	1890	1892	1892	1891	1878	1891
1894	1890	1896	1890	1891	1899	1891	1891	1891	1892	1890	1890	1891	1892
1890	1895	1876	1889	1891	1891	1892	1890	1890	1891	1897	1892	1891	1890

2.4 The following table contains data on the number of cycles a mold-spring, which supports molds in a forge press, lasts before breaking. Prepare a histogram, calculate the average and standard deviation, and find the proportion of springs failing before the average. (Based on McGinty 2000.)

143	248	67	49	47	51	249	267	81	5	37	125	95	17	43	5	72	202	63	167	60	27	10
200	101	5	90	46	315	54	233	31	189	68	3	30	53	135	13	56	49	13	36	133	7	16
120	3	69	122	13	25	3	74	22	15	36	32	5	106	173	44	146	20	2	29	32	7	325
226	152	74	116	73	128	3	427	30	85	240	93	12	111	7	74	43	36	229	177	58	406	151
141	116	199	240	37	22	47	115	65	169	43	150	52	46	24	11	60	101	79	5	12	47	

2.5 Mathematical Models for Describing Populations

The histogram gives a picture of where a population is located and how much it is dispersed. The numerical measures, the average, and the standard deviation give a quantitative evaluation of the location and dispersion of a population. Often, we have to make predictions about the proportion of a population below a given value, above a given value, or between two given values. To make such predictions, it is convenient to use a mathematical model to represent the form of the distribution. There are several such models available for this purpose. Such models are idealized mathematical functions to represent the graph of the frequency distribution. These models are called the "probability distribution functions." Many such functions have been developed by mathematicians to model various forms of frequency distributions encountered in the real world. The normal, Poisson, binomial, exponential, gamma, and Weibull are some of the names given to such mathematical models. Many of these are useful as models for populations encountered in quality engineering. An understanding of what these models are, which model would be suitable to represent what population or process, how they are used to make predictions about process conditions, and how they are helpful in solving quality-related problems are the topics discussed next.

To discuss these models of distributions, we need to define several terms, including probability, random variable, and probability distribution function. We begin with the definition of probability.

2.5.1 Probability

Before the term "probability" can be defined, a few preliminary definitions are needed. An experiment is a clearly defined procedure that results in observations. A single performance of an experiment is called a trial, and each trial results in an outcome, or observation. The experiments we deal with here are called random experiments, because the outcome in any one trial of the experiment cannot be predicted with certainty, but all possible outcomes of the experiment are known. The set of all possible outcomes of a random experiment is called the "sample space" and is denoted by S. Each element of a sample space is called a sample point. Some examples of experiments and their sample spaces are given below.

Experiment	Sample Space
1. Toss a die, and observe the number	S_1: {1, 2, 3, 4, 5, 6}
2. Toss two coins, and observe the faces on both	S_2: {HH, HT, TH, TT}
3. Toss two coins, and count the number of heads	S_3: {0, 1, 2}
4. Pick a sample of 10 bulbs from a box of 200, and count the number of defectives	S_4: {0, 1, 2,…,10}

An "event" is a subset of the sample space such that all the elements in it share a common property. An event can be specified by the common property or by enumerating all the elements in it. The events are labeled using the capital letters A, B, C, and so on. Given below are some examples of events defined in some sample spaces.

Experiment	Sample Space	Example Events
1. Toss a die and, observe the number	{1, 2, 3, 4, 5, 6}	A: {Number less than 4} = {1, 2, 3}
2. Toss two coins, and observe the faces	{HH, HT, TH, TT}	B: {At least one head} = {HH, HT, TH}
3. Toss two coins, and count the number of heads	{0, 1, 2}	C: {No head} = {0}
4. Pick a sample of 10 bulbs, and count the number of defectives (D)	{0, 1, 2,…,10}	E: {$D < 3$} = {0, 1, 2}

2.5.1.1 Definition of Probability

The probability of an event is a number between 0 and 1 that indicates the likelihood of occurrence of the event when the associated experiment is performed. The probability of an event that cannot occur is 0, and the probability of an event that is certain to occur is 1.0.

We use the notation $P(A)$ to denote the probability of the event A. Thus, the definition of probability in notations is:

$$0 \le P(A) \le 1.$$

$P(\Phi) = 0$ (where Φ denotes the null event, i.e., the event that cannot occur.)
$P(S) = 1$ (because when an experiment is performed, any one of the outcomes must occur.)

This is just the definition of the term "probability." Next, we will see how to compute the probability of events occurring when experiments are performed.

2.5.1.2 *Computing the Probability of an Event*

There are two basic methods for computing the probability of events:

1. Method of analysis of experiment
2. Method of relative frequency

The first method is used when we know, from an understanding of the experiment, the relative chance for occurrence of the outcomes. The second method is used when we do not know the relative chance for the occurrence of the outcomes. Use of these methods is illustrated in the examples below. A third method, which involves subjectively assigning probability values to events based on one's prior experience with such events, is also available. Such probabilities are used in a branch of statistics known as "Bayesian statistics." Bayesian methods are not discussed in this book.

2.5.1.2.1 *Method of Analysis*

Step 1: Formulate the sample space of the experiment.

Step 2: Assign weights to each of the elements in the sample space, such that the weights reflect the chance for occurrence of the elements, the weights are nonnegative, and the weights of all elements add to 1.0.

Step 3: Calculate probability of an event A, $P(A)$, as the sum of the weights of the elements in A.

Example 2.4

A card is drawn from a deck. What is the probability that the card has a number and not a picture?

Solution

Step 1: Formulate S:

S = {hearts number, hearts picture, clubs number, clubs picture, diamond number, diamond picture, spade number, spade picture}.

Step 2: Assign weights to the elements. We could assign the following weights to the outcomes from our knowledge of the relative occurrence of the outcomes:

{9/52,	4/52,	9/52,	4/52,	9/52,	4/52,	9/52,	4/52}
S = {HN,	HP,	CN,	CP,	DN,	DP,	SN,	SP}

Step 3: Calculate $P(A)$:

Event A: {the card is a number} = {HN, CN, DN, SN}
Therefore, $P(A) = 36/52 = 9/13$.

Example 2.5

A loaded die has even numbers that are twice as likely to show as the odd numbers. What is the probability that the number that shows is less than 5 when such a die is thrown?

Solution

Step 1: $S = \{1, 2, 3, 4, 5, 6\}$.
Step 2: Let w be the weight that represents the chance of an odd number. Then, $2w$ is the weight of an even number. The total for the weights of all the elements is $9w$, and this should be equated to 1.0, which means that $w = 1/9$. Therefore, the weights for the six outcomes are:

$$\{1/9, 2/9, 1/9, 2/9, 1/9, 2/9\}$$

Step 3: Event A: (Number is less than 5) = {1, 2, 3, 4}:

$$P(A) = 1/9 + 2/9 + 1/9 + 2/9 = 6/9 = 2/3$$

2.5.1.2.2 A Special Case

If the sample space of an experiment consists of elements that are all equally likely, then:

$$P(A) = \frac{\text{Number of elements in } A}{\text{Number of elements in } S}$$

It is easy to see how this result is true; it follows from the analysis method above. If there are k outcomes in the sample space, all equally likely, then each element has a weight equal to $1/k$. If event A has a number of elements in it, then $P(A) = a/k$, which is the ratio of number of elements in A to number of elements in S.

Example 2.6

When a fair coin is tossed three times, what is the probability that all tosses will show the same face?

Solution

The sample space S: {HHH, HHT, HTH, THH, HTT, THT, TTH, TTT}. If A is the event that all tosses have the same face, then A = {HHH, TTT}.

Are the elements in S all equally likely? Yes. Just take the example of tossing a fair coin two times with the sample space S = {HH, HT, TH, TT}. A little bit of reflection will show that if the two outcomes in one trial are equally likely, then the four outcomes of the two trials are also all equally likely. This can be extended to simultaneous performance of any number of trials, so the eight outcomes in the above sample space are all equally likely. Therefore, $P(A) = 2/8 = 1/4$.

Note: For any experiment, the sample space can be written in a few different ways. Whenever possible, it is advisable to write the sample space as made up of elements that are all equally likely. The probability of events in that sample space can then be computed using this simple formula.

2.5.1.2.3 Method of Relative Frequency

The method of relative frequency is used when the experimenter cannot assign weights to the outcomes in S from knowledge of the experiment. The probability can then be determined through experimentation:

Step 1: Perform the experiment a certain number of times—say, N—and find the number of times the event A occurs—say, n.

Step 2: Calculate the relative frequency of the event A:

$$f_A = \frac{n}{N}. \text{ Then, } P(A) = \lim_{N \to \infty} f_A$$

That is, $P(A)$ is given by f_A, provided that N is large.

Example 2.7

A survey of 100 students at Bradley University's College of Engineering showed the following:

	From Illinois Outside Chicago	From Chicago	From Other Areas	Total
Boys	13	55	10	78
Girls	10	10	2	22
Total	23	65	12	100

What is the probability that a student (picked at random) from Bradley's College of Engineering comes from Chicago? What is the probability that a randomly selected student is a girl and is from Illinois outside Chicago?

Solution

Here, the experiment has been performed 100 times and the outcomes recorded. The relative frequency of any event can be calculated out of the 100 trials. For example, the relative frequency of a student being from Chicago is 65/100. Can this relative frequency be taken as the probability? In other words, is the number of trials ($N = 100$) large enough for the relative frequency to be used as the probability of the events?

When an experiment is repeated a number of times, the relative frequency of an event tends to become a constant after a certain number of repetitions. When the repetitions begin yielding constant relative frequencies, we say that the number of trials is large enough. Then, the relative frequency can be used as probability of the event. In this case, $N = 100$ should be large enough (based on a statistician's judgment) to make the relative frequencies constant. Therefore,

P(A student comes from Chicago) = 65/100 = 0.65, and

P(A student is a girl and is from Illinois outside Chicago) = 10/100 = 0.1.

The two methods above would help in computing the probability of simple events of the kind seen in the above examples. However, we often have to deal with more complex events. For example, consider this problem: A box contains 20 pencils of which 4 are defective and rest are good. If a sample of 4 pencils is drawn from this box, what is the probability that there is no more than one defective in the sample? The methods discussed above for calculating probability are not adequate to find an answer to this; we need to know some additional theorems in order to be able to compute the probability of such complex events.

2.5.1.3 *Theorems on Probability*

2.5.1.3.1 *Addition Theorem of Probability*

If A and B are any two events in a sample space, (i.e., A and B are two possible events when an experiment is performed) then the probability of A or B occurring:

$$P(A \cup B) = P(A) + P(B) - P(A \cap B)$$

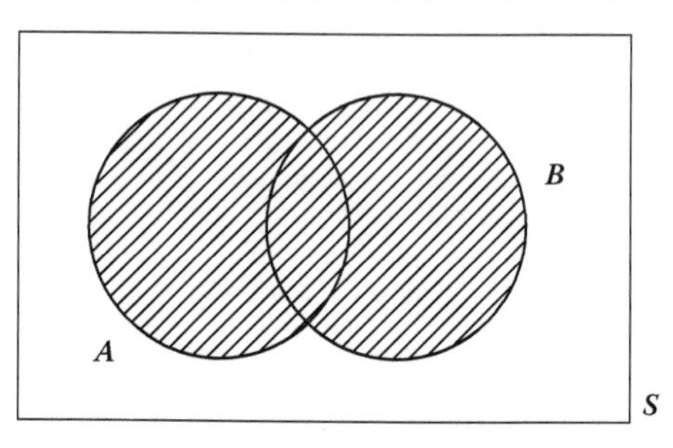

To see how the theorem is true, consider the following. The Venn diagram shown in the sketch above shows the relationship of the events with respect to each other and to the sample space. With reference to the diagram, $P(A \cup B)$ is given by the sum of probabilities of elements inside A or B (cross-hatched area). We can get the $P(A \cup B)$ from the sum $P(A) + P(B)$, but this sum includes the probability of the elements in $(A \cap B)$ twice. Therefore, if we subtract $P(A \cap B)$ once from $P(A) + P(B)$, we will get $P(A \cup B)$.

Corollary

If A and B are mutually exclusive—that is, there are no common elements between them—then $P(A \cap B) = 0$, and $P(A \cup B) = P(A) + P(B)$.

Example 2.8

When a pair of dice is thrown, what is the probability that the numbers 5 or 6 will show on either of the dice?

Solution

1,1	1,2	1,3	1,4	1,5	1,6
2,1	2,2	2,3	2,4	2,5	2,6
3,1	3,2	3,3	3,4	3,5	3,6
4,1	4,2	4,3	4,4	4,5	4,6
5,1	5,2	5,3	5,4	5,5	5,6
6,1	6,2	6,3	6,4	6,5	6,6

A B S

We first construct the sample space of the experiment as in the sketch above:
A: {one of the numbers is 5} and *B*: {one of the numbers is 6}

$$P(A)=11/36 \quad P(B)=11/36 \quad P(A\cap B)=2/36$$

$$P(A\cup B)=P(A)+P(B)-P(A\cap B)=\frac{11}{36}+\frac{11}{36}-\frac{2}{36}=\frac{20}{36}=\frac{5}{9}$$

Example 2.9

When two dice are thrown, what is the probability that the total is less than 4 or that one of the numbers is 4?

Solution

1,1	1,2	1,3	1,4	1,5	1,6
2,1	2,2	2,3	2,4	2,5	2,6
3,1	3,2	3,3	3,4	3,5	3,6
4,1	4,2	4,3	4,4	4,5	4,6
5,1	5,2	5,3	5,4	5,5	5,6
6,1	6,2	6,3	6,4	6,5	6,6

A B S

With respect to the sample space shown above:

A: (total < 4), then $P(A) = 3/36$
B: (one of the numbers is 4), then $P(B) = 11/36$

A and *B* are mutually exclusive—that is, there are no common elements between *A* and *B*. Hence,

$$P(A\cup B)=\frac{3}{36}+\frac{11}{36}=\frac{14}{36}$$

2.5.1.3.2 The Extension of the Addition Theorem

When there are three events—A, B, and C, as shown in the sketch below—the addition theorem becomes:

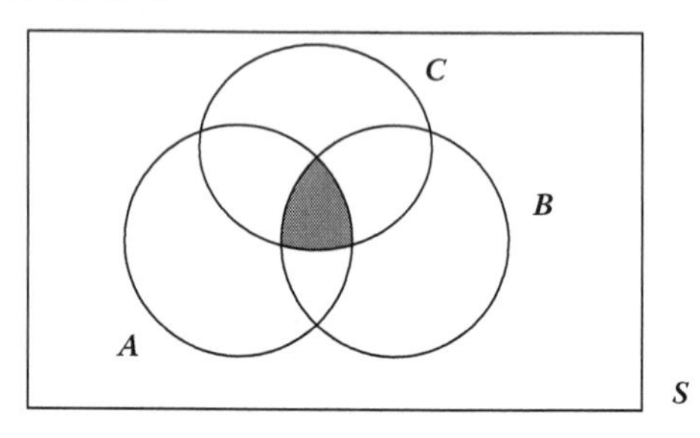

$$P(A \cup B \cup C) = P(A) + P(B) + P(C) - P(A \cap B) - P(A \cap C)$$
$$-P(B \cap C) + P(A \cap B \cap C)$$

Similar extension can be made for four, five, or any number of events; and the expression on the right-hand side becomes more and more complex. When the events are all mutually exclusive, however, the extension is simple for any number of events: if $A_1, A_2, \ldots, A_k$ are all mutually exclusive events, then:

$$P(A_1 \cup A_2 \cup \cdots \cup A_k) = P(A_1) + P(A_2) + \cdots + P(A_k).$$

2.5.1.3.3 Complement Theorem of Probability

Events A and A^c are said to be complement to each other if they are mutually exclusive and together make up the sample space (see the sketch below).

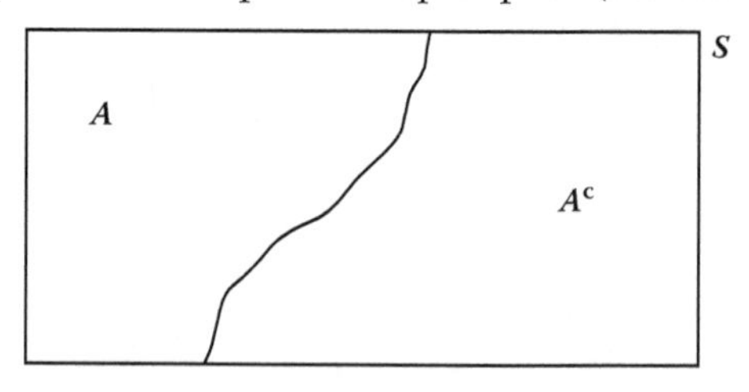

That is, if $(A \cap A^c) = \varnothing$, and $(A \cup A^c) = S$, then,

$$P(A^c) = 1 - P(A).$$

To see how the theorem is true: by definition $(A \cup A^c) = S$; therefore $P(A \cup A^c) = P(S) = 1.0$. Because they are mutually exclusive, $P(A \cup A^c) = P(A) + P(A^c)$. Thus, $P(A) + P(A^c) = 1.0$. Hence, $P(A^c) = 1 - P(A)$.

Example 2.10

When a coin is tossed six times, what is the probability that at least one head appears?

Solution

S = (HHHHHH, HHHHHT,..., TTTTTH, TTTTTT)

There are $2^6 = 64$ equally likely outcomes in S. Therefore,

P(at least 1 head) = 1 − P(no head) = 1 − (1/64) = 63/64.

Sometimes it is easier to compute the probability of an event that is the complement of the event of interest. The above theorem is useful in such cases.

2.5.1.3.4 Theorems on the Joint Occurrence of Events

When we have to find the probability of the joint occurrence of two events—that is, the probability of *A and B* occurring—we need to use one of the multiplication theorems given below, depending on whether or not the events are independent. First we need to define the independence of events; we start with defining conditional probability.

2.5.1.3.5 Conditional Probability

Conditional probability can be interpreted as follows: when an experiment is performed, the event B occurs. Given that B has occurred, what is the probability that A also occurs? That probability is called the conditional probability of "A given B" and is written as P(A | B). This can be seen in the sketch below to be the proportion of sample points in *B* that are also in *A*.

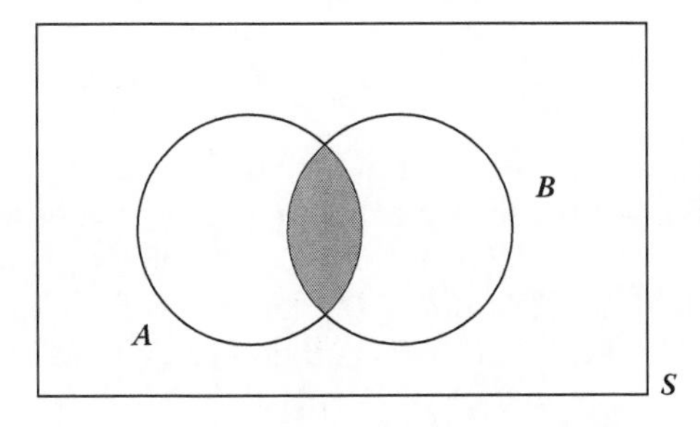

It is obtained as the ratio of the probability of both *A* and *B* occurring to the probability of event *B* occurring alone. In terms of notations,

$$P(A \mid B) = \frac{P(A \cap B)}{P(B)}.$$

Example 2.11

Find the probability that when two dice are thrown, the total equals 6, given that one of the numbers is 3.

Solution

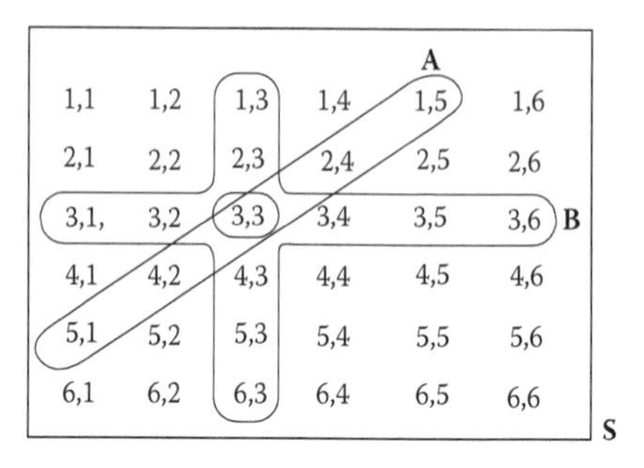

With reference to the sketch above,

$$A : \{\text{total equals } 6\}$$

$$B : \{\text{one of the numbers is } 3\}$$

We have to find $P(A|B)$:

$$P(A|B) = \frac{P(A \cap B)}{P(B)} = \frac{1/36}{11/36} = 1/11$$

Notice that $P(A \cap B)$ and $P(B)$ are calculated with respect to the original sample space of the experiment. Notice also, that out of the 11 sample points in B, only one is also in A. Thus, $P(A|B)$ can be obtained as the proportion of sample points in B that are also in A.

2.5.1.3.6 Independent Events

Two events in a sample space are said to be independent if the occurrence of one does not affect the probability of occurrence of the other. Independence of events can be defined using conditional probabilities as follows:

If A and B are such that $P(A|B) = P(A)$, then A and B are said to be independent. We can show that if $P(A|B) = P(A)$, then $P(B|A) = P(B)$.

Sometimes, the independence of events will be obvious; at other times, independence has to be verified. When independence is not obvious, we can use the above definition to verify if independence exits.

Example 2.12

Toss a pair of dice. Let E_1 be the event that the sum of the numbers is 6 and E_2 be the event that the sum is 7. Let F be the event when the first number is 3.

a. Are E_1 and F independent?
b. Are E_2 and F independent?

The events are identified in the sample space shown in the sketch below.

Solution

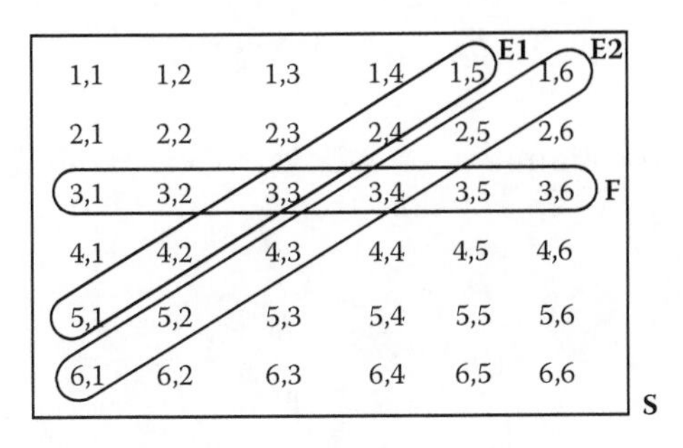

a. E_1 and F are independent if $P(E_1|F) = P(E_1)$.
With reference to the sketch above,

$$P(E_1 \mid F) = \frac{1}{6}, \qquad P(E_1) = \frac{5}{36}$$

because $P(E_1|F) \neq P(E_1)$, E_1 and F are *not* independent.

b. E_2 and F are independent if $P(E_2|F) = P(E_2)$. With reference to the sketch above,

$$P(E_2 \mid F) = \frac{1}{6}, \quad P(E_2) = \frac{1}{6}$$

because $P(E_2|F) = P(E_2)$, E_2 and F *are* independent.

Although E_1 and E_2 seem to be similar events, they apparently have different relationships with the event F. The meaning of the above result is that the total of the two numbers being 7 is independent of the number on the first die, whereas the total being 6 is not independent of the number on the first die. If the number on the first die is 6, then the total cannot be equal to 6.

Statistical independence is an important concept, and the above example, adapted from Walpole and Myers (1978), illustrates the meaning of independence. How we calculate the probability for the joint occurrence of two events depends on whether they are independent or not. The following theorems are used to find the probability for the joint occurrence of events when they are independent and when they are not.

2.5.1.3.7 The Multiplication Theorems of Probability

If A and B are any two events in a sample space, then:

$$P(A \cap B) = P(A \mid B)P(B) = P(B \mid A)P(A)$$

This result follows from the definition of conditional probabilities $P(A|B)$ and $P(B|A)$.

If A and B are independent, then:

$$P(A \cap B) = P(A)P(B)$$

This follows from the definition of independence. The following two examples illustrate the use of these theorems.

Example 2.13

A box contains seven black balls and five white balls. If two balls are drawn with replacement—that is, the ball drawn is put back after observing its color—what is the probability that both balls drawn are black?

Solution

Let B_1 be the event that the first ball is black. Then, $P(B_1) = 7/12$.

Let B_2 be the event that the second ball is black. Because the first ball is replaced, $P(B_2) = 7/12$.

It is easy to see that B_1 and B_2 are independent, because the outcome in the first pick has no effect on the outcome in the second pick. Hence,

$$P(B_1 \cap B_2) = P(B_1)P(B_2) = \frac{7}{12} \times \frac{7}{12} = \frac{49}{144}$$

Example 2.14

A box contains seven black balls and five white balls. If two balls are drawn without replacement—that is, the ball drawn is not put back after observing its color—what is the probability that both balls drawn are black?

Solution

Let B_1 be the event that the first ball is black. Then, $P(B_1) = 7/12$.

Let B_2 be the event that the second ball is black. B_1 and B_2 are not independent, because $P(B_2)$ depends on what happens in the first pick. We have to use conditional probability:

$$P(B_2 \mid B_1) = 6/11$$

$$P(B_1 \cap B_2) = P(B_2 \mid B_1) \times P(B_1) = \frac{6}{11} \times \frac{7}{12} = \frac{42}{132}.$$

In some situations, the probability of an event of interest is known only when conditioned on the occurrence of other events. The theorem of total probability that follows makes use of conditional probabilities and gives a very useful result when only conditional information is available.

2.5.1.3.8 The Theorem of Total Probability

Let $B_1, B_2, \ldots, B_k$ be partitions of a sample space S such that $(B_1 \cup B_2 \cup \ldots \cup B_k) = S$ and $(B_i \cap B_j) = \varnothing$ (null set) for any pair i and j (see diagram below). The partitions are mutually exclusive events that jointly make up the sample space. If A is an event of interest in the same sample space, then:

$$P(A) = P(A \mid B_1)P(B_1) + P(A \mid B_2)P(B_2) + \cdots + P(A \mid B_k)P(B_k)$$

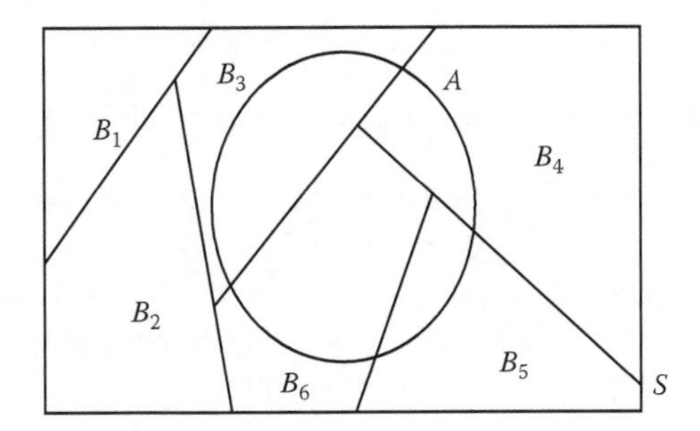

The individual product terms on the right-hand side of the above equation gives the joint probabilities of A with each of the partitions (see sketch above). When they are added together, the sum gives the total unconditional probability of A.

Example 2.15

In a college of engineering, the majors are distributed as follows: 26% electrical, 25% mechanical, 18% civil, 12% industrial, and 19% manufacturing. It is known that 5% of electrical, 10% of mechanical, 8% of civil, 45% of industrial, and 4% of manufacturing majors are female. If a student is picked at random in this college, what is the probability that the student is female?

Solution

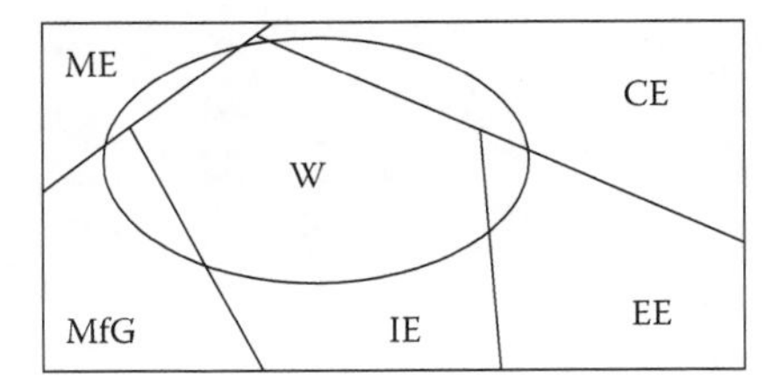

The majors form a partition of all the students in the college. The probabilities of the partitions are (see sketch above):

$$P(EE) = 0.26, P(ME) = 0.25, P(CE) = 0.18, P(IE) = 0.12, P(MfE) = 0.19.$$

The conditional probabilities that the student is female, given that the student is in a particular major, are:

$$P(W|EE) = 0.05, P(W|ME) = 0.10, P(W|CE) = 0.08,$$
$$P(W|IE) = 0.45, P(W|MfE) = 0.04.$$

The unconditional probability that the student is female is:

$$P(W) = P(W|EE)P(EE) + (W|ME)P(ME) + P(W|CE)P(CE)$$
$$+ P(W|IE)P(IE) + P(W|MfE)P(MfE) = (0.05)(0.26) + (0.10)(0.25)$$
$$+ (0.08)(0.18) + (0.45)(0.12) + (0.04)(0.19) = 0.114$$

Note that this probability represents the proportion of female students in the whole college. We are able to find the proportion of female students in the college from the information on their proportions within majors.

2.5.1.4 Counting the Sample Points in a Sample Space

Sometimes, finding the number of sample points in a sample space may be a challenge. The following methods provide help in those situations.

2.5.1.4.1 The Multiplication Rule

If an operation can be performed in n_1 ways and another operation can be performed in n_2 ways, then the two operations can be performed together in $n_1 \times n_2$ ways.

The proof of this result can be seen in the tree diagram shown below. For each way of performing the first operation, there are n_2 ways of performing the second operation. Therefore, both operations can be performed together in $n_1 \times n_2$ ways.

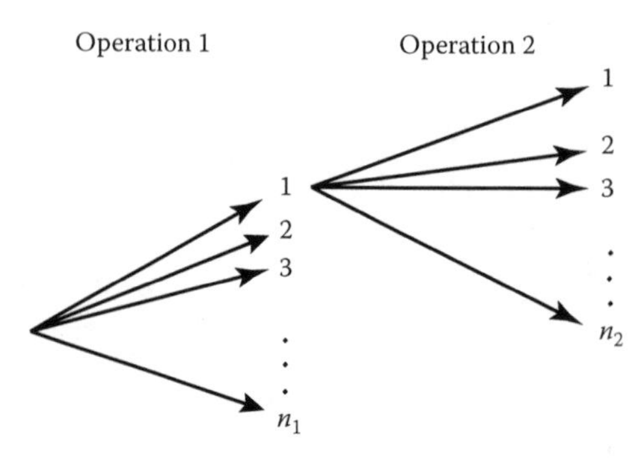

More generally, if Operation 1 can be performed in n_1 ways, Operation 2 in n_2 ways, …, and Operation k in n_k ways, then the k operations can be performed together in $n_1, n_2, \ldots,$ and n_k ways.

Example 2.16

There are 5 boys and 4 girls in a group. Teams of two are to be formed with 1 boy and 1 girl. How many such teams are possible?

Solution

This problem can be looked at as filling two boxes, one with a boy and the other with a girl. Then, we want to find the number of ways in which both boxes can be filled together. There are 5 ways of filling the first box with a boy and 4 ways of filling the second with a girl:

B	G
5	4

Both boxes can be filled together in $5 \times 4 = 20$ ways. Thus, there are 20 ways of choosing a team of two consisting of 1 boy and 1 girl, out of the 5 boys and 4 girls.

Example 2.17

How many four-lettered words are possible from the 5 letters E, F, G, H, and I, if each letter is used only once? How many of them will end with a vowel?

Solution

This problem again can be looked at as filling four boxes:

1	2	3	4

$$2 \times 3 \times 4 \times 5 = 120$$

Starting with the fourth box, there are 5 possible ways of filling this box. After the fourth box is filled, there are 4 possible ways of filling the third box, and so on. There are 120 ways of filling the four boxes together and, therefore, 120 four-lettered words are possible from the 5 letters.

If the word has to end with a vowel, then the last box can be filled in only 2 ways with E or I. Thus, the number of ways in which the four boxes can be filled with this restriction is 48.

1	2	3	4

$$2 \times 3 \times 4 \times 2 = 48$$

2.5.1.4.2 Permutations

A permutation is an arrangement of all or part of a given set of objects. For example, the three objects a, b, and c can be permuted as follows, if all of them are taken together. So there are six permutations of the three objects taken all at the same time.

$$abc, acb, bac, bca, cab, cba$$

Similarly, there are six permutations of the three objects taken two at a time:

$$ab, ac, ba, bc, ca, cb$$

In both of these examples, we wrote down all of the permutations and counted them. Suppose there are a large number of objects—say, 20—and it is not possible to write down all of the possible permutations and count them. The following theorem helps.

2.5.1.4.3 Theorem on Number of Permutations

The number of permutations of n distinct objects taken r at a time denoted by nPr, is given by

$${}_nP_r = \frac{n!}{(n-r)!}, r \leq n$$

When $r = n$, ${}_nP_n = n!$

The above result can be obtained by considering the formation of a permutation as filling r boxes with n objects:

r		$r-1$		$\cdots$		2		1	No. of boxes
□		□		$\cdots$		□		□	
$(n-r+1)$	×	$(n-r+2)$	×	$\cdots$	×	$(n-1)$	×	n	No. of ways of filling boxes

So, the number of ways of filling all the r boxes together is:

$$(n-r+1)(n-r+2)\cdots(n-1)n = \frac{n!}{(n-r)!}$$

Example 2.18

How many starting lineups are possible with a team of 10 basketball players?

Solution

Because the arrangement (or position) of the players is important in a lineup, the number of lineups is given by:

$$_{10}P_5 = \frac{10!}{(10-5)!} = \frac{10!}{5!} = 10 \times 9 \times 8 \times 7 \times 6 = 30{,}240$$

2.5.1.4.4 Combinations

A combination is a group of a certain number of objects taken from a given collection of objects. Here, unlike in the permutations, no attention is paid to the arrangement or the relative location of the objects in the group. For example, *abc* and *acb* are different permutations, but they are one and the same combination. Permutations are arrangements, but combinations are just groupings.

2.5.1.4.5 Theorem on Number of Combinations

The number of combinations of n distinct objects taken r at a time written as $\binom{n}{r}$ is given by:

$$\binom{n}{r} = \frac{n!}{r!(n-r)!}, \quad \text{for } r \leq n.$$

This result follows from the result for the number of permutations. Each group of r objects, when permuted within the group, can produce $r!$ permutations, so the number of combinations multiplied by $r!$ should be equal to the number of permutations. That is,

$$_nP_r = \binom{n}{r} r!$$

Therefore, $\binom{n}{r} = \frac{_nP_r}{r!} = \frac{n!}{(n-r)!r!}$

Example 2.19

How many different teams of 5 can be formed from a group of 10 players?

Solution

$$\text{Number of teams of 5 out of 10} = \binom{10}{5} = \frac{10!}{5!5!} = 252.$$

So, we see that 252 teams (combinations) of 5 can be formed out of 10 players, and 30,240 lineups (permutations) are possible out of these teams.

Example 2.20

How many different committees of three can be formed with 2 women and 1 man out of a group of 4 women and 6 men?

Solution

Look at this problem as having to fill two boxes, one with 2 women and the other with 1 man:

W	M
2	1

The box with women can be filled in $\binom{4}{2}$ ways and the box with men in $\binom{6}{1}$ ways.

The two boxes together can be filled in: $\binom{4}{2}\binom{6}{1} = \frac{4!}{2!2!} \times \frac{6!}{1!5!} = 6 \times 6 = 36$ ways.

Example 2.21

A box contains 20 pencils, of which 16 are good and 4 are defective.

a. How many samples of 4 pencils can be drawn from the box?
b. How many samples of 4 are possible with exactly 1 defective?
c. If a sample of 4 is drawn from the box, what is the probability that exactly 1 is defective?
d. What is the probability that a sample of 4 will have no more than 1 defective?

Solution

a. Number of samples of 4 out of $20 = \binom{20}{4} = \frac{20!}{4!16!} = 4845.$

b. Number of samples with exactly 1 defective $= \binom{16}{3} \times \binom{4}{1}$

$$= \frac{16!}{3!13!} \times 4 = 2240.$$

c. $P(1 \text{ defective}) = \frac{\binom{16}{3}\binom{4}{1}}{\binom{20}{4}} = \frac{560 \times 4}{4845} = 0.462.$

d. $P(\text{No more than 1 defective}) = P(0 \text{ defective or } 1 \text{ defective})$

$$= P(0 \text{ defective}) + P(1 \text{ defective})$$

$$= \frac{\binom{16}{4}\binom{4}{0} + \binom{16}{3}\binom{4}{1}}{\binom{20}{4}}$$

$$= \frac{1820 + 2240}{4845} = \frac{4060}{4865} = 0.838.$$

2.5.2 Exercise in Probability

2.5 An experiment consists of choosing a student at random in a college of engineering and asking his or her major (ME, EE, CE, IE, or MfE), whether he or she has taken a course in statistics (0 = no, 1 = yes), and, if the answer is no, asking whether he or she intends to take a course in statistics before graduation (0 = no, 1 = yes, 2 = not sure).

a. What is the sample space?

b. Identify the event where the student is not an IE and intends to take a course before graduation.

2.6 An experiment consists of throwing two dice and observing the numbers on both. An event E occurs if the difference between the two numbers is greater than 3. Write the sample space, and identify the elements in E.

2.7 An experiment consists of picking a student and measuring the circumference of the head, X, in inches, and their IQ, Y. If $X > 20$ and $100 \leq Y \leq 200$, formulate the sample space. Identify the elements in the event $\{X > 24 \text{ and } Y < 120\}$. (Use a graph.)

2.8 An urn has 4 black balls and 4 white balls. If 2 balls are drawn from it *with* replacement, what are the possible outcomes?

a. Write the sample space [hint: Mark each ball as B1, B2,..., W1, W2,...).

b. Identify the elements in the event: {Both balls are black}.

2.9 An urn has 4 black balls and 4 white balls. If 2 balls are drawn from it *without* replacement, what are the possible outcomes?

a. Write the sample space [hint: Mark each ball as B1, B2,..., W1, W2,...).

b. Identify the elements in the event: {Both balls are black}.

2.10 Three sides of a die are painted red, 2 sides yellow, and 1 side blue. What is the probability that when this die is thrown, the face that comes up is not red?

2.11 An urn has 4 black balls and 4 white balls. If 2 balls are drawn from it *with* replacement, what is the probability that both are black?

2.12 An urn has 4 black balls and 4 white balls. If 2 balls are drawn from it *without* replacement, what is the probability that both are black?

2.13 When two dice are thrown, what is the probability that the number on the first die is 4 or that the total is greater than 7?

2.14 When two dice are thrown, what is the probability that the difference between the two numbers is at least 2?

2.15 In a class where there are a total of 50 students, including both graduate and undergraduates, 25 are graduate students, 10 are IEs, and 5 are IE graduate students. If a student is picked at random from this class, what is the probability that the student is an IE or a graduate student?

2.16 In an electoral district, 35% of the electorate are African-American voters, and 50% are women. If 15% of the voters are African-American women, what percentage of the voters are neither African-American nor women?

2.17 In a shopping population, 20% are teenagers. Among the teenagers, 50% are girls.

a. If a shopper is selected at random, what is the probability that the shopper is a teenage girl?

b. If teenage girls are 25% of all women, what proportion of the shopping population are women?

2.18 When three dice are tossed, what is the probability that all three will not show the same number?

2.19 **a.** An unfair coin has $P(H) = 1/5$ and $P(T) = 4/5$. If this coin is tossed twice, what is the probability of getting exactly one head?

b. If this coin is tossed 5 times, what is the probability that 2 of the 5 tosses will result in heads?

2.20 **a.** In how many different ways can a 9-question, true-or-false examination be answered without regard to being right or wrong?

b. If a student answers an examination as above at random, what is the probability that he or she will have all 8 correct?

2.21 In how many ways can 4 boys and 3 girls be seated in a row if no 2 boys or girls should sit next to each other?

2.22 An urn contains 15 green balls and 12 blue balls. What is the probability that two balls drawn from the urn *with* replacement will both be green?

2.23 An urn contains 15 green balls and 12 blue balls. What is the probability that two balls drawn from the urn *without* replacement will both be green?

2.24 Three defective items are known to be in a container containing 30 items. A sample of 4 items is selected at random without replacement.

- **a.** What is the probability that the sample will contain no defectives?
- **b.** What is the probability that the sample will contain exactly 2 defectives?
- **c.** What is the probability that the number of defectives in the sample will be 2 or less?

2.25 A high-rise tower built by a developer contains 200 condominium units. The QA department of the developer has to approve the tower before turning it over to sales. The QA department will select a random sample of 8 units and inspect them. If more than three major defects are found in any unit, they will reject the unit as defective. If more than 2 of the 8 inspected units are defective, the entire tower will be rejected. If 10 of the 200 units are known to have more than three major defects, what is the probability that the tower will be rejected?

2.26 A manufacturer receives a certain part from four vendors in the following percentages: Vendor A = 28%, Vendor B = 32%, Vendor C = 18%, and Vendor D = 22%. Inspection of incoming parts reveals that 2% from Vendor A, 1.5% from Vendor B, 2.5% from Vendor C, and 1% from Vendor D are defective. What percentage of the total supplies received by the manufacturer is defective?

2.27 A candidate running for U.S. Senate in the state of Illinois found in a survey that 75% of registered Democrats, 5% of the registered Republicans, and 60% of independents would vote for him. If the voters in the state are 54% registered Democrats, 26% registered Republicans, and the rest independents, what percentage of the overall vote can the candidate expect to get?

2.5.3 Probability Distributions

2.5.3.1 Random Variable

A random variable is a variable that assumes for its values the outcomes of a random experiment. A random variable takes only real numbers for its values. If an experiment produces outcomes that are not in numbers—that is, its outcomes are in notations such as HHH or TTTH—then a random variable is used to produce numbers to represent the outcomes. Random variables are named or labeled using the capital letters X, Y, Z, and so on. The set of all possible values of a random variable X is called its "range space" and is denoted by R_X. As in the case of the sample space, events can be defined in the range space, and their probabilities can be determined. A few examples of random variables are given below.

1. The number that shows when a die is thrown.
2. The number of heads obtained when a coin is tossed three times.
3. The number of bad widgets in a sample of 20.
4. The number of coin tosses needed to get five heads in a row.
5. The weight of sugar in a 1-lb. sugar bag.
6. The amount of snowfall in January in Peoria, Illinois.
7. The life of a car battery in hours.

Random variables are of two kinds:

1. Discrete random variables, and
2. Continuous random variables.

A discrete random variable is one that takes a finite (or countably infinite) number of possible values. A continuous random variable is one that takes an infinite number of possible values; it takes values in an interval. The first three of the above examples are discrete random variables, the fourth example is a discrete random variable that takes a countably infinite number of values, and the final three are continuous random variables.

Random variables are used to represent populations with variability in them. For example, we say: let X be the height of students in a university to mean that the values of heights in the population are the possible values of the random variable X. The random variable thus makes the population with variability in it a mathematical entity.

We come across several random variables in the natural and man-made worlds. Often, decisions have to be made involving these variables, and therefore knowledge of their behavior is necessary. For example, a car dealer wants to know the number of cars of a certain model to be ordered for the next month. The demand for the cars in any month is a random variable and the dealer would want to know how the demand behaves. In other words, the dealer would like to know the possible values of the random variable and how likely each value is to occur. The dealer needs the probability distribution of the random variable "demand."

Many examples used in this and later chapters will show how the outcomes of many processes we deal with in quality engineering are random variables and how knowledge of their behavior is necessary for predicting the process outputs. That knowledge is obtained by modeling the process outputs using distributions appropriate to those processes. Predictions thus made would enable evaluation of the capability of the processes to meet customer requirements. If the predictions presage inadequacy, the same models would provide directions for making improvements so that adequate capabilities can be achieved.

Although a random variable may seem to behave in a haphazard or chaotic manner to an ordinary observer, statisticians have found that they follow certain patterns and that the behavior can be captured in models. These models, represented by formulas, graphs, or tables, are called "probability distributions." Because the two types of random variables mentioned above are mathematically different, one being continuous and the other noncontinuous, they need two different types of models. The model used to describe a discrete random variable is called the "probability mass function," and that used to describe a continuous random variable is called the "probability density function." First, we will discuss the probability distribution of a discrete random variable, the probability mass function.

2.5.3.2 *Probability Mass Function*

If X is a discrete random variable, a function $p(x)$ with the following properties is defined as the probability mass function (*pmf*) of the random variable:

1. $p(x) \geq 0$ for all values of x
2. $\sum_x p(x) = 1$
3. $p(x) = P(X = x)$

[Note: The uppercase letter X is used to represent the name of the random variable and the lowercase letter x is used to represent a value assumed by X.]

In words, a nonnegative function, denoted by $p(.)$ is used to describe a discrete random variable. The function value of $p(.)$ at any possible value of the random variable gives the probability that the random variable takes that particular value, and the function values add to 1.0 when summed over all possible values of the random variable. Again, such a function is called the probability mass function (*pmf*) of the discrete random variable.

Example 2.22

A random variable X denotes the number of tails when a coin is tossed three times. Find its probability mass function.

Solution

The possible values of the random variable, or the range space of X, R_X: (0, 1, 2, 3).

We can find the probability of events in R_X by identifying their equivalent events in the sample space. An equivalent event in the sample space is made up of all those elements that will be mapped by the random variable onto the event of interest in the range space. The probability of the event in R_X is the same as the probability of the equivalent event in sample space.

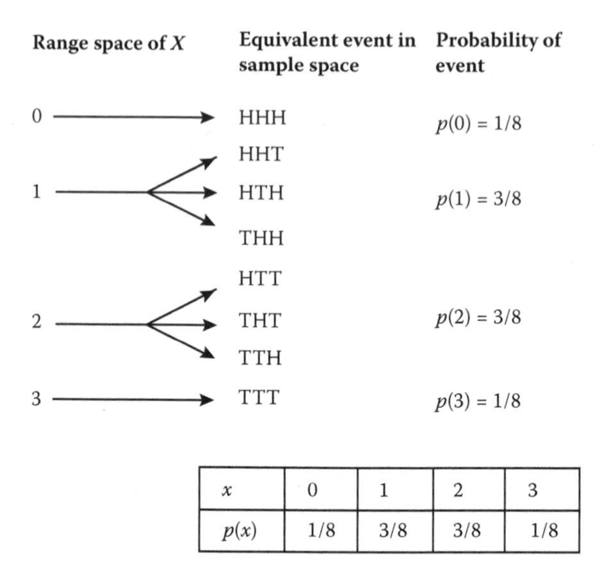

x	0	1	2	3
$p(x)$	1/8	3/8	3/8	1/8

FIGURE 2.9a
Calculating the probabilities of events in a range space.

The probabilities of the random variable taking different values in the example are found as follows:

Figure 2.9a shows the elements of the range space, their equivalent events in the sample space, and how the probabilities of the events in the range space are obtained.

We have obtained the probability mass function of the random variable X. The function $p(x)$ satisfies all the required properties, and the probability distribution is presented in a table as part of Figure 2.9a. It can also be represented in a graph, as shown in Figure 2.9b, which is called a "probability histogram." Or, it can be represented in a closed form:

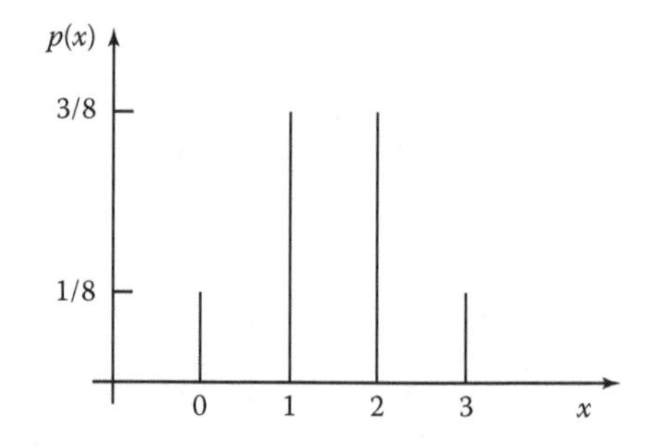

FIGURE 2.9b
Graphical representation of a probability mass function (*pmf*) of Example 2.22.

$$p(x) = \frac{\binom{3}{x}}{8}, \quad x = 0,1,2,3$$

The closed form expression is concise and convenient. By any form, however, $p(x)$ gives the probability of the random variable taking the possible values and, thus, describes the behavior of the random variable.

2.5.3.3 Probability Density Function

The method used above to describe a discrete random variable will not work in the case of a continuous random variable, because the continuous random variable takes an infinite number of possible values. The probability that a continuous random variable equals exactly any one of the infinite possible values is zero, so the probabilities of individual values cannot be tabulated. Hence, we must consider the probability of a continuous random variable taking values in an interval.

If X is a continuous random variable, then a function $f(x)$ is defined with the following properties and is called the "probability density function" (*pdf*) of X:

1. $f(x) \geq 0$ for all values of x
2. $\int_x f(x)dx = 1$
3. $P(a \leq X \leq b) = \int_a^b f(x)dx$

In words, a positive-valued function denoted by $f(.)$ is used to describe a continuous random variable. The function integrates to 1.0 over all possible values of the random variable and its integral value over any interval gives the probability that the random variable lies in that interval. Such a function is called the probability density function (*pdf*) of the continuous random variable.

Example 2.23

A random variable X is known to have the following *pdf*:

$$f(x) = \begin{cases} 0.01x, & 0 \leq x < 10 \\ 0.01(20 - x) & 10 \leq x \leq 20 \\ 0, & \text{otherwise} \end{cases}$$

a. Verify if $f(x)$ is a valid *pdf*
b. Find $P(5 \leq X \leq 10)$

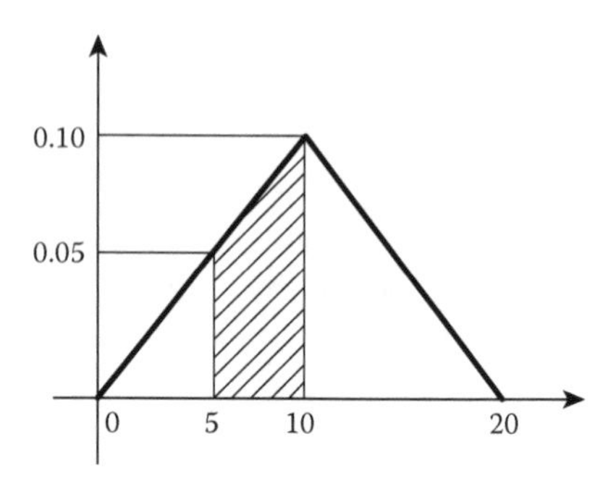

FIGURE 2.10
Graph of a probability density function (*pdf*).

Solution

Figure 2.10 shows the graph of the function.

a. The function is positive valued:

$$\int_0^{20} f(x)\,dx = \text{area under the curve over all possible values of } X$$
$$= \text{area of the triangle } = (1/2)(20)(0.1) = 1.$$

b. $P(5 \le X \le 10) = \int_5^{10} f(x)\,dx = $ area under the "*curve*" between 5 and 10

$$= \frac{0.05 + 0.1}{2} \times 5 = \frac{0.75}{2} = 0.375.$$

In this problem, the areas could be computed easily from basic geometry. If the graph of the function is not a simple geometric figure, however, then integration has to be used in order to compute the areas.

2.5.3.4 The Cumulative Distribution Function

When we use mathematical models to describe populations, it is often convenient to have another function, called the "cumulative distribution junction" (*CDF*), which is defined as $F(x) = P(X \le x)$. $F(x)$ for any x gives the probability that the random variable takes values starting from the lowest possible value up to and including the given value x.

If X is a discrete random variable with *pmf* $p(x)$, then

$$F(x) = P(X \le x) = \sum_{t \le x} p(t)$$

If X is a continuous random variable with *pdf* $f(x)$, then

$$F(x) = P(X \leq x) = \int_{t \leq x} f(t)\,dt$$

The *CDF* $F(x)$ is a step function for a discrete random variable, which increases in steps with the increases occurring at the values assumed by X. If X is a continuous random variable, then $F(x)$ increases continuously. Figure 2.11 shows the shapes of the *CDF* of discrete and continuous random variables.

As will be seen later, the availability of $F(x)$ for various probability distributions, often conveniently tabulated, is very helpful when proportions in populations below a given value, above a given value, or between two given values are to be computed. With the help of such tables we can avoid repeated summation and integration of probability distribution functions.

2.5.3.5 The Mean and Variance of a Distribution

A probability distribution function chosen appropriately for a random variable completely describes the behavior of the random variable. In the case of some distributions, however, the description can be accomplished adequately using one or two (or sometimes three) numerical measures. These measures are referred to as "characteristics of the distributions." The mean and the variance are the two most important of these characteristics:

1. Mean:

   If X is a discrete random variable with *pmf* $p(x)$, then the mean is defined as:

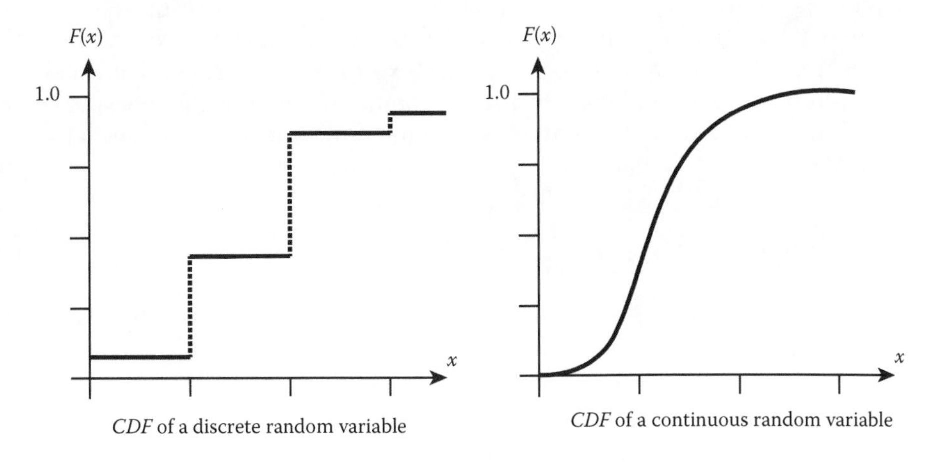

FIGURE 2.11
Graph of the cumulative distribution function (*CDF*).

$$\mu_X = \sum_x xp(x)$$

If X is a continuous random variable with *pdf* $f(x)$, then the mean is defined as:

$$\mu_X = \int_x xf(x)dx$$

2. Variance:

   If X is a discrete random variable with *pmf* $p(x)$, then the variance is defined as:

$$\sigma_X^2 = \sum_x (x - \mu_X)^2 p(x)$$

   If X is a continuous random variable with *pdf* $f(x)$, then the variance is defined as:

$$\sigma_X^2 = \int_x (x - \mu_X)^2 f(x)dx$$

From the definition of the mean μ_X, we can see that the mean is a weighted average, which represents the center of gravity of a distribution and indicates where a distribution is located on the x-axis. The variance σ_X^2 is the weighted average of the squared deviations of the values of the variable from its mean. It says how dispersed the distribution is about the mean; the larger the value of σ_X^2, the more variability there is in the distribution. The standard deviation of a distribution is the (positive) square root of the variance. Thus, the standard deviation σ_X is also a measure of the variability of a distribution. The standard deviation is commonly used as the measure of variability in quality engineering because its unit is the same as that of the variable.

Example 2.24

If X represents the number of heads when a coin is tossed three times, find the following:

a. μ_X
b. σ_X^2
c. $F(x)$
d. $P(X \leq 1)$
e. $P(1 < X \leq 3)$.

Solution

From an earlier example, the distribution of X is given by:

x	0	1	2	3
$p(x)$	1/8	3/8	3/8	1/8

a. $\mu_X = \sum_x xp(x) = 0\times\frac{1}{8}+1\times\frac{3}{8}+2\times\frac{3}{8}+3\times\frac{1}{8}=\frac{12}{8}=\frac{3}{2}$.

b. $\sigma_X^2 = \sum_x (x-\mu_x)^2 p(x) = \left(0-\frac{3}{2}\right)^2\times\frac{1}{8}+\left(1-\frac{3}{2}\right)^2\times\frac{3}{8}+\left(2-\frac{3}{2}\right)^2\times\frac{3}{8}$

$+\left(3-\frac{3}{2}\right)^2\times\frac{1}{8}=\frac{9}{4}\times\frac{1}{8}+\frac{1}{4}\times\frac{3}{8}+\frac{1}{4}\times\frac{3}{8}+\frac{9}{4}\times\frac{1}{8}=\frac{3}{4}$.

c. $F(x)$:

x	$x<0$	$0\le x<1$	$1\le x<2$	$2\le x<3$	$X\ge 3$
$F(x)$	0	1/8	4/8	7/8	$8/8=1.0$

The $F(x)$ is a step function for this discrete random variable. For example, $F(x)=1/8$ from $x=0$ until just before $x=1$. At $x=1$, it jumps to 4/8 and remains at that value until just before $x=2$. At $x=2$, it jumps to 7/8, and so on.

d. $P(X\le 1)$ can be read off the table as $F(1)=4/8$

e. $P(1<X\le 3)=P(X\le 3)-P(X\le 1)=F(3)-F(1)=1.0-0.5=0.5$

Example 2.25

A random variablen X has *pdf* $f(x)=\begin{cases}2x, & 0\le x\le 1\\ 0, & \text{otherwise}\end{cases}$. Find the following:

a. μ_X
b. σ_X^2
c. $F(x)$
d. $P(X\le 0.3)$
e. $P(0.1<X\le 0.3)$.

Solution

a. $\mu_X = \int_x x(2x)\,dx = 2\frac{x^3}{3}\Big|_0^1 = \frac{2}{3}$

b. $\sigma_X^2 = \int_x (x-\mu_X)^2(2x)\,dx = \frac{1}{18}$

c. $F(x) = \int_0^x 2t\,dt = t^2\Big|_0^x = x^2,\ 0 \le x \le 1$

Therefore,

$$F(x) = 0,\ x < 0$$

$$= x^2,\ 0 \le x \le 1$$

$$= 1,\ x > 1.$$

d. $P(X \le 0.3) = F(0.3) = 0.3^2 = 0.09$

e. $P(0.1 < X \le 0.3) = F(0.3) - F(0.1) = 0.3^2 - 0.1^2 = 0.09 - 0.01 = 0.08$

2.5.4 Some Important Probability Distributions

We just defined the generic probability distribution functions, one for describing a discrete random variable and another to describe a continuous random variable. These are idealized mathematical functions with certain properties. Many such mathematical functions have been proposed to describe random variables encountered in various situations. A few that are most useful in describing random variables encountered in quality engineering are discussed below. The nature of these distributions, the context in which they are useful, and their important characteristics are examined. Specifically, we will look next at:

1. The binomial distribution.
2. The Poisson distribution.
3. The normal distribution.

The binomial and Poisson are discrete distributions, whereas the normal is a continuous distribution. The chi-squared distribution and the *t* distribution are two continuous distributions that will be discussed later in this chapter while discussing inference methods. The exponential distribution, which is used to describe random variables that represent the life of products, will be discussed in Chapter 3 when discussing reliability methods.

2.5.4.1 The Binomial Distribution

A random variable X is said to have the binomial distribution if its probability distribution (the probability mass function) is given by:

$$p(x) = \binom{n}{x} p^x (1-p)^{n-x},\quad x = 0,1,\ldots,n.$$

The binomial distribution has two parameters. The parameters of a distribution are the quantities that need to be specified in order to complete the description of the distribution. The above expression for $p(x)$ contains two unknown quantities, n and p, which are the parameters of this distribution. We will use the notation $X \sim Bi(n, p)$ to indicate that the random variable X is binomially distributed with parameters n and p.

This distribution is the model for a random variable that represents the number of "successes" out of n independent trials when each trial can result in one of two possible outcomes—"success," or "failure." The probability of success in a trial is denoted by p and the probability of failure by $(1 - p)$. The random variable takes values 0, 1,..., n, because when the trial is repeated n times, the number of successes can be 0, 1, 2,..., or n.

We do not attempt to prove here that the above expression is, indeed, the distribution of the random variable representing the number of successes in n independent trials. That derivation is available in textbooks on probability and statistics, such as Hines and Montgomery (1990) or Hogg and Craig (1965). We also will not show the proof or derivation of any other distribution in this chapter. We will show how they are used as models of random variables of interest to us. A few examples of binomial random variables are given below.

1. X: the number of heads when a fair coin is tossed 10 times: (A fair coin is one that has $P(H) = P(T) = 0.5$.)

$$X \sim Bi(10, 1/2)$$

2. Y: the number of baskets a ballplayer makes in 12 free throws if her average is 0.4:

$$Y \sim Bi(12, 0.4)$$

3. W: the number of defectives in a sample of 20 taken from a (large) lot having 2% defectives:

$$W \sim Bi(20, 0.02)$$

In each of the above experiments the individual trials must be independent in order to use the binomial model to represent the "number of successes" in the n trials. In the first example above, the trials are obviously independent because the outcome in one toss of the coin will not affect the outcomes in the other tosses. The number of heads out of n tosses of a coin is a perfect example of a binomial random variable. In the second example, it might be reasonable to assume that the trials are independent, although they may not be strictly so. The assumption of independence is necessary in order to use the binomial model in this case. In the third example, the independence requirement will not be met if the sample is chosen from a small lot, as then the individual picks (made without replacement) will not be independent. As the lot size becomes large compared to the sample size, however, the assumption of independence becomes more valid. Hence, the binomial model can be used in the third example only when the sampling is done from large lots. For practical purposes, a lot is considered to be "large" if its size is 30 or larger. The following example shows the usefulness of the binomial distribution as a model for describing random variables of this kind.

Example 2.26

A sample of 12 bolts is picked from a production line and inspected. If the production process is known to produce 2% defectives, what is the probability that the sample will have exactly 1 defective? What is the probability that there will be no more than 1 defective?

Solution

Let X represent the number of defectives in the sample. Then, assuming the process produces a large population,

$$X \sim Bi(12, 0.02)$$

$$p(x) = \binom{12}{x}(.02)^x(.98)^{12-x}, \quad x = 0, 1, \ldots, 12$$

$$P(X=1) = p(1) = \binom{12}{1}(.02)^1(.98)^{11} = 12(.02)(.98)^{11} = 0.192.$$

$$P(\text{no more than 1 defective}) = P(X \le 1) = p(0) + p(1)$$

$$= \binom{12}{0}(0.02)^0(.98)^{12} + \binom{12}{1}(.02)^1(.98)^{11}$$

$$= 0.784 + 0.192 = 0.976.$$

The probability is 0.976 (it is almost a certainty) that a sample of 12 drawn from a 2% lot will have 1 or less defectives.

2.5.4.1.1 *The Mean and Variance of a Binomial Variable*

If $X \sim Bi(n, p)$, it can be shown, using the definition for mean and variance, that:

$$\mu_X = \sum_{x=0}^{n} x\binom{n}{x}p^x(1-p)^{n-x} = np$$

$$\sigma_X^2 = \sum_{x=0}^{n} (x-\mu_X)^2\binom{n}{x}p^x(1-p)^{n-x} = np(1-p).$$

Thus, the mean of a binomial random variable with parameters n and p equals np, and the variance is $np(1 - p)$. The standard deviation of the binomial variable is $\sqrt{np(1-p)}$.

Example 2.27

If samples of 12 bolts are drawn repeatedly from a production line having 2% defectives, what will be the average number of defectives in the samples? What will be the standard deviation of the number of defectives in the samples?

Solution

$$X \sim Bi(12, 0.02)$$
$$\mu_X = 12 \times 0.02 = 0.24$$
$$\sigma_X = \sqrt{np(1-p)} = \sqrt{12 \times 0.02 \times 0.98} = 0.485$$

This means that if samples of 12 are repeatedly taken from the 2% lot, the number of defectives per sample, in the long run, will average at 0.24. The variability in the number of defectives will be given by $\sigma = 0.485$.

2.5.4.2 The Poisson Distribution

A random variable X is said to have the Poisson distribution if its probability distribution (probability mass function) is given by:

$$p(x) = \frac{e^{-\lambda}\lambda^x}{x!}, \qquad x = 0, 1, 2, \ldots$$

The Poisson distribution has one parameter denoted by λ. We will use the notation $X \sim Po(\lambda)$ to denote that the random variable has a Poisson distribution with the parameter λ.

The Poisson variable takes values from zero to infinity, which is often used as a clue for recognizing Poisson variables. The possible values of the variable are countable and could be anywhere from zero to a very large number.

The Poisson distribution has been found to be a good model to describe random variables that represent counts that can take values anywhere from zero to infinity. The following are examples of Poisson random variables.

1. Number of knots per sheet of plywood.
2. Number of blemishes per shirt.
3. Number of pinholes per square-foot of galvanized steel sheet.

4. Number of accidents per month in a factory.
5. Number of potholes per mile of a city road.

2.5.4.2.1 The Mean and Variance of the Poisson Distribution

If $X \sim Po(\lambda)$, then the mean of X can be shown to be:

$$\mu_X = \sum_{x=0}^{\infty} x \frac{e^{-\lambda}\lambda^x}{x!} = \lambda$$

and the variance can be shown to be

$$\sigma_X^2 = \sum_{x=0}^{\infty} (x - \mu_X)^2 \frac{e^{-\lambda}\lambda^x}{x!} = \lambda$$

Notice that the mean and the variance of a Poisson random variable are equal. This is a unique property of the Poisson distribution. Also, the mean of a Poisson distribution is equal to its parameter. This means that if the average value of a Poisson variable is known, then its distribution is immediately defined, and the probability of various events relating to the random variable can be found.

Example 2.28

A typist makes, on average, three mistakes per page. What is the probability that the page he/she types for a typing test will have no more than one mistake?

Solution

Let X be the number of mistakes per page. Then,

$$X \sim Po(3)$$

$$\Rightarrow p(x) = \frac{e^{-3}3^x}{x!}, \quad x = 0, 1, \ldots$$

$$P(\text{no more than 1 defect}) = P(X \le 1) = p(0) + p(1) = e^{-3}\left[\frac{3^0}{0!} + \frac{3^1}{1!}\right]$$

$$= e^{-3}[4] = 0.199$$

Example 2.29

If a typist makes, on average, three mistakes per page, what would be the standard deviation of mistakes per page in the long run by this typist?

Solution

$$X \sim Po(3)$$

$$\sigma_X = \sqrt{3} = 1.732$$

2.5.4.3 The Normal Distribution

A random variable X is said to have a normal distribution if its probability distribution (probability density function) is given by:

$$f(x) = \frac{1}{\sigma\sqrt{2\pi}} e^{-\frac{1}{2}\left(\frac{x-\mu}{\sigma}\right)^2}, \qquad -\infty < x < \infty, \ \sigma > 0,$$

where the two quantities, μ and σ, are the unknown parameters of the distribution. We use the notation $X \sim N(\mu, \sigma^2)$ to denote that the random variable X has normal distribution with parameters μ, and σ^2. By specifying values for μ and σ^2, the distribution of a normal random variable is completely described. The graph of the density function is as shown in Figure 2.12a. (We usually try to graph the shape of a *pdf* to gain an understanding of how it behaves.) The graph of the normal *pdf*, which is known as the "normal curve," has certain special properties:

1. It is symmetric with respect to a vertical line at $x = \mu$.
2. It is asymptotic with respect to the x-axis on both sides of the vertical line.
3. The maximum value of $f(x)$ occurs at $x = \mu$.
4. The two points of inflexion occur at σ distances on each side of μ.

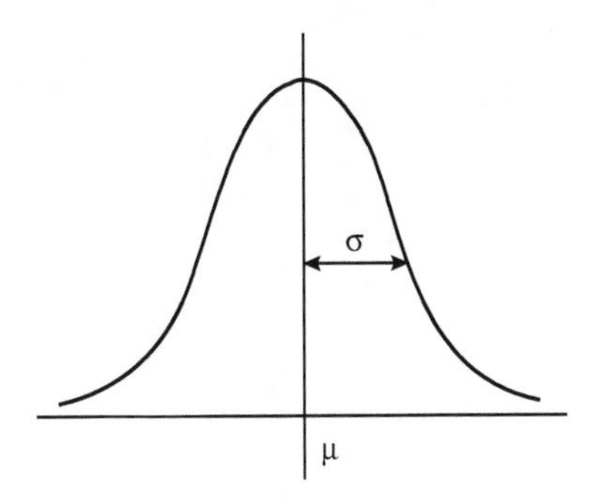

FIGURE 2.12a
Graph of the normal probability density function.

It can be shown that:

$$\int_x f(x)\, dx = 1 \qquad \text{[Area under the curve } = 1.0]$$

$$\int_x xf(x)\, dx = \mu \qquad \text{[Mean of the distribution} = \mu]$$

$$\int_x (x-\mu)^2 f(x)\, dx = \sigma^2 \qquad \text{[Variance of the distribution } = \sigma^2].$$

Notice that of the two parameters of the normal distribution, one equals its mean and the other its variance.

We frequently come across normally distributed random variables in quality engineering. Most measurements, such as the length of bolts, diameter of bores, strength of wire, or weight of parcels, are examples of measurements that can be expected to follow the normal distribution. Often, we would know that a random variable (a population) has the normal distribution with a known average and a known variance. We will be interested in finding the proportion of the population that lies in a given interval.

As an example, we may know that the net weight of nails in 20-lb. boxes is normally distributed with a mean of 20 lb and a variance of 9 lb^2. That is, if the random variable X represents the weights, then $X \sim N(20, 9)$. We may want to find the proportion of the boxes that have a net weight of less than 15 lb, the lower specification limit for net weights; that is, we want $P(X \le 15)$. This probability is given by the area below 15 under the curve defined by the function

$$f(x) = \frac{1}{3\sqrt{2\pi}} e^{-\frac{1}{2}\left(\frac{x-20}{3}\right)^2}$$

Thus, $$P(X \le 15) = \int_{-\infty}^{15} \frac{1}{3\sqrt{2\pi}} e^{-\frac{1}{2}\left(\frac{x-20}{3}\right)^2} dx. \quad \text{(See Figure 2.12b.)}$$

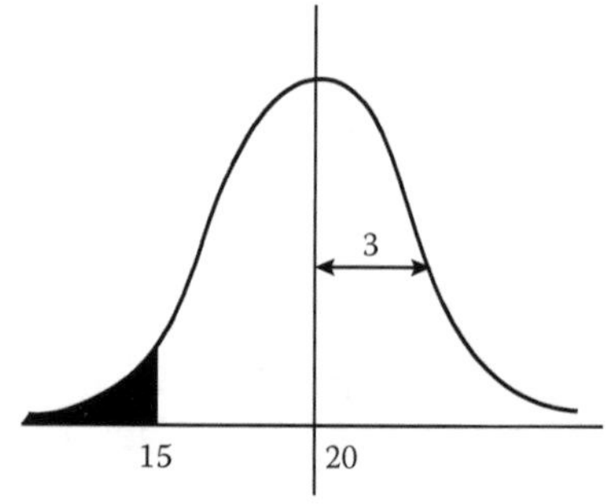

FIGURE 2.12b
Finding the proportion in a normal population.

It is not easy to find this area using calculus unless numerical methods of integration (with the help of a computer program) are used. A different approach is taken to find the required probability using the standard normal distribution, which is defined below.

2.5.4.3.1 The Standard Normal Distribution

A random variable that is normally distributed with $\mu = 0$ and $\sigma^2 = 1$ is called the "standard normal variable." The standard normal variable is denoted by a unique label, Z. Its distribution is called the "standard normal distribution." Thus,

$$Z \sim N(0, 1)$$

Its *pdf* is given by:

$$\phi(z) = \frac{1}{\sqrt{2\pi}} e^{-\frac{1}{2}z^2} \qquad -\infty < z < \infty$$

And its *CDF* is given by:

$$\Phi(z) = \int_{-\infty}^{z} \phi(t)\, dt$$

The notations $\phi(z)$ and $\Phi(z)$ are specifically assigned to the *pdf* and the *CDF*, respectively, of the standard normal distribution. The *CDF*, $\Phi(z)$, which represents the area under the standard normal curve from $-\infty$ up to any z (see Figure 2.12c), has been tabulated for many z values, as in Table A.1 in the Appendix. This table is called the "standard normal table," or simply the "normal table."

There is a relationship that relates any normal distribution to the standard normal distribution. This relationship can be used to find the areas under any normal distribution by converting the problem into one of the standard normal

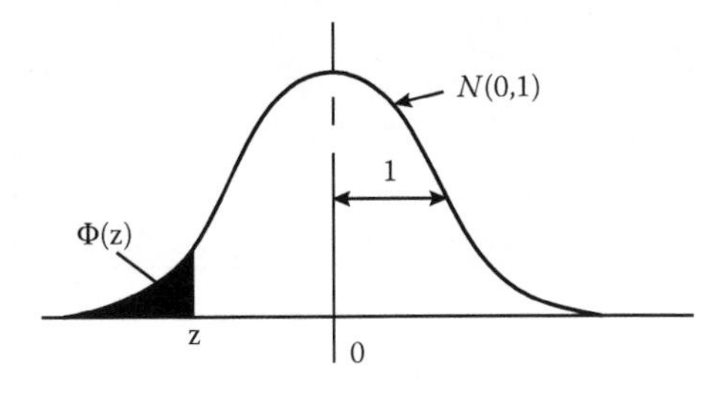

FIGURE 2.12c
The cumulative probabilities in the standard normal distribution.

distribution. First, we will solve some examples to see how the standard normal table is used to find areas under the standard normal distribution.

Example 2.30

If $Z \sim N(0, 1)$:

a. Find $P(Z \leq 2.62)$.
b. Find $P(Z \leq -1.45)$.
c. Find $P(Z > 1.45)$.
d. Find $P(-1.5 \leq Z \leq 2.5)$.
e. Find t such that $P(Z < t) = 0.0281$.
f. Find s such that $P(Z > s) = 0.0771$.
g. Find k such that $P(-k \leq Z \leq k) = 0.9973$.

Solution

Sketches in Figures 2.13a to 2.13g have been made to help identify the areas under the standard normal curve.

a. From the normal table, $P(Z \leq 2.62) = \Phi(2.62) = 0.9956$.

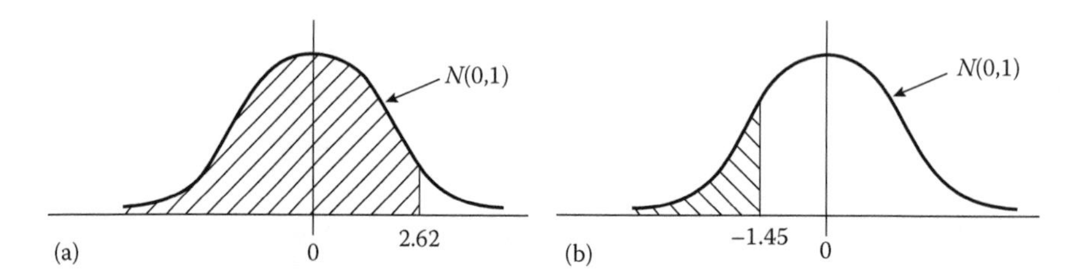

FIGURE 2.13a,b
(a) Area in the standard normal distribution for Example 2.30a. (b) Area in the standard normal distribution for Example 2.30b.

b. From the normal table, $P(Z \leq -1.45) = \Phi(-1.45) = 0.0735$.

c. $P(Z > 1.45) = 1 - P(Z \leq 1.45) = 1 - \Phi(1.45) = 1 - 0.9265 = 0.0735$.
[Note: $P(Z > 1.45) = P(Z < -1.45)$ because of the symmetry of the normal curve.]

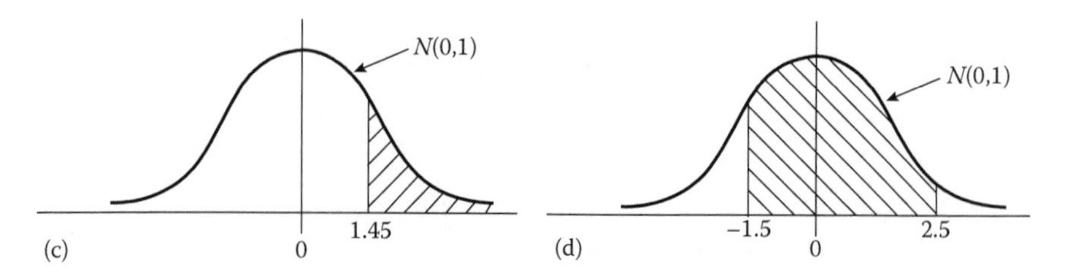

FIGURE 2.13c,d
(c) Area in the standard normal distribution for Example 2.30c. (d) Area in the standard normal distribution for Example 2.30d.

d. $P(-1.5 \leq Z \leq 2.5) = P(Z \leq 2.5) - P(Z \leq -1.5) = \Phi(2.5) - \Phi(-1.5) = 0.9938 - 0.0668 = 0.9270$.

e. From the normal table, $t = -1.91$.

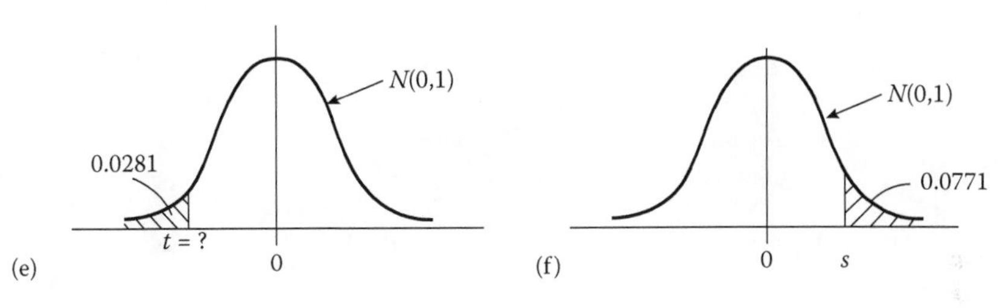

FIGURE 2.13e,f
(e) Area in the standard normal distribution for Example 2.30e. (f) Area in the standard normal distribution for Example 2.30f.

f. In other words, we have to find s such that $P(Z \leq s) = 1 - 0.0771 = 0.9229$. From the normal table $s = 1.425$.

g. Because of symmetry, $P(Z \leq -k) = P(Z \geq k) = 0.00135$.
We have to find k such that $\Phi(-k) = 0.00135$. From the normal table, $k = 3$.

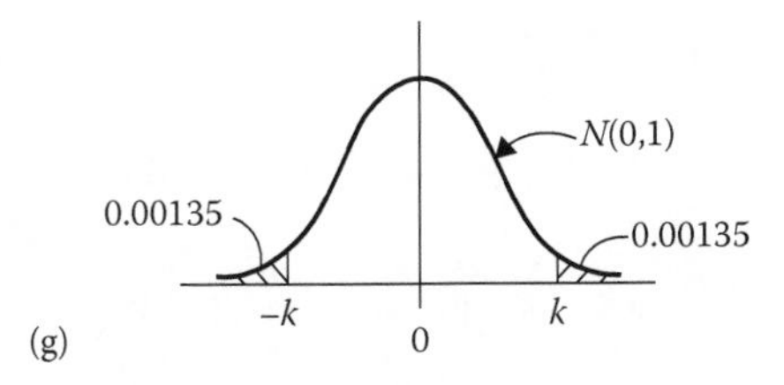

FIGURE 2.13g
Area in the standard normal distribution for Example 2.30g.

The following theorem provides the relationship between any normal distribution and the standard normal distribution.

Theorem

$$\text{If } X \sim N(\mu, \sigma^2), \text{ then } \frac{X-\mu}{\sigma} \sim N(0,1).$$

According to this theorem, if a random variable is normally distributed, then a function of it, $(X - \mu)/\sigma$, has the standard normal distribution. (The proof of this theorem, which employs methods of finding the distributions of functions of random variables, can be found in books on probability and statistics, such as Mood et al. 1974.) The examples below show how this relationship is used in finding the areas under any normal curve.

Example 2.31

Consider a random variable $X \sim N(2.0,\ 0.0025)$. That is, $\mu_x = 2.0$, $\sigma_x^2 = 0.0025 \Rightarrow \sigma_x = 0.05$.

a. Find $P(X \le 1.87)$.
b. Find $P(X > 2.20)$.
c. Find $P(1.9 \le X \le 2.1)$.
d. Find t such that $P(X \le t) = 0.05$.
e. Find s such that $P(X > s) = 0.05$.
f. Find k such that $P(\mu - k\sigma \le X \le \mu + k\sigma) = 0.9973$.

Solution

a. $$P(X \le 1.87) = P\left(\frac{X - 2.0}{0.05} \le \frac{1.87 - 2.0}{0.05}\right)$$

$$= P(Z \le -2.6) = \Phi(-2.6) = 0.0047$$

b. $$P(X > 2.2) = P\left(Z > \frac{2.2 - 2.0}{0.05}\right) = P(Z > 4.0) = \Phi(-4.0) = 0.0$$

[For all practical purposes, take $P(Z \le -3.5) = \Phi(-3.5) = 0$.]

c. $$P(1.9 \le X \le 2.1) = P\left(\frac{1.9 - 2.0}{0.05} \le Z \le \frac{2.1 - 2.0}{0.05}\right)$$

$$= \Phi(2) - \Phi(-2) = 0.9772 - 0.0228 = 0.9544$$

d. We need to find t such that $P\left(Z \le \frac{t - 2.0}{0.05}\right) = 0.05$

From Normal tables, $\frac{t - 2.0}{0.05} = -1.645$

$$\Rightarrow t = -1.645(0.05) + 2.0 = 1.918$$

e. We need to find s such that $P\left(Z > \frac{s - 2}{0.05}\right) = 0.05$

$$\Rightarrow \frac{s - 2}{0.05} = 1.645 \Rightarrow s = 2.082$$

f. $$\Rightarrow P\left(\frac{\mu - k\sigma - \mu}{\sigma} \le Z \le \frac{\mu + k\sigma - \mu}{\sigma}\right) = 0.9973$$

$$\Rightarrow P(-k \le Z \le k) = 0.9973$$

$$\Rightarrow k = 3$$

The result from Example 2.31f means that for any normal distribution, 99.73% of the population falls within 3σ distance on either side of the mean. Using a similar analysis, it can be shown that 95.44% of a normal population lies within 2σ distance on either side of the mean, and 68.26% within 1σ distance on either side of the mean.

2.5.4.3.2 Application of the Normal Distribution

Example 2.32

The diameters of bolts that are mass produced are known to be normally distributed with a mean of 0.25 in. and a standard deviation of 0.01 in. Bolt specifications call for 0.24 ± 0.02 in.

a. What proportion of the bolts is outside the specifications?
b. If the process mean is moved to coincide with the center of the specifications, what proportion will be outside the specifications?

Solution

Let D be the diameter of the bolts. Then, $D \sim N(0.25, 0.01^2)$.

Note that the variance, not the standard deviation, is written as the parameter of the distribution in the statement above. Note also the language we use. When the random variable represents the population of bolt diameters, the probability the random variable lies below a given value is used as the proportion of the population that lies below the given value.

a. The upper and lower specifications are at 0.22 and 0.26, respectively. We need:

$$P(D < 0.22) + P(D > 0.26) = P\left(Z < \frac{0.22 - 0.25}{0.01}\right) + P\left(Z > \frac{0.26 - 0.25}{0.01}\right)$$

$$= P(Z < -3) + P(Z > 1)$$

$$= \Phi(-3) + 1 - \Phi(1) = 0.00135 + 0.1587 = 0.16.$$

That is, 16% of the bolts are outside specification. Figure 2.14a shows the process in relation to the specification limits.

b. When the process mean is adjusted to coincide with the specification mean: $D \sim N(0.24, 0.01^2)$ [assuming the standard deviation remains the same]

$$P(D < 0.22) + P(D > 0.26) = P\left(Z < \frac{0.22 - 0.24}{0.01}\right) + P\left(Z > \frac{0.26 - 0.24}{0.01}\right)$$

$$= 2.(0.0228) = 0.0456.$$

That is, 4.56% will be outside specification. Figure 2.14b shows the process centered with respect to the specifications.

Example 2.32 brings home an important message for a quality engineer. Centering a process with respect to the specifications will often lead to a considerable reduction in out-of-spec products. In many situations, centering may require only some simple adjustment—such as adjusting a tool setting to raise or lower the process average. Further reduction in out-of-spec diameters may have to come from reducing the variability.

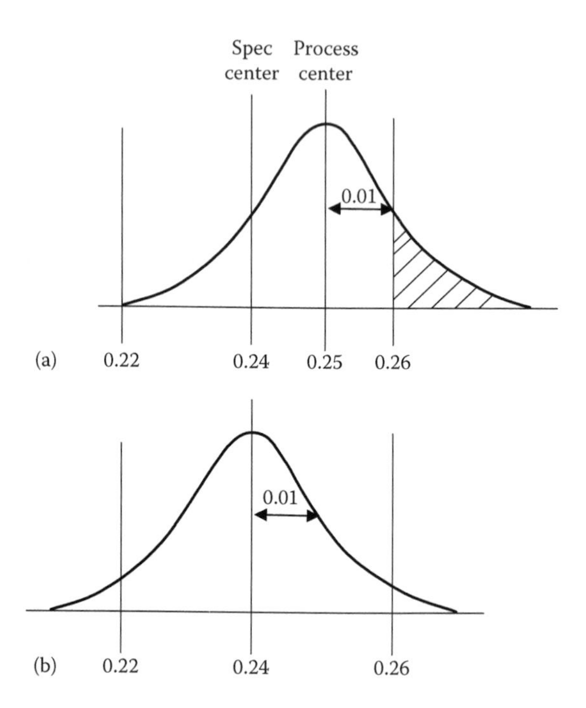

FIGURE 2.14
The process making diameters of bolts before and after centering.

Example 2.33

A battery manufacturer gives warranty to replace any battery that fails before four years. The life of the manufacturer's batteries is known to be normally distributed, with a mean of five years and a standard deviation of 0.5 year.

a. What percentage of the batteries will need replacement within the warranty period?
b. What should be the standard deviation of the battery life if no more than 0.5% of the batteries should require replacement under warranty?

Solution

Let X be the random variable that denotes the life of the batteries in years. Then, $X \sim N(5, 0.5^2)$.

a. The proportion of the batteries failing before four years is calculated as:

$$P(X < 4) = P\left(Z < \frac{4-5}{0.5}\right) = P(Z < -2.0) = 0.0228$$

That is, 2.28% of the batteries will have to be replaced during the warranty period.

b. Let σ' be the new standard deviation. Then, we need to find the value of σ' such that $P(X < 4) = 0.005$

$$\text{i.e., } P\left(Z < \frac{4-5}{\sigma'}\right) = 0.005$$

$$\Rightarrow \frac{4-5}{\sigma'} = -2.575 \text{ (from Normal tables)}$$

$$\Rightarrow \sigma' = \frac{1}{2.575} = 0.3883.$$

The standard deviation should be reduced from 0.5 to 0.39; a 23% reduction in variability is needed.

Example 2.33 is a situation in which changing the center of the process, or increasing the average life of the batteries, cannot be accomplished by a mere change of a tool setting or some other simple means. Improving battery life may require research and invention of a new process or new technology. However, improved results on battery life can be achieved by reducing the variability—in other words, by making the batteries more uniform—possibly through controlling the process variables at more consistent levels.

2.5.4.4 Distribution of the Sample Average $\overline{X}$

The sample average is a random variable. Suppose, for instance, that we take samples of size 16 repeatedly from a population and calculate the average from each sample, these averages will not all be the same. Each time a sample is taken, the $\bar{x}$ obtained from the sample is a value assumed by the random variable $\overline{X}$. If $\overline{X}$ is a random variable, then it should have a distribution, a mean, and a variance. The following result gives the distribution of $\overline{X}$ from samples taken from a normal population.

Theorem

$$\text{If } X \sim N(\mu, \sigma^2), \text{ then } \overline{X}_n \sim N\left(\mu, \frac{\sigma^2}{n}\right)$$

This result says that if samples of size n are taken repeatedly from a population that is normally distributed and the averages are computed from those samples, then the averages will be normally distributed, with a mean equal to the mean of the parent population and variance smaller than the population variance, depending on the sample size.

Example 2.34

Diameters of bolts are known to be normally distributed, with a mean of 2.0 in. and a standard deviation of 0.15 in. If a sample of nine bolts is drawn and their diameters are measured, what is the probability that the average diameter from the sample will be less than 2.1 in.?

Solution

Let X denote the diameter of bolts:

$$X \sim N(2.0, 0.15^2).$$

Then the sample averages from the samples of size $n = 9$ will be distributed as:

$$\overline{X}_9 \sim N\left(2.0, \frac{0.15^2}{9}\right)$$

That is,

$$\mu_{\overline{X}} = 2.0$$

$$\sigma^2_{\overline{X}} = \frac{0.15^2}{9} = \left(\frac{0.15}{3}\right)^2 = 0.05^2$$

$$\sigma_{\overline{X}} = 0.05.$$

We need $P(\overline{X} < 2.1)$. To convert this problem into one of the standard normal distribution, we have to subtract the mean of the random variable $\overline{X}$, $\mu_{\overline{X}}$, from both sides of the inequality and divide both sides by the standard deviation of the random variable $\overline{X}$, $\sigma_{\overline{X}}$. Therefore,

$$P(\overline{X} \leq 2.1) = P\left(Z \leq \frac{2.1 - 2.0}{0.05}\right) = P(Z \leq 2) = 0.9772$$

That is, 97.72% of the averages will be less than 2.1 in.

2.5.4.5 The Central Limit Theorem

The previous theorem gave the distribution of $\overline{X}$ when the population is normally distributed. What if the population is not normal or its distribution is not known? The central limit theorem (CLT) gives the answer.

Let a population have any distribution with a finite mean and a finite variance. That is, let $X \sim f(x)$, with mean $= \mu$ and variance $= \sigma^2$. Then,

$$\overline{X}_n \xrightarrow[n \to \infty]{} N\left(\mu, \frac{\sigma^2}{n}\right)$$

The CLT says that no matter what the population distribution is, the sample averages tend to be normally distributed if the sample size is large. It is known that this happens even for sample sizes as small as four or five if the population distribution is not extremely nonnormal. This is the reason statistical methods based on $\overline{X}$ are known to be robust with respect to the normality assumption for the population. In other words, if a method is based on $\overline{X}$, then even if an assumption of normality for the population is required, the conclusions from the method will remain valid even if the assumption of normality for the population is not quite valid. The confidence intervals and tests of hypotheses about population means—which we will discuss in the next section—as well as the control charts for measurements discussed in Chapters 4 and 5, have this robustness.

Example 2.35

Samples of size four are taken from a population whose distribution is not known but its mean μ and variance σ^2 are known to be 2.0 and 0.0225, respectively. The $\overline{X}$ values from the samples are plotted sequentially on a graph. Two limit lines are to be drawn, with the population mean at the center and the limit lines on either side, equidistant from the center, so that 99.73% of the plotted $\overline{X}$ values will fall within these limits. Where will the limit lines be located?

Solution

$$X \sim N(2.0, 0.0225) \quad \Rightarrow \overline{X} \sim N(2.0, 0.0225/4)$$

That is,

$$\mu_{\overline{X}} = 2.0 \quad \text{and} \quad \sigma_{\overline{X}} = \sqrt{\frac{(0.0225)}{4}} = \frac{0.15}{2} = 0.075$$

To include 99.73% of the $\overline{X}$ values, the limits must be located at 3-sigma distance from the mean of $\overline{X}$, where the "sigma" is $\sigma_{\overline{X}} = 0.075$. (Refer to Example 2.31f.) Therefore, the limits must be at 2.0 ± 3(0.075)—that is, at 1.775 and 2.225, respectively.

2.5.5 Exercise in Probability Distributions

2.28 If Y denotes the difference between two numbers when two dice are thrown, find its *pmf*, mean, and variance.

2.29 An urn contains two white balls and two black balls. If X denotes the number of black balls in a sample of three balls drawn from it without replacement, find the *pmf*, mean, and variance of X.

2.30 A random variable X has the following *pdf*. Find $P(X < 4)$, as well as the mean and variance of X.

$$f(x) = \begin{cases} \dfrac{2(1+x)}{27}, & 2 \le x \le 5 \\ 0, & \text{otherwise} \end{cases}$$

2.31 A random variable has the following *pdf*. Find the value of a so the function is a valid *pdf*.

$$f(x) = \begin{cases} ax^3, & 0 \le x \le 1 \\ 0, & \text{otherwise} \end{cases}$$

Also, find $F(x)$, $F(1/2)$, $F(3/4)$, $P(1/2 \le X \le 3/4)$, and the mean and variance of X.

2.32 A plane has four engines. Each engine has a probability of failing during a flight of 0.3. What is the probability that no more than two engines will fail during a flight? Assume that the engines are independent.

2.33 A consignment of 200 brake cylinders is known to contain 2% defectives. The customer uses a sampling rule according to which a sample of eight units will be taken. If the sample contains no defectives, the consignment will be accepted; otherwise, the consignment will be rejected. What is the probability that this consignment of 200 cylinders will be accepted using the sampling rule?

2.34 A chemical plant experiences an average of three accidents per month. What is the probability that there will be no more than two accidents next month?

2.35 In a foundry, the castings coming out of a mold line have an average of three gas holes that can be salvaged. A customer will reject a casting if it has more than six such holes. If a lot of 200 castings is sent to the customer in a month, approximately how many of them would you expect to be rejected?

2.36 Let $X \sim N(10, 25)$. Find $P(X \le 15)$, $P(X \ge 12)$, and $P(9 \le X \le 20)$.

2.37 Diameters of bolts are normally distributed with $\mu = 2.02$ in. and $\sigma = 0.02$ in. If the specifications for the diameter are at 2.0 ± 0.06 in., find:

a. The proportion of diameters below the lower specification.

b. The proportion above the upper specification.

c. The proportion in specification.

2.38 Let $X \sim N(5, 4)$. Find b such that $P(X > b) = 0.20$.

2.39 The thickness of gear blanks produced by an automatic lathe is known to be normally distributed, with a mean of 0.5 in. and a standard deviation of 0.05 in. If 10% of the blanks are rejected for being too thin, where is the lower specification located?

2.40 Let $X \sim N(5, 9)$. Find the values of a and b such that $P(a \le X \le b) = 0.80$ if the interval (a, b) is symmetrical about the mean.

2.41 The life of a particular type of battery is normally distributed, with a mean of 600 days and a standard deviation of 49 days. What fraction of these batteries will survive beyond 586 days?

2.42 Automatic fillers are used to fill cans of cooking oil, which have to meet a minimum specification of 10 oz. To ensure that every can meets this minimum, the company has set a target value for the process average at 11 oz. The standard deviation of the amount of oil in a can is known to be 0.20 oz.

a. At the process average of 11 oz, what percentage of the cans will contain less than 10 oz of cooking oil? Assume that the amount of oil in cans is normally distributed.

b. Assuming that virtually all values of a normal distribution fall within a distance of 3.5σ from the mean, find the minimum value to which the process average can be lowered so that virtually no can will have less than 10 oz?

2.6 Inference of Population Quality from a Sample

As mentioned earlier, we often have to draw conclusions about the quality of a population from the quality observed in a sample. In statistical terms, knowing the quality of a population is equivalent to knowing the distribution of the population along with its parameter values. If the population in question has the normal distribution, which we can expect with regard to many quality characteristics, then the values of the two parameters, the mean μ and the variance σ^2, must be known or estimated.

As mentioned in Section 2.4 of this chapter, if we can get a large size sample (≥ 50) from such a population, then the $\overline{X}$ and S^2 obtained from the sample can be used as the estimates of μ and σ^2, respectively. If, however, only a small size sample is available, then the sampling error, or the variability in $\overline{X}$ and S^2, inherited from the variability of the population, will be too large, thus rendering them useless as estimates. For these situations, statisticians have devised methods for making inferences about population parameters—these are called "inference procedures."

In this section, we discuss two such procedures for normally distributed populations: the method of creating confidence intervals for population parameters, and the method of hypothesis testing to verify if the hypotheses, or claims, made about population parameters are valid. In addition, we also discuss the methods for verifying the distribution of a given population, especially when that population is expected to follow the normal distribution. These include a graphical method using the normal probability paper and an analytical method called the "goodness-of-fit" test. First, a few definitions relating to the inference procedures are necessary.

2.6.1 Definitions

A sample of n units drawn from a population produces n observations $\{X_1, X_2, \ldots, X_n\}$. This set of n observations is identified as "the sample" for further mathematical work. It is easy to see that each element of the sample is a random variable having the same distribution as the population. In addition, we assume that the observations are independent. Thus, a sample of size n is a collection of n independent, identically distributed (I.I.D.) random variables. When we pick a sample (of physical units) in practice and take measurements on them, we get one set of observations $\{x_1, x_2, \ldots, x_n\}$ on the I.I.D. random variables. A statistic is a function of the random variables constituting the sample. The sample average $\overline{X}$, range R, standard deviation S, the largest value X_{max}, and so on, are examples of statistics. A statistic is a random variable, because it is a function of random variables. For example, $\overline{X}$ is a random variable, and when we take a sample and calculate the average of the observations, we obtain one observation $\bar{x}$ of the random variable $\overline{X}$.

We said earlier that when we want to know the quality of a population, we in fact want to know the values of the two parameters, assuming the population is normal. The exact values of these parameters are never known, although they exist. All we can do is to estimate their values from samples. When we want to estimate a parameter, we choose one of the statistics from the sample as the *estimator* for the parameter. A few statistics may be available as candidates for being chosen as the estimator. For example, $\overline{X}$ or $\tilde{X}$, the median, can be used as the estimator for μ, the mean of a normal population; and the sample standard deviation S, or the sample range R can be used as the estimator for σ, the standard deviation of the normal population. Certain criteria are used to select a "good" estimator from the available candidates. Being *unbiased* is one of the desirable properties of an estimator. An estimator is said to be unbiased if its expected value, or the average value (over many samples), equals the parameter being estimated. Using the notation $E(.)$ to denote the expected value, or the long-run average value of a random variable, it can be shown for normal populations that:

$$E(\overline{X}) = \mu$$

$$E(\tilde{X}) = \mu$$

$$E(S^2) = \sigma^2$$

In the last expression above, $S^2 = \sum (X_i - \overline{X})^2 / n - 1$, with $(n - 1)$ used in the denominator, as only then is it an unbiased estimator for σ^2. If there was n in the denominator instead of $(n - 1)$, then it would not be an unbiased estimator for σ^2. Also, note that S^2 is unbiased for σ^2 but that S is not an unbiased estimator for σ. Furthermore, the range R is not an unbiased estimator for σ. (All these results have proofs that can be found in books on probability and statistics, such as Guttman et al. 1982).

If several unbiased estimators are among the candidates, then the one with the least amount of variability would be preferred. Such an estimator is called the "minimum variance unbiased" (MVUB) estimator. The statistics $\overline{X}$ and S^2 are known to be such MVUB estimators for μ and σ^2, respectively, for a normal population. That is the reason these estimators with such "good" properties are used in the methods discussed below for making inferences about μ and σ^2 of normal populations.

The value of an estimator computed from a single sample is called a "point estimate." The point estimate, especially from a small sample, is of no value because it is just one observation of a random variable. Its value will vary from sample to sample and cannot be trusted as the true value of the parameter. Therefore, we resort to interval estimation, in which we create an interval in such a way that there is certain level of confidence—say, 95% or 99%—that the parameter we are seeking lies in the interval. Such an interval is called a "confidence interval."

2.6.2 Confidence Intervals

Suppose we want to estimate the mean μ of a normal population using sample average $\overline{X}$ as the estimator. A sample from the population is taken, and the value of $\overline{X}$ is computed. This observed value of $\overline{X}$, denoted by $\overline{x}$, is a point estimate for μ. Using this point estimate, an interval $(\overline{x} - k, \overline{x} + k)$ is created such that $P(\overline{x} - k, \leq \mu \leq \overline{x} + k) = 1 - \alpha$, where $(1 - \alpha)$ is called the "confidence coefficient." The interval $(\overline{x} - k, \overline{x} + k)$ is a $(1 - \alpha)100\%$ confidence interval (CI) for μ. The value of k is determined as shown below from the distribution of the estimator $\overline{X}$.

2.6.2.1 CI for the μ of a Normal Population When σ Is Known

The estimator is $\overline{X}$. On the assumption that the population has $N(\mu, \sigma^2)$, as stated earlier in the chapter, the $\overline{X}$ has $N(\mu, \sigma^2/n)$, where n is the sample size. Then, $(\overline{X} - \mu)/(\sigma/\sqrt{n}) \sim N(0,1)$, from which the following statement can be made:

$$P\left(-z_{\alpha/2} \leq \frac{\overline{X} - \mu}{\sigma/\sqrt{n}} \leq z_{\alpha/2}\right) = 1 - \alpha,$$

where $z_{\alpha/2}$ is a number such that $P(Z > z_{\alpha/2}) = \alpha/2$. (See Figure 2.15).

The terms on the left-hand side of the above equation can be rearranged to give the statement

$$P\left(\overline{X} - z_{\alpha/2}\frac{\sigma}{\sqrt{n}} \leq \mu \leq \overline{X} + z_{\alpha/2}\frac{\sigma}{\sqrt{n}}\right) = 1 - \alpha$$

(In making the rearrangement, we have used the principle that when both sides of an inequality are multiplied by a constant or the same quantity is

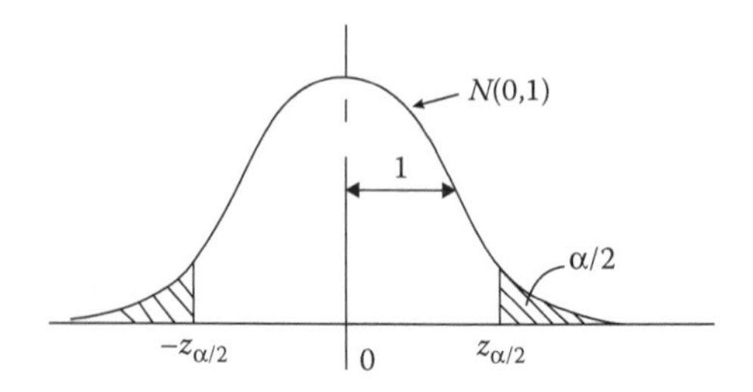

FIGURE 2.15
Definition of $z_{\alpha/2}$.

added or subtracted from both sides of an inequality, the inequality remains unchanged.) This equation, indeed, provides the CI for μ with (1 − α) as the confidence coefficient. Therefore, we can make the following statement: A (1 − α)100% CI for μ of a population that is normally distributed with a known standard deviation σ is given by:

$$\left[\bar{x} - z_{\alpha/2}\frac{\sigma}{\sqrt{n}},\ \bar{x} + z_{\alpha/2}\frac{\sigma}{\sqrt{n}}\right]$$

where $z_{\alpha/2}$ is the number that cuts off α/2 probability at the upper tail of the standard normal distribution. In this model we have assumed that the value of the population standard deviation σ is known. We can, however, build another model for the situation when σ is not known by using the same procedure.

Example 2.36

A random sample of four bottles of fabric softener was taken from a day's production and the amount of turbidity in them was measured as: 12.6, 13.4, 12.8, and 13.2 ppm, respectively. Find a 99% CI for the mean turbidity of the population of bottles filled in this line on that day. It is known that turbidity measurements are normally distributed with a standard deviation of σ = 0.3.

Solution

$(1-\alpha) = 0.99 \Rightarrow \quad \alpha/2 = 0.005 \Rightarrow \quad z_{0.005} = 2.575$ (from normal tables)

$\bar{x} = 52/4 = 13.0 \quad \sigma = 0.3$

99% CI for $\mu = \left[13.0 - 2.575\left(0.3/\sqrt{4}\right), 13.0 + 2.575\left(0.3/\sqrt{4}\right)\right] = [12.61, 13.39]$

2.6.2.2 Interpretation of CI

The CI in the above example is interpreted as follows: If 99% confidence intervals are set up, as above, from samples of size four on 100 days, then on

99 out of the 100 days the true mean that we are seeking will lie inside those intervals, and on 1 day the true mean will not lie within the interval.

2.6.2.3 CI for μ When σ Is Not Known

In the above model of CI for μ, we assumed that the population standard deviation σ was known. Suppose the population standard deviation is not known; then, σ is replaced with S, the sample standard deviation, in the statistic $(\overline{X}-\mu)/(\sigma/\sqrt{n})$ to obtain a new statistic $(\overline{X}-\mu)/(S/\sqrt{n})$. This new statistic is known to have the t distribution with $(n - 1)$ degrees of freedom. The CI is created using this new statistic and its distribution, the t distribution.

The t distribution is a symmetrical distribution with a mean of zero and a shape that resembles the standard normal distribution except for heavier tails. It has one parameter and is called the "degrees of freedom" (*df*), and the shape of the distribution and the thickness of the tails depend on the value of the parameter *df*. When the *df* is 30 or larger, the shape approaches that of the standard normal distribution. Tables, such as Table A.2 in the Appendix, tabulate percentiles of the t distribution for various degrees of freedom. The tabled values are $t_{\alpha,\nu}$, where α is the probability the tabled value cuts off at the upper tail of the t distribution with ν degrees of freedom. (See Figure 2.16.) The formula for the CI is obtained following the same procedure as was outlined for the previous model.

A (1 − α) 100% CI for μ of a population that is normally distributed, when σ is not known, is given by:

$$\left[\overline{x}-t_{\alpha/2,n-1}\frac{s}{\sqrt{n}},\ \overline{x}+t_{\alpha/2,n-1}\frac{s}{\sqrt{n}}\right],$$

where $\overline{x}$ is the sample average and s is the sample standard deviation from a sample of size n, and $t_{\alpha/2,n-1}$ is such that $P(t_{n-1} > t_{\alpha/2,n-1}) = \alpha/2$. That is, $t_{\alpha/2,n-1}$ is the number that cuts off α/2 probability at the upper tail of the t distribution with $df = n - 1$.

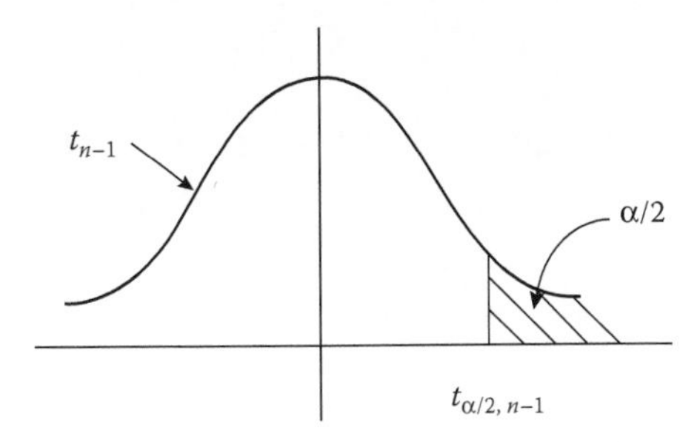

FIGURE 2.16
Example of a t distribution.

Example 2.37

Four sample measurements of the turbidity in bottles of fabric softener filled in a day on a filling line were as follows: 12.6, 13.4, 12.8, and 13.2 ppm, respectively. Set up a 99% CI for the average turbidity in all the bottles of fabric softener filled that day. Assume a normal distribution. The σ is not known.

$$\alpha/2 = 0.005 \quad \bar{x} = 13.0$$

$$n-1 = 3 \quad t_{0.005,3} = 5.841 \text{ (from the } t \text{ table)}$$

$$s = 0.366 \text{ (calculated from the sample observations)}$$

$$\text{99\% confidence interval} = \left[13.0 \pm 5.841\left(0.366/\sqrt{4}\right)\right] = [11.93, 14.06].$$

Note that the CI in the above example is wider compared to that in the σ-known case. The wider a CI becomes, the less precise it is in estimating the parameter. In this case, the loss of precision is due to a lack of information on the population standard deviation.

2.6.2.4 CI for σ^2 of a Normal Population

When setting CI for the variance of the population σ^2, the sample variance $S^2 = \Sigma(X_i - \bar{X})^2/n-1$ is used as the estimator. The fact that the statistic $(n-1)S^2/\sigma^2$ has the χ^2 (chi-squared) distribution with $(n - 1)$ degrees of freedom is made use of in obtaining the confidence interval for σ^2.

The χ^2 distribution is a positive-valued distribution with a single parameter, *df*. The shape of a χ^2 distribution depends on the value of the parameter *df*, and percentiles of the χ^2 distribution for various values of *df* are available in tables such as Table A.3 in the Appendix. The tabled values are $\chi^2_{\alpha,\nu}$, where α is the probability the tabled value cuts off at the upper tail of the χ^2 distribution with ν degrees of freedom. (See Figure 2.17.)

A $(1 - \alpha)$ 100% CI for the variance σ^2 of a normal population is given by:

$$\left(\frac{(n-1)s^2}{\chi^2_{\alpha/2,n-1}}, \frac{(n-1)s^2}{\chi^2_{1-\alpha/2,n-1}}\right),$$

where s^2 is the sample variance, n the sample size, and $\chi^2_{\alpha,n-1}$ is such that $P\left(\chi^2_{n-1} > \chi^2_{\alpha,n-1}\right) = \alpha$.

Example 2.38

Set up a 99% CI for the standard deviation σ of the turbidity in bottles of cloth softener if a sample of four bottles gave the following measurements: 12.6, 13.4, 12.8, and 13.2 ppm, respectively. Assume normality for turbidity in bottles.

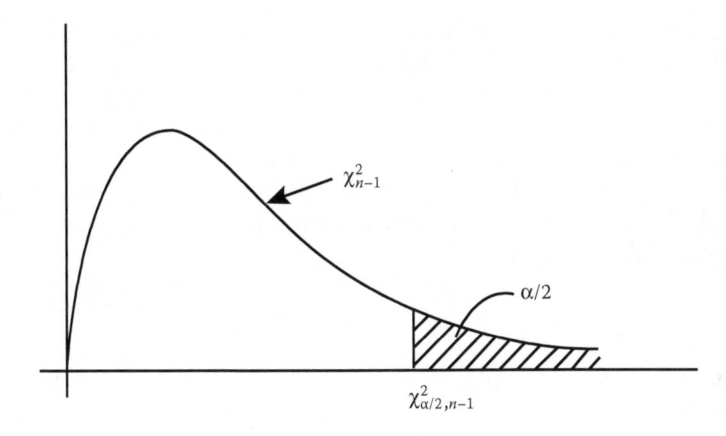

FIGURE 2.17
Example of a chi-squared distribution.

Solution

First, we find the CI for σ^2, and then obtain the CI for σ:

$$\alpha/2 = 0.005 \quad (n-1) = 3 \quad s^2 = 0.133 \text{ (calculated from the sample).}$$

We have to calculate the limits: $\left[\dfrac{3(0.133)}{\chi^2_{0.005,3}}, \dfrac{3(0.133)}{\chi^2_{0.995,3}}\right]$

From the χ^2 tables: $\chi^2_{0.005,3} = 12.838 \quad \chi^2_{0.995,3} = 0.0717$

$$99\% \text{ CI for } \sigma^2 = \left[\frac{(3)(0.133)}{12.838}, \frac{(3)(0.133)}{0.0717}\right] = [0.031, 5.56]$$

$$99\% \text{ CI for } \sigma = \left[\sqrt{.031}, \sqrt{5.56}\right] = [0.176, 2.36].$$

Now, suppose we want a 95% CI for σ in the above example:

$$\alpha/2 = 0.025 \quad \chi^2_{0.025,3} = 9.348 \quad \chi^2_{0.975,3} = 0.216$$

$$95\% \text{ CI for } \sigma^2 = [0.042, 1.847]$$

$$95\% \text{ CI for } \sigma = [0.205, 1.359].$$

We see that the smaller the confidence coefficient, the narrower the CI becomes.

We have so far seen some examples of using confidence intervals to estimate population parameters. The discussion was restricted to models for estimating the mean and variance of a normal population—the two cases that are encountered most commonly in quality engineering work. Other models for estimating parameters of other distributions follow the same principles used in creating the models above. Models exist to set up confidence intervals for the fraction defectives in a population, using the fraction defectives obtained from a sample. Other models exist for the CI for the difference between two population means when the standard deviations are known, as well as when standard deviations are not known. These models are used to compare the means of two populations. When such models are needed, the reader should refer to the books on probability and statistics listed as references at the end of this chapter.

2.6.3 Hypothesis Testing

Hypothesis testing is another inference procedure in which a hypothesis proposed about the value of a population parameter is tested for its validity, based on the information obtained from a sample taken from the population. The hypothesis could also be about the relationship of parameters of two populations, such as whether they are equal or one of them is larger than the other. This method will be used when someone makes a claim about a population parameter and the correctness of that claim has to be verified. Suppose, in the case of the fabric softener, that the manufacturer claims that the average turbidity is no more than, for example, 10 ppm. The hypothesis testing can then be used to verify such a claim.

In this procedure, two hypotheses are proposed; one is called the "null hypothesis," denoted by H_0; and the other is called the "alternative hypothesis," denoted by H_1. Generally, H_0 and H_1 are complementary to each other in the sense that if one is accepted, then the other is rejected, and vice versa. The claim made about a parameter is usually included in H_1 for reasons that are explained a little later.

After the hypotheses have been set up, a sample is taken from the population in question and the value of an appropriate test statistic is computed from the sample results. For each testing situation, a suitable test statistic is chosen in such a way that it relates a sample estimator to the population parameter in question and the distribution of the test statistic is known.

A statistical test could result in any one of the four possible events shown in Figure 2.18. Of these four, two of the events lead to errors in conclusion. These errors are designated as Type I and Type II errors, as shown in Figure 2.18, and can be summarized as follows:

Outcomes from a statistical test		The test declares H_0	
		True	Not True
In reality H_0 is	True	OK	**Error - Type I**
	Not True	**Error - Type II**	OK

FIGURE 2.18
Possible outcomes of a statistical test.

- A Type I error is said to occur if H_0 is declared not true when, in fact, it is true.
- A Type II error is said to occur if H_0 is declared true when, in fact, it is not true.
- The probability of a Type I error is denoted by α and is called the "level of significance."
- The probability of a Type II error is denoted by β, and $(1 - \beta)$ and is called the "power" of the test.

When a test is designed using a certain test statistic, the probability of these errors occurring can be calculated from the knowledge of the distribution of the test statistic. More importantly, the test can be designed such that the probability of an error does not exceed a specified value.

Designing a test involves setting up a *critical region* (CR) in the distribution of the test statistic. The CR is the region in the distribution of the test statistic, such that if the observed value of the test statistic (from a sample) falls in this region, it will lead to rejection of H_0 in favor of H_1. Identification of the CR is the major part of test design, and it depends on how H_0 and H_1 are set up. We will see below how the CR is identified for different hypotheses scenarios.

The steps involved in hypothesis testing can be summarized as follows:

1. Set up H_0 and H_1.
2. Choose a level of significance.
3. Choose an appropriate test statistic.
4. Identify the CR.
5. Select a sample from the population, and compute the observed value of the test statistic.
6. If the observed value falls in the CR, reject H_0 in favor of H_1; otherwise, do not reject H_0.
7. Interpret the results.

Examples are provided below to illustrate the method of using hypothesis testing to draw inference about population parameters.

2.6.3.1 Test Concerning the Mean μ of a Normal Population When σ Is Known

The hypotheses are:

H_0: $\mu = \mu_0$ (hypothesize that the mean equals a number μ_0).
H_1: $\mu > \mu_0$ (if the mean is not equal to μ_0, it must be greater than μ_0).

It should be pointed out that in the above set up, although H_0 has only the equality sign (=), the inequality sign (≤) opposite to that in the H_1, is implied, because only then will H_0 and H_1 complement each other.

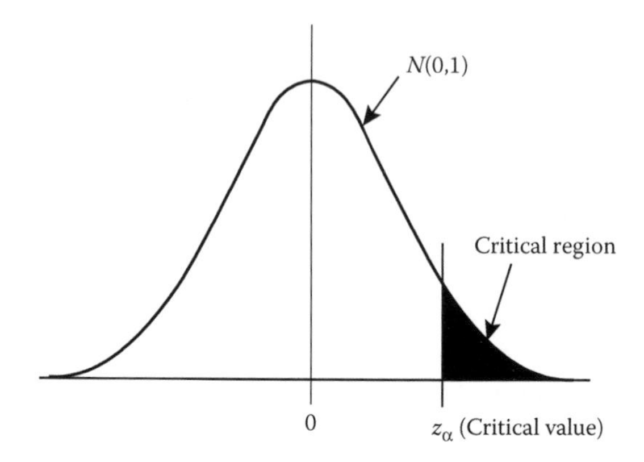

FIGURE 2.19
Critical region in the distribution of the test statistic.

The test statistic is:

$$\frac{\bar{X}-\mu_0}{\sigma/\sqrt{n}} \sim N(0,1)$$

Note that this test statistic relates the sample average $\bar{X}$, the estimator, to the population mean μ through its known distribution, the Z distribution ($N(0, 1)$).

The alternate hypothesis determines where the CR is located. In this case, the CR is chosen as the set of all observed values of the test statistic that are greater than z_α, where z_α is such that $P(Z > z_\alpha) = \alpha$. The following is the logic used in choosing this critical region.

If H_0 is true, then the observed value of the test statistic will be a value from the Z distribution, most probably a value near its mean, zero. If, on the other hand, H_1 is true, then the observed value will be a large value falling far to the right of zero, on the higher side, as shown in Figure 2.19. We need to draw a line somewhere to determine how large is too large in order to decide if H_0 should be rejected in favor of H_1. Wherever the line is drawn, there will be a probability that H_0 is rejected when, in fact, it is true, because the random variable Z can assume very large values (or very small values) even if its mean is zero, as the normal curve is asymptotic with respect to the x-axis. We draw the line in such a way that this probability is not more than a small number, α. So, the line is drawn at z_α, where z_α cuts off α probability at the upper tail in the Z distribution. The z_α is the critical value of the observed statistic. The values in the distribution beyond z_α constitute the critical region.

Example 2.39

A supplier of cotton rope claims that their new product has an average strength greater than 10 kg. A sample of 16 rope pieces gave an average of 10.2 kg. If the

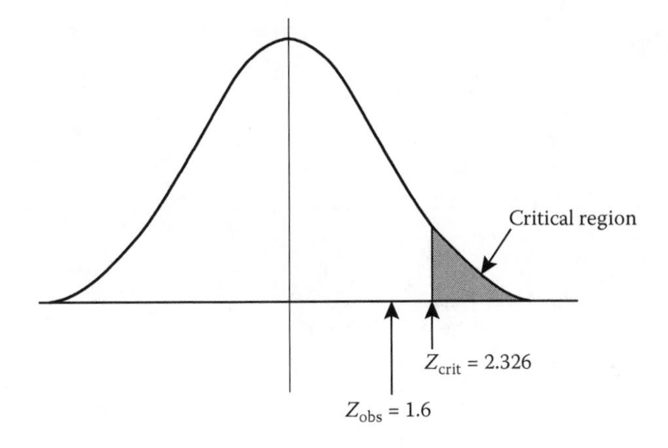

FIGURE 2.20
Critical region for testing claim about cotton rope strength.

standard deviation of the strength is known to be 0.5 kg, test the hypothesis that $\mu = 10$ vs. $\mu > 10$. Use $\alpha = 0.01$.

Solution

$$H_0 : \mu = 10$$
$$H_1 : \mu > 10$$

The hypotheses have been set up in such way that the claim is included in the alternate hypothesis.

$$\text{The test statistic is: } \frac{\overline{X} - 10}{\sigma/\sqrt{n}} \sim z$$

The CR is: $Z_{obs} > z_\alpha = z_{.01} = 2.326$.
That is, all values of Z_{obs} greater than 2.326 will lead to the rejection of H_0, as shown in Figure 2.20.

$$\text{The observed value of the test statistic from sample } Z_{obs} = \frac{10.2 - 10}{0.5/\sqrt{16}} = 1.6.$$

The Z_{obs} is not in the CR, and H_0 is not rejected. The mean strength of the population of cotton ropes is not greater than 10 kg; the supplier's claim is not valid.

This is an example of a "one-tailed test," so-called because the CR is located on one tail of the distribution of the test statistic. So, the alternate hypotheses of the kind H_1: $\mu > \mu_0$ and H_1: $\mu < \mu_0$ will need the one-tailed test.

2.6.3.2 Why Place the Claim Made about a Parameter in H_1?

The hypotheses for the above example could have been set up in two possible ways, as in Sets 1 and 2 below:

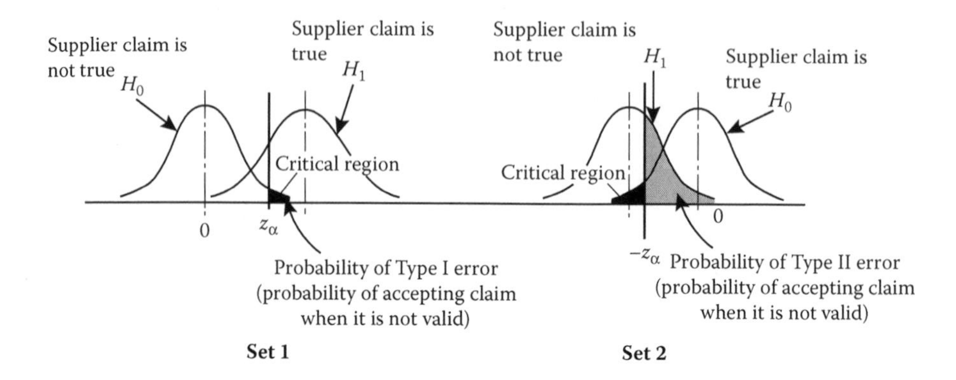

FIGURE 2.21
Two possible sets of hypotheses to verify a claim.

Set 1	Set 2
$H_0 : \mu \le 10$	$H_0 : \mu \ge 10$
$H_1 : \mu > 10$	$H_1 : \mu < 10$

For both sets, the test statistic is: $\dfrac{\overline{X}-10}{\sigma/\sqrt{n}} \sim Z.$

The supplier's claim is included in H_1 in Set 1, whereas the claim is included in H_0 in Set 2 (with the implied > sign as part of H_0). The CR corresponding to each set is shown in Figure 2.21.

In order to understand the difference between the two sets, let us focus on the event that the supplier's claim is accepted when it is not true, which is an important event from the point of view of a customer and should be controlled. In Set 1, this event happens when H_0 is rejected when it is true. This is the Type I error in this set up and its probability is limited to α. Also, it does not depend on where the distribution of the supplies is located. In the case of Set 2, this event occurs when H_1 is true and it is not accepted. This is the Type II error in this set up and its probability depends on where the distribution of the supplies is located. It could be a small value or it could be a very large value, depending on the distribution of the supplies. This lack of predictability of the probability of accepting the supplier's claim when it is not true makes Set 2 unreliable. Hence the choice of Set 1 where the claim is placed in the alternate hypothesis is preferable.

2.6.3.3 The Three Possible Alternate Hypotheses

We mentioned that the selection of the CR depends on the alternate hypothesis. Three possible scenarios that are encountered in practical testing and the applicable hypotheses for each are shown in Figure 2.22. The particular set

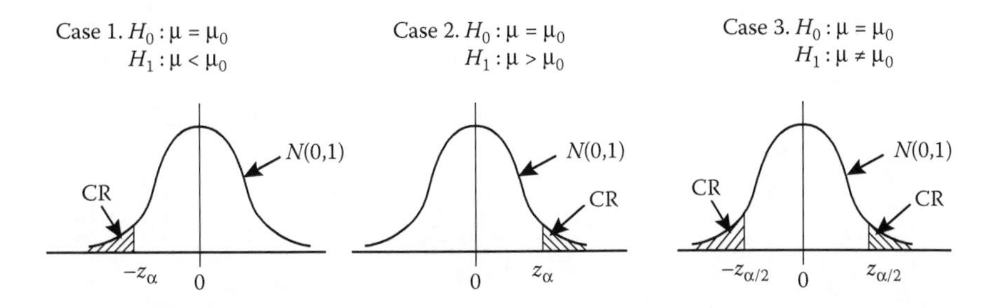

FIGURE 2.22
Alternate hypotheses and corresponding critical regions (CR).

of hypotheses chosen for a given situation will depend on where the experimenter believes the mean will be if it is not at the hypothesized value.

The test statistic will be the same for all three situations, and the location of the CR in each case will change according to the alternate hypothesis, as shown in Figure 2.22.

In Case 1, the experimenter believes that if the mean is not equal to the hypothesized value μ_0, it should be less than that value. The choice of H_1 reflects this belief. The H_0 will be rejected if the observed value of the test statistic is too small, and the CR determines how small is too small. In Case 2, the experimenter has reasons to believe that if the mean is not equal to μ_0, it should be larger than μ_0. The H_0 will be rejected if the observed value of the test statistic is too large and the CR determines how large a value of test statistic is too large. In both cases, the risk of α of rejecting H_0 when it is not true is assigned to the side of the distribution where the mean is expected to fall if not at the hypothesized value. In Case 3, the experimenter has no idea which side of the μ_0 the mean will be if it is not at μ_0. The H_1 reflects this, and the CR determines how far away on either side from zero the value of the test statistic should be before H_0 is rejected. Here, the risk of α is divided equally on either side of the distribution, because the mean could fall on either side if not at the hypothesized value. Cases 1 and 2 are one-tailed tests, and Case 3 is a two-tailed test.

Example 2.39 above is an example under Case 2. Examples of Cases 1 and 3 will be found in the next section.

2.6.3.4 *Test Concerning the Mean μ of a Normal Population When σ Is Not Known*

When σ is not known, the sample standard deviation S is used instead of σ in the test statistic. This new statistic is known to have the t distribution with $(n - 1)$ degrees of freedom. That is,

$$\frac{\overline{X} - \mu_0}{S/\sqrt{n}} \sim t_{n-1}.$$

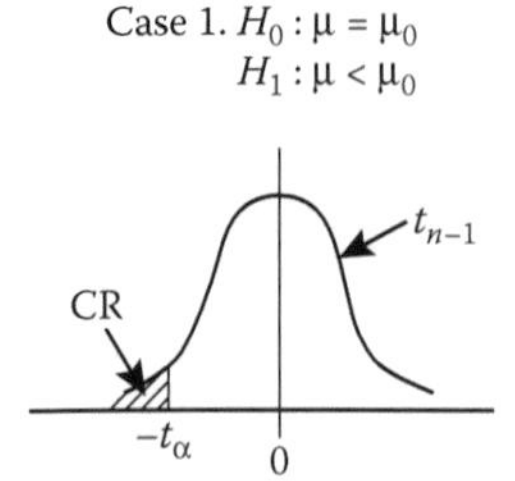

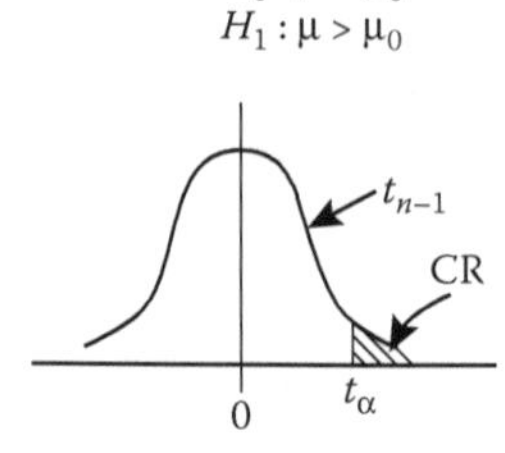

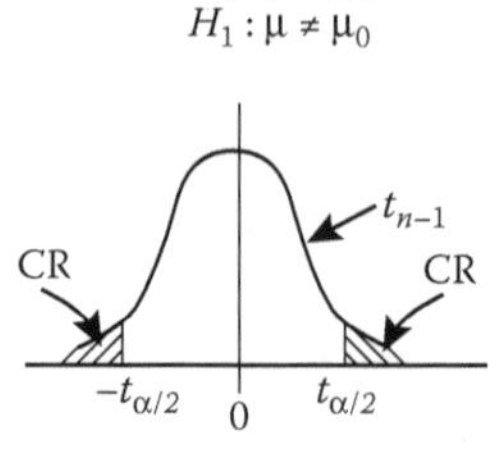

FIGURE 2.23
Critical regions for tests for μ when σ is not known.

There are, again, three possible alternate hypotheses. The critical regions corresponding to the three alternate hypotheses are shown in Figure 2.23.

Example 2.40

The amount of ash in a box of sugar should be less than 2 g according to a producer's claim. The lab analysis of a sample of five boxes gave the following results: 1.80, 1.92, 1.84, 2.02, and 1.76 g, respectively. Does the sample show evidence that the mean ash content in the boxes is less than 2 g, as claimed by the producer? Use $\alpha = 0.05$.

Solution

$H_0 : \mu = 2 \quad H_1 : \mu < 2$ (the claim is in the alternate hypothesis).

$\bar{x} = 1.868 \quad s = 0.104$ (computed from sample observations).

The test statistic is: $\dfrac{\bar{X} - 2}{S/\sqrt{5}} \sim t_4$.

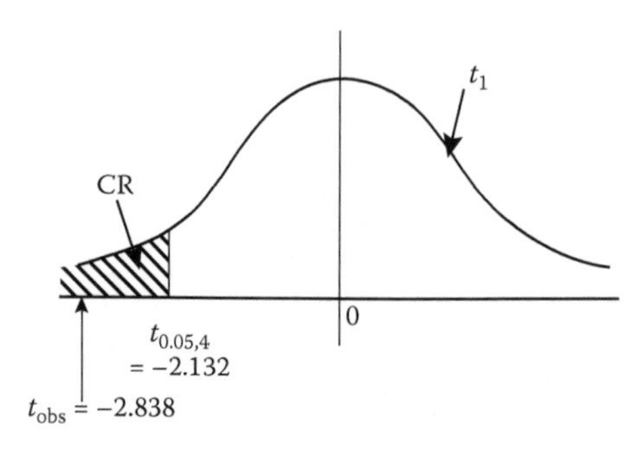

FIGURE 2.24
Location of critical region for ash in sugar boxes.

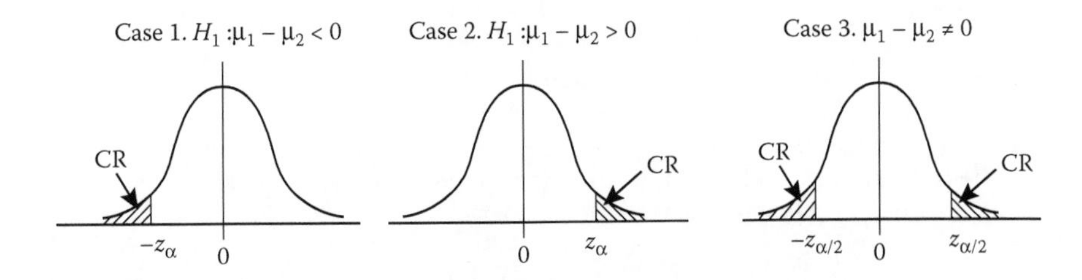

FIGURE 2.25
Critical regions for hypotheses about difference of two means.

The CR is made up of values of $t_4 < -t_{0.05,4} = -2.132$ (from the t table, Table A.2). That is, reject H_0 if the observed value of the test statistic is less than −2.132. (See Figure 2.24.)

The observed value of t_4 from the sample is: $t_{obs} = \dfrac{1.868 - 2}{0.104/\sqrt{5}} = -2.838$.

Thus, t_{obs} is in CR; therefore, reject H_0. The mean ash content is less than 2 g; the producer's claim is valid.

2.6.3.5 Test for Difference of Two Means When σs Are Known

This test model is useful when two populations have to be compared with respect to their mean values. For example, the mean life of bulbs from one manufacturer may have to be compared with the mean life from another manufacturer.

$$\text{Test Statistic}: \frac{(\overline{X}_1 - \overline{X}_2)}{\sqrt{\sigma_1^2/n_1 + \sigma_2^2/n_2}} \sim Z$$

H_0: $\mu_1 - \mu_2 = 0$ (no difference between the two population means).

Again, there are three possible alternate hypotheses, and the CR corresponding to the three cases are shown in Figure 2.25.

Example 2.41

A random sample of 10 male workers in a factory earned an average annual salary of \$35,000, and a random sample of 8 female workers earned an average of \$34,600. The standard deviations of the (population of) salaries of male and female workers are known to be \$1200 and \$1800, respectively. Test the hypothesis that male and female workers in this factory earn equal pay against the alternate hypothesis that they do not. Use $\alpha = 0.01$.

Solution

Male Workers	Female Workers
$\overline{x}_1 = 35{,}000$	$\overline{x}_2 = 34{,}600$
$\sigma_1 = 1200$	$\sigma_2 = 1800$
$n_1 = 10$	$n_2 = 8$

$$H_0 : \mu_1 - \mu_2 = 0$$
$$H_1 : \mu_1 - \mu_2 \neq 0$$

where μ_1 and μ_2 are, respectively, the means of male and female earnings.

The two-sided H_1 is chosen because the experimenter has no reason to believe the difference in average is positive or negative.

$$\text{Test statistic} : \frac{(\bar{X}_1 - \bar{X}_2)}{\sqrt{\sigma_1^2/n_1 + \sigma_2^2/n_2}}$$

$$\text{CR: values of } Z_{obs} > z_{\alpha/2} \text{ or } Z_{obs} < -z_{\alpha/2}$$

$$z_{\alpha/2} = z_{0.005} = 2.575 \text{ (from the normal tables)}$$

$$Z_{obs} = \frac{400}{\left[\frac{1200^2}{10} + \frac{1800^2}{8}\right]^{1/2}} = \frac{400}{740} = 0.541.$$

Z_{obs} is not in the critical region (see Figure 2.26); hence, do not reject H_0. Evidence from samples does not show a significant difference between the salaries of male and female workers.

We have seen only some basic models in hypothesis testing to get an idea of how this statistical procedure works and where it can be used. There are models available that can be used to test hypotheses regarding the difference between two means when population standard deviations are not known and sample sizes are small. We then use a test statistic that has a t distribution. The model to test the hypothesis concerning population variance uses a test statistic that has a χ^2 distribution. The model to test the hypothesis

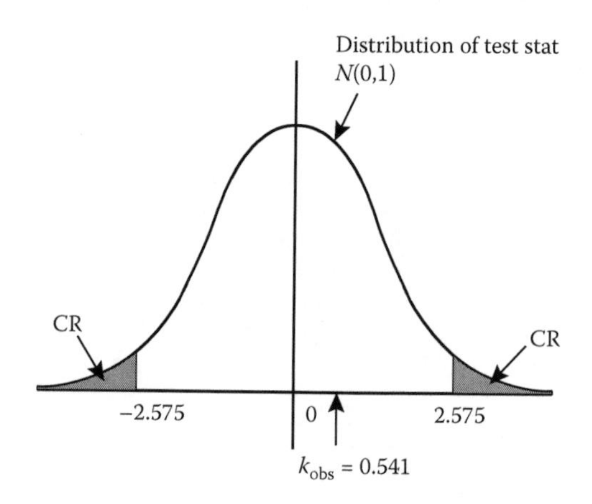

FIGURE 2.26
Critical region for testing the equality of the salary of male and female workers.

about the ratio of two variances uses a test statistic that has a F distribution. All these models assume that the populations concerned have a normal distribution. There are testing procedures that do not require an assumption of normality for the population; these are known as "distribution-free" or "nonparametric" tests. Any book in statistics will give further details on these methods.

We made an assumption in the foregoing discussions of confidence intervals and hypothesis testing that the populations concerned were normal. We will make similar assumptions when using other statistical tools discussed in other chapters in this book. Although we may be justified in making this assumption in many situations, we also may want to verify, on occasions, if this assumption is valid.

Tests exist for verifying the type of distribution a population follows based on the sample data drawn from it. These tests examine how well a set of data fits a hypothesized distribution, and they can be used to verify any distribution—such as exponential, normal, or Weibull. We will describe below the methods as they are used to verify the normality of populations.

2.6.4 Tests for Normality

A simple, informal method to verify the normality of a population is to take a large sample ($n \geq 50$) from the population and draw a histogram from the sample data. If the histogram follows the symmetrical bell shape, then it is verification that the population in question follows the normal distribution. This, however, is a very subjective, approximate method, and it is not recommended when large samples are not available. Two other, more formal methods, one graphical and the other analytical, are described below.

2.6.4.1 Use of the Normal Probability Plot

Normal probability plotting involves plotting the cumulative percentage distribution of the data on specially designed normal probability paper (NPP). The NPP is designed such that if the data came from a normal population, the cumulative distribution of the data will plot as a straight line. Conversely, if the cumulative distribution of the data plots as a straight line on an NPP, then we conclude that the data set comes from a normal population. The procedure is described below using an example.

Table 2.3 contains information taken from Table 2.2 on the fill-weights of deodorant in cans. Figure 2.27 shows the cumulative percentage distribution from Table 2.3 plotted on a commercially available NPP (from www.Weibull.com). The cumulative percentages have been plotted against the upper limit of each of the cells. We see that the plotted points all seem to fall on a straight line, indicating that the data come from a normal population. If a line is drawn approximately through "all" the plotted points, the resulting

TABLE 2.3
Frequency Distribution and Cumulative Frequency Distribution

	Cell Limits	Midpoint	Frequency (α_i)	% Frequency	% Cumulative frequency
1	160–180	170.5	1	1	1
2	181–200	190.5	3	3	4
3	201–220	210.5	6	6	10
4	221–240	230.5	9	9	19
5	241–260	250.5	22	22	41
6	261–280	270.5	34	34	75
7	281–300	290.5	15	15	90
8	301–320	310.5	5	5	95
9	321–340	330.5	4	4	99
10	341–360	350.5	1	1	100
		Total	100	100	

line represents the cumulative distribution of the population of fill-weights. From this line, the mean μ and the standard deviation σ of the population can be estimated as follows:

1. Using the notation X_p as the p-th percentile of the population, the estimate for the mean $\hat{\mu} = X_{50}$. This is based on the fact that the mean is the 50th percentile for a normal distribution.
2. The estimate for standard deviation is $\hat{\sigma} = (X_{84} - X_{16})/2$. This formula for the standard deviation comes from the fact that there is a distance of (approximately) 2σ between X_{16} and X_{84} in any normal distribution. The paper used already has markings for X_{16} and X_{84}.

Thus, for the example,

$$\hat{\mu} = 264 \qquad \hat{\sigma} = \frac{298 - 232}{2} = 33.$$

2.6.4.2 Normal Probability Plot on the Computer

Most statistical software packages provide a routine to make the normal probability plot for a given set of data. Figure 2.28 shows a normal probability plot made by the Minitab software. The computer software plots each point in the data against its cumulative probability, the cumulative probabilities for individual values being estimated by the mean rank of the value in the data. To calculate the mean rank, the data are first arranged in ascending order. Then, the mean rank of the observation that has the i-th rank in the data is $i/(n + 0.5)$, where n is the total number of observations in the data. The mean ranks of a few values for the data in Table 2.1 are shown in Table 2.4. The cumulative probabilities also are estimated by the median rank, which is calculated as $(i - 0.3)/(n + 0.4)$.

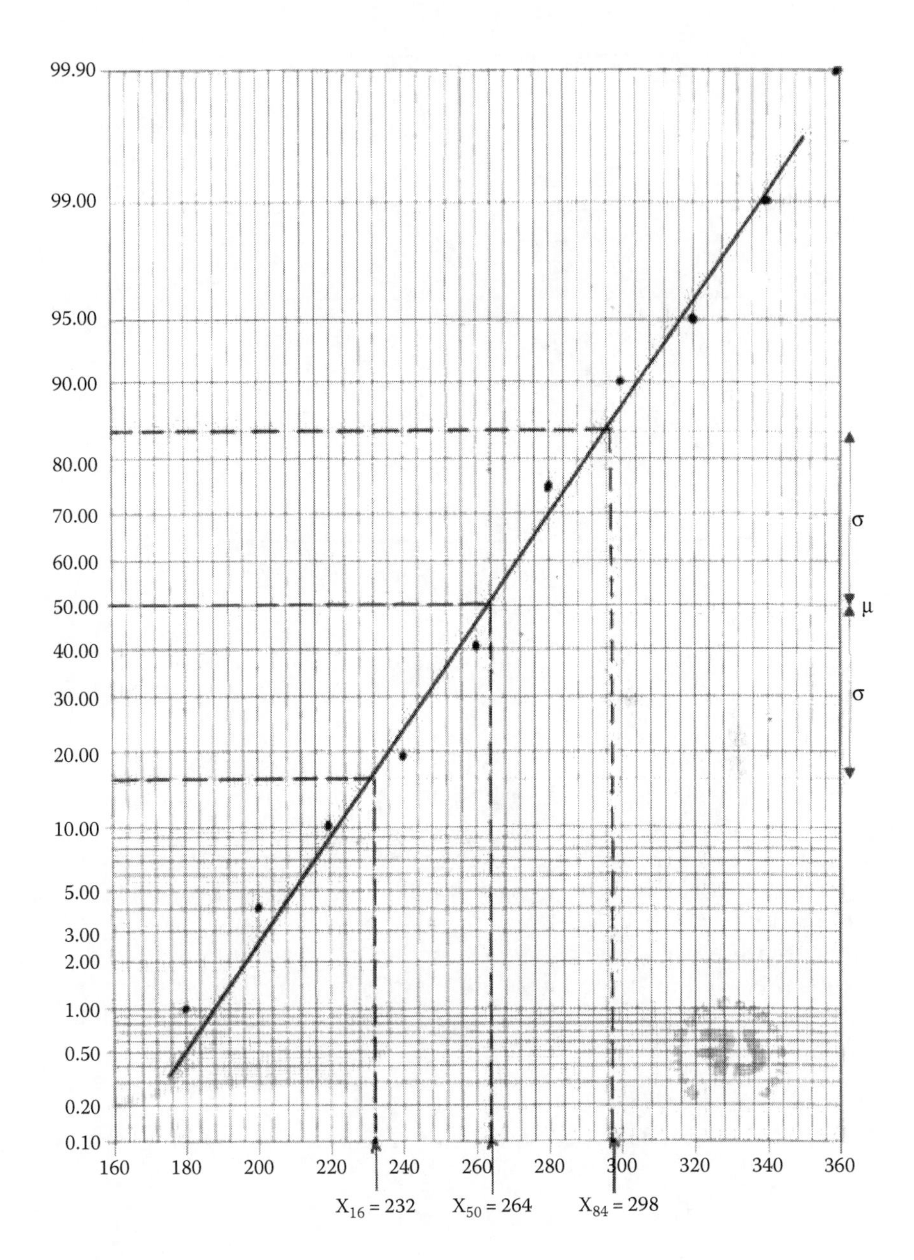

FIGURE 2.27
Example of a normal probability plot using commercially available normal probability paper (From www.Weibull.com).

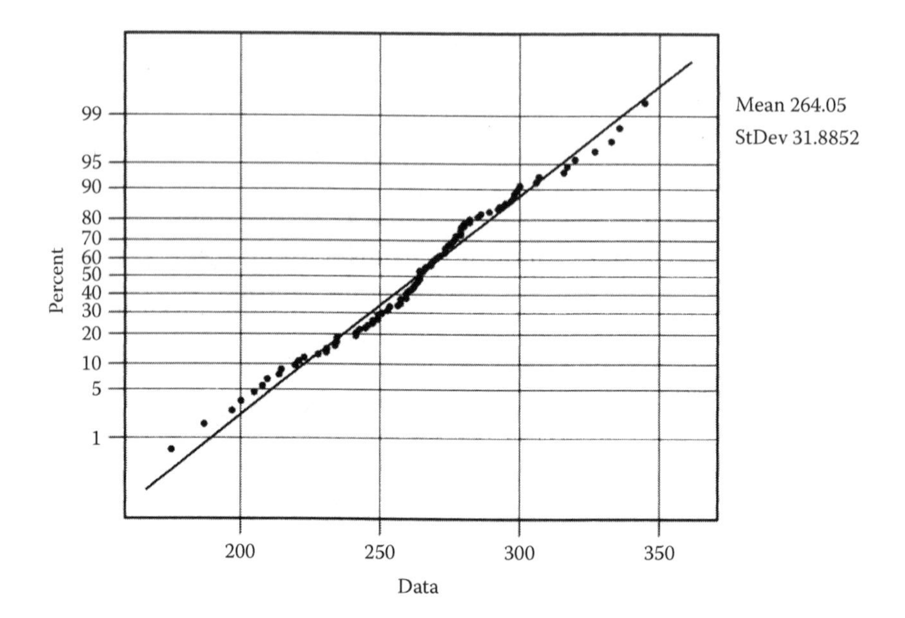

FIGURE 2.28
Normal probability plot produced by Minitab software for data in Table 2.1.

The normal probability plot provides a subjective test to determine if a population has a normal distribution. Whether the plots all fall on a straight line has to be determined by eye judgment, and different people may come to different conclusions, especially in marginal cases. In addition, there is the subjectivity in drawing the line to pass through "all" the plotted points.

A quantitative evaluation of how well a data set fits a hypothesized distribution is provided by a goodness-of-fit test. Several such tests—such as the χ^2, Kolmogorov–Smirnov, and the Anderson–Darling tests—are available.

TABLE 2.4

Cumulative Distribution of Individual Data Values for Normal Probability Plot

Value	Rank in Data	Mean Rank (×100)
175	1	0.995
187	2	1.99
197	3	2.985
⋮	⋮	⋮
328	98	97.71
337	99	98.51
346	100	99.50

2.6.4.3 A Goodness-of-Fit Test

A goodness-of-fit test generally compares the actual cumulative distribution of the data with that of the hypothesized distribution and obtains a quantity to measure the deviation of the actual from the hypothesized. This is done by computing the deviation of the actual distribution from the hypothesized distribution at several points within the range of data and then computing the total of the deviations. If the total deviation is too large, then the hypothesis that the data came from that hypothesized distribution is rejected. The difference among the various methods lies in how the total deviation is calculated and how the total deviation, a statistic, is distributed. The observed value of the total deviation is compared with the critical percentile of its distribution corresponding to a chosen α value. If the observed value is larger than the critical percentile, it is considered too large and the hypothesis that the data came from the hypothesized distribution is rejected. Otherwise the hypothesis is not rejected.

As an example, the χ^2 goodness-of-fit test is done by grouping the data into cells, as for making a histogram, and then calculating a quantity $\Sigma_i (a_i - e_i)^2/e_i$ as a measure of the total deviation. In this case, a_i represents the actual number of observations in each cell, and e_i represents the expected number of observations in each cell if the distribution were normal, with a hypothesized mean μ and standard deviation σ. This quantity, a statistic, is known to follow approximately the χ^2 distribution with $(p - 1)$ degrees of freedom (Hogg and Ledolter 1987), where p is the number of cells into which the data have been tallied. The observed value of this quantity is compared with the critical value $\chi^2_{\alpha, (p-1)}$. If the observed value is larger than the critical value $\chi^2_{\alpha, (p-1)}$, then the hypothesis that the data came from the hypothesized normal distribution is rejected.

If, however, any parameter of the hypothesized distribution was estimated from data, then the *df* of the χ^2 statistic is reduced by the number of parameters estimated. If q parameters were estimated, then the χ^2 statistic will have $df = (p - q - 1)$.

The calculations for the χ^2 goodness-of-fit test for the example are shown in Table 2.5, where the expected frequencies are calculated for the hypothisis

H_0: The population has $N(264, 32^2)$.

The values of the mean and standard deviation for the hypothesized distribution are the values estimated by $\overline{X}$ and S of the data. For example, the expected percentage frequency in the interval 241 to 260 is calculated as:

$$P(240.5 \le X \le 260.5) = \Phi\left(\frac{260.5 - 264}{32}\right) - \Phi\left(\frac{240.5 - 264}{32}\right)$$

$$= \Phi(-0.11) - \Phi(-0.73) = 0.2235 \text{ or } 22.35\%.$$

[Note: We have used in the above calculation the half-step correction to properly distribute the area under the normal curve lying between 240 and 241, and again between 260 and 261.]

TABLE 2.5

Frequency Distribution, Cumulative Frequency Distribution, and Calculation of χ^2 Statistic

	Cell Limits	Midpoint	Frequency (a_i)		% Frequency	% Cumulative Frequency	% Expected Frequency (e_i)		$\frac{(e_i - a_i)^2}{e_i}$
1	161–180	170.5	1		1	1	0.372		
2	181–200	190.5	2	9	2	3	1.85	8,32	0.056
3	201–220	210.5	6		6	9	6.17		
4	221–240	230.5	10		10	19	14.21		1.247
5	241–260	250.5	18		18	37	22.35		0.846
6	261–280	270.5	35		35	72	24.12		4.91
7	281–300	290.5	17		17	89	17.82		0.038
8	301–320	310.5	6		6	95	9.02		
9	321–340	330.5	4	11	4	99	3.13	12.9	0.278
10	341–360	350.5	1		1	100	0.745		
		Total	100		100		100		7.375

Some of the cell frequencies had to be grouped to meet the requirement of the χ^2 test that the expected frequency in each cell is at least five. The total of the last column—7.375, which is the observed value of the χ^2 statistic—is compared with the critical value $\chi^2_{0.05,3} = 9.348$. (The *df* for the χ^2 statistic is three because there are p = six cells and q = two estimated parameters.) Because the observed value is not greater than the critical value, H_0 is not rejected, and we conclude that the population has the hypothesized normal distribution.

2.6.5 The *P*-Value

If a computer program is used for hypothesis testing, in addition to calculating the observed value of the test statistic, the computer program also gives the *P*-value corresponding to the observed value. What is the *P*-value? How is it to be interpreted?

It is easy to explain the meaning of the *P*-value using the one-tailed test as an example. Referring to Figure 2.29a, the *P*-value is the probability in the distribution of the test statistic that lies beyond the observed value, away from values favoring H_0. We can see from Figure 2.29a, if the *P*-value is smaller than the chosen α, it means that the observed value of the test statistic lies in the critical region, and so H_0 should be rejected. If the *P*-value is larger than α, the observed value is not in the critical region and the H_0 is not to be rejected.

If the test is a two-tailed test, the probability in the distribution of the test statistic lying beyond the observed value should be compared with $\alpha/2$, as shown in Figure 2.29b. Hence, the computer program calculates the *P*-value as two times the probability in the distribution of the test statistic lying

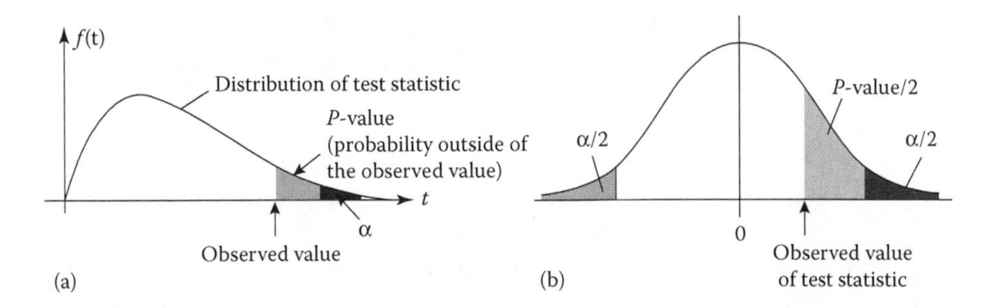

FIGURE 2.29
Meaning of the *P*-value.

outside the observed value. The *P*-value is then directly compared with α. So, the rules to use with the *P*-value are as follows:

$$\text{If } P\text{-value} \leq \alpha, \text{ reject } H_0.$$
$$\text{If } P\text{-value} > \alpha, \text{ do not reject } H_0.$$

These rules are the same whether it is a one-tailed or a two-tailed test.

The *P*-value is defined by some (e.g., Walpole et al. 2002) as the "lowest level of significance, at which the observed value of the test statistic is significant." This means that if the level of significance of the test had been chosen equal to the *P*-value obtained, then the observed value would be declared significant. A little reflection will convince the reader that this definition of the *P*-value is the same as the probability (the "*P*" in the *P*-value really stands for probability) lying in the distribution of the test statistic beyond the observed value.

We should note that only computer programs can compute the *P*-values if the distribution of the test statistic is continuous. The *P*-value cannot be computed manually with any accuracy even for tests involving the normal distribution. Calculating the *P*-value manually is impossible when other distributions, such as the *t* or χ^2 distributions, are involved.

2.6.6 Exercise in Inference Methods

2.6.6.1 Confidence Intervals

2.43 A random sample of 12 specimens of core sand in a foundry has a mean tensile strength of $\bar{X} = 180$ psi. Construct a 99% CI on the mean tensile strength of the core sand under study if the standard deviation of the tensile strength σ is 15 psi. Assume the strength is normally distributed.

2.44 A random sample of nine electric bulbs was chosen from a production process and tested until failure. The lives in hours were 1366,

1372, 1430, 1246, 1449, 1268, 1408, 1468, and 1502, respectively. Give a 95% CI for the average life of the bulbs produced in the process if it is known that the life is normally distributed with a standard deviation of 100 hours.

2.45 A random sample of 20 cans of hair spray was found to have an average of $\bar{X} = 1.15$ oz of concentrate. The standard deviation of the concentrate obtained from the sample was $s = 0.25$ oz. Find a 99% CI on the mean quantity of concentrate in the cans.

2.46 A random sample of nine electric bulbs was chosen from a production process and tested until failure. The lives in hours were 1366, 1372, 1430, 1246, 1449, 1268, 1408, 1468, and 1502, respectively. Give a 90% CI for the mean life of the bulbs produced in the process if it is known that the life is normally distributed but the population standard deviation is not known.

2.47 A random sample of 15 bronze rods gave the following measurements of diameter in millimeters: 6.24, 6.23, 6.20, 6.21, 6.20, 6.28, 6.23, 6.26, 6.24, 6.25, 6.19, 6.25, 6.26, 6.23, and 6.24. Assuming that the shaft diameter is normally distributed, construct a 99% CI for the mean and variance of the shaft diameter. What is the 95% CI for the standard deviation?

2.48 The following data come from strength of steel specimens in thousands of psi: 77.8, 78.7, 58.6, 43.7, 67.6, 46.9, 82.1, 90.8, 67.9, 76.9, 56.6, 74.8, 72.0, 82.9, and 81.9. Find the 95% CI for the mean and standard deviation of strength of steel in this population if the population is assumed to be normal.

2.6.6.2 Hypothesis Testing

2.49 The yield of a chemical process is being studied. The variance of the yield is known from previous experience as $\sigma^2 = 5$ (percentage^2). The past six days of plant operation have resulted in the following yields (in percentages): 92.5, 91.6, 88.75, 90.8, 89.95, and 91.3. Is there reason to believe that the yield is less than 90%? Use $\alpha = 0.05$. Assume that the yield is normal.

2.50 The shelf life of a battery is of interest to a manufacturer. A random sample of 10 batteries gave the following shelf lives in days: 108, 122, 110, 138, 124, 163, 124, 159, 106, and 134. Is there evidence that the mean shelf life is greater than 125 days? Assume that the shelf life is normal. Use $\alpha = 0.05$.

2.51 Two machines are used for filling bottles with a type of liquid soap. The filled volumes can be assumed to be normal, with standard deviations of $\sigma_1 = 0.015$ and $\sigma_2 = 0.018$. A random sample is taken from the output of each machine, and the following volume checks in milliliters were obtained. Verify if both machines are filling equal volumes in the bottles. Use $\alpha = 0.10$.

Machine 1		Machine 2	
16.03	16.01	16.02	16.08
16.04	15.96	15.97	16.04
16.05	15.98	15.96	16.02
16.05	16.02	16.01	16.01
16.02	15.99	15.99	16.11

2.52 A survey of alumni of two engineering programs in a college gave the following data on their salaries after five years. Is there evidence to show that the average salaries of the graduates from the two programs are different? Assume that both the salaries are normally distributed and have a standard deviation of $12,000. Use $\alpha = 0.10$.

IE Graduates	ME Graduates
66,500	64,600
58,200	58,900
80,400	76,900
48,500	45,800
64,500	70,100
68,300	69,700
77,300	77,900
84,900	52,300

2.6.6.3 Goodness-of-Fit Test

2.53 Using the χ^2 goodness-of-fit test, test the data in Exercise 2.1 for normal distribution. Compare the results from a test for normality of the same data using computer software.

2.54 Using the χ^2 goodness-of-fit test, test the "Before Improvement" data in Exercise 2.2 to verify if the data come from a normal distribution.

2.7 Mini-Projects

Mini-Project 2.1

An industrial engineering department has implemented some changes to the way they teach engineering statistics to their majors during the junior year. The new method makes extensive use of computer software, with

PowerPoint presentations as opposed to deriving relationships and calculating results on the blackboard. The students take a class in quality control after their junior year and are required to take a test on fundamentals of engineering statistics in the QC class—after a brief review. The data below show the scores in the review test for the recent years.

The scores in spring 2001 and spring 2003 reflect the effect of the new teaching methods. Analyze the data, and verify if the new methods have, indeed, improved the students' understanding of the fundamentals of engineering statistics. If you wish, you may use any computer software to analyze the data. A good analysis, as you might know, includes both graphical and analytical methods. Make any assumptions that are necessary. Verify your assumptions if necessary.

Spring 1999		Spring 2000		Spring 2001		Spring 2003	
100	86	100	60	100	70	97	60
100	84	95	58	95	68	90	57
96	80	93	53	95	66	87	57
94	78	90	40	91	63	80	53
92	76	88		89	61	77	53
92	74	83		89	52	63	50
90	70	78		88	48	63	50
90	62	78		84	46	63	47
88	56	75		84	45	63	43
86	46	73		79	43	63	37
86		68		73	34	63	
86		63		73		60	

Mini-Project 2.2

The data below represent the tensile strength of core sand (sand mixed with binders and baked, from which foundry cores are made) obtained from two different testing machines in a foundry. Each test yields three readings from three specimens made from the same sand. The foundry is currently using the Detroit (D) tester and is planning to purchase the Thwing—Albert (TA) tester, because the D testers are getting old and cannot be repaired or replaced. The foundry wants to know if the results from the new TA tester are about the "same" as the results from the old tester for the same sand. Many of the current specifications for sand are based on the results of the D tester and they want to know if the specifications would hold good for test results from the new tester. In the absence of a definitive answer, there is debate in the foundry as to whether to continue using the old specification or reset all new specifications with the new tester.

An experiment was conducted to obtain test results over several types of sand using both the old and new tester and the results were recorded as shown in the table below. It is necessary to determine if the testers give "equal" measurements for all sand types in all ranges of strength.

Analyze the data, and provide suitable answers to the company. Make any assumptions you need, and state them clearly. You can also verify the assumptions you make. Remember, a good analysis includes use of both graphical and analytical tools, and we do not conclude if two populations are equal or not equal until we have made a significance test about their parameters.

Sand Batch	Trial	D Tester			TA Tester		
		Specimen 1	Specimen 2	Specimen 3	Specimen 1	Specimen 2	Specimen 3
1	1	340	360	375	359	351	345
	2	350	335	345	348	329	348
2	1	225	210	250	254	255	247
	2	255	245	230	244	253	202
3	1	220	250	245	243	251	235
	2	240	225	230	241	251	204
4	1	230	230	255	246	247	237
	2	220	235	245	246	248	241
5	1	430	445	450	441	445	441
	2	400	435	435	445	442	443

Mini-Project 2.3

Given below in the tables are monthly statistics on maximum and minimum temperatures in Peoria, Illinois, gathered from historical records maintained in the Weather Underground website (www.wunderground.com). The data represent the monthly maximum/minimum of the daily maximums/minimums. For example, in January 1950, the maximum among the daily maximums was 68°F and the minimum among the daily maximums was 46.

Based on these data, is there evidence that global warming is affecting the weather in this midwestern city in the United States? You can pick any stream(s) of data from the tables that you think would help in answering the question. As always, a good analysis would include the use of both graphical and analytical methods. Any declaration of the existence or nonexistence of difference between populations should come out of a significance test, which will take into account both the average and variability in the sample data. Use only the methods you have learned from Chapter 2.

1950		Jan	Feb	Mar	Apr	May	Jun	Jul	Aug	Sep	Oct	Nov	Dec
Max	Max	68	46	73	80	90	93	91	88	84	86	80	53
	Min	46	36	46	55	68	75	72	68	68	64	51	32
Min	Max	16	10	19	36	54	60	66	66	59	55	7	12
	Min	3	–9	3	21	35	45	52	44	39	34	0	–5

1960		Jan	Feb	Mar	Apr	May	Jun	Jul	Aug	Sep	Oct	Nov	Dec
Max	Max	59	45	70	82	82	88	91	91	93	80	71	64
	Min	46	33	48	62	62	70	71	75	72	59	53	46
Min	Max	16	12	10	36	41	66	73	73	60	44	30	5
	Min	0	1	−9	21	32	52	52	54	46	21	16	−11
1970		**Jan**	**Feb**	**Mar**	**Apr**	**May**	**Jun**	**Jul**	**Aug**	**Sep**	**Oct**	**Nov**	**Dec**
Max	Max	57	55	60	86	87	93	97	90	90	80	57	69
	Min	35	32	51	68	66	73	75	71	73	62	45	42
Min	Max	0	3	30	37	53	60	69	66	61	46	19	24
	Min	−18	−5	17	25	35	52	51	55	37	30	7	7
1979		**Jan**	**Feb**	**Mar**	**Apr**	**May**	**Jun**	**Jul**	**Aug**	**Sep**	**Oct**	**Nov**	**Dec**
Max	Max	34	39	70	75	87	91	89	91	88	86	71	59
	Min	25	28	57	59	66	70	70	75	68	66	46	45
Min	Max	2	0	25	37	57	69	72	64	66	41	27	18
	Min	−22	−17	6	19	37	48	51	45	41	25	16	−2
1990		**Jan**	**Feb**	**Mar**	**Apr**	**May**	**Jun**	**Jul**	**Aug**	**Sep**	**Oct**	**Nov**	**Dec**
Max	Max	62	63	80	84	79	90	93	91	93	82	73	59
	Min	44	36	60	63	59	73	75	73	72	61	55	35
Min	Max	28	19	30	39	50	66	64	71	55	44	39	7
	Min	14	6	19	21	37	48	55	51	37	28	19	−2
2000		**Jan**	**Feb**	**Mar**	**Apr**	**May**	**Jun**	**Jul**	**Aug**	**Sep**	**Oct**	**Nov**	**Dec**
Max	Max	62	54	**	**	79	90	93	91	93	82	73	59
	Min	39	43	**	**	59	73	75	73	72	61	55	35
Min	Max	17	28	**	**	50	66	64	71	55	44	39	7
	Min	−2	10	**	**	37	48	55	51	37	28	19	−2
2010		**Jan**	**Feb**	**Mar**	**Apr**	**May**	**Jun**	**Jul**	**Aug**	**Sep**	**Oct**	**Nov**	**Dec**
Max	Max	48	41	79	84	91	91	92	94	88	85	73	46
	Min	35	33	51	64	67	75	77	76	68	57	48	25
Min	Max	8	20	36	51	50	75	79	75	62	45	36	14
	Min	−11	0	22	32	35	57	58	54	43	27	20	−1

Mini-Project 2.4

The quality of air in any place in the U.S. is measured by Air Quality Index (AQI), which indicates how polluted the air is in that place. It is computed from the measurements of five major pollutants: ground-level ozone, particulate matter, carbon monoxide, sulfur dioxide, and nitrogen oxide. Based on the amount of each pollutant in the air, the AQI assigns a numerical value to air quality, and it can be interpreted as follows according to the U.S. Environmental Protection Agency (EPA): 0 to 50 (good); 51 to 100 (moderate); 101 to 150 (unhealthy for sensitive groups); 151 to 200 (unhealthy); 201 to 300 (very unhealthy); 301 to 500 (hazardous). An AQI of 100 is considered "acceptable" by the EPA even if it be a moderate health concern for a very small number of people. When it exceeds 100, the air could cause harm to people's health in different degrees, depending on their sensitivity.

Given below are data on air quality of some selected cities in the U.S. recorded between 1990 and 2000. The data represent the number of days the AQI exceeded 100 in a given city, and in a given year. Obviously there is some variability in the readings from year to year. Is it possible to draw meaningful conclusions from these data? Making sense out of measurements is the common challenge to the engineering statistician.

You have to propose one or more meaningful hypotheses such as: cities on the East Coast have cleaner air than those on the West Coast, or cities near the northern border are more livable than those near the southern border, and so on. Then pick suitable data from the table below and analyze them to verify if your hypotheses are valid. State your conclusions in a way that would be meaningful to a couple who want to retire in an air-friendly city in the U.S. They would like to have some alternatives to chose from. Use a map of the U.S. if you are not familiar with the locations of the cities.

Cities	1990	1991	1992	1993	1994	1995	1996	1997	1998	1999	2000
Atlanta, Georgia	42	23	20	36	15	35	25	31	50	61	26
Baltimore, Maryland	29	50	23	48	41	36	28	30	51	40	16
Boston, Massachusetts	7	13	9	6	10	8	2	8	7	5	1
Chicago, Illinois	4	25	6	3	8	23	7	9	10	14	0
Cleveland, Ohio	10	23	11	17	24	27	18	13	22	21	5
Dallas, Texas	24	2	12	14	27	36	12	20	28	23	20
Denver, Colorado	9	6	11	6	2	3	0	0	7	3	2
Detroit, Michigan	11	27	7	5	11	14	13	11	17	15	3
El Paso, Texas	19	7	10	7	11	8	7	4	6	6	3
Houston, Texas	51	36	32	27	38	65	26	47	38	50	42
Kansas City, Missouri	2	11	1	4	10	23	10	17	15	5	10
Los Angeles, California	173	168	175	134	139	113	94	60	56	27	48
Miami, Florida	1	1	3	6	1	2	1	3	8	5	0
Minneapolis/St. Paul, Minnesota	4	2	1	0	2	5	0	0	1	0	0
New York, NY	36	49	10	19	21	19	15	23	17	24	12
Philadelphia, Pennsylvania	39	49	27	62	37	38	38	38	37	32	18
Phoenix, Arizona	12	11	11	15	10	22	15	12	14	10	10
Pittsburgh, Pennsylvania	19	21	9	13	19	25	11	21	39	23	4
St. Louis, Missouri	23	24	14	9	32	36	20	15	23	29	14

(*continued*)

Cities	1990	1991	1992	1993	1994	1995	1996	1997	1998	1999	2000
San Diego, California	96	67	66	59	46	48	31	14	33	16	14
San Francisco, California	0	0	0	0	0	2	0	0	0	0	0
Seattle, Washington	9	4	3	0	3	0	6	1	3	1	1
Washington, DC	25	48	14	52	22	32	18	30	47	39	11

Source: U.S. Environmental Protection Agency, Office of Air Quality Planning & Standards.

References

Bilgin, M., and B. Frey. May 1999. "Applying SPC at a Candy Manufacturing Co." Unpublished project report, IME 522, IMET Department, Bradley University, Peoria, IL.

Garrett, L. May 2003. "Statistical Model of the Game Time of a Major League Baseball Game." Unpublished project report, IME 526, IMET Department, Bradley University, Peoria, IL.

Guttman, I., S. S. Wilks, and J. S. Hunter. 1982. *Introductory Engineering Statistics*. 3rd ed. New York: John Wiley.

Guyer, K., and A. Lalwani. December 1994. "Analysis of Plastometric Whiteness in the G-198 PVC Resin." Unpublished project report, IME 522, IMET Department, Bradley University, Peoria, IL.

Hines, W. W., and D. C. Montgomery. 1990. *Probability and Statistics in Engineering and Management Science*. 3rd ed. New York: John Wiley.

Hogg, R. V., and A. T. Craig. 1965. *Introduction to Mathematical Statistics*. 2nd ed. New York: Macmillan.

Hogg, R. V., and J. Ledolter. 1987. *Engineering Statistics*. New York: Macmillan.

McGinty, D. May 2000. "Die Spring Life Analysis." Unpublished project report, IME 526, IMET Department, Bradley University, Peoria, IL.

Mood, M. M., A. G. Graybill, and D. C. Boes. 1974. *Introduction to the Theory of Statistics*. 3rd ed. New York: McGraw-Hill.

Walpole, R. E., and R. H. Myers. 1978. *Probability & Statistics for Engineers & Scientists*. 2nd ed. Upper Saddle River, NJ: Prentice Hall.

Walpole, R. E., R. H. Myers, S. L. Myers, and K. Ye. 2002. *Probability & Statistics for Engineers & Scientists*. 7th ed. Upper Saddle River, NJ: Prentice Hall.

3

Quality in Design

3.1 Planning for Quality

3.1.1 The Product Creation Cycle

The scheme of activities that are performed from the time a product is conceived to the time the product is made and delivered to the customer can be described as a *product creation cycle*. Such a scheme, adapted from the "quality planning timing chart" described in the *Reference Manual of Advanced Product Quality Planning and Control Plan*, published jointly by the big three automakers (Chrysler LLC, Ford Motor Company, and General Motors Corporation 2008a), is presented in Figure 3.1. The modern approach to making quality products requires that quality issues be addressed and the necessary quality activities performed throughout this product creation cycle.

The product creation cycle in Figure 3.1 shows the activities divided into the following six stages:

1. Product planning
2. Product design and development
3. Process design and development
4. Product and process validation
5. Production
6. Feedback, assessment, and corrective action

These stages are not sequential in the sense that the activities overlap. The quality activities performed in the stages of product planning, product design, process design and product and process validation are collectively called "quality planning activities," and those performed during the production stage are called "quality control activities." The activities relating to quality performed during the planning stages, along with their objectives, the tools employed, and the outcomes achieved, are covered in this chapter.

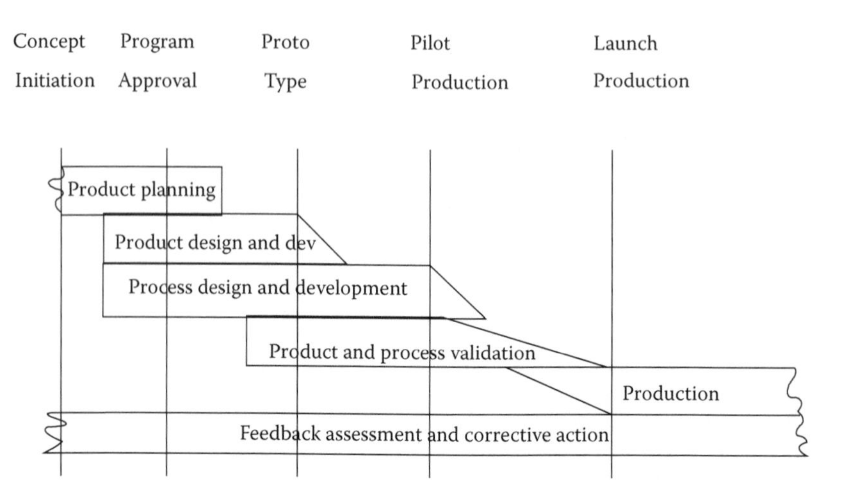

FIGURE 3.1
The product creation cycle. (Reprinted from Chrysler LLC, Ford Motor Company, and General Motors Corporation. 2008a. *Advanced Product Quality Planning (APQP) and Control Plan—Reference Manual*, 2nd edition. Southfield, MI: A.I.A.G. With permission.)

The major tools employed during the quality planning stage are:

- *Customer surveys*—used to find the needs of the customers.
- *Quality function deployment*—used for translating customer needs into product features.
- *Failure mode and effects analysis*—used to proof the product and process designs against possible failures.
- *Basic principles of reliability*—needed to define, specify, measure, and achieve reliability in products.
- *Design of experiments*—used to select product characteristics and process parameters to obtain desired product and process performance.
- *Tolerancing*—used to determine the economic limits of variability for product characteristics and process parameters.

3.2 Product Planning

This is the first stage of planning for a product, when the major features for the product are determined. If the product is a car, features such as horsepower, body style, transmission type, safety standards, fuel consumption, and so on, are determined at this stage. If the product is a lawnmower, such major features as engine horsepower, deck size, and whether it will be self-propelled or self-starting will be determined. Quality and reliability goals for the product are also established at this stage.

The quality goals are chosen in terms of several aspects of quality, such as performance, safety, comfort, appearance, and so on, to meet the needs of the customer after ascertaining what their needs and expectations are in each of these areas. Reliability goals are set in terms of length of failure-free operation of the product—again based on customer preferences—cost constraints, and prevailing competition.

The quality and reliability goals selected at this stage drive the specifics of design activities during the next stage, in which product design details are worked out and process design is undertaken to proceed side by side with product design. Detailed drawings for parts and subassemblies are also prepared, along with bills of materials in this stage. A preliminary process flow chart is prepared, including the choice of material and machinery for making the product. Preparation of the flow chart, among other things, will help to determine if existing technology will be adequate or if new technology is needed to make the product. The bills of materials will help in deciding what parts will be produced and what parts will be procured.

Quality planning activities are generally performed by a cross-functional team called the "quality planning team." This team is comprised of representatives from various functional areas, such as marketing, product engineering, process engineering, material control, purchasing, production, quality, and customer service. Supplier and customer representatives are included when appropriate. Quality planning begins with finding the needs of the customer.

3.2.1 Finding Customer Needs

Finding customer needs is often referred to as "listening to the voice of the customer." The customers' voice can be heard in several ways:

1. Surveying of past and potential customers.
2. Listening to focus groups of customers, such as chain stores, dealers, or fleet operators.
3. Collecting information from the history of complaints and warranty services.
4. Learning from the experiences of cross-functional team members.

Deciding on which approach should be used in a given situation depends on the nature of the product, the amount of historical information already available, and the type of customer being served. For example, industrial customers may have to be approached differently from the general public, and customers for cars must be approached differently from customers for toys. The customer survey is the most commonly used method and is often combined with other methods, such as interviewing a focus group. Thus, the customer survey is an important tool in assessing the needs of the customer.

3.2.1.1 Customer Survey

A typical customer survey attempts to establish the customers' needs and the level of importance that customers attach to the different needs. When appropriate, information is elicited on how much the customers favor a competitor's product and why. The survey could be conducted by phone interview, personal interview, or mail; each method has its advantages and disadvantages. For example, direct contact of customers, either by phone or in person, could generate information that the design team might not have even thought of asking (Gryna 1999). Mailed surveys will mostly produce answers to prepared questions. These surveys may be cheaper to administer, whereas the direct contact may be expensive. In either case, the quality planning team should have a prepared survey instrument or questionnaire.

Often, the survey tools used for measuring the customers' satisfaction with an existing product can be used for projecting what the customers would want in a new design, or a new product. Designing a survey starts with identifying the attributes that customers might look for in the product. The customers are asked to express the level of their desire for the chosen attributes, usually on a scale of one to five. Tables 3.1 and 3.2 show examples of customer surveys; one is for a tangible product and the other is for an intangible service.

Designing a customer survey is a science in itself. There are good references (Hayes 1998; Churchill 1999) that provide guidance on preparing a survey instrument. The following discussion covers some of the important fundamentals of creating a customer survey.

The list of characteristics on which the customer's rating is requested has to be drawn by people with a good understanding of the product being planned and the customer being served. The items in this list must be relevant, concise, and unambiguous. Each item should contain only one thought and be easy to understand. For new products or customers, the list should have room for the addition of items by the customer. Usually, a five-point (Likert-type) scale is chosen for the customer to express the strength of their requirement as the Likert-type format is known to give more reliable results than a true-false response (Hayes 1998). The questionnaire should have a brief introduction, as shown in Tables 3.1 and 3.2, to let the customer know the objective of the questionnaire and how the questionnaire is to be completed.

The survey instrument must be reliable. A survey is reliable when the survey results truly reflect the preferences of the customer. One way to measure this reliability is to give the survey to the same set of sample customers twice, with an intervening delay, and evaluate the correlation between the

TABLE 3.1

Customer Requirement Survey for a New Book in Quality Engineering

A new book is planned in quality engineering to be published by an international publisher. The book will present the statistical and managerial tools of quality in an integrative manner. We want to find out from you (a possible customer) the topics that you would like covered in the book and their level of importance to you. Some of the topics have been anticipated and listed below, but there is room for additions. Please rate the topics on a scale of 1 (not needed) to 5 (highly desirable) to express your priority. Please also rate a book that you may be currently using in terms of how the requirements are met by that book. Please identify (if you would) the book you are currently using.

Information about yourself (mark the most suitable one): (1) I am a professor teaching quality subjects _____(2) I am a professional engaged in quality work _____(3) I am a student interested in quality topics.

	Required in New Book					Level of Satisfaction in Current Book				
Statistical Methods										
1. Rigor in mathematics	1	2	3	4	5	1	2	3	4	5
2. Fundamentals of prob. & stat.	1	2	3	4	5	1	2	3	4	5
3. Statistical process control methods	1	2	3	4	5	1	2	3	4	5
4. Acceptance sampling methods	1	2	3	4	5	1	2	3	4	5
5. Design of experiments	1	2	3	4	5	1	2	3	4	5
6. Regression analysis	1	2	3	4	5	1	2	3	4	5
7. Reliability principles	1	2	3	4	5	1	2	3	4	5
____________	1	2	3	4	5	1	2	3	4	5
Management Topics										
1. Supply chain management	1	2	3	4	5	1	2	3	4	5
2. Team approach	1	2	3	4	5	1	2	3	4	5
3. Strategic planning	1	2	3	4	5	1	2	3	4	5
4. Principles of management	1	2	3	4	5	1	2	3	4	5
5. Organizational theory	1	2	3	4	5	1	2	3	4	5
6. QFD	1	2	3	4	5	1	2	3	4	5
____________	1	2	3	4	5	1	2	3	4	5
Quality Management Systems										
1. ISO (QS) 9000 standards	1	2	3	4	5	1	2	3	4	5
2. Baldridge Award criteria	1	2	3	4	5	1	2	3	4	5
3. Six Sigma system	1	2	3	4	5	1	2	3	4	5
____________	1	2	3	4	5	1	2	3	4	5
Quality Improvement Tools										
1. PDCA/Breakthrough methods	1	2	3	4	5	1	2	3	4	5
2. Magnificent seven	1	2	3	4	5	1	2	3	4	5
3. Benchmarking	1	2	3	4	5	1	2	3	4	5
4. Use of computer software	1	2	3	4	5	1	2	3	4	5
5. FMEA	1	2	3	4	5	1	2	3	4	5
____________	1	2	3	4	5	1	2	3	4	5

TABLE 3.2

Customer Satisfaction Questionnaire for a Bank

To better serve you, we would like to know your opinion about the quality of our products and services. Please indicate the extent to which you agree or disagree with the following statements about the service you received from (company name). Some of the statements are similar to one another to ensure that we accurately determine your opinion concerning our service. Circle the appropriate number against each question using the scale below.

1—I strongly disagree with this statement (SD).
2—I disagree with this statement (D).
3—I neither agree nor disagree with this statement (N).
4—I agree with this statement (A).
5—I strongly agree with this statement (SA).

	SD	D	N	A	SA
1. I waited a short period of time before getting help.	1	2	3	4	5
2. The teller completed my transactions in a short time.	1	2	3	4	5
3. The financial consultant was available to schedule me at a good time.	1	2	3	4	5
4. The teller greeted me in a pleasant way.	1	2	3	4	5
5. The teller listened carefully to me when I was requesting a transaction.	1	2	3	4	5
6. The bank lobby was quiet and comfortable.	1	2	3	4	5
7. The teller did not pay attention to what I told him/her.	1	2	3	4	5
8. The waiting line was too long when I arrived.	1	2	3	4	5
9. The teller took his/her own time to complete my transaction.	1	2	3	4	5
10. The financial consultant was not available at a time convenient to me.	1	2	3	4	5
11. The teller was very personable.	1	2	3	4	5
12. The lobby was noisy and cold.	1	2	3	4	5
13. Additional comments.					

two sets of responses. A high correlation between the two sets of responses will indicate high reliability of the questionnaire. Another way of evaluating this reliability is to include in the questionnaire two questions for each attribute, worded differently. The correlation between the responses for the same attribute is then evaluated. A high correlation would indicate high reliability of the questionnaire. The example questionnaire from the bank in Table 3.2 contains questions of this kind.

A survey result becomes more reliable with improvements in the clarity and relevance of the questions. It also becomes more dependable with an increase in the number of questions (Hayes 1998), which, of course, has to be balanced with relevance so that whole survey instrument remains concise.

Once the survey instrument is prepared, the plan for administering the survey should be made. Because of the cost and time constraints, surveying the entire population is not possible except in small populations, and statistical sampling techniques are required. Two commonly used sampling techniques are:

1. Simple random sampling
2. Stratified random sampling

In simple random sampling, the sample is chosen from the entire population such that each customer in the population has an equal chance of being included in the sample. In stratified random sampling, the population is first divided into several strata, based on some rational criteria—such as sex, age, income, or education—and simple random sampling is done within each stratum. Stratified sampling provides better precision (i.e., less variability) in the estimates for a given sample size compared to the simple random sampling. It also provides estimates of customer preferences for each of the strata, which may provide additional useful information for some product designers.

The size of the sample is determined by the confidence level needed so that the error in estimates does not exceed a chosen value. If $(1 - \alpha)100\%$ level of confidence is required that the error in the results is not more than $\pm e$, then the sample size n is given by the formula:

$$n = \left(\frac{z_{\alpha/2} \times s}{e}\right)^2,$$

where $z_{\alpha/2}$ is the value that cuts off $\alpha/2$ probability on the upper tail in the standard normal distribution and s is the estimate for standard deviation of the scores, known from previous, similar surveys.

Example 3.1

Suppose we want to determine the sample size for a survey that requires the expression of customer preference on the one-to-five scale. Suppose also that the confidence level needed is 95% that the error in the estimates does not exceed ±0.1, and the estimate for standard deviation of the scores from previous, similar surveys is 0.5. Find the sample size needed.

Solution

Because $(1 - \alpha) = 0.95$, $\alpha = 0.05$, and $Z_{\alpha/2} = Z_{0.025} = 1.96$,

$$n = \left(\frac{Z_{0.025} \times s}{e}\right)^2 = \left(\frac{1.96 \times 0.5}{0.1}\right)^2 \cong 96.$$

This means that at least 96 customers must be surveyed in order to obtain the results within an error of ±0.1 with 95% confidence.

The above discussion included the basics of conducting a customer survey to make the reader aware of the issues involved. This may even be adequate to make some simple surveys. For more complete information on sample surveys, the reader is referred to the references cited earlier.

The next important quality-related function in product planning is translating the customers' voice into design parameters or design features of the product. A formal tool used in this translation is called the "quality function deployment."

3.2.2 Quality Function Deployment

The quality function deployment (QFD) is the method used to select the design features of a product in a manner to satisfy the expressed preferences of the customers. This method, which originated in Japan, also helps in prioritizing those features and picking the most important ones for special attention further along the design process. The major component of the QFD method is a matrix created with the customer's preferences on the rows and the design features selected to meet those preferences on the columns (Figure 3.2). The intersecting cell between a column and a row is used to record the strength of the relationship, or how well the chosen feature will meet the corresponding customer preference. These relationships are determined by the collective judgment of the product planning team and are recorded in the matrix using notations that express the strength of those relationships. Some design features may satisfy more than one customer preference, while other design features may have a strong relationship with some customer preferences but a weak relationship with others. Three different notations are generally used to indicate weak, strong, and very strong relationships, as shown in Figure 3.2.

The purpose of making this matrix is to identify those design features that are most important from the point of view of meeting customer preferences. A design feature that has strong relationships with several customer preferences will come out ahead of a feature that has only weak relationships with a few customer preferences. This process of prioritizing the design features based on their contribution to satisfying customer preferences is done using some simple arithmetic calculations.

Numerical values are assigned to the customer preferences to signify how strongly the customer prefers one requirement relative to the other. These preference numbers are the ones obtained from the customers through the customer survey, and are indicated in the column immediately next to the customer requirements. These numbers are multiplied by the numerical equivalent of cell entries that express the strength of relationships between customer preferences and design features. The total of these resulting products for any design feature represents how important that feature is in terms of satisfying the needs of the customer. These totals (obtained column-wise) are entered at the bottom of each column,

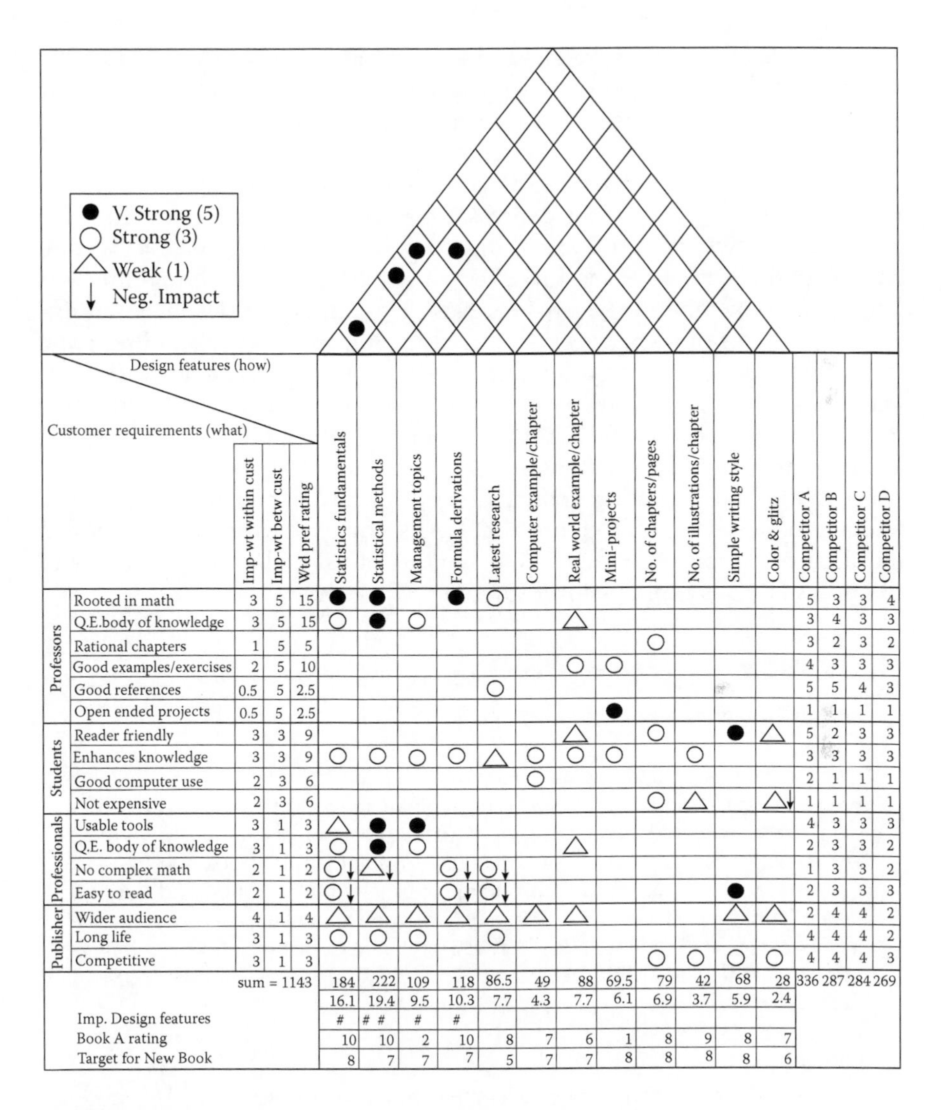

FIGURE 3.2
Example of a house of quality. (From www.QIMacros.com.)

thus assigning a score for each design feature. Those with the highest scores are the most important features from the customer's point of view. These features are further studied with regard to the advantage they may provide in comparison to one or more competitors. The design features that are important from the view of satisfying customer preferences, and those that provide certain advantages over competitors, are identified and given special treatment in the new design. This is the basic function of the QFD method.

The matrix of customer preference versus design feature is topped by a triangular matrix, which shows how the design features are related among one another. Knowledge of these relationships helps in alerting the designer to changes that may occur in other design features while making changes to one of them. The QFD method also provides for studying how a competitor's product fares against the expressed preferences of the customers. This provision—which helps in evaluating competitors and, thus, enables a comparison of the features of the new product with those of the competitors—is known as "benchmarking." This enables the identification of the strengths and weaknesses of the competitor's product. It also helps in building upon their strengths or winning a competitive edge by satisfying a need the competitors have not addressed, as shown in the example below. Different users of the QFD method also add many other details to the main function in order to suit their individual products.

The matrices, when put together, look so much like the picture of a house with a roof, windows, and doors, that the assembly of the matrices is called the "house of quality" (HOQ). The procedure of using the QFD method and creating a HOQ is explained in Example 3.2.

Example 3.2

QFD for Designing a Book

This example relates to designing a book on quality engineering to be used mainly by engineering professors in classes for engineering majors. It would be desirable if the book would also be of use to quality professionals in industry to meet their training needs. It will be an additional advantage if the book can be used in classes for other majors as well, such as business, mathematics, and science, because publishers like books that appeal to as wide an audience as possible. The features of the book that would satisfy the preferences of the various customers must be identified and prioritized.

Solution

The first task for a design team is to identify who the customers are. In this case, the author is the "planning team," and anyone who will be impacted by the design is a customer. Thus, the professors, students, professionals, and publisher are the customers, and they all have their own preferences. A list of these preferences is first generated based on the experience of the author and enquiry among his colleagues and students. The customers and their needs are shown on the left-hand side of the HOQ (Figure 3.2). Notice that the items in this list are in the language of the customers.

3.2.2.1 Customer Requirements and Design Features

Next, the preference numbers reflecting the relative importance of the various customer preferences are determined. When a product has multiple customers—as in this case—a set of importance-weights, which are used to represent the importance of one customer relative to others, is first generated by the planning team. A total of 10 points is distributed among the customers, based on their relative importance. In this case a weight of five is assigned to professors, three to students, one to professionals, and one to the publisher. The numbers expressing the relative importance of the preferences within each customer are then obtained from those expressed by the customers in the customer surveys. A total of 10 points is distributed among the preferences within each customer.

For any customer preference, the product of the importance-weight of the particular preference within the customer and the importance-weight of the customer (relative to other customers) is obtained as the indicator of how important the preference is overall. The computation to arrive at the importance-weight of the customer preferences is done on the columns next to the column that lists the customer preferences.

Next, a list of design features that meet the customer preferences identified in the previous step is developed. This list is in the language of the designer. These design features must be measurable. Note that there is at least one design feature that responds to each of the customer requirements. The strength of relationships among the customer preferences and design features are then estimated by the design team and are marked on the matrix. As mentioned earlier, three levels of strength are used—weak, strong and very strong. The strength notations generally indicate that an increase in the measure of the design feature will provide increased satisfaction of the particular customer preference. Some design features may help in meeting a preference of one customer but may work against a preference of another customer. Where a design feature affects a customer requirement inversely, such an affect is indicated by a down-arrow (↓) next to the strength notation.

3.2.2.2 Prioritizing Design Features

For each cell in the relationship matrix, the product of the numerical equivalent of the strength relationship and the importance-weight of the corresponding customer preference is obtained and added column-wise, and the total is placed under each column in the row at the bottom of the relationship matrix. (The relationships with down-arrows also make a positive contribution in this step when we are determining the importance of a design feature. Their negative significance is taken into account later, when we are determining target values for the design features of the new design.) The numbers in this row at the bottom of the central matrix represent the importance of the design features in meeting key customer requirements. For this example, the following numerical equivalents for the strength relationships have been used: weak = 1, strong = 3, and very strong = 5. This is the usual scale employed, but other scales, such as (1, 3, 9) instead of (1, 3, 5), are also sometimes used.

The numbers obtained for each of the design features are then "normalized" using the formula $y_j = 100(x_j/\Sigma x_j)$, where y_j is the normalized score and x_j is the raw score for the j-th design feature. These normalized scores represent the relative importance of a given design feature among all the design features. These relative importance scores (called the "normalized contributions" of the design features) are used to prioritize the design features for further deployment. Usually, three or four features with top-ranking normalized scores will be chosen as the most important design features. For the example, the top-ranking features are identified with a (#) mark below the normalized scores, with the top-most feature being identified with a (##) mark.

3.2.2.3 Choosing a Competitor as Benchmark

Also shown on the right-hand side of the HOQ is the assessment on how well competitor books fare with regards to the established customer preferences. For this example, four books—A, B, C, and D—are identified as competitors for the new design. These books are evaluated by the planning team and assigned numbers on a scale of 1 to 5 to represent their ability to meet the established customer preferences. The products of these numbers and the importance-weights of customer requirements are added and the total is shown at the bottom of the column for each competitor. These numbers represent how well a competitor's book satisfies the customer preferences. The competitor with the largest of these numbers is the best-in-class, and it is chosen as the benchmark. For this example, the benchmark is Competitor A.

3.2.2.4 Targets

For the design features that have been prioritized as the most important, targets are selected for the new product based on a comparison with the benchmark. The benchmark is first evaluated by the planning team, and scored on a scale of 1 to 10 to reflect how well it has handled the chosen design features. These scores are shown in a row below the row containing the normalized contributions of design features of the new product. These numerical scores for the benchmark provide the basis for selecting the target for the new product. The new product will then have the important design features at targets chosen based on a comparison with the benchmark.

For the book example, the design features "statistics fundamentals," "statistical methods," "management topics," and "formula derivations" are identified as the most important. The targets for these features in the new book will be chosen by taking into account the numerical sores the benchmark secured for the features. From the scores assigned for the different features, we notice that this competitor is lacking in one important design feature, "management topics," which can be taken advantage of in the new design. The targets for each design feature of the new book are chosen by keeping in mind how the design features satisfy the customer requirements and how these features interact among one another. These target values are shown in the row below the row displaying the feature scores of the benchmark, Competitor A. These target numbers are relative numbers, related to the scores of the corresponding features of Competitor A. For example, the number of statistical methods covered in the new book will be about seven-tenths of those covered in Competitor A. The target for the statistical methods is made smaller than the competitor's in order to balance out the increase in the target for the number of management topics so that the total size of the new book will still be comparable with that of the benchmark.

A simple example of designing a book was used above to describe the QFD methodology. The reader can imagine the level of details needed for a product like a refrigerator or a car. The above example, however, illustrates the important principles involved in using the QFD methodology for identifying the needs of the customer and designing a product to satisfy those needs. Several good books (e.g., Cohen 1995; Akao 1990) are available for further study.

At the planning stage, the selection of major design features of a product includes the selection of quality and reliability goals. The customers are asked for their quality and reliability preferences for the particular grade level of the product. Their needs expressed in this regard through past complaints are also gathered. Suitable design features are then incorporated to respond to these needs. A basic understanding of the principles of reliability is needed for a design engineer—and the other members of the planning team—to understand customer needs in this respect, and to be able to respond to them. Knowledge regarding how reliability is defined, measured, and specified is necessary. The engineers should also know how reliability goals are chosen and achieved in products. The following discussion on the fundamentals of reliability is given to meet this objective.

3.2.3 Reliability Fundamentals

Reliability refers to the ability of a product to perform without failure over a period of time. It is related to the length of life of a product before a failure occurs. This length of life, simply referred to as "life," is a random variable in the sense that in a given population, although the units are all built by the same process, the life of different units will vary. Those life values are viewed as values of a random variable, usually denoted as T.

The variability in the life variable can be described by a frequency distribution, and this frequency distribution can be obtained from data collected on lives of sample units if the product already exists. The frequency distribution of a future product can be projected based on similar past models. An example of a frequency distribution is shown in Figure 3.3. This frequency distribution can also be represented by a mathematical function, which we call the probability density function (*pdf*) of the random variable T. (We will

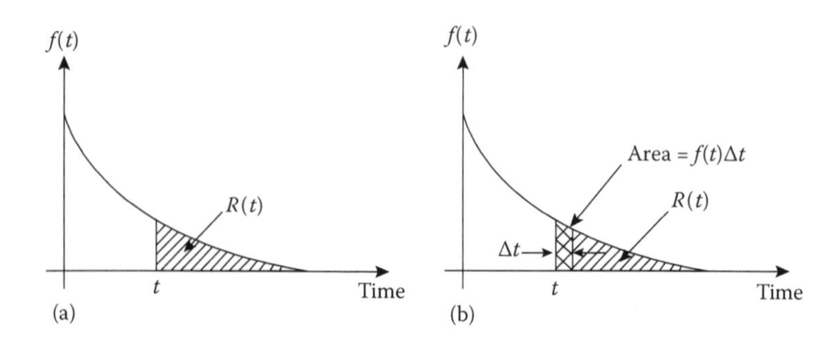

FIGURE 3.3
Frequency distribution of life and definition of reliability.

assume that T, which generally represents life in hours, days, or months, is a continuous variable.) The frequency distribution of the life variable is the basic information necessary to assess the reliability of a product. The cumulative distribution function of the life variable is called its "life distribution."

3.2.3.1 Definition of Reliability

Reliability is expressed as a function of time. The reliability of a product at time t, denoted as $R(t)$, is defined as the probability that the product will not fail before the time t, under a stated set of conditions. In notations,

$$R(t) = P(T > t).$$

This probability can be seen in Figure 3.3 as the area under the curve above t. This probability also represents the proportion in the population that survives beyond time t. If $f(t)$ is the *pdf* of T, then

$$R(t) = \int_t^\infty f(x)\,dx.$$

Also, it can be seen that $R(t) = 1 - F(t)$, where $F(t)$ is the cumulative distribution function (*CDF*) of T.

3.2.3.2 Hazard Function

An important definition relating to reliability is that of the hazard function, which is also known as the "instantaneous failure rate" or "mortality rate" function. The hazard function, denoted by $h(t)$, represents the rate at which survivors at time t will fail in the instant immediately following time t. In other words, $h(t)$ will provide answers to questions such as: What proportion of the refrigerators that are five years old will fail in the next year? Or, what proportion of engines that have run for 100,000 miles will fail in the next mile? The $h(t)$ is obtained as described below.

With reference to Figure 3.3b, the proportion of the population surviving beyond time t is $R(t)$. The proportion of these survivors failing in the interval Δt immediately following t is:

$$\frac{f(t)\times \Delta t}{R(t)}$$

The rate at which they fail in the interval Δt is

$$\frac{f(t)\times \Delta t}{R(t)\times \Delta t}.$$

We want the rate at the instant immediately following time t—that is, we are looking to find the rate when $\Delta t \rightarrow 0$. That is, we want:

$$\lim_{\Delta t\rightarrow 0}\frac{f(t)\times \Delta t}{R(t)\times \Delta t}=\frac{f(t)}{R(t)}.$$

Therefore,

$$h(t)=\frac{f(t)}{R(t)}.$$

The $h(t)$ is not a probability; instead, it reflects the susceptibility of a product to failure when it has reached a certain age or when it has worked certain hours or run a certain distance. It can be obtained once the failure distribution is known either as an empirical function obtained from the failure data of sample units, or as a mathematical function projected on the basis of the analysis of historical failure data. The study of the $h(t)$ of a product over time can reveal quite a bit of information about the reliability of the product at different stages of its life. The $h(t)$ for a given product can be an increasing function of t, a decreasing function of t, or a constant that is independent of t. The $h(t)$ could behave differently—increasing, decreasing, or remaining constant—at different periods of life even for the same product. The following example shows how $h(t)$ can be calculated from empirical data obtained from the failure times of sample units of a product.

Example 3.3

Table 3.3 shows the data on the life of 1000 compressors giving the number that failed in each time interval. Draw the frequency distribution, reliability, and failure rate curves as functions of time.

TABLE 3.3

Life distribution, Reliability, and Failure Rate Calculations for Compressors

(1) Interval (Month)	(2) No. of Failures	(3) Frequency of Failures in Interval	(4) Cumulative Frequency of Failures	(5) Reliability at End of Interval	(6) Failure Rate at End of Interval
0–0	0	0.00	0.00	1.000	0.0347
$0 < t \leq 10$	347	0.347	0.347	0.653	0.0093
$10 < t \leq 20$	61	0.061	0.408	0.592	0.0117
$20 < t \leq 30$	69	0.069	0.477	0.523	0.0166
$30 < t \leq 40$	87	0.087	0.564	0.436	0.023
$40 < t \leq 50$	101	0.101	0.665	0.335	0.031
$50 < t \leq 60$	103	0.103	0.768	0.232	0.044
$60 < t \leq 70$	101	0.101	0.865	0.135	0.074
$70 < t \leq 80$	97	0.097	0.966	0.034	0.100
$80 < t \leq 90$	34	0.034	1.000	0.00	—
Total	1000	1.000			

Solution

The calculations for frequency distribution, reliability, and failure rate are shown in Table 3.3. In this table, Column 3 gives the frequency distribution of failures in each interval, and Column 6 gives the failure rate at the instant following each interval. While the numbers in Column 3 are the proportion that fail in each interval out of the original sample of 1000, the numbers in Column 6 represent the proportion of those that survive past the interval, failing in the instant (month) immediately following the interval.

To illustrate the computation of failure rate, the failure rate of the compressors that are 20 months old is calculated in Table 3.3 as:

$$h(20) = \frac{f(20)}{R(20)} = \frac{\text{Rate of failures per month following the 20th month}}{\text{Reliability at the 20th month}}$$

$$= \frac{0.069/10}{0.592} = 0.0117 \text{ units/month.}$$

This means that out of the compressors that have worked for 20 months, 1.17% will fail within a month. This failure rate increases to about 10% for the 80-month-old compressors. These compressors have had a larger rate of failure when they were newer (<10 months old). Figure 3.4 shows the graph of the frequency distribution of failure times, reliability, and failure rate for the compressors.

3.2.3.3 The Bathtub Curve

Failure rate curves have been used to study the life behavior of many different types of equipment. Figure 3.5 (a to d) shows a few different types of failure rate curves experienced by different types of equipment. There are equipments whose failure rate increases with time, decreases with time, or remains

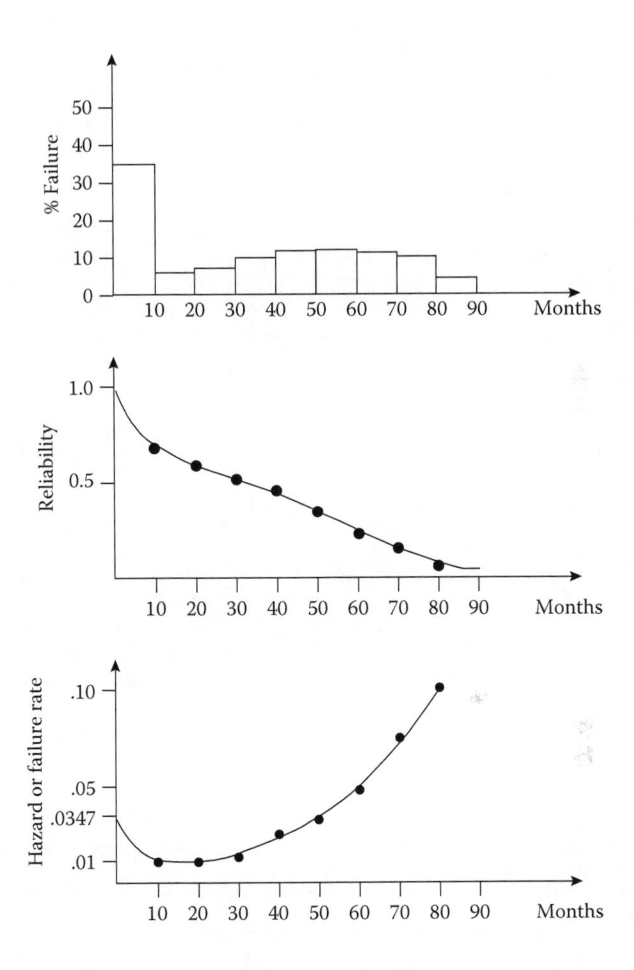

FIGURE 3.4
Example of failure distribution, reliability and failure rate functions.

constant over time. The one failure rate curve that seems to describe the failure rate behavior of a variety of complex equipment is shown in Figure 3.5d. Because of its shape, it is called the "bathtub curve." This curve shows how the failure rate for this type of equipment changes differently at various periods of the product's life. Based on the nature of change in the failure rate, the life of such equipment can be divided into three major periods.

Period A

Period A represents a time of decreasing failure rate, at the end of which the rate tends to become constant. The decrease in failure rate is caused by defective units in the population failing early and being repaired or removed from the population. This region is known as the "infant mortality" region. The causes

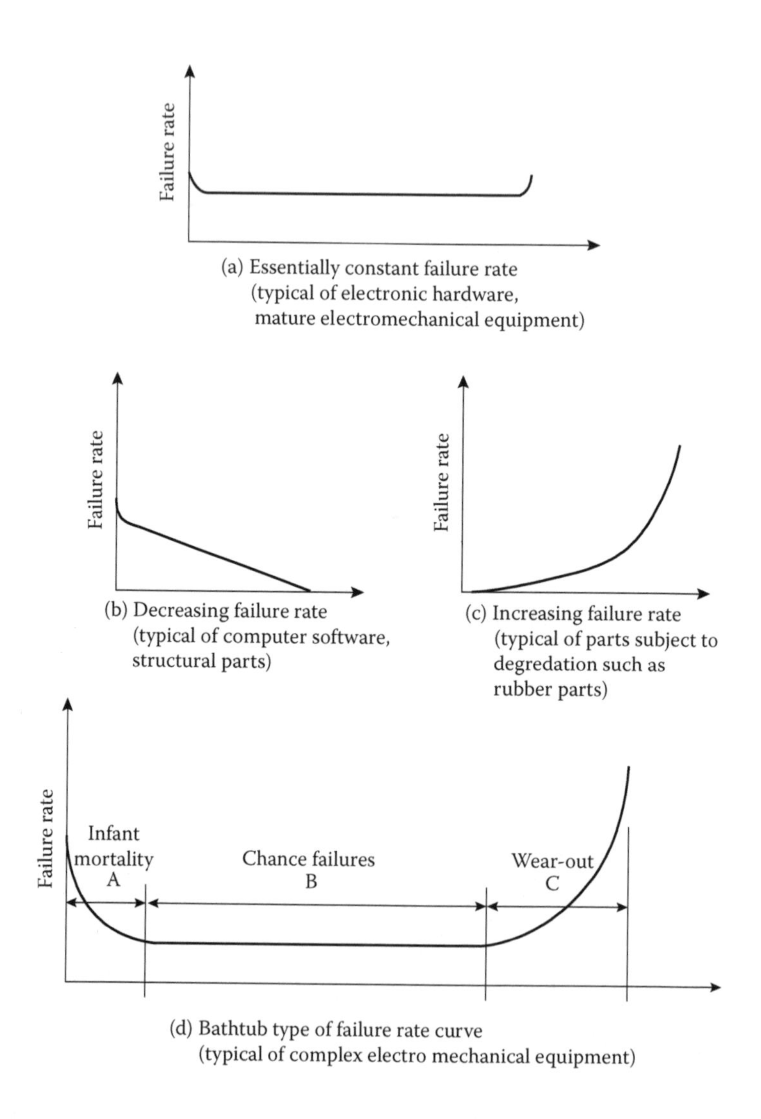

FIGURE 3.5
Examples of failure rate curves.

for failure in this period, which are called "early failures," are mainly poor material or bad workmanship, such as use of a wrong clip, a missing washer, or a bolt not tightened enough. They are not caused by any design weakness. A high early failure rate or long period of infant mortality would indicate an inadequate process control during manufacture or assembly. The length of this period determines the length of burn-in that is required—the time for which the product must be run in the factory before shipping in order to avoid customers experiencing the early failures. A design engineer would want to keep the early failure rate small and the infant mortality period short.

Period B

Period B represents a time of constant failure rate, at the end of which the rate starts to increase. Here, failures occur not because of manufacturing defects but because of "accidents" caused by chance loads exceeding the design strength. As will be seen later, a constant failure rate implies that the life distribution of units in this period follows exponential law. This region is known as the period of "chance failures," or "useful life." The latter name is derived from the fact that the product performs during this period without failure, except for accidents, giving its most useful performance during this period.

Period C

Period C is a time of increasing failure rate because of parts wearing out. Failures occur in this period as a result of fatigue, aging, or embrittlement. This region is known as the "wear-out" period. Knowledge of when wear-out begins helps in planning replacements and overhauls so that wear-out can be delayed and the useful life extended.

3.2.3.4 Distribution of Product Life

The above discussion of the failure rate behavior of products and the phenomenon of the bathtub curve brings out the fact that the distribution of the life variable could change over the life of a product. At least three distinctly different periods can be identified for most electromechanical equipment, in which the life distribution follows different characteristics. We would want to know the distribution pattern of life in the different periods of a product's life. The distribution characteristics have to be studied by gathering data on the failure times of the product and then checking which distribution fits the data best. Some of the candidates include the exponential, Weibull, log-normal, and gamma distributions. The properties of these distributions, and how their fit to a given set of data should be determined, are discussed in books on reliability—such as Ireson and Coombs (1988), Krishnamoorthi (1992), and Tobias and Trindade (1995)—which are cited at the end of this chapter. Of all the distributions employed to model the life variable, however, the exponential distribution is the most common as many parts and products are known to follow the exponential law during a major part of their lives. Exponential distribution is to the life variable what the normal distribution is to other measurable quality characteristics. A brief description of the exponential distribution follows.

3.2.3.5 The Exponential Distribution

The function form, or the *pdf*, of the exponential distribution is given by:

$$f(t) = \lambda e^{-\lambda t}, \ t \geq 0$$

The exponential random variable takes only positive values and the distribution has one parameter, λ. We write $T \sim \text{Ex}(\lambda)$ to indicate that a random variable T has exponential distribution with parameter λ. The distributions shown in Figure 3.3 do, in fact, represent the shape of an exponential distribution.

If $T \sim \text{Ex}(\lambda)$, then:

$$CDF\ F(t) = P(T \le t) = \int_0^t \lambda e^{-\lambda x} dx = 1 - e^{-\lambda t},\ t \ge 0$$

$$R(t) = 1 - F(t) = e^{-\lambda t},\ t \ge 0$$

$$h(t) = \frac{f(t)}{R(t)} = \frac{\lambda e^{-\lambda t}}{e^{-\lambda t}} = \lambda, \text{ a constant independent of } t$$

The mean of the distribution is:

$$\mu_T = \int_0^\infty t\lambda e^{-\lambda t} dt = \frac{1}{\lambda}$$

The variance of the distribution is:

$$\sigma_T^2 = \int_0^\infty (t - \mu_T)^2 \lambda e^{-\lambda t} dt = \frac{1}{\lambda^2}$$

And the standard deviation is:

$$\sigma_T = \frac{1}{\lambda}$$

We can see from the above that the mean and the standard deviation of an exponentially distributed random variable are equal. We can also see that a product with exponential failure times has a constant failure rate equal to the value of the parameter of the distribution, λ. This failure rate can be estimated from the failure data from sample units, using the following formula:

$$\hat{\lambda} = \frac{\text{Number of failures}}{\text{Total number of hours of running time}}$$

3.2.3.6 Mean Time to Failure

The mean of the distribution, or the average life of all units in the population, is called the "mean time to failure" (*MTTF*) and is equal to the reciprocal of the failure rate λ. The term *MTTF* is used for products that have only one life (i.e., those that are not repairable). For products that are repairable, the term "mean time between failures" (*MTBF*) is used to denote the average time between repairs. The *MTTF* (or the *MTBF*) has a special significance when the life distribution is exponential. Then, as shown above, it is equal to the reciprocal of the failure rate λ, the single parameter of the distribution. This means that knowledge of the *MTTF* alone provides information about the lives of the entire population. The evaluation and prediction of all measures relating to reliability can be made once the *MTTF* is known.

It should be pointed out, however, that the *MTTF* does not have the same significance when the life distribution is not exponential. For example, if the Weibull distribution (another popular model for life variables) is used to model the life of a product, then knowledge of the *MTTF* is not adequate to define the life distribution. Instead, the two parameters of the Weibull distribution must be estimated.

Example 3.4

A production machine has been in operation for 2200 days. If it broke down three times during this period, what is the estimate for the failure rate and the *MTBF* of the machine? Assume that the time between failures is exponential.

Solution

$$\hat{\lambda} = 3/2200 = 0.00136 \text{ failures/day.}$$

Therefore, $MTBF = 2200/3 = 733.3$ days.

Example 3.5

Twelve fan belts were put on test, of which four failed at 124, 298, 867, and 1112 hours, respectively. The remaining belts had not failed by 1200 hours, when the test was stopped. What is the estimate for the failure rate and the *MTTF* if the failure time can be assumed to be exponential?

Solution

This is an example of censored data, in which the test was stopped before all the test units failed. The estimate for λ is calculated taking into account the fact that those units that did not fail, called "run-outs," survived until the test was ended. So the fact that the eight units that did not fail each ran for 1200 hrs is counted in calculating the total hours of running time. Thus,

$$\hat{\lambda} = \frac{4}{124 + 298 + 867 + 1112 + 8 \times 1200} = 0.00033 \text{ failures/hour,}$$

and the *MTTF* is $\hat{\mu} = 1/\hat{\lambda} = 3000.25$ hours.

Example 3.6

The life of some engine seals is known to have the exponential distribution, with $MTTF = 12{,}000$ hours.

a. What is the reliability of the seals at 2000 hours?
b. What proportion of the seals would have failed by 8000 hours?

Solution

Let T represent the life of the seals, then,

$$f(t) = \lambda e^{-\lambda t}, \quad t \geq 0$$

where $\lambda = 1/12{,}000$. Therefore,

$$f(t) = \frac{1}{12{,}000} e^{-\frac{1}{12{,}000}t}, \quad t \geq 0$$

$$F(t) = \int_0^t f(x)dx = 1 - e^{-\lambda t} = 1 - e^{-\frac{t}{12{,}000}}, \quad t \geq 0$$

$$R(t) = 1 - F(t) = e^{-\frac{t}{12{,}000}}, \quad t \geq 0.$$

a. $R(2{,}000) = e^{-\frac{2{,}000}{12{,}000}} = e^{-0.167} = 0.846.$
b. $F(8{,}000) = 1 - e^{-\frac{8{,}000}{12{,}000}} = 1 - e^{-0.667} = 0.487.$

From the above discussion, we see that reliability is a function of time and is quantified as the probability that a product will survive beyond a given time. Reliability can also be interpreted as the proportion of the population that will survive beyond the given time. The failure rate expressed as a function of age represents the susceptibility of a product to failure at a given age, and it provides another measure of reliability. The reliability of a product can be evaluated if its life distribution is known. The life distribution can be obtained from empirical data on the failure times of sample units. It can also be modeled using a distribution function that is chosen to "fit" the historical failure data of the product. Probability distributions (such as the exponential, Weibull, log-normal, and gamma) are commonly used to model life variables.

The most commonly used mathematical model for describing life variables, however, is the exponential distribution. It has one parameter, λ, called the "failure rate." This failure rate is a constant and is independent of age. If a product life is exponential, then its reliability can be measured using its failure rate or its reciprocal *MTTF*. The failure rate, or *MTTF*, can then be used for setting reliability goals and monitoring reliability achievements. We have also seen how knowledge about the behavior of the failure rate over time, expressed as a bathtub curve, can be used to understand—and possibly enhance—a product's reliability.

3.2.3.7 Reliability Engineering

The discipline of applying reliability principles to evaluating, predicting, and achieving reliability in products is called "reliability engineering."

Reliability goals are first chosen at the overall system level. For example, if a manufacturer produces trucks, the reliability goal is first chosen for the entire truck. This reliability is then apportioned to the components so that by working together, the component reliabilities will contribute to achieving the desired system reliability. The apportioning of the component reliability is done with an understanding of the relationship of the components to the system. First, the functional relationship of the components to the system is expressed in what is called a "reliability block diagram" (RBD). The system reliability is then calculated from component reliabilities using procedures that are documented in books on reliability (e.g., Tobias and Trindade 1995). Apportioning the system reliability to the components is done by trial and error until component reliabilities, which have been assigned taking into account the capabilities of individual components, would achieve the desired system reliability. At the end of the apportioning exercise, the reliability required of components will be known. For example, in the case of the truck, the individual reliability requirements of the engine, transmission, controls, and so on, will be known. The reliability requirement of purchased components will be written as requirements in the purchase contracts. Achieving reliability goals for manufactured components will be the responsibility of the design engineers and the quality planning team.

The reliability of current designs can be estimated either from field failure data or from laboratory test data. From these, the gap between the required and actual reliability can be obtained for each component, and from these estimates will emerge a few critical components that must have their reliability improved in order to attain the system reliability goals. Reliability improvement can be accomplished by studying the failure rate behavior and failure mechanisms of parts and subassemblies. If failures are happening in early life, process controls should be implemented, or improved. Variability in process parameters must be reduced and poor workmanship must be avoided. If the failures result from wear-out, it may be because seals, belts, and hoses that have increasing failure rates. Better material, better tolerances, and improved maintenance will delay the wear-out failures. If the failure rate is high during the useful life, load-strength studies (see Chapter 4 in O'Connor 1985) will help in identifying opportunities to minimize "accidental" failures. Also, designs can be made more robust; that is, less susceptible to failure due to changes in environment over which the user has no control.

This is done by optimal choice of product parameters and their tolerances. The issues relating to the choice of product parameters and their tolerances are discussed next as part of the product design. Note that any significant improvement in reliability has to come from design changes.

3.3 Product Design

Product design consists of two stages: first, the overall parameters are chosen; and second, the details of the parameters are worked out. Engineering drawings and specifications are created, and prototype testing is done for validating the design. As indicated in Figure 3.1, process design is undertaken even as the product design progresses. Trial runs are made for validating both the product and the process design. The major quality-related activities at this stage are:

1. Parameter design
2. Tolerance design
3. Failure mode and effects analysis
4. Design for manufacturability study
5. Design reviews

The objectives, procedures, and outcomes of each of these activities are explained below.

3.3.1 Parameter Design

Parameter design, in the context of product design, refers to selecting the product parameters or those critical characteristics of the product that determine its quality and enhance the product's ability to meet its intended use and satisfy the customer.

At the end of the QFD exercise, the major design features of the product and their target values would have already been decided. For example, if the product is a lawnmower, the planning team would have chosen the performance target as: mowing an average yard of about 10,000 sq. ft. in less than one hour. They must then decide the product parameters, such as blade size, blade angle, engine horsepower, speed of rotation, deck height, chute angle, and so on, in order to accomplish the target performance. Most product designers would have initial values for these parameters based on their experience with previous models. The question to be answered is whether these initial values are good enough to meet the new target or if they need to be changed to attain the new target for the new model. Often, the answer has to be found through experimentation—that is, by trying different values for the product parameters, measuring corresponding performances, and choosing the set of parameters that give the desired performance. Thus, in the lawnmower example, an experiment has to be conducted with the objective of finding the best set of values for the product parameters that will enable the cutting of a yard that is 10,000 sq. ft. in size in one hour or less.

A vast body of knowledge exists on how to perform experiments efficiently so that the required information about the product performance, vis-à-vis the product parameters, is obtained with the minimum amount of experimental work. This branch of statistics, referred to as the "design of experiments," (D.O.E.), was pioneered by Sir Ronald Fisher, the English statistician, who in early 1920s was researching the selection of the best levels of inputs, such as fertilizer, seed variety, amount of moisture, and so on, to maximize the yield from agricultural fields. The designed experiments were subsequently used profitably in industrial environments to optimize the selection of product parameters during product design, and process parameters during process design. The Japanese engineer Dr. Genichi Taguchi propagated the philosophy that experiments must be used for selecting the product and process variables in such a way that the performance of the product or process will be "robust." By this, he meant that the selection of parameters should be such that the performance of the product will not be affected by various noise or environmental conditions to which the product or process may be subjected.

The basics of designed experiments are discussed below. The objective here is to impress upon readers the need for experimentation when choosing product and process parameters, and alert them to the availability of different experimental designs to suit different occasions. Details on how experiments should be conducted and their results analyzed are also provided for some popular designs. By the end, it is hoped that readers will be able to appreciate the value of designed experiments in the context of product or process design, perform some basic experiments, and analyze the data from them. Readers will also be better prepared to explore more advanced designs when the need for them arises. The discussion below relates to two simple, but important, designs that are used in industrial experimentation, which are known as 2^2 and 2^3 factorial designs. Some additional designs in the 2^k family are included in Chapter 5.

3.3.2 Design of Experiments

An experiment is designed to study the effect of some input variables, called "factors," on a response, which may be the performance of a product or output of a process. The factors can be set at different levels, and the product performance or the process output could change depending on the levels at which the different factors are kept. The design of the experiment involves choosing the relevant factors, selecting the appropriate levels for them, and determining the combinations of the factor levels, called the "treatment combinations," at which the trials will be conducted. The design also determines the number of times that the trials will be repeated with each treatment combination in order to obtain a measure of the random variability present in the results. In addition, the design will specify the sequence in which the trials should be run, and is usually accompanied by a procedure for analyzing the data from the trials and drawing conclusions.

Sometimes, an experimenter is concerned with only one factor and wants to determine the best level of that factor to achieve the desired level of a response. In such a case, an experiment would be conducted by running trials at different levels of that one factor. Such an experiment is known as a "one-factor experiment." More often, though, we will be dealing with situations where several factors are influencing a response, which is a quality characteristic of a product or output of a process, and we have to find out how the different factors, individually and jointly, affect the response. With this being the more common situation, we will focus here on the multifactor experiments.

There are many multifactor experimental designs to suit the varying situations in which experiments have to be run. We will discuss below one type of design called the "2^k factorial" design, in which k number of factors, each with two levels, are studied to learn of their effect on a response. These designs are very useful in the selection of product and process parameters and are considered to be "workhorse" designs in industrial experimentation. We will discuss the 2^2 and 2^3 designs in this chapter, which, being simple designs, are useful in explaining the concepts and terminology of experimental design. The more general 2^k design is discussed in Chapter 5.

3.3.2.1 2^2 Factorial Design

A factorial experiment is an experiment in which each level of one factor is combined with each level of every other factor to obtain all possible treatment combinations at which trials are to be performed. The 2^2 design is a factorial experiment involving two factors, each at two levels, and is explained below using an example.

Example 3.7

Consider the case of the lawnmower mentioned earlier. The response is the time needed to cut a yard of 10,000 sq. ft. The possible factors would be blade diameter, blade angle, rotation speed, engine horsepower, deck height, chute size, chute location, along with a few others. Suppose that for a new design of the lawnmower, only changes to two factors—the blade angle and the deck height—are being considered for achieving the desired performance. The two levels for the factor "blade angle" are 12° and 16°, and the two levels for the factor "deck height" are 5 in. and 7 in. The best combination of blade angle and deck height for achieving the desired performance is to be determined.

Solution

All the possible treatment combinations for the factorial experiment can be represented graphically, as shown in Figure 3.6. When two factors each have two levels, there are four possible treatment combinations. This experimental set up, or design, is called a 2^2 factorial design.

The 2^2 factorial design shown in Figure 3.6 can also be represented in a table, as shown in Table 3.4. In this table, the lower level of each factor is designated by a (–) sign and the higher level by a (+) sign. Notice the simple pattern of the signs in the columns under each factor: the signs alternate one at a time in the column of Factor A and they alternate two at a time in the column of Factor B. This enables the creation of the design easily and quickly. The design is to run the experiment at the four treatment combinations numbered one to four. For example, in Treatment Combination 1, both the factors will be at low levels; and in Treatment Combination 2, the Factor A is at its higher level and Factor B is at its lower level. In addition, the design recommends that two trials be run at each treatment combination to obtain two replicates. The results from the two replicates will then be averaged, and the average will be used as the response from the treatment combination.

The number of replicates needed is determined based on the variability expected from the individual trials as well as the cost associated with, and

Factors	Levels	
	Low(–)	High(+)
A: Blade angle	12	16
B: Deck height	5 in.	7 in.

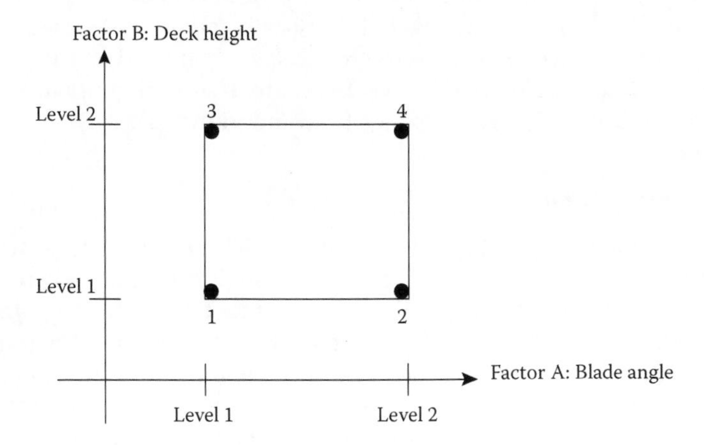

FIGURE 3.6
Graphical representation of a 2^2 factorial experimental design.

TABLE 3.4

A 2^2 Factorial Experimental Design in a Table

Treatment Combination	Factor A (Angle)	Factor B (Height)	Response (Min.)		
			Replicate 1	Replicate 2	Average
1	–	–	72	74	73
2	+	–	47	49	48
3	–	+	59	61	60
4	+	+	83	83	83

time available for, repeating the trials. At least two replicates are necessary, as will be seen later, to obtain an estimate for the experimental error, or the variability in the response because of noise. However, when the number of factors is large, (>3), there are ways of obtaining estimates for experimental error without multiple replications of the trials. For two- and three-factor experiments, at least two replicates must be performed.

"Noise" refers to the environmental conditions that are not deliberately changed as part of the experiment, but changes in them may influence the experimental results. For example, in the case of the lawnmower, we would like to perform the experiment with different treatment combinations on the same piece of yard so that things other than the experimental factors will remain the same for all treatment combinations. For obvious reasons, however, it is not possible to do this, and we have to repeat the experiment with different treatment combinations on different yards (called the experimental units), which may all be similar yards of "equal" size. When different yards are used for different treatment combinations, there will be some difference in the yards because of small differences in size, slopes, weeds, wet patches, and so on. These will cause variability in the results outside of the effects due to the factor levels being investigated. Replicates allow us to evaluate this variability, and the averages from the replicates when used as the response from the treatment combinations will contain less variability from such noise, compared to the single readings from individual trials.

3.3.2.2 Randomization

Just as noise or extraneous factors can cause variability in replicates that are run with the same treatment combination, noise factors can also affect outcomes between two treatment combinations, either by masking or adding to the factor effects. For example, dry grass in the afternoon may require less time to cut than moist grass in the morning. The morning/afternoon effect will add to, or subtract from, the real effect of the factors; therefore, care should be taken to avoid the morning/afternoon effect influencing the real factor effects in any systematic manner. To remove—or at least minimize—this effect of noise, the trials are randomized. In other words, the trials are not run in the order indicated in the design; instead, they are run in a randomly chosen order. Randomizing prevents the noise factors from affecting the trials systematically and minimizes the influence of noise factors on the trial outcomes.

In the example of the lawnmower, the eight trials can be randomly assigned to eight different yards (of approximately 10,000 sq. ft. in area and similar in all other aspects), and the eight trials can be run in a random sequence, using one operator. The experiment will then be considered to be a "completely randomized" experiment. On the other hand, suppose there is some restriction on the number of trials that can be run in one day—for example, only four trials are possible in a day. In such a case, it may make sense to run the four trials from one replication on one day, and the four trials from the other replication on another day. This way, if there is a "day effect," it will affect all four treatment combinations in one replication in the same way. It is possible then to separate this day effect from the rest of the noise by using a suitable analysis technique and, thus, minimize the amount of variability that is not accounted for. This is an example of blocking, in which the trials are run in two blocks, with the days being the blocks. The trials within a block are run in a randomized sequence. Such an experiment will be called a "randomized complete block" design, with the term "complete" being used to indicate that each block includes one full, or complete, replication. When a full replication cannot be accommodated inside a block, we will have a "randomized incomplete block" design. This gives an idea of why so many different designs have been developed to accommodate the different circumstances of experimentation. We will discuss here only the randomized complete design. More details on the design and analysis of experiments with blocks can be found in the references (e.g., Hicks and Turner 1999) cited at the end of this chapter.

3.3.2.3 Experimental Results

Suppose the eight trials of the above experiment are run in a completely randomized manner and the results from the trials are as shown in Table 3.4. The results are presented in the graph in Figure 3.7, with the respective corners of the square representing the treatment combinations. It is easy to see that Treatment Combination 2 produces the best result, requiring the least amount of time to cut the given-size yard. Looking at the difference between the two replicates at each treatment combination (see Table 3.4), not much variability can be seen between the replicates, indicating that the experimental error, or unexplained variability, is almost not there. This means that there is not much noise; therefore, the signal is clear, and it is easy to see the best treatment combination.

The results, however, do not come out this clear from many real experiments. There may be much difference in the results from replicates of the same treatment combination, and the results for the various treatment

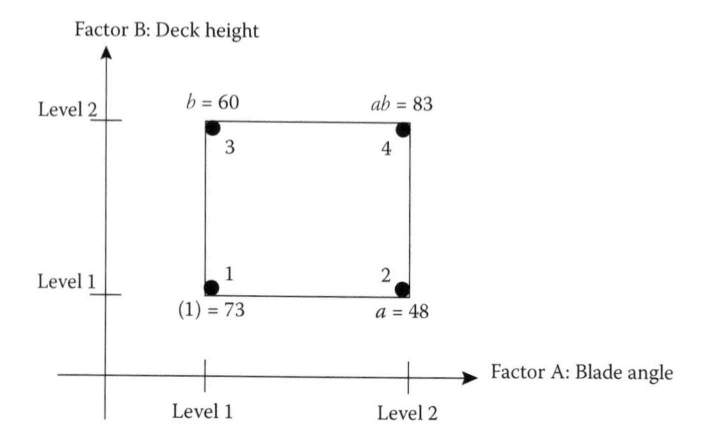

FIGURE 3.7
Results from a 2^2 experiment (see Table 3.5 for treatment combination codes).

combinations may not be far apart to give a clear-cut choice. The results from the trials of treatment combinations may even overlap. When the differences among the results from the treatment combinations are not far apart, showing obvious differences due to factors, we would then want to know whether the differences in the results are, indeed, true differences because of factors or just experimental variability. In such a case, we would have to calculate the effect of the individual factors and the effect of interaction among the factors, and use a statistical technique to determine if these effects are significant.

3.3.2.4 Calculating the Factor Effects

The data from the experiment under discussion are rearranged in Table 3.5 to facilitate the calculation of the effects. This table has a new column,

TABLE 3.5
Calculating Effects of a 2^2 Factorial Experimental Design

	Design Columns		Calculation Column	Response (Min.)		
Treatment Combination Code	**Factor A (Angle)**	**Factor B (Height)**	**Interaction (AB)**	**Replicate 1**	**Replicate 2**	**Average**
(1)	–	–	+	72	74	73
a	+	–	–	47	49	48
b	–	+	–	59	61	60
ab	+	+	+	83	83	83
Contrast	–2	22	48			
Effect	–1	11	24			

with the heading "interaction," added to those from Table 3.4. The original columns with (–) and (+) signs are named "design columns," and the new column is called the "calculation column." The treatment combinations are identified by new codes: "(1)" for the treatment combination in which both factors are at low level, "*a*" for the treatment combination in which Factor A is at high level and Factor B at low level, "*b*" for the treatment combination in which Factor B is at high level and Factor A at low level, and "*ab*" for the treatment combination in which both factors are at high level. These codes represent the average response from the respective treatment combinations in the formulas that are given below for calculating effects. The graphic representation of the design in Figure 3.7 includes these new notations to help readers follow the development of formulas for computing factor effects. (Please note that the codes we use here represent the averages from the treatment combinations. Some authors use the codes to represent the totals from the treatment combinations, so the formulas given here may look different from theirs.)

3.3.2.5 Main Effects

Factors A and B are called the main factors in order to differentiate them from interactions that also arise as outcomes of experiments. The effect caused by a main factor, called a "main effect," is calculated by subtracting the average response at the two treatment combinations where the factor is at the lower level from the average response at the treatment combinations where the factor is at the higher level. For example, the average of the responses of treatment combinations where Factor A is at the higher level is:

$$\frac{ab+a}{2}$$

and the average of the responses of treatment combinations where Factor A is at the lower level is:

$$\frac{(1)+b}{2}$$

The difference between these two averages gives the effect of Factor A and is denoted as A. So,

$$A=\frac{-(1)+a-b+ab}{2}$$

Similarly,

$$B=\frac{-(1)-a+b+ab}{2}$$

So, for the example,

$$A = \frac{-73 + 48 - 60 + 83}{2} = -1$$

$$B = \frac{-73 - 48 + 60 + 83}{2} = 11$$

These effects can be interpreted as follows: if the blade angle (Factor A) is changed from its lower level of 12° to the higher level of 16°, then the mowing time decreases by 1 minute, and if the deck height (Factor B) is changed from 5 in. to 7 in., then the mowing time increases by 11 minutes.

3.3.2.6 Interaction Effects

Interaction between two factors exists if the effect of the two factors acting together is much more, or much less, than the sum of the effects caused by the individual factors acting alone. It is necessary to detect the existence of interaction between factors, because when significant interaction exists, the main effects calculations are rendered suspect. The interaction effect between two factors also has practical meaning and helps in understanding how the factors work together.

The interaction effect between Factors A and B in the two-factor experiment is calculated as follows: take the average of the responses from the treatment combinations where both factors are at the high and both are at the low level; this is the average of the responses at the two ends of the leading diagonal of the square in Figure 3.7. Then take the average of the responses from the treatment combinations where one factor is at the high level and the other is at the low level; this is the average of the responses at the two ends of the other diagonal in Figure 3.7. Subtract the latter from the former; the difference is the interaction effect caused by increasing A and B simultaneously, denoted as the *AB* interaction.

For the example, to get the *AB* interaction, find the average of the treatment combinations where both factors are at the high and at the low level:

$$\frac{(1) + ab}{2}.$$

Then, find the average of the treatment combinations where one factor is at the high and one factor is at the low level:

$$\frac{a + b}{2}.$$

The difference between the averages is the *AB* interaction:

$$AB = \frac{(1) - a - b + ab}{2}.$$

For this example,

$$AB = \frac{73 - 48 - 60 + 83}{2} = 24.$$

This means that increasing both the blade angle and the deck height together increases the mowing time much more than the sum of the effects from increasing them individually. Considerable interaction between Factors A and B exists in this case.

The above formula for the interaction effect will give a value of zero if the response from increasing the two factors together equals the sum of the responses from increasing them individually. When this happens, we say the joint effect is additive. Thus, we can see that interaction exists when the joint effect is *not* additive.

3.3.2.7 A Shortcut for Calculating Effects

The product of the column of treatment combination codes and the column of signs under any factor is called a "contrast" of that factor. For example, the product of the column of codes and column of signs under Factor A = [–(1) + a – b + ab] is called *Contrast A*. The effect of a factor can be obtained from its contrast. The general rule for calculating an effect is:

$$\text{Effect of a factor} = \frac{\text{Contrast of the factor}}{2^{k-1}},$$

where k is the number of factors in the experiment.

Calculations of the effects using this formula have been made in Table 3.5 and the results are shown in the bottom rows of the table. Incidentally, the order of arrangement of the treatment combinations in Table 3.5 is called the "standard order." We should note that this order ((1), a, b, and ab for the 2^2 design) has enabled the use of a simple method of creating the design and a shortcut for calculating the effects. Therefore, whenever a design is created, the standard order should be used. This standard order also lends itself to be expanded easily while designing higher order designs, as will be explained later while discussing the 2^3 design.

3.3.2.8 Determining the Significance of Effects

Next, we have to determine if the effects are significant. The analysis of variance (ANOVA) method is normally used to answer this question. The ANOVA method is explained in Chapter 5; here, we will use a quick and simple method (adapted from Hogg and Ledolter 1987) for evaluating the significance of factors using confidence intervals. The method consists of first obtaining an estimate for the experimental error and then an estimate for the standard error of the factor effects. (The standard deviation of an average is called the "standard error," and because a factor effect is an average, its standard deviation is called its standard error.) Then, approximate 95% confidence intervals are established for each of the main and interaction effects using the estimated standard error. If any of the confidence intervals includes zero, then the corresponding effect is *not* significant. If, on the other hand, any confidence interval does not include zero, then the corresponding effect *is* significant.

To explain the method of obtaining an estimate for the experimental error, we lay out the results of the experiment as shown in Table 3.6. In this table, the term y_{ijk} represents the observation from the k-th replicate of the treatment combination, with Factor A at the i-th level and Factor B at the j-th level.

We assume there are n observations (from n replicates) in each of the treatment combinations located in each cell. The quantity $\bar{y}_{ij.}$ represents the average from each cell, $\bar{y}_{i..}$ the average of each row, $\bar{y}_{.j.}$ the average of each column, and $\bar{y}_{...}$ the average of all observations.

Take, for example, the cell (1, 1), where both factors are at Level 1. The quantity $(1/(n-1))\Sigma_{k=1}^{n}(y_{11k}-\bar{y}_{11.})^2$ obtained from observations of this cell represents the variability among the n observations, all made at the same treatment combination represented by the cell (1, 1). If there is no experimental error, then all the values in this cell would be the same, and the value of the above quantity would be zero. If its value is not zero, then it represents an estimate for the error variance from this cell. There are four such estimates for experimental error in this 2^2 experiment, and if they are pooled together, the quantity $S^2 = (1/4(n-1))\Sigma_{i=1}^{2}\Sigma_{j=1}^{2}\Sigma_{k=1}^{n}(y_{ijk}-\bar{y}_{ij.})^2$ gives a better estimate for the variance of the experimental error. The general formula for estimating the error variance with k factors in an experiment is $S^2 = (1/2^k(n-1))\Sigma_{i=1}^{k}\Sigma_{j=1}^{2}\Sigma_{k=1}^{n}(y_{ijk}-\bar{y}_{ij.})^2$ The standard error of an effect can then be estimated as described below.

Notice that in Table 3.6 the effect of Factor A, for example, is the difference between the two row averages $\bar{y}_{1..}$ and $\bar{y}_{2..}$. If σ^2 is the variance of the individual observations in the cells, then the variance of each of the row averages is $\sigma^2/2n$ because $2n$ observations go into each of the averages. Because the effect is the difference of the two row averages, the variance of

TABLE 3.6

Layout of Data From a 2^2 Factorial Experiment

	Factor B		
Factor A	**Level 1**	**Level 2**	**Average**
Level 1	$y_{111}, y_{112}, \ldots, y_{11n}$ average: $\bar{y}_{11.}$	$y_{121}, y_{122}, \ldots, y_{12n}$ average: $\bar{y}_{12.}$	$\bar{y}_{1..}$
Level 2	$y_{211}, y_{212}, \ldots, y_{21n}$ average: $\bar{y}_{21.}$	$y_{221}, y_{222}, \ldots, y_{22n}$ average: $\bar{y}_{22.}$	$\bar{y}_{2..}$
Average	$\bar{y}_{.1.}$	$\bar{y}_{.2.}$	$\bar{y}_{..}$

the effect is $2(\sigma^2/2n) = \sigma^2/n$, because the variance of the difference (or sum) of two (independent) random variables is equal to the sum of the variances of the random variables. The estimate of the variance of an effect, therefore, is S^2/n, and the estimate for the standard deviation of an effect is $\sqrt{S^2/n}$. The general formula for calculating the standard error (*s.e.*) of an effect with k factors is: $\sqrt{S^2/2^{k-2}n}$, which reduces to $\sqrt{S^2/n}$ for the present case, with $k = 2$.

The value of the *s.e.* for the example is calculated as:

$$s^2 = \frac{1}{4}[(72-73)^2 + (74-73)^2 + (47-48)^2 + (49-48)^2$$

$$+ (59-60)^2 + (61-60)^2 + 0 + 0] = \frac{6}{4} = 1.5$$

$$s.e. = \sqrt{\frac{1.5}{2}} = 0.87.$$

An approximate 95% confidence interval for an effect is given by:

$$\text{Factor effect} \pm 2(s.e.),$$

where *s.e.* is the standard error of the effect. Therefore, for the example, the approximate 95% confidence intervals for the effects are:

$$\text{Factor A} = -1 \pm 2(0.87) = [-2.74, 0.74]$$

$$\text{Factor B} = 24 \pm 2(0.87) = [22.26, 25.74]$$

$$\text{Interaction effect AB} = 11 \pm 2(0.87) = [9.26, 12.74]$$

From these confidence intervals, we can see that there is no significant effect due to Factor A because 0 lies in that interval. The effects of Factor B and the AB interaction are significant because 0 does not lie in those intervals. The existence of interaction can also be detected in the graph of the response drawn against factor levels, as shown in Figure 3.8. The graph shows that the response changes in one way (decreases) between the two levels of Factor A when Factor B is at Level 1 but changes in the opposite way (increases) when Factor B is at Level 2. If there is no interaction, the lines will be parallel and will not cross each other.

When such interaction exists, the calculation of main effects becomes meaningless, as averaging done to obtain the effect of a main factor hides the fact that the response behaves differently at the different levels of the other factor. In such situations, we simply look at the responses at the treatment combinations

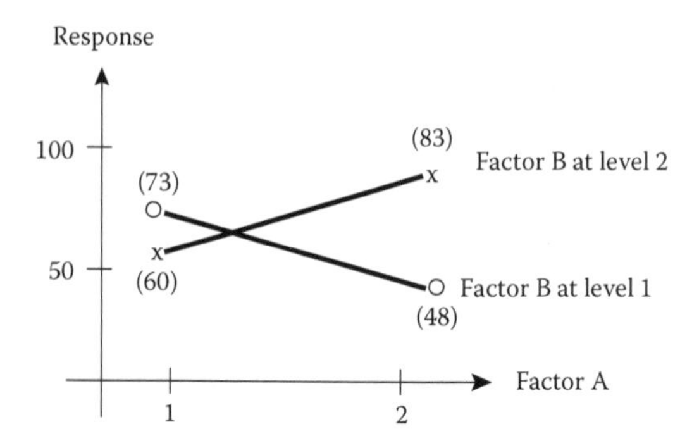

FIGURE 3.8
Graph of response against factor levels.

at which the trials were run and then choose the treatment combination that produces the best response. In this example, the combination with Factor A at the higher level and Factor B at the lower level produces the least time for mowing and, therefore, provides the best combination of the parameters considered. This method of selecting the best combination of levels from those at which trials were conducted is referred to as "running with the winner," meaning that no further analysis of the data is made to investigate responses at other possible treatment combinations in the factor level space. The method of exploring the treatment combination space for the best response using a model built out of the results of the experiment, known as "response surface methodology," is explained later in another section of this chapter.

3.3.2.9 The 2^3 Design

The 2^3 design will be used when there are three factors affecting a response and each factor is studied at two levels. We will illustrate this design using another lawnmower example.

Example 3.8

In this example, we are experimenting with another mower model and are considering one additional factor, chute location, along with blade angle and deck height. Chute location is a different kind of factor in that it is not measurable, as the other two factors are. The chute location is called a "qualitative factor," whereas the other two are termed "quantitative factors." There are two possible locations for the chute: one location is (arbitrarily) identified as the "lower level" and the other as the "higher level." The response, again, is the time in minutes required to cut a yard of 10,000 sq. ft. area. There are eight possible treatment combinations for the 2^3 design, and they are shown graphically in Figure 3.9 and in Table 3.7. We will need to find the main and interaction effects and check if they are significant.

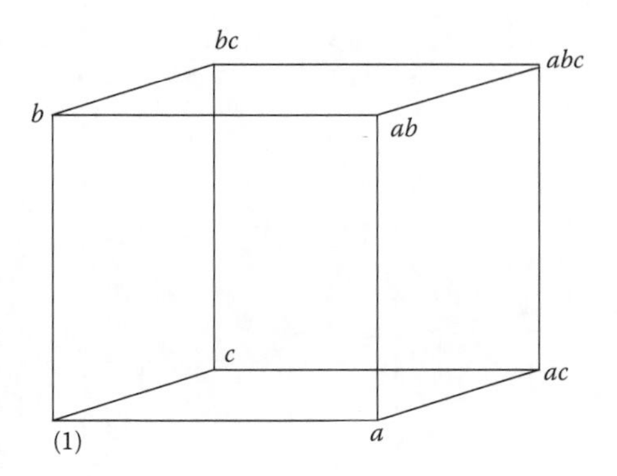

FIGURE 3.9
Graphical representation of the 2^3 design.

Solution

First the treatment combinations are arranged in the standard order, as in Table 3.7.

The standard order for the 2^3 design is created as an extension of the standard order for the 2^2 design, as shown below:

Standard order for the 2^2 design:

(1) *a* *b* *ab*

Standard order for the 2^3 design:

(1) *a* *b* *ab* *c* *ac* *bc* *abc*

The order for the 2^2 design is first written down and is then multiplied by the code for the third factor. The result is appended to the original order of the 2^2. The above rule can be extended to any number of factors. For example, if we want the standard order for the 2^4 design, the order would be:

(1)	*a*	*b*	*ab*	*c*	*ac*	*bc*	*abc*
d	*ad*	*bd*	*abd*	*cd*	*acd*	*bcd*	*abcd*

The order for the 2^3 design is multiplied by the code for the fourth factor, and the result is appended to the order for the 2^3 design.

Once the standard order for the 2^3 design is determined, the design is obtained by placing the (–) and (+) signs alternatively in the column for Factor A, placing (–, –) and (+, +) signs alternatively in the column for Factor B, and placing (–, –, –, –) and (+, +, +, +) signs in the column for Factor C. This completes the design of the experiment. Because the columns under Factors A, B, and C are used in creating the design in this fashion, they are called the "design columns." The signs under the interaction columns are obtained as the product of the corresponding individual factor columns. For example, the column of signs for the AB interaction is the product of the

TABLE 3.7

The 2^3 Design

Treatment Combination Code	Design Columns			Calculation Columns						
	Factor A (Blade Angle)	Factor B (Deck Height)	Factor C (Chute Location)	Interaction AB	Interaction AC	Interaction BC	Interaction ABC	Replicate 1	Replicate 2	Average
(1)	–	–	–	+	+	+	–	78	82	80
a	+	–	–	–	–	+	+	73	75	74
b	–	+	–	–	+	–	+	75	79	77
ab	+	+	–	+	–	–	–	69	71	70
c	–	–	+	+	–	–	+	86	82	84
ac	+	–	+	–	+	–	–	82	80	81
bc	–	+	+	–	–	+	–	89	85	87
abc	+	+	+	+	+	+	+	72	76	74
Contrast	–29	–11	25	–11	–3	3	–9			
Effect	–7.25	–2.75	6.25	–2.75	–0.75	0.75	–2.25			

columns of signs under Factors A and B. The column of signs for the ABC interaction is the product of columns AB and C. The interaction columns are grouped together and referred to as "calculation columns" in Table 3.7.

With the data arranged as in Table 3.7, the factor effects and interaction effects can be calculated using the following formula:

$$\text{Effect of a factor} = \frac{\text{Contrast of the factor}}{2^{k-1}}.$$

For this example, when $k = 3$, the effect of a factor $= \dfrac{\text{Contrast of the factor}}{4}$.

This convenient format for creating the design for a 2^k factorial experiment and obtaining the contrasts for factor and interaction effects is called the "table of contrast coefficients" (Box, Hunter, and Hunter 1978).

Continuing with the example, the main and interaction effects are:

$$A = \frac{-80+74-77+70-84+81-87+74}{4} = -7.25$$

$$B = \frac{-80-74+77+70-84-81+87+74}{4} = -2.75$$

$$C = \frac{-80-74-77-70+84+81+87+74}{4} = 6.25$$

$$AB = \frac{+80-74-77+70+84-81-87+74}{4} = -2.75$$

$$AC = \frac{+80+74-77-70-84-81+87+74}{4} = -0.75$$

$$BC = \frac{+80+74-77-70-84-81+87+74}{4} = 0.75$$

$$ABC = \frac{-80+74+77-70+84-81-87+74}{4} = -2.25$$

These calculations can be conveniently made on Table 3.7 and the results recorded in the two bottom rows, as shown in the table. To check if the effects are significant, we first estimate the experimental error by pooling the variances from within each treatment combination using the formula: $S^2 = \dfrac{1}{2^k(n-1)} \sum_{i=1}^{k} \sum_{j=1}^{2} \sum_{k=1}^{n} (y_{ijk} - \bar{y}_{ij.})^2$, which gives:

$$s^2 = \frac{1}{8}\left\{(78-80)^2 + (82-80)^2 + \cdots + (72-74)^2 + (76-74)^2\right\} = 5.75,$$

with a standard error of an effect of:

$$\sqrt{\frac{S^2}{2^{k-2}n}} = \sqrt{\frac{S^2}{4}} = \sqrt{\frac{5.75}{4}} = 1.2.$$

The approximate 95% confidence intervals for the effects are:

$$\begin{aligned}
\text{C.I. for } A &= -7.25 \pm 2(1.2) = [-9.65, -4.85] \\
\text{C.I. for } B &= -2.75 \pm 2(1.2) = [-5.15, -0.35] \\
\text{C.I. for } C &= 6.25 \pm 2(1.2) = [3.85, 8.65] \\
\text{C.I. for } AB &= -2.75 \pm 2(1.2) = [-5.15, -0.35] \\
\text{C.I. for } AC &= -0.75 \pm 2(1.2) = [-3.15, 1.65] \\
\text{C.I. for } BC &= 0.75 \pm 2(1.2) = [-1.65, 3.15] \\
\text{C.I. for } ABC &= -2.25 \pm 2(1.2) = [-4.65, 0.15]
\end{aligned}$$

3.3.2.10 Interpretation of the Results

Looking at the confidence intervals, we see that all three main effects, *A*, *B*, and *C*, are significant because there is no 0 included in any of those intervals. The *AB* interaction is also significant. No other interaction is significant. The responses from the trials are presented on the design cube in Figure 3.10. If not for the significant *AB* interaction, we could have pursued the analysis further to locate the optimal combination of factor levels that gives the best product performance. Because of the significant *AB* interaction we have to limit our search for the best treatment combination to those at which trials were performed.

From the graph in Figure 3.10, we see that the lower-level chute location has uniformly smaller mowing time. This can be seen at the four corners of the side of the cube nearest to the viewer, with (1), *a*, *b*, and *ab* treatment

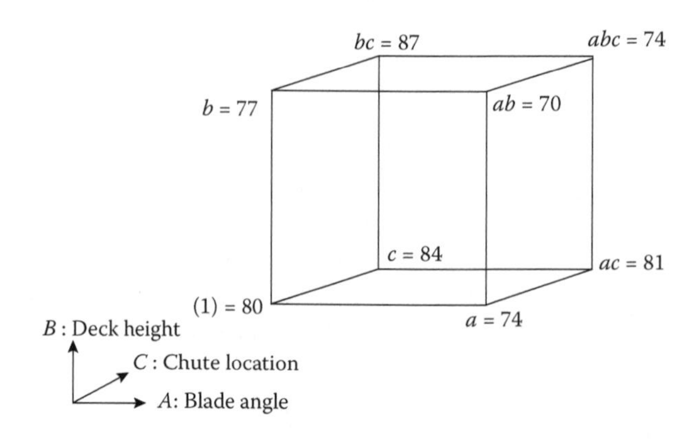

FIGURE 3.10
Results from a 2^3 experiment.

combinations at the corners. On this side of the cube, we can see that an increase in levels of Factors A and B decreases the mowing time, and together they reduce the mowing time even further. All this points to the corner with the Treatment Combination *ab*, where the mowing time is 70 minutes. That is the best choice of parameter values, so the best choice of parameters would be as follows: blade angle high, deck height high, and chute location low.

3.3.2.11 Model Building

One of the important aspects of post-experiment data analysis is model building and exploring the factor-level space for optimal factor-level combinations that produce performances even better than those obtained in the experimental trials. A mathematical model is postulated from the test results to represent the response as a function of the factor levels. This model is then used to predict the response at any factor-level combination—including those at which the experiment was not run. Using these predicted values, the factor-level combination that produces the best response can be located. This line of investigation, searching on the response surface for the combination of treatment levels that produces the best possible response, is called the "response surface methodology." The reader is referred to books on the design of experiments, such as Box et al. (1978) for more details on this method.

A model for the above three-factor experiment would be:

$$y = \beta_0 + \beta_1 x_1 + \beta_2 x_2 + \beta_{12} x_1 x_2 + \beta_3 x_3 + \beta_{13} x_1 x_3 + \beta_{23} x_2 x_3 + \beta_{123} x_1 x_2 x_3 + \varepsilon$$

which proposes that a value of the response y from the experiment is made up of an overall mean β_0 modified by effects from Factors 1, 2, and 3 (for Factors A, B, and C, respectively) and their interactions, and an error term ε. For 2^3 designs, the estimate for β_0 is the overall average of the responses from all eight treatment combinations, and estimates for β_1, β_2,... are equal to half the effects of factors A, B, ... (Montgomery 2001). Thus, for the experiment with the lawnmower, the model would be:

$$\begin{aligned} y = 78.4 - 3.652x_1 - 1.375x_2 + 3.125x_3 - 1.375x_1x_2 \\ - 0.375x_1x_3 + 0.375x_2x_3 - 1.125x_1x_2x_3 \end{aligned}$$

This model can now be used to predict the value of the response corresponding to any combination of values for the three factors. Such predicted values for the response can be plotted to identify contours, peaks, and valleys in the response surface, thus helping to locate the best combination of factor levels that optimizes the response.

The above model can also be used to verify if the assumptions regarding the error in the responses—that they are independent and follow $N(0, \sigma^2)$—are true. For this, the error term for a treatment combination is estimated as the difference between the actual value obtained from the experiment for the treatment combination and the predicted value from the above model. This difference, called the "residual," is calculated for all observed values of the response and is then analyzed for assumptions of normality and independence. The residuals are also plotted against different factor levels and studied to see if they remain constant over the levels of the factors, for all factors.

We have discussed above an important design (the 2^3 factorial) that is a very useful design in industrial experimentation. If only three parameters are considered by the experimenter as being important in determining the level of a quality characteristic, this design can be used to determine the optimum levels at which these parameters must be set in order to obtain the desired level of the quality characteristic. At least two replicates are needed if an estimate of the experimental error is to be obtained in order to determine the significance of the effects. When the number of factors is more than three—say, for example, that it is four—the number of treatment combinations becomes 16. If two replicates are needed, the number of trials needed is 32, which is too large considering the time and other resources needed to complete the trials. Statisticians have therefore created methods to estimate the experimental error even with one replicate. These are discussed in Chapter 5.

When the number of factors becomes even larger—for instance, seven—the number of treatment combinations becomes $2^7 = 128$, which is prohibitively high in terms of practical considerations. Then, a fraction—say, one-half or one-fourth—of the trials needed for a full factorial is run, with the trials being chosen judiciously to give all useful information while sacrificing some information that may not have much practical importance. These are called "fractional factorial designs."

The 2^k designs and the fractional designs derived from them are known as "screening designs" and are used to find out which subset of all factors in a given context are important in terms of their influence on the response. Another experiment is then run with the chosen subset, usually consisting of three or four factors, to identify the best combination of levels of the chosen subset of factors. Some additional discussion of the designs, including the fractional factorials, can be found in Chapter 5, and the reader is referred to references given at the end of this chapter for further details on experimental design.

3.3.2.12 Taguchi Designs

Taguchi designs, which have become very popular among engineers, are of the fractional factorial type, in which the designs are provided as orthogonal arrays, which are columns of signs for factors. Whereas traditional designs handle noise by randomizing the experimental sequence and expecting that their effect is thus neutralized, Taguchi designs seek to deliberately vary the noise factors and study their effect on the response. More specifically, the best factor level combination is chosen as the one at which the effect of the noise factors is minimum. This is done by using what is called the "signal-to-noise" ratio, which roughly equals the ratio of an effect to its standard error. The factor level combination that produces the largest value for the signal-to-noise ratio will be the best treatment combination. As mentioned earlier, fractional factorials save on experimental work but result in the loss of some information. When choosing fractional factorials, one must be careful not to lose important information for the sake of economy of experimental work; otherwise the results may provide inadequate or even false information. At this point, we wish to make it clear that the Taguchi designs are advanced designs that should be used only by those who understand the fundamentals of designed experiments and are aware of the advantages and shortcomings of using these designs.

3.3.3 Tolerance Design

Tolerance design, or tolerancing, is the process of determining the allowable variations in a product characteristic around the selected targets so that the product can be produced at a competitive cost while meeting the customer's needs for limited variability.

The provision of tolerance recognizes that the variability in a part or product characteristic is unavoidable, but should be kept within limits. For a producer, a narrower or tighter tolerance means an increased production cost. For the customer, who could be the end-user or the operator in the next assembly line, tolerance design affects how the product looks and performs, or how easily it can be assembled in the next higher assembly. It may also determine the length of life or the reliability of the final product. All these have economic value for the customer. Therefore, tolerancing is an economic issue for both the customer and the producer, and a gain for one may result in a loss for the other. Thus, it has to be decided by a suitable trade-off.

The tolerance provided for a characteristic depends on the manufacturing process by which the characteristic is created. Therefore, tolerancing for characteristics should be conducted after the manufacturing processes have been chosen.

3.3.3.1 Traditional Approaches

Traditionally, tolerancing has been done based on the experience of designers regarding what has worked for them satisfactorily. Experienced designers have documented the tolerances that have worked well for different manufacturing processes. These selections have also been considerably influenced by what tolerance a particular process is capable of achieving. For example, standard tolerance tables, such as those given in the *Standard Handbook of Machine Design* (Shingley and Mischke 1986), provide tolerances for various processes. The tolerance for drilled holes, for example, is given as:

Drill size (in.)	Tolerance (in.)
0.2660–0.4219	+0.007, –0.002
0.4375–0.6094	+0.008, –0.002, and so on

The tolerance for bored holes is given as:

Diameter (in.)	Tolerance (in.)
0.2510–0.5000	+0.001, –0.001
0.5010–0.7500	+0.001, –0.001, and so on

These tolerances will be adequate and appropriate to use for a new product if the production machinery used for producing the characteristic has the same capability as those assumed in obtaining the standard tolerances. Here, "capability" refers to the variability around a target with which a machine or a process is capable of producing the characteristic. This capability is measured using the standard deviation (σ) of the characteristic produced by the machine or process. In general, these tolerances are calculated as $T \pm 3\sigma$, where T is the target and the 3σ spread is chosen on the assumption that the characteristic follows the normal distribution.

The subject of process or machine capability is discussed in more detail in Chapter 4, in which the measures for quantifying process capability, such as C_p and C_{pk}, are defined. These indices illustrate how well a machine or process is capable of holding the process variability within a required set of limits. For now, we only note that the tolerance selected for a characteristic should be related to the capability of the machine or process by which the characteristic is produced. For design engineers to assign rational, feasible tolerances, they should have information on the capabilities of the machines and processes that are used to create the characteristics. When the capability of a process to produce a characteristic is known through the standard deviation—say, σ_0—of the characteristic, then a reasonable tolerance for the characteristic would be $T \pm 3\sigma_0$, where T is the nominal or target.

When design engineers choose tolerances without knowing if the available processes will be able to hold those tolerances, that is when the production people are frustrated and the tolerances lose their meaning and are likely to be ignored. If tighter tolerances than what the current processes are capable of are needed, process improvements must be made to reduce the output variability before tighter tolerances are demanded of them.

3.3.3.2 Tolerancing According to Dr. Taguchi

Traditional tolerancing, which provides an upper and lower specification limit (USL and LSL, respectively) for a characteristic around a target, is based on the premise that the fitness of the product is good as long as that

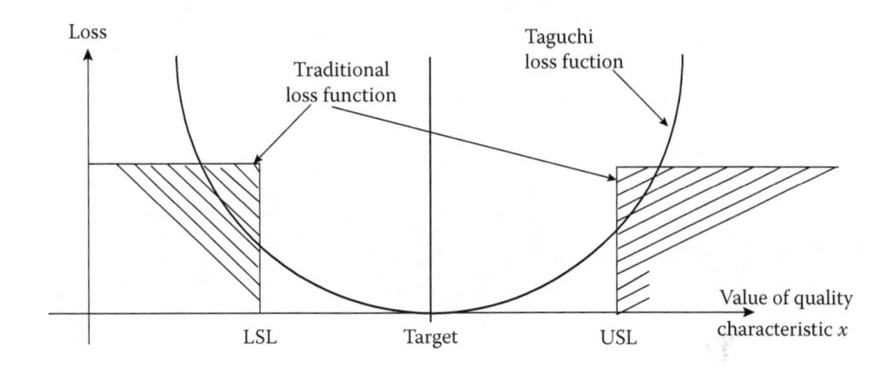

FIGURE 3.11
Traditional and Taguchi's views of loss from a characteristic not on target.

characteristic is at the target or anywhere within the limits specified, and that the product suddenly becomes unfit when the characteristic exceeds these limits. In other words, the losses arising from the characteristic are zero when it is on target or within the limits and become unacceptable as soon as the limits are crossed. This is shown graphically in Figure 3.11 as the "traditional loss function."

According to Dr. Taguchi, the loss is zero only when the characteristic is on target, and it increases parabolically as the value of the characteristic moves away from the target, as indicated by the "Taguchi loss function" in Figure 3.11. Doctor Taguchi provided models for computing the losses, such as $L(x) = k(x - T)^2$, where x is the value of the characteristic, T is its target, and k is a constant for the characteristic. The value for k can be determined by collecting data on loss at a few values of the characteristic. This form of quantifying the loss also enables a determination of the optimal tolerance limits to obtain the correct trade-off between the cost of producing a tolerance and the gains resulting from it. For more on this, refer to DeVor, Chang, and Sutherland (1992).

3.3.3.3 Assembly Tolerances

Many situations arise in industrial settings, where assembly tolerances have to be calculated from the component tolerances. For example, the track for a bulldozer is made up of 120 links. The tolerances of the links are known, and we may want to know the tolerance that the assembled track will meet. Another example would be finding the tolerance on the clearance between a bore and a shaft given the individual tolerances on the bore and shaft diameters. In situations like these, when we need the tolerance on the total (or difference) of certain characteristics, we use the root sum of squares (RSS) formula.

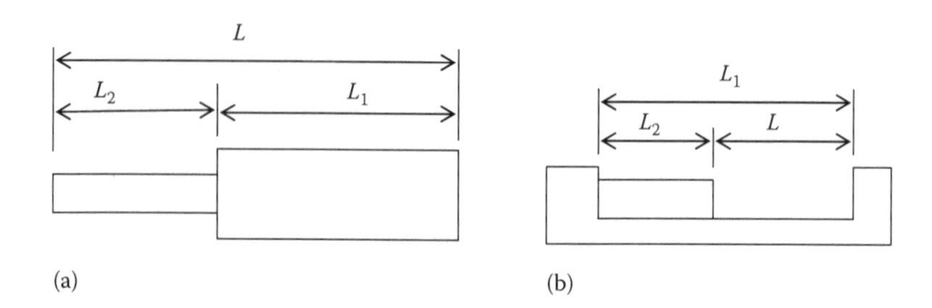

FIGURE 3.12a, b
Assemblies made of components.

3.3.3.4 The RSS Formula

Suppose L_1 and L_2 are the lengths of two components (see Figure 3.12a) and $L = L_1 + L_2$. Let t_1 and t_2 be the tolerances on L_1 and L_2, respectively. Then the tolerance on L is given by $t_L = \sqrt{t_1^2 + t_2^2}$. We need the assumption that L_1 and L_2 are independent. Similarly, in the Figure 3.12b $L = L_1 - L_2$, and if t_1 and t_2 are the tolerances on L_1 and L_2 respectively, then, assuming independence between L_1 and L_2, the tolerance on L is also given by $t_L = \sqrt{t_1^2 + t_2^2}$. Notice that the "+" sign inside the radical of both formulas irrespective of whether L is the sum or the difference of L_1 and L_2.

This is based on the following results for linear combinations of normally distributed random variables. Suppose that $X_1 \sim N(\mu_1, \sigma_1^2)$ and $X_2 \sim N(\mu_2, \sigma_2^2)$, and that X_1 and X_2 are independent. It can then be shown (Hines and Montgomery 1990) that:

$$(X_1 + X_2) \sim N(\mu_1 + \mu_2, \sigma_1^2 + \sigma_2^2),$$
$$(X_1 - X_2) \sim N(\mu_1 - \mu_2, \sigma_1^2 + \sigma_2^2).$$

Note that the variances add up for both the sum and the difference of the random variables. Hence, the standard deviation of the sum (or the difference) of the random variables is:

$$\sigma_{X_1 \pm X_2} = \sqrt{\sigma_1^2 + \sigma_2^2}$$

Tolerance is usually taken as $\pm 3\sigma$, or some multiple of the standard deviation, for both the components and the assembly, thus:

$$t_{X_1 \pm X_2} = \sqrt{t_1^2 + t_2^2}$$

More generally, if $X_1, X_2, \ldots, X_n$ are n independent random variables, each of which is normally distributed with means $\mu_1, \mu_2, \ldots, \mu_n$, and variances $\sigma_1^2, \sigma_2^2, \ldots, \sigma_n^2$, then $Y = \pm X_1 \pm X_2 \pm \cdots \pm X_n$ has normal distribution with mean $\mu_Y = \pm\mu_1 \pm \mu_2 \pm \cdots \pm \mu_n$ and variance $\sigma_Y^2 = \sigma_1^2 + \sigma_2^2 + \cdots + \sigma_n^2$. Therefore the more general RSS formula can be stated as follows:

Suppose that $L_1, L_2, \ldots, L_n$ are lengths of n independent components, and their assembly length is given by $L = \pm L_1 \pm L_2 \pm \cdots \pm L_n$. If $t_1, t_2, \ldots, t_n$ are the tolerances on the individual lengths, then the tolerance on the total length t_L is given by:

$$t_L = \sqrt{t_1^2 + t_2^2 + \cdots + t_n^2}.$$

This rule can also be used for other measurements such as weights, heights, and thicknesses.

Example 3.9

A track for a tractor is made up of 120 links. Each link is produced with a nominal length of 12 in. and a tolerance of $t = \pm 0.125$ in.

a. If the links meet their tolerance, what tolerance will the tracks meet? Assume that the links are independent and normal.
b. If the specification for the total length of track is 1440 ± 1 in., what proportion of the track lengths will be out of specification? Assume that the link lengths are independent and normal, and meet their tolerance. The tolerance is chosen as three-times the standard deviation.
c. If the tolerance on the total track length should not be more than ± 1 in., what should be the maximum tolerance on the links?

Solution

a. If t is the tolerance on the total track, then $t = \sqrt{120 \times 0.125^2} = 1.37$ in. Hence, the tracks will meet the specification 1440 ± 1.37 in.
b. The current tolerance on links is $t = \pm 0.125$. Therefore, the current standard deviation of links is $\sigma_L = 0.125/3 = 0.0417$. If the link lengths are normally distributed, then the track length, which is the sum of the link lengths, is also normally distributed, with $\mu_T = 1440$ in. and $\sigma_T = \sqrt{120 \times 0.0417^2} = 0.4568$. If the specification for the track is [1439,1441], then the proportion outside specification

$$= 1 - P\left(\frac{1439 - 1440}{.4568} \le Z \le \frac{1441 - 1440}{.4568}\right)$$
$$= 1 - P(-2.19 \le Z \le 2.19) = 1 - \Phi(2.19) + \Phi(-2.19)$$
$$= 1 - 0.9857 + 0.0143 = 0.0286$$

Therefore, about 2.9% of the track lengths are outside specification.

c. If t is the tolerance on the links, then:

$$1 = \sqrt{120 \times t^2}$$

$$t = 1/\sqrt{120} = 0.091 \text{ in.}$$

Therefore, the specification for links should be 12 ± 0.091 in.

Example 3.10

An assembly (as shown in Figure 3.12c) calls for a tolerance on the "gap." Assume that the parts A, B, C, and D are manufactured independently and have the nominal and tolerance in inches, as shown.

a. What is the expected tolerance on the gap?
b. If the customer requires a tolerance of ±0.02 on the gap, what proportion of the assemblies will be outside specification? Assume that the lengths of the parts A, B, C, and D are independent and normally distributed. Also assume that the tolerances are based on the 3σ rule.

Solution

a. The nominal value of the gap is:

$$L_{gap} = L_D - [L_A + L_B + L_C] = 4.5 - [1.0 + 1.5 + 1.75] = 0.25.$$

The tolerance for the gap is:

$$t_{gap} = \sqrt{t_A^2 + t_B^2 + t_C^2 + t_D^2} = \sqrt{0.01^2 \times 3 + 0.02^2} = 0.026.$$

The gap will be able to meet a specification: 0.25 ± 0.026.

b. $\sigma_A = \sigma_B = \sigma_C = 0.01/3 = 0.0033$
$\sigma_D = 0.02/3 = 0.0066$

Therefore, $\sigma_{gap} = \sqrt{0.0033^2 \times 3 + 0.0066^2} = 0.0087.$

The specification limits for the gap are given as 0.25 ± 0.02. The proportion out of specification is: $1 - P[0.23 < \text{Gap} < 0.27]$, which can be found as follows. Since the average of the gap is 0.25 and $\sigma_{gap} = 0.0087$,

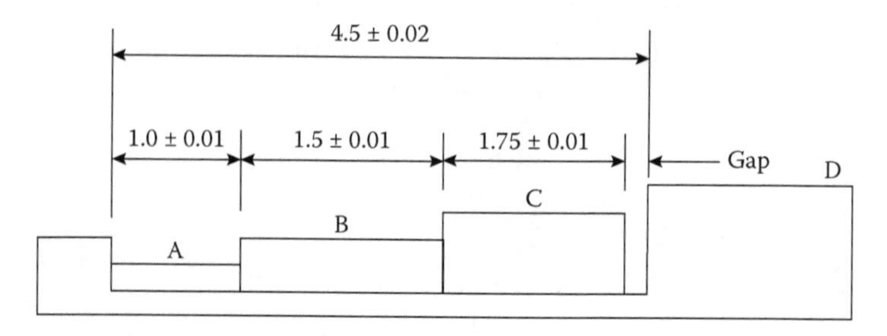

FIGURE 3.12c
Example of setting assembly tolerance.

$$\text{Proportion out of spec} = 1 - P\left(\frac{-0.02}{0.0087} \le Z \le \frac{0.02}{0.0087}\right)$$
$$= 1 - [\Phi(2.3) - \Phi(-2.3)]$$
$$= 1 - 0.9786 = 0.0214.$$

Therefore, 2.14% of the assemblies will have gaps that are either too small or too large.

3.3.3.5 **Natural Tolerance Limits**

"Specification limits" or "tolerance limits" usually refer to the limits of variability that a customer has imposed based on where and how the product is to be used. They might have been given by the customer or chosen by the designer based on his or her assessment of the customer's requirements.

We may sometimes want to know what tolerance limits a process is capable of meeting in its current condition. Such tolerance limits, which reflect the current capability of a process, are called the "natural tolerance limits" (NTLs) of the process. The NTLs are chosen as 3σ tolerance as $\hat{\mu} \pm 3\hat{\sigma}$ where $\hat{\mu}$ and $\hat{\sigma}$ are the estimates for the mean and standard deviation of the process, respectively. If data are available from a random sample ($n \ge 50$), then the NTLs are chosen from the estimates of the process mean and process standard deviation as $(\bar{X} \pm 3S)$. If data are available from $\bar{X}$ & R-control charts, then NTLs are chosen as $(\bar{\bar{X}} \pm 3\bar{R}/d_2)$, as explained in Chapter 4.

The NTLs are also useful when studying the capabilities of processes with respect to customer specifications. If the NTLs fall within the specifications given by the customer, then the process has good capability; otherwise, the capability of the process should be improved.

The NTLs are most useful when setting tolerances for process variables and in-process product variables. The latter refer to the characteristics of the material or subassemblies of a product at the intermediate stages of production, which the customer will not see. For example, the thickness of rubber sheets that are used in layers for building a tire is an important variable in tire production. This thickness must be uniform over all the sheets used in all the layers of the tire for the process to produce quality tires. However, the customer, as the user of the tires, will have no interest in knowing the target and tolerance for the thickness of the rubber sheets. The process engineer will choose the tolerance for the sheets as the NTLs of the sheet-making process.

3.3.4 Failure Mode and Effects Analysis

Failure mode and effects analysis (FMEA) is a technique that is used to investigate all possible weaknesses in a product design and prioritize the weaknesses in terms of their potential to cause product failure. This prioritization enables taking corrective action on the most critical weaknesses in order to reduce the overall likelihood of failure of the product. When this technique is used at the stage of evaluating the design of a product, it is called the "design FMEA." This same technique can be used in the design of a process that produces products. Then, the method will help in identifying and prioritizing the modes of failures in the process that might cause poor quality in the products. This is called the "process FMEA."

The FMEA is typically done by a multifunctional quality planning team. The design engineer who is responsible for the product design will have the major responsibility for gathering the team and conducting the design FMEA. The FMEA study should start with a block diagram of the product in question that shows the functional relationship of the parts and assembly. The FMEA is illustrated using an example, where it is used for evaluating a product design.

Figure 3.13 shows a design FMEA for the front doors of an automobile. The analysis relates to the design features to prevent failure of the door panel from rust and corrosion. Referring to the identifying letters shown in the figure, Items A through G relate to information on the product name, designer name, date, reference number, and so on, which are meant for documenting traceability. Items a_1 and a_2 give a full description of the part and its functional requirement. Item b is a potential failure mode, one of several modes to be considered.

Item c is the description of the possible effects of the failure mode in using the product. Item d is the "severity" column, where the severity of the effect that the failure mode will cause to the next component (subsystem, or the end-user) is recorded. Severity is estimated on a scale of 1 to 10, with 10 being the greatest severity. If the failure of the part because of the mode in question will result in unsafe vehicle operation—thus causing injury to the customer—then the severity rating will be 10. If the failure will have no effect on the proper functioning of the assembly and is not likely to cause any inconvenience to the customer, then the rating will be 1. In this example, a severity rating of 5 had been given, possibly because the rust may result in the poor functioning of locks and other hardware in the door. Item e is for a special classification by automotive manufacturers to identify those characteristics that may require special process control in subsequent processing of the part. Such characteristics may relate to emission control, safety, or other government ordinances.

Item f lists the possible causes that are responsible for the failure mode in question. Item g is the "occurrence" column, where the frequency of occurrence of the failure due to the different causes is rated on a scale of 1 to 10. The occurrence of a cause relates to the number of failures that may occur because of the particular cause during the design life of the equipment (e.g., 100,000 miles for a car). One possible scale: rate the occurrence as 10 if failure occurs in one in every two vehicles from this cause and rate the occurrence as 1 if the occurrence is less than one in one million vehicles. The rating for

POTENTIAL
FAILURE MODE AND EFFECTS ANALYSIS
(DESIGN FMEA)

—— System
—— Subsystem
—— Component B
Model years(s)/Programs(s) D
Core team G

Design Responsibility C
Key Date E

FMEA Number A
Page ______ of ______
Prepared by: H
FMEA Date (Orig.) F

Item / Function	Require ment	Potential failure mode	Potential Effect(s) of failure	Severity	Classification	Potential Cause(s) of failure	Current design				RPN	Recommended Action	Responsibility & Target completion date	Action resilts				
							Controls prevention	Occurence	Controls Detection	Detection				Action taken Completion date	Severity	Occurrence	Detection	RPN
Front Door LH HOHX-0000-A	Maintain integrity of inner door panel	Integrity breach allows envlron access of inner door panel	Corroded interior lower door panels Deterloraled life of door leading to: .Unsallsfactory appearance due to rust through paint over time. .Impaired function of interior door hardware	6		Upper edge of protective.Wax application specified for inner door panels is too low	Design requirements (#31268) and best practice (BP 3455)	3	Vehicle durability test, T-118 (7)	7	105	Laboratory accelerated corrosion test	A.Tale Body Engineer OX 09 03	Based on test results (test no. 148-1) upper edge spec raised 126 OX 09 30	7	2	3	30
						Insufficient wax thickness specified	Design requirements (#31268) and best practice (BP 3455)	3	Vehicle durability test, T-118 (7)	7	105	Laboratory accelerated corrosion test	A.Tale Body Engineer OX 09 03	Test results (Test No. 1481) show specified thickness is adequate. OX 09 30	5	2	3	30
												Design of Experiments (DOE) on Wax thickness	J. Smythe body engineer OX 10 18	DOE shows 25% variation in specified thickness is acceptable OX 10 25	6	2	3	30
						Inappropriate wax formulation specified	Industry standard MS-1893	2	Physical and chemical lab test - report No. 1265 (5) vehicle durability test. T-118 (7)	5	50	None						
						Corner design prevents spray equip from reaching all areas		5	Design aid with non-functioning spray head (8) vehicle durability test. T-118 (7)	7	175	Team evaluation using production spray equipment and specified wax	T.Edwards body engineer and assy ops Ox 11 15	Based on test: 3 addillonalvent holes provided in affected areas (error-proofed) OX 12 15	5	1	1	5
	SAMPLE					Insufficient room between panels for spray head access		4	Drawing avaluation of spray head access (4) vehicle durability test T-118 (7)	4	80	Team evaluation using design aid buck and spray head	Bodt Engineer and assy ops OX 11 15	Evaluation showed adequate access OX 12 15	5	2	4	40
a1	a2	b	c	d	e	f	h	g	h	i	j	k	l	m	----n----			

FIGURE 3.13
Example of a design FMEA study. (Reprinted from Chrysler LLC, Ford Motor Company, and General Motors Corporation. 2008b. *Potential Failure Mode and Effects Analysis (FMEA)—Reference Manual*. 4th edition. Southfield, MI: A.I.A.G. With permission.)

occurrence can be educated estimates based on data obtained from previous model experience or warranty information. The rating standard must have the agreement among the planning team and should be consistent.

Item h shows the current control activities pertaining to the cause of the failure mode that are already incorporated in the design. The information on current activities is taken into account while rating the failure mode in the next column, Item i on "detectability." If the current control activity is certain to detect the failure due to a particular mode, then the rating will be 1. If the current activities will not—or cannot—detect the failure, then the detectability rating will be 10. Item j, the "risk priority number" (RPN), is the product of ratings for severity (S), occurrence (O), and detectability (D).

$$\text{RPN} = S \times O \times D$$

The RPN will range from 1 to 1000 and is used to prioritize the various modes of failure and—within each mode—the various causes of failure. The larger the RPN is for a particular cause or mode, then the greater the risk that failure will occur due to that cause or mode. Organizations set threshold limits for RPN to take action when the RPN exceeds the limit. Whenever an RPN is encountered that is greater than the threshold number, suitable action is taken to reduce the RPN below the threshold value. Items k through m show what action is planned and who has the responsibility to see that action is completed. Teams also set time schedules to have the actions completed by. Item n is the new RPN for the failure mode and respective causes after the recommended actions have been completed.

Example 3.11 shows the analysis for only one failure mode along with the causes contributing to that failure mode. For any product or a component, however, there will be a number of failure modes, and each one must be analyzed as described above and the high RPN modes must be addressed with a suitable response.

The FMEA technique helps in identifying possible failure modes, and possible causes of those failure modes, and enables prioritizing them on a rational basis to select those that need to be addressed. It provides a format for a design engineer to think together with a team, proactively anticipate possible failures, and implement improvements to the design. Thus, the FMEA method:

1. Facilitates concurrent design of products, because it brings several functions together
2. Contributes to a better quality product
3. Keeps the customer in focus during the design activities
4. Reduces future redesigns
5. Provides documentation of the improvement activities

3.3.5 Concurrent Engineering

Concurrent engineering (CE) is the design approach that calls for the design activity to include several related functions, such as manufacturing engineering, production, quality, marketing, customer service, customers, and suppliers, in order to obtain their inputs early in the design process and incorporate them into the product design. This is in contrast to older models followed in design work, termed "sequential engineering" (SE), in which the designers worked in isolation, prepared designs and passed on the blueprints to manufacturing engineers "over the wall," who in turn would make the process designs sitting in their cells and then pass them on over the wall to the production people. If any drawbacks were discovered by the production people regarding manufacturability or any other aspect of the design, the documents would travel the same course backward, over the walls, to the designers, who would make the appropriate corrections and then send the designs back to production over the same course. Such back-and-forth transmission of documents caused delays, made the changes costly, extended the time for completion of the designs, and delayed the production and delivery of the product to the market.

Under the CE model, representatives from all the functions related to design work are assembled in a team, the members of which review the designs from their individual perspectives as the design is being prepared, and provide timely feedback. The designers would incorporate necessary changes based on the feedback before finalizing the drawings. The manufacturing engineers, for example, would start on process designs, using the preliminary information provided to them; and while the product design is still under preparation, they would suggest changes to the product design if they encountered any difficulties with the original design. Thus, the product design and process design can proceed concurrently, with information being shared mutually. This type of design activity—which contributes to the simultaneous progression of product design, process design, quality planning, and other related activities—is called CE. This approach has produced some dramatic results (Clausing 1994; Syan and Menon 1994; Skalak 2002), including:

1. Improved product quality
2. Reduced product cost
3. Reduced time to market
4. Reduction in number of redesigns
5. Reduced cost of design
6. Improved customer satisfaction
7. Increased profitability
8. Improved team spirit among employees

Of all the benefits, the most cited is the reduced time to market, which is the time between the definition of the product to the delivery of the first unit. Studies (Skalak 2002) have shown that the manufacturers of products, from telephones to airplanes, have recorded 30% to 80% reductions in time to market through the use of CE principles. The oft-quoted example of success in the use of CE concepts is the story of designing the Taurus/Sable models used by Ford between 1980 and 1985. In using the new approach, the Ford design team not only achieved a reduction in development time, fewer engineering changes, a reduction in time to market, and an improvement in quality of product, they designed the car that was declared one of the world's 10 best cars for three years in a row, between 1986 and 1988 (Ziemke and Spann 1993).

CE is also a very important component of the Toyota Production System, also called Lean Production Method. How Toyota improved the quality of their cars, reduced the time to market, and reduced the total hours for design using CE methods in the 1980s is legend (Womack, Jones, and Roos 2007).

The CE methodology makes use of several tools, such as QFD, design and process FMEA, designed experiments, reliability analysis, design for manufacturability and assembly, and design reviews, along with computer-aided tools for design and integration. An extensive computer-based communication network for exchanging data, text, and drawings among the members of the design team is an important component of CE work.

Many of the tools mentioned above are covered in this chapter as well as in other chapters of this book. Design for manufacturability and design reviews are explained below.

3.3.5.1 Design for Manufacturability/Assembly

"[Design for manufacturability/assembly] represents a new awareness of the importance of design as the first manufacturing step" (Syan and Swift 1994). This study is meant to make the design engineers conscious of the limitations of the manufacturing and assembly processes, and to make the design manufacturable. The availability of technology, capability of processes to hold tolerances, capability of measuring instruments with acceptable measurement variability, and restrictions on material handling are some of the issues that design engineers should consider in making their designs. Guidelines have been proposed for achieving manufacture-friendly and assembly-friendly designs that will result in high-quality, low-cost products. These guidelines include suggestions such as:

1. Design for a minimum number of parts
2. Develop a modular design
3. Design parts for multiple uses
4. Design parts for ease of fabrication
5. Design for top-down assembly
6. Maximize part symmetry

7. Use simple designs with minimal complexity
8. Provide lead-in chamfers
9. Provide for simple handling and transportation
10. Avoid parts that will tangle

Systematic procedures, some of them computerized, can be used by the design team to evaluate the manufacturability and ease of assembly of designs. Feedback to team members from personnel dealing with manufacturing, quality, and materials would help greatly in making designs that minimize the difficulties and cost of manufacturing or assembly. Formal design reviews that are made at suitable times during the design process further help in accomplishing this goal.

3.3.5.2 Design Reviews

Design reviews are regularly scheduled meetings of a review team organized by design engineers. These meetings include representatives from manufacturing, quality, materials, suppliers, and customers, to review designs and monitor progress. The reviews are made to verify:

1. Functional requirements to meet customer needs
2. Quality and reliability goals
3. Design FMEA study results
4. Design for manufacturability and ease of assembly
5. Prototype test results
6. Final assembly tolerances

Design reviews are typically conducted in three stages: one at the preliminary stage before detailed designs, one after prototype testing, and one before launching to full production. Design reviews are not to be considered as second-guessing the designer's work. The reviews are meant to channel all contributions from people related to the design for making the best possible design, one which will meet the needs of the customer and enable the most efficient production of the product.

3.4 Process Design

Manufacturing engineers design the processes to convert the raw material into the finished product. As mentioned earlier, they begin their work even when the product design is still progressing, and they complete it after the product design is completed. Among the outputs of the process design are several quality-related outcomes:

1. Process flow chart
2. Process parameter selection
3. Floor plant layout
4. Process FMEA
5. Process control plan
6. Process instructions
7. Packaging standards
8. Preliminary process capability studies
9. Product and process validation

3.4.1 The Process Flow Chart

The process flow chart (PFC) is a schematic representation of the operations, or activities, starting from raw material and leading to the production of the final product. The PFC is an important planning document used by manufacturing engineers in selecting appropriate machinery, methods, and measurement tools. It provides a perspective on the flow of activities in a production process and, thus, facilitates planning for quality-related activities at appropriate junctures in the course of production. Therefore, it forms the central tool in making the quality plan for a product.

The PFC is an important communication tool among the process designers, including the quality planners, and a convention exists for the use of symbols to denote the various activities involved in a process:

An ellipse for the start of a process or subprocess

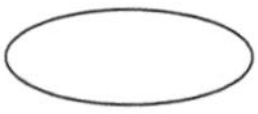

A circle for an operation

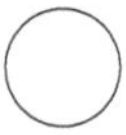

A diamond for a decision point

An arrow for transportation

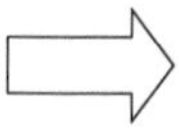

An inverted triangle for storage

A square for inspection

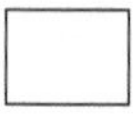

And a large "D" for delay

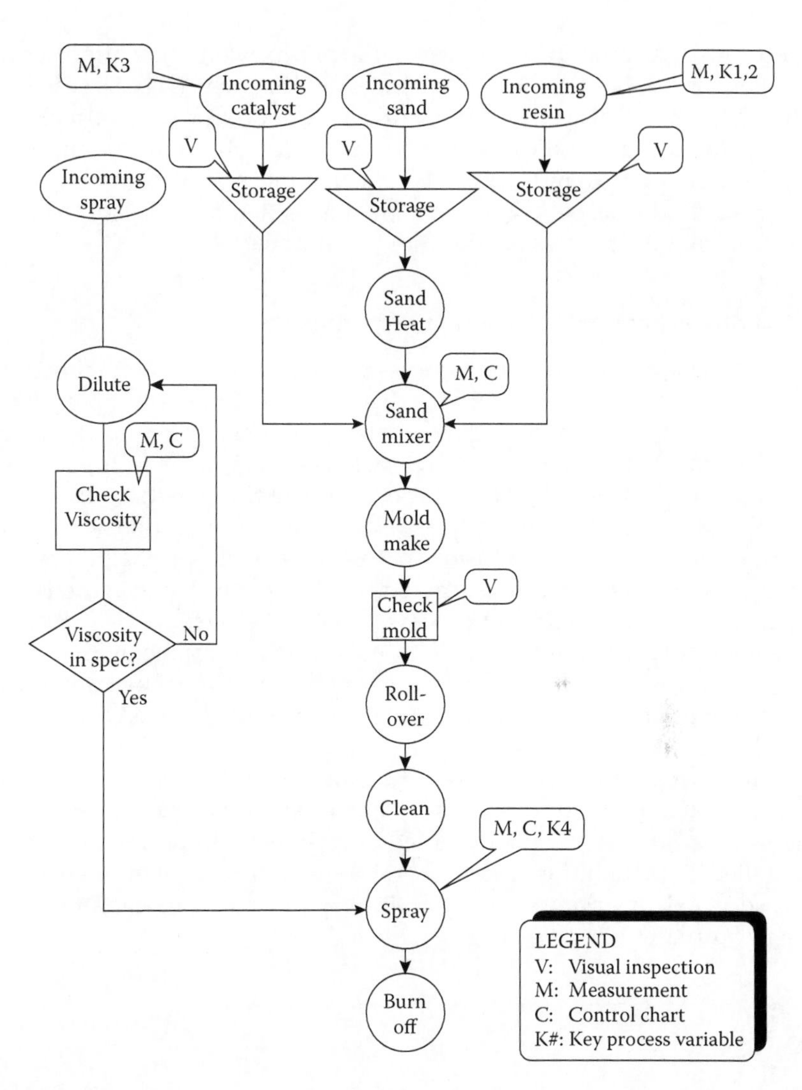

FIGURE 3.14
A process flow chart for making a foundry mold.

The quality planning team describes the checks and control activities to be performed during production at suitable points in the flow process chart. The square inspection boxes mark the major control activities to be performed. It could be a quality characteristic of the product to be checked at certain critical points during production, or it could be a critical process variable that influences an important product characteristic that must be controlled. Determining when to measure a product characteristic, or a process variable, is an important decision in planning for quality. One rule is to verify the condition of the product before any critical or expensive operation is to be performed. Another rule is to verify the product condition after a critical characteristic has been created. The process FMEA, which is discussed later in this chapter, will also disclose the critical product or process variables that must be checked, and indicate where in the process they must be checked.

Figure 3.14 is an example of a PFC for making molds for castings from sand, resin, and catalyst. Notice how the inspection and control to be performed during the process are indicated on the chart.

3.4.2 Process Parameter Selection: Experiments

Selecting the levels at which the process variables are to be maintained during the production of a product is perhaps the most important part of process design in terms of assuring the quality of the final product. In practice, this selection is often made based on previous experience with the same or a similar process, which might have been arrived at through formal, or—most probably—informal, seat-of-the-pants experimentation carried out by individuals sometime during the history of the process. The production of unacceptable goods causing waste in production shops or rejection and return by customers is often the result of process variables selected in an *ad hoc* manner, without experimentation, or with unplanned experiments that do not generate the best combination of process variables. Experimentation using designed experiments is the best course to determine the best combination of levels of process variables.

The following case study shows, in some detail, how an experiment is conducted on a production process, and gives an idea of the logistical problems involved in executing a designed experiment and of how one should be handled. Although this case study relates to an experiment performed to improve an existing process, many of the steps are also applicable to designing a new process.

Case Study 3.1

This case study relates to choosing process variables for making molds in a foundry that will result in castings of good quality. The castings in question were experiencing a defect called "blisters," most of which would not surface until the castings were machined in the customer's machine shop. The blisters had to be eliminated to keep the customer from taking the business to a competitor.

The process improvement team, consisting of tooling engineers (process designers), production, quality, and customer-service people, determined in a brainstorming session that the following three process variables would be the factors in the experiment to improve the existing process. The team also determined the levels for the factors as shown:

Factors	Levels	
	Low	High
A: Drying oven temp.	350°F	400°F
B: Sand for port-cores	Silica	Lake
C: Flow-off vents	Yes	No

A 2^3 factorial design was chosen with eight treatment combinations having a sample size $n = 20$ at each of the treatment combinations. Note that all the factors in this experiment are qualitative factors. The oven temperature is measurable; however, because of other restrictions, the two levels indicated were the only levels possible. The response in the experiment is the proportion of defective castings found in the samples, which is an attribute response. By taking a large sample size $n = 20$ ($n > 20$ would be too expensive in this process), the team expected that the proportion of defectives found in the sample would be representative of the proportion of defectives in the process.

In this case, complete randomization of the units was not possible because of the many restrictions resulting from the automatic production line. For example, the oven could not be turned up and down at will. When a temperature change is made in the oven, it may take about 30 minutes for the oven to stabilize at the new temperature. Therefore, it was desirable to keep changes in the oven condition to a minimum. Also, the molds could not be taken off the conveyor in order to be given a particular treatment. The molds should travel on the conveyor in the sequence in which they were produced and then transported in the same sequence to the oven and to the mold line. However, as will become apparent in the plan shown below, the experimental units assigned to any treatment combination were assembled at different, random times, and this provided a sort of natural, random assignment of units to the treatment combinations. In addition, this also resulted in a random sequencing of the treatment combinations.

In this process, the molds were made and the proper port-cores assembled into the mold (one of the factors) in a cell at one location. The molds were then transported to another location, where they were dipped in a wash and dried in an oven—the air temperature of which was controlled at the design level (another factor). The dried molds were then taken to the mold conveyor line, where the venting was done (the third factor) before the iron was poured into the molds. The logistics required adequate manpower to tag the experimental molds and follow them through the various processes, making sure that they received the appropriate treatments at the proper workstations. The plan shown in Figure 3.15 and Table 3.8 helped in planning the logistics of the experiment. The plan

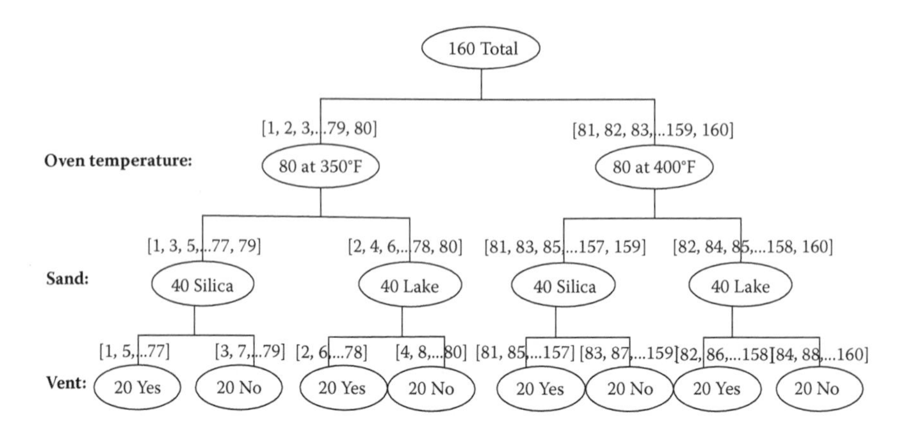

FIGURE 3.15
Assigning experimental units to treatment combinations.

makes use of the serial numbers assigned to the molds, which are written outside and etched inside the molds in order to provide traceability between a mold and its casting. The mold numbers are shown in square parentheses in Figure 3.15. In this plan, we should note that each level of a factor is combined with each level of every other factor.

A total of 160 molds were needed for the experiment, and they were divided into groups of 20 and assigned to the different treatment combinations, as shown in Figure 3.15. Copies of Table 3.8, which was prepared out of Figure 3.15, were handed to the workstation operators so that they could provide the correct treatment to the molds when the molds arrived at their workstations.

TABLE 3.8

Treatments Given to the Experimental Units

Mold	Oven Temperature °F	Port-core	Vent
1	350	Silica	Yes
2	350	Lake	Yes
—	—	—	—
20	350	Lake	No
21	350	Silica	Yes
—	—	—	—
40	350	Lake	No
41	350	Silica	Yes
—	—	—	—
80	350	Lake	No
81	400	Silica	Yes
—	—	—	—
—	—	—	—
159	400	Silica	No
160	400	Lake	No

Copies of Table 3.8 were given to operators in a meeting that was organized to explain the details of the experiment and the importance of making sure the right treatment is given to the molds when they arrived at their workstation. When all the molds were prepared according to the plan, iron was poured into them with the utmost consistency regarding iron temperature and other pouring practices. (When the process output is tested against certain factors, it is absolutely essential that other factors that might affect the output are controlled at constant levels as far as is practical.) When the iron cooled and the castings were cleaned the next day, the castings were cut at appropriate places and were evaluated for blisters, and the numbers of defective castings in each of the treatment combinations were tallied. The proportions of defectives at each treatment combination were computed and the results presented in a design cube, as shown in Figure 3.16.

Figure 3.16 shows that a treatment combination exists that produced 0% defectives, whereas all others at which trials were conducted produced some defectives. The "current practice," which is represented by the treatment combination where all factors are at the low level, produced a 39% defect rate. Although the quality engineers were hesitant to conclude that the treatment combination showing 0% defectives had indeed produced no defectives, because of the possible variability in the outcome, the production people wanted to run the process immediately at the newly discovered treatment combination having the potential to produce "perfect" quality. A trial run was ordered for one full day at the "best" treatment combination. When the trial run confirmed the experimental results, the process designer changed the process specs for regular production. The blister problem disappeared.

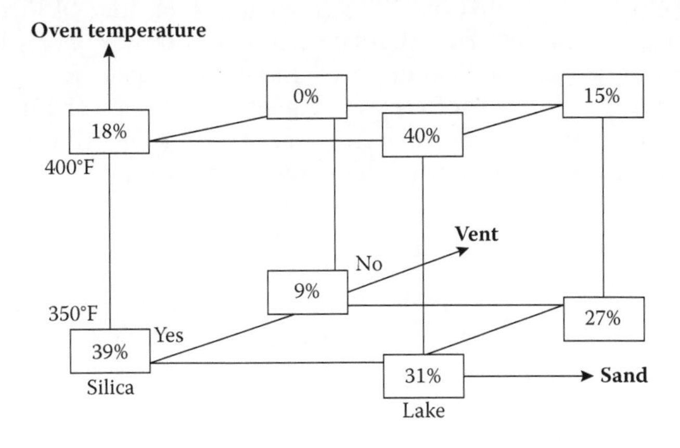

FIGURE 3.16
Results from an experiment to choose process variables.

In the case study above, no further analysis was made after finding the treatment combination that provided the best possible result that could be expected. If such clear-cut results are not forthcoming and the results must be analyzed for the effect of individual factors or their interactions, then calculation of the effects can be made, as was done in the example of the lawnmower design earlier on in this chapter. Verifying the significance of an effect, however, has to be done keeping in mind that the response here is an attribute, not a measurement, and that there are no replicates. The best approach would be to assume that the proportion of defectives among the treatment combinations are normally distributed on the assumption that the sample sizes are large, and then make a normal probability plot of the effects, as discussed in Chapter 2. The significant effects will show as outliers and be identified as such. Adjustments to the process design can then be made accordingly to obtain the desired results.

The purpose of presenting the above case study is, firstly, to emphasize the need for designing processes using statistically designed experiments; and secondly, to note that the responsibility of a quality engineer in running designed experiments does not end with drawing up the design on paper and then handing it to the production personnel, giving them the responsibility to run the experiment. There may be occasions when the production personnel can conduct the experiment themselves because of simple process configuration, and changing factor levels meant only changing a few dial settings.

For example, if the experiment involves changing the temperature, pressure, and amount of water added to a chemical reactor and then measuring an output at the end of the reaction period, this experiment can be entrusted to production operators. On the other hand, if the process configuration is such as in the case study above, where treatments to the experimental units are given at different places by different people, several opportunities exist for making mistakes because of logistics and lack of communication. The quality engineer should participate in the execution of such an experiment and make sure that participants both understand the objectives of the experiment and follow the instructions meticulously. Then, if mistakes occur, timely detection will prevent large-scale rejection of data or making erroneous conclusions.

3.4.3 Floor Plan Layout

The facilities engineers prepare the floor plan layout to facilitate the smooth flow of material and subassemblies toward the final assembly. The quality planning team has to verify that sufficient room has been assigned for inspection stations, sample collection and transmittal, control chart location, computer terminals for data gathering and display, and suitable fixtures for displaying work instructions and inspection standards. Some inspection and instrument calibration may need to be done under controlled environmental conditions. The team should make sure that the plant layout makes suitable provision for these requirements.

3.4.4 Process FMEA

The process FMEA is performed to assess the effect of possible process failure modes on the quality of the product. How the FMEA is made has been explained earlier in this chapter while discussing its use in product design. A process FMEA must be performed for any new process, or for a current process scheduled to produce a new product, in order to make certain that possible failures are considered and suitable remedial actions are implemented. The process FMEA is also a good tool to use while investigating the root causes for defectives produced in an existing process.

An example of a process FMEA is shown in Figure 3.17. The items identified by letters in the figure have the same meaning as those for the design FMEA, except that the modes and effects of failure now belong to a process. This process FMEA relates to one failure mode—the lack of sufficient wax coating on the inside panel of a car door, causing rusting of the door and poor functioning of the door hardware. Note that FMEA analysis has been used to discover the causes of the failure mode and prescribe solutions to rectify the cause, which, when implemented, result in smaller values for the RPN.

3.4.5 Process Control Plan

The process control plan is a working document prepared using the information from the PFC, FMEA, design reviews, and other tools employed for the process design. The information from these documents is transferred in a manner that can be easily used by the operating personnel. The process control plan is, in essence, the consolidation of all process design information that can be handed off to the operating personnel for use. Thus, it becomes the source of information for process operators for creating all the product characteristics and keeping them in control. The process control plan is also a "living document" in the sense that it is continuously updated (by the process engineers) based on the changing requirements of the customer, experience gained with the processes, or solutions discovered in resolving previous problems.

3.4.6 Other Process Plans

3.4.6.1 Process Instructions

Process instructions are documents describing how each operation in the process should be performed. Process instructions and control plans complement each other in defining how operations are to be performed. The process

POTENTIAL
FAILURE MODE AND EFFECTS ANALYSIS
(PROCESS FMEA)

Item: B
Model years(s)/Programs(s) D
Core team G

Process Responsibility C
Key Date E

FMEA Number A
Page ____ of ____
Prepared by: H
FMEA Date (Orig.) F

Process step / Function	Requirement	Potential failure mode	Potential Effect(s) of failure	Severity	Classification	Potential Cause(s) of failure	Current process: Controls prevention	Occurence	Controls Detection	Detection	RPN	Recommended Action	Responsibility & Target completion date	Action resilts: Action taken Completion date	Severity	Occurrence	Detection	RPN
Op to: manual application of wax inside door pannel	Corner inner door, lower surfaces with wax to specification thickness	Insufficient wax coverage over specified surface	Allowa integrity breach of inner door panel Connraded interox lover door panels Deterloraled life of door leading to: .Unsallsfactory appearance due to rust through paint over time. .Impaired function of interior door hardware	7		Manually insurial sprary head not inserted ler enough	None	8	Variable check for dim Visual check for coverage	5	200	Add positive depth stop to sprayer	Mig engineering by 0X 10 15	Stop added, sprayer checked	7	2	5	70
												Automato spraying	Mig engineering by 0X 12 15	Rejected due to complexity of different doors on fly same line				
						Spray head clogged -vennocity ico high -temperature lea kpw -Pressure loo low	Tool appay all start-up and after idle products and preventative maintenance program to clean heeds	5	Variable check for dim thickness Visual check for coverage	5	175	Use design of exporiments (DDE) on viscosity vs. temperature vs. pressure.	Mig engineering by 0X 10 01	Temp and press units wore determined and line controls have been installed- control cleats show process is in control cpic-1.85	7	1	5	35
						Spray head deformed due to inspect	Proventation maintanance program to maintain feects	2	Visual check for film thickness Visual check for coverage	5	70	None						
	SAMPLE					Spray lime sufficient	None	5	Operator instructions. Lo sampling (visual) check coverage of artical access	7	245	Install spray linear	Maintenance xxxxxxxxxx	Automatic spray linner installed- operator sinris spray, spitor controls shut off control chanels show process is in control-cpic-2.03	7	1	7	49
		Executive wax coverage over specified surface																
a1	a2	b	c	d	e	f	h	g	h	i	j	k	l	m	n			

FIGURE 3.17
Example of a process FMEA. (Reprinted from Chrysler LLC, Ford Motor Company, and General Motors Corporation. 2008b. *Potential Failure Mode and Effects Analysis (FMEA)—Reference Manual*. 4th edition. Southfield, MI: A.I.A.G. With permission.)

instructions specify the optimal sequence in which motions should be made in order to accomplish a job. The control plans give the quality-critical information. The process instructions are also useful when training new operators on the job. These have to be prepared out of information obtained from PFCs, control plans, and the recommendations of equipment manufacturers. The experience gained in past operations by the operators must be used in finalizing the instructions.

3.4.6.2 Packaging Standards

Designing the packaging for the final product is an important activity in quality planning. The packaging design should ensure that the product performance will not be altered during handling and shipment to the customer. The quality planning team should make sure that adequate attention is paid to designing, testing, and specifying the packaging as part of the quality plan.

3.4.6.3 Preliminary Process Capabilities

The meaning of process capability and the method of measuring process capabilities are explained in detail in Chapter 4. Process capability refers to the ability of a process to produce a desired product characteristic within the variability allowed by (internal or external) customer specifications. At this stage of planning, the quality planning team should make certain that the processes have been chosen to produce the product characteristics within the allowable variability. Often, a pilot study may be necessary to evaluate the capability of the chosen process to perform the operations for a new product with acceptable capabilities. Such studies would give estimates for the expected capabilities of the new process and also reveal opportunities for improving those capabilities.

3.4.6.4 Product and Process Validation

This is the step before the product is launched into production, when the process is tried out to check if it is capable of producing the product to meet the customer's needs. The process will be tested at production-quantity levels, using production workers and production tooling, in the normal production environment. The following are the quality-related outcomes expected from such a trial run:

- Process capability results
- Measurement system evaluation
- Product/process approval
- Feedback, assessment, and corrective action

3.4.6.5 Process Capability Results

These are measured during the validation run and are compared with expected or targeted benchmarks. Changes should be made to the process if any shortfall occurs in the expected process capabilities or product quality. Changes to process variables or additional process controls may have to be implemented to achieve the target capabilities specified by customers. For measurements, the capabilities are generally specified using C_p and C_{pk}. Customers usually demand C_p and C_{pk} values of 1.33 or greater. For attributes, capabilities are specified in defects per thousand (dpt) or parts per million (ppm). The target capability varies with the type of product, type of attribute, and needs in subsequent processes or use by the customer.

3.4.6.6 Measurement System Analysis

The requirements for measuring instruments in terms of accuracy, precision, resolution, and so on, and how the instruments are evaluated against these requirements, are described in Chapter 4. The objective of measuring system analysis is to ensure that the production function has adequate capability to measure and verify if the process variables and product characteristics are on target and within specifications. The instruments and measuring schemes must be checked for their adequacy in meeting the process needs. Gage R&R studies (see Chapter 4) may have to be conducted, which will reveal an instrument's capability for repeatability and reproducibility. If these capabilities are not adequate then repairing the hardware or training the operators to follow consistent procedures in using the instrument would help to improve them. If an instrument's capability cannot be improved, however, it may have to be replaced with a more capable instrument.

3.4.6.7 Product/Process Approval

Product/process approval refers to testing the product from the preproduction trial and verifying that the product meets the design specifications. Characteristics such as strength, chemistry, temperature rise, and fuel consumption must be evaluated from the trial run and certified as to meeting the expected standards.

3.4.6.8 Feedback, Assessment, and Corrective Action

This is the stage when the effectiveness of all the quality planning work in previous stages is verified. The objective is to gather all the information together and assess the capabilities of the processes for producing the product characteristics to meet customer satisfaction. The capabilities of both measurements, as well as attributes, must be evaluated periodically, and opportunities to make improvements, to reduce variability further, must be explored. Additional effort must be made to satisfy the customer through help in using the product in the customer's environment. Customer feedback should be obtained in order to review and revise the original design to improve their satisfaction even more. The customer's requirements on delivery and service must be sought and fulfilled. Whenever needed, the customer's participation in the design and planning activities must also be encouraged.

3.5 Exercise

3.5.1 Practice Problems

3.1 A customer survey containing 11 questions was administered twice, with a gap of three months in between, to a sample group of 30 customers. The questions were to be answered in the scale of one to five. The survey results (average responses for each of the 11 questions) are shown below. Based on these results, comment on the reliability of the survey instrument.

Question no.	1	2	3	4	5	6	7	8	9	10	11
First survey	4.2	2.7	3.9	2.8	4.8	5.0	1.8	3.1	4.1	2.2	3.7
Second survey	3.8	3.6	2.9	1.7	4.1	4.6	2.6	3.8	4.8	3.1	3.3

3.2 A customer survey contains two sets of questions, with each set containing questions similar to those in the other set but worded differently. The average scores received for the questions, with similar questions paired together, are given below. Is the questionnaire reliable?

Set 1	4.2	3.8	3.2	4.9	4.0	2.8	1.8	2.2	4.3	3.9
Set 2	2.7	4.5	1.6	3.2	1.6	4.3	2.0	3.5	4.0	4.1

3.3 Determine the sample size for a customer survey if the allowable error in the estimates should not be more than ±0.3 and the desired confidence level is 95%. The standard deviation of the individual responses is known to be 0.7.

3.4 A survey is to be conducted among the alumni of an educational institution to obtain impressions of their experience when they were in school. A questionnaire has been prepared that asks for their response against questions on a scale of one to five. If the standard deviation of the scores from similar past surveys is 0.7, and the present results are desired within an error of ±0.2 with a 95% confidence level, how many alumni should be surveyed?

3.5 The following data represent the life of electric bulbs in months. Check what distribution the data follow by preparing a histogram and checking its shape. Determine the *MTTF*. (*MTTF* for a population that has normal distribution is given by X_{50} and that for an exponential distribution is given by $X_{63.2}$.)

260	195	28	83	41	39	11	128	61	7
19	400	4	72	36	2	200	107	41	43
25	151	36	21	82	97	88	113	7	30
3	158	21	11	84	20	112	2	66	191
9	155	11	187	6	0	65	66	153	81

3.6 Calculate the percentage frequency of failures, reliability, and failure rate for the following set of data on failure of a production line, and draw the graphs of these functions in the manner of Figure 3.4.

Interval (days)	10–20	20–30	30–40	40–50	50–60	60–70
No. of failures	5	17	28	24	18	8

3.7 The following data represent the failure times in years of a sample of 10 batteries of a certain make. The numbers shown with a superscript (+) are not failure times; they show that those batteries were still working by the time that the experiment was concluded. Calculate the *MTTF* and failure rate of these batteries. Assume that the battery life follows an exponential distribution.

1.0	2.3	4.4	5.4	6.1	11.8	12.9	13.7	13.7^{+}	13.7^{+}

3.8 The time to failure of a certain make of automobile transmissions is known to follow the exponential distribution, with a mean of 100,000 miles. If these transmissions carry a warranty for 75,000 miles, then:

a. What proportion of the transmissions will need repair before the warranty expires?

b. If no more than 4% of the population can be replaced economically under warranty, what warranty time would be economically feasible?

3.9 A brand of washing machine is known to fail according to an exponential distribution, with a failure rate of 1 in 12,000 hours of operation. If the manufacturer gives a five-year warranty, what proportion of the washing machines will need warranty service? A machine is used for an average of 600 hours per year.

3.10 In a study of a machining operation, the amount of vibration caused on the job was measured as the response with tool angle (15° vs. 20°) and speed of rotation (2600 vs. 3000 rpm) as the factors. The following data were collected from two replicates of the experiment.

Analyze the results of the experiment, and choose the combination of tool angle and speed that produces the least amount of vibration. Calculate the main and interaction effects and determine if they are significant.

		Speed	
		1 (2600)	2 (3000)
Tool angle	1 (15°)	18.2, 14.4	14.5, 15.1
	2 (20°)	24.0, 22.5	41.0, 36.3

3.11 In an experiment conducted to understand the effects of percentage moisture and percentage clay on the compactability of sand, a quality characteristic, the following results were obtained. Calculate the effect of each factor and their interaction. Also, verify if the effects are significant.

		Moisture	
		3%	6%
Clay	1%	48, 44	56, 50
	3%	42, 46	52, 49

3.12 The following data come from an experiment to find the optimal combination of process parameters to obtain the best yield from a process. The parameters were Factor A, Factor B, and Factor C. The response, obtained in two replicates, was the number of pieces produced per run. Calculate the main and interaction effects, and estimate the standard error of the effects. Create the confidence intervals, and comment on the results.

	(1)	*a*	*b*	*ab*	c	*ac*	*bc*	*abc*
Replicate 1	226	322	350	553	442	406	608	399
Replicate 2	318	432	346	475	457	375	504	415

3.13 The compressive strength of concrete is known to be affected by three factors: percentage fines in the aggregate, ratio of cement to aggregate (C & A ratio) in the mix, and percentage by volume of water added to the mix. The following results were obtained in an experiment to determine the best levels of these factors to obtain the maximum compressive strength. Two trials were performed at each treatment combination. Find the effects of the factors and the significance of the factors, and determine the best combination of the factor levels.

		C&A Ratio—Low		C&A Ratio—High	
		Moisture—Low	Moisture—High	Moisture—Low	Moisture—High
% Fines	Low	528, 520	580, 640	720, 640	780, 708
	High	480, 540	510, 590	690, 710	720, 740

3.14 Twenty-pound boxes of nails are packaged in cartons, with 12 boxes per carton. Each 20-lb. box has a tolerance limit of ±0.5 lb. What is the tolerance on the carton weight? Assume that the box weights are independent and are normal.

3.15 For a shaft-bearing assembly, the shaft diameters were found to have an average of 2.2 in., with a standard deviation of 0.12 in. The bearing bore diameters were found to have an average of 2.1 in., with a standard deviation of 0.07 in. What proportion of the assemblies will have a clearance of less than 0.01 in.? Assume that the shaft and bearing dimensions are independent.

3.16 The total weight of a cereal box includes the weight of the cereal and the weight of the cardboard box. The total weight of the filled boxes is known to meet the specification 20 ± 1.0 oz. If the empty boxes meet the specification 4 ± 0.5 oz, does the net weight of the cereal meet the tolerance 16 ± 0.8 oz?

3.17 Hundred-pound bags of alcohol (the solid, industrial variety) are stacked 25 per skid for shipping. The skids are weighed before shipping to check that the bag weights meet the customer specifications. The customer specifications call for 100 ± 4 lb per bag, so the material handler at the stacking operation uses 2500 ± 100 lb as the specification for the skid weight (less the weight of the wooden skid). The customer complains of receiving bags that are underweight. What would be the correct specification for the skid weight, assuming that the bag weights are independent? (The company uses this procedure as a quick way of checking the bag weights.)

3.18 The amount of time a maintenance person is out on a service call, including travel, is 120 ± 25 minutes. The travel time between jobs is 60 ± 20 minutes. What is the average and the "tolerance" of the service time? Assume that the travel time and the service times are normal and independent.

3.5.2 Mini-Projects

Mini-Project 3.1

Create a customer survey instrument to find out from the group of IE (ME/EE/CE or any other) majors how satisfied they are with the services they receive in an IE (ME/EE/CE or any other) department in both academic and nonacademic areas. Alternatively, create a customer survey instrument to find out from users of a university library their level of satisfaction with the several services provided by the library.

Mini-Project 3.2

Based on the survey instrument created in Mini-Project 3.1, construct a HOQ, and choose the design features that you will use to enhance the services provided by the IE (ME/EE/CE or any other) department or the library. Assume some reasonable data for responses.

Mini-Project 3.3

This project is meant to verify the theoretical basis of the root sum of squares formula used for obtaining assembly tolerance from parts tolerance. The theoretical results in the simplest form can be expressed as follows:

$$\text{If } X_1 \sim N(\mu_1, \sigma_1^2) \text{ and } X_2 \sim N(\mu_2, \sigma_2^2),$$
$$\text{then } (X_1 + X_2) \sim N(\mu_1 + \mu_2, \sigma_1^2 + \sigma_2^2),$$
$$\text{and } (X_1 - X_2) \sim N(\mu_1 - \mu_2, \sigma_1^2 + \sigma_2^2).$$

In other words, the sum of the two random variables has the mean equal to the sum of the two means. The standard deviation of the sum equals $\sqrt{\sigma_1^2 + \sigma_2^2}$, not $(\sigma_1 + \sigma_2)$. Similarly, the difference of the two random variables has the mean equal to the difference of the means. The standard deviation of the difference equals $\sqrt{\sigma_1^2 + \sigma_2^2}$, not $\sqrt{\sigma_1^2 - \sigma_2^2}$, nor $(\sigma_1 - \sigma_2)$.

Using Minitab, generate 50 observations from two normal distributions, one with mean 30 and standard deviation 3, and the other with mean 40 and standard deviation 4, and store the data in two columns. Add the numbers in each row to obtain 50 observations from the random variable, which is the sum of the above two random variables. Draw the histogram of all three sets of data, and calculate their means and standard deviations. What is the relationship between the mean of the two original distributions and that of the distribution of the "sum"? What is the relationship between the standard deviation of the two original distributions and the standard deviation of the "sum"?

Repeat the above experiment, except now calculate the difference of the two original observations to obtain the column of differences. Draw its histogram, and calculate the average and standard deviation of the "difference" and compare them with those of the original distributions.

You could also choose any mean and any standard deviation for the original two distributions instead of what is suggested.

References

Akao, Y., ed. 1990. *Quality Function Deployment: Integrating Customer Requirements into Product Design*. Cambridge, MA: Productivity Press.

Box, G. E. P., W. G. Hunter, and J. S. Hunter. 1978. *Statistics for Experimenters*. New York, NY: John Wiley.

Chrysler LLC, Ford Motor Company, and General Motors Corporation. 2008a. *Advanced Product Quality Planning (APQP) and Control Plan—Reference Manual*, 2nd edition. Southfield, MI: A.I.A.G.

Chrysler LLC, Ford Motor Company, and General Motors Corporation. 2008b. *Potential Failure Mode and Effects Analysis (FMEA)—Reference Manual*. 4th edition. Southfield, MI: A.I.A.G.

Churchill, G. A., Jr. 1999. *Marketing Research: Methodological Foundations*. 7th ed. Chicago: Dryden Press.

Clausing, D. 1994. *Total Quality Development*. New York, NY: ASME Press.

Cohen, L. 1995. *Quality Function Deployment*. Reading, MA: Addison-Wesley Longman.

DeVor, R. E., T. H. Chang, and J. W. Sutherland. 1992. *Statistical Quality Design and Control—Contemporary Concepts and Methods*. New York, NY: Macmillan.

Gryna, F. M. 1999. "Market Research and Marketing." Section 18 in *Juran's Quality Handbook*. 5th ed. J. M. Juran and A. B. Godfrey, eds. New York, NY: McGraw Hill.

Hayes, B. E. 1998. *Measuring Customer Satisfaction*. 2nd ed. Milwaukee, WI: ASQ Quality Press.

Hicks, C. R., and K. V. Turner. 1999. *Fundamental Concepts in the Design of Experiments*. 5th ed. New York, NY: Oxford University Press.

Hines, W. W., and D. C. Montgomery. 1990. *Probability and Statistics in Engineering and Management Science*. 3rd ed. New York, NY: John Wiley.

Hogg, R. V., and J. Ledolter. 1987. *Engineering Statistics*. New York, NY: Macmillan.

Ireson, W. G., and C. F. Coombs, Jr., eds. 1988. *Handbook of Reliability Engineering and Management*. New York, NY: McGraw-Hill.

Krishnamoorthi, K. S. 1992. *Reliability Methods for Engineers*. Milwaukee, WI: ASQ Quality Press.

Montgomery, D. C. 2001. *Introduction to Statistical Quality Control*. 4th ed. New York, NY: John Wiley.

O'Connor, P. 1985. *Practical Reliability Engineering*. 2nd ed. New York, NY: John Wiley.

Shingley, J. E., and C. R. Mischke. 1986. *Standard Handbook of Machine Design*. New York, NY: McGraw-Hill.

Skalak, S. C. 2002. *Implementing Concurrent Engineering in Small Companies*. New York, NY: Marcel Dekker.

Syan, C. S., and U. Menon, eds. 1994. *Concurrent Engineering—Concepts, Implementation, and Practice*. London: Chapman & Hall.

Syan, C. S., and K. G. Swift. 1994. "Design for Manufacture." In *Concurrent Engineering—Concepts, Implementation, and Practice*, C. S. Syan and U. Menon, eds. London: Chapman & Hall.

Tobias, P. A., and D. C. Trindade. 1995. *Applied Reliability*. 2nd ed. New York, NY: Van Nostrand Reinhold.

Womack, J. P., D.T. Jones, and D. Roos. 2007. *The Machine that Changed the World*. New York, NY: Free Press.
Ziemke, M. C., and M. S. Spann. 1993. "Concurrent Engineering's Roots in the World War II Era." In *Concurrent Engineering—Contemporary Issues and Modern Design Tools*, H. R. Parsaei and W. G. Sullivan, eds. London: Chapman & Hall.

4

Quality in Production—Process Control I

In the last chapter, we discussed the quality methods that are employed in the product planning, product design, and process design stages of the product creation cycle. In this chapter, we discuss the quality methods used during the production stage of the cycle. The major tools used in the production stage are the control charts, of which several different types exist to handle the different types of product characteristics and process variables encountered in production and service environments. The three major types of charts needed to monitor the three major types of variables are discussed here. The measures for evaluating process capability, or ability of a process to produce products within specifications, are defined in this chapter, as are the methods for assessing the capability of measuring instruments and measurement systems.

The topics in process control that are considered essential for performing the day-to-day process control work in the production or service environments are included in this chapter. Some of the advanced topics in process control, such as the charts needed for special situations, evaluating the performance of control charts and characteristics of a six sigma process, are covered in the next chapter.

4.1 Process Control

Process control is the most important component of the quality effort during manufacture, and its objective is to proactively evaluate the condition of a process and control it so as to prevent the production of defective units.

Any productive work can be viewed as a process consisting of machinery, manpower, methods, and measuring schemes, as shown in Figure 4.1. The environment can sometimes play a significant part in the production activity and can influence the quality of the output; therefore, it should also be taken into account while studying the process. It is possible that measures may have to be incorporated into the process in order to prevent changes in the environment from affecting the process output in any negative way.

A process receives inputs—mainly the material, parts, or subassemblies in a production process—and then processes them in a sequence of operations using machinery, tools and manpower, and delivers an output. Each operation of the process has to be performed under certain selected conditions,

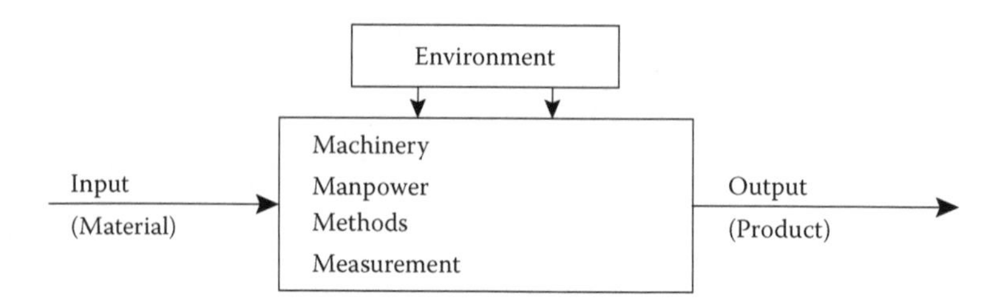

FIGURE 4.1
The scheme of a process.

such as speed, feed, temperature, pressure, and so on. These are called the "process parameters." The levels at which the process parameters are maintained during production determine the levels at which the product characteristics are obtained, which, in turn, determine the quality level of the product. The target levels for the process parameters must be selected using experiments to yield the desired levels of the product characteristics. The selected parameters must be maintained consistently at the chosen levels during production in order to obtain the product characteristics consistently at the desired levels. Maintenance of the process parameters at the chosen levels is accomplished using the control charts. These are also used to monitor the consistency of the level of the product characteristics, or the process output.

The term "process parameter" is used in quality literature to indicate both the variables of a process that influence the product characteristics, as well as the quantities that define the distribution of a process, such as μ and σ of a normal distribution. Where there is a chance for confusion as to which of the two we are referring—the term "process variables" will be used to refer the former and the term "process parameter" will be used to indicate the characteristics of a distribution.

4.2 The Control Charts

Dr. Walter Shewhart, working in early 1920s at the Bell Telephone Laboratories in Princeton, New Jersey, proposed the procedures that later came to be known as the control charts. According to him (Shewhart 1931), the variability in a product characteristic, or process parameter can be from two sources:

1. From a "stable system of *chance causes*," meaning the aggregate of small, unavoidable variability arising from natural differences in material, manpower, machinery, instruments, and the environment, that should be expected.

2. From *"assignable causes,"* meaning the causes arising from a specific occurrence, such as a broken tool, pressure surge, or temperature drop that might occur unexpectedly.

These two sources of variability were later renamed by Dr. Deming as "common cause" variability and "special cause" variability, respectively. The former is an integral part of the process and cannot be eliminated at reasonable cost. Therefore, it must be accepted as part of the process. The latter produces disturbances to the process and usually increases the variability beyond acceptable levels, and so must be discovered and eliminated. Dr. Shewhart proposed the control chart as a means of differentiating between the condition of the process when it is subject only to the natural, chance causes, and when it is affected by one or more assignable causes.

4.2.1 A Typical Control Chart

A control chart typically has a centerline (CL) and two control limits—an upper control limit (UCL) and a lower control limit (LCL), such as those shown in Figure 4.2. These limits represent the limits for chance-cause variability and are computed using data drawn from the process. (We will see later how these limits are calculated so that they represent the limits for the chance-cause variability.) Samples are taken from the process at regular intervals, and a measure that reflects the quality of the product, or process, is computed out of the sample observations. The values of the measure are plotted on a chart, onto which the limit lines have been drawn to a suitable scale. If the values from all the samples taken during a period of time lie within the limits, the process is said to be "in-control" for that time period. However, if the value from any of the sample plots lie outside either one of the limits, the

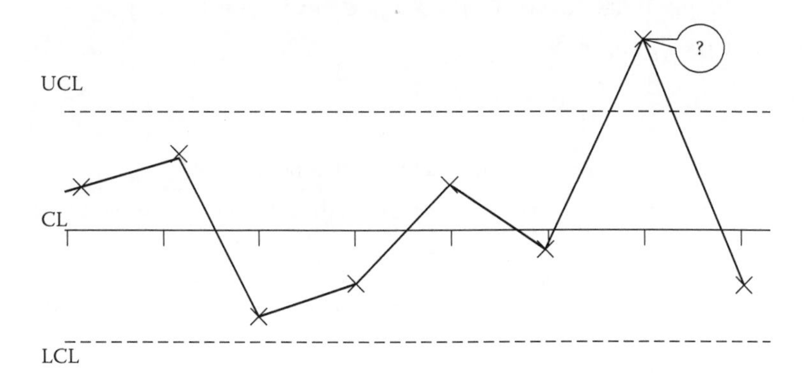

FIGURE 4.2
A typical control chart.

process is said to be "not-in-control" during that time period. When the process is not-in-control, one or more assignable causes, or special causes, must have occurred, and action must be taken to discover and eliminate them.

The measure to be computed from sample observations depends on what we desire to control in the process. Suppose that the average value of a product characteristic is to be controlled; the average of that characteristic from the sample observations will be then plotted. If the variability in the characteristic is to be controlled, then the standard deviation, or range, or a similar measure suitable to estimate the variability, will be computed and plotted. The control chart takes the name of the measure that is plotted. Table 4.1 shows examples of measures (these are sample statistics) that are plotted to control various process parameters. There are other control charts that use different sample statistics to control these or other process parameters. However, those listed in Table 4.1 are the most commonly used and will be discussed in detail below. The context in which these are used, the method of computing their limits, and the manner of interpreting the results will also be explained.

Many advantages are realized when processes are controlled using these control charts, which are also referred to as "statistical process control" (SPC) tools. Essentially, they contribute to reducing the variability in the process parameters and product characteristics, and enable the production of products of consistent quality. The benefits arising from implementing control charts on processes include:

- Avoidance of defectives by producing products right the first time
- Reduction of waste, and increase in throughput
- Satisfied customers, and improved customer relationships
- Better knowledge of the processes and their capabilities
- Improved worker morale because of the satisfaction derived from fruitful results
- Improved image for the producer, and better market share
- Improved profitability

TABLE 4.1

Examples of Control Charts Used to Control Different Parameters

Process Parameter to be Controlled	Sample Statistic Plotted	Name of the Chart
Process average (μ)	Sample average ($\overline{X}$)	$\overline{X}$-chart
Process variability (σ)	Sample range (R), or sample standard deviation (S)	R-chart S-chart
Process proportion defectives (p)	Sample proportion defectives (P)	P-chart
Process defects per unit (c)	Sample defects per unit (C)	C-chart

4.2.2 Two Types of Data

As stated in Chapter 2, the data from sample observations are classified into either of two categories:

1. Measurement data
2. Attribute data

Measurement data result from measurements taken on a continuous scale, such as height, weight, thickness, and so on, and result in observations such as 142.7 in., 1.28 kg, and 68 cft. Attribute data come from inspections performed based on attributes such as taste, feel, and eye judgment. Attribute data can also come from gaging, in which a product is classified into categories such as good/bad, tight/loose, or flat/not flat. Attribute data are usually in counts, proportions, or percentages, such as three defects per printed circuit, two of eight too small, or 10% too tight. The reader will recognize that the measurement data come from observations made on continuous random variables and the attribute data come from observations made on discrete random variables. It is necessary to recognize the type of data one has to deal with in a given situation in order to determine the type of control chart to use. This is because the limits for the charts are computed based on the distribution that the data came from, and the distributions for the two types of data are different.

4.3 Measurement Control Charts

The measurement control charts are designed to control measurements, with the assumption that those measurements follow the normal distribution. When a measurement is to be controlled, both its average and its variability must be controlled. A measurement will fall outside its specification limits if its average moves to a location other than the target, its variability increases beyond the specified range, or both (see Figure 4.3).

For normal populations, the sample average and sample measures of variability, such as sample variance and sample range, are statistically independent—a result proved in books in mathematical statistics. This means that controlling the sample average will not control the process variability and vice versa. Hence, two control charts are needed: one that uses the sample average to control the process mean, and another that uses a sample measure of variability to control the standard deviation of the process.

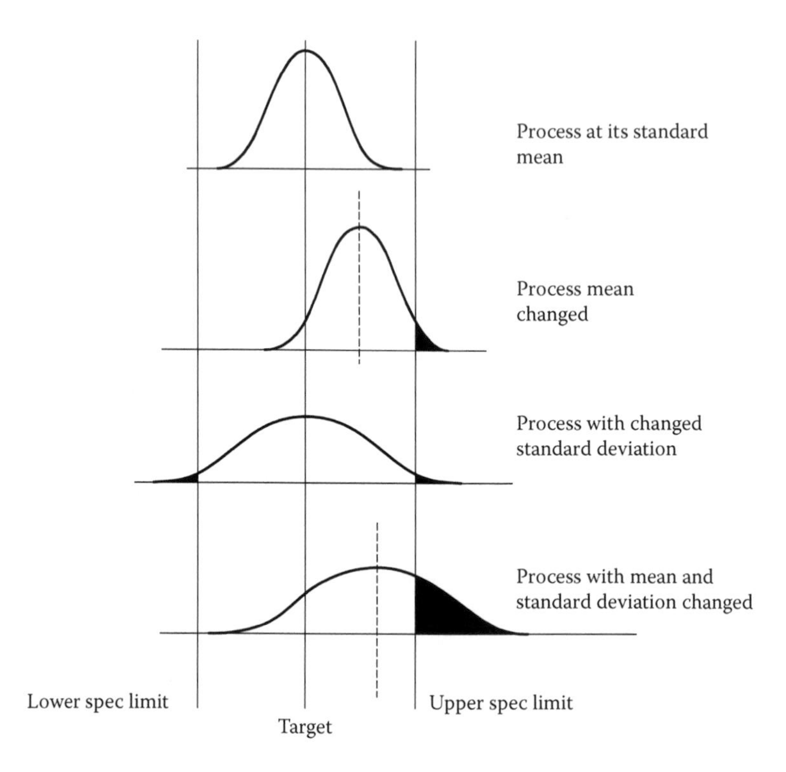

FIGURE 4.3
Process conditions when process mean and/or standard deviation change.

4.3.1 $\bar{X}$- and *R*-Charts

The most popular combination used to control a measurement are $\bar{X}$- and *R*-charts: the former to control the process mean and the latter to control the process variability. Both are made from the same samples and are generally used together. Samples, usually of size four or five, are taken from the process at regular intervals, and the sample average $\bar{X}$ and the sample range R are computed. These values are then plotted on graphs where the limits calculated from the following formulas have been drawn:

Control limits for *R*-chart	Control limits for $\bar{X}$-chart
$\mathrm{UCL}(R) = D_4\bar{R}$	$\mathrm{UCL}(\bar{X}) = \bar{\bar{X}} + A_2\bar{R}$
$\mathrm{CL}(R) = \bar{R}$	$\mathrm{CL}(\bar{X}) = \bar{\bar{X}}$
$\mathrm{LCL}(R) = D_3 R$	$\mathrm{LCL}(\bar{X}) = \bar{\bar{X}} - A_2\bar{R}$

In the above formulas, $\bar{\bar{X}}$ and $\bar{R}$ are the averages of at least 25 sample averages and sample ranges, respectively, which have been obtained from the process being controlled. A_2, D_3, and D_4 are factors chosen based on sample size

TABLE 4.2

Factors for Calculating Limits for Variable Control Charts

n	A	A_2	A_3	B_3	B_4	c_4	D_1	D_2	D_3	D_4	d_2	d_3
2	2.121	1.880	2.659	0	3.267	0.798	0	3.686	0	3.267	1.128	0.853
3	1.732	1.023	1.954	0	2.568	0.886	0	4.358	0	2.574	1.693	0.888
4	1.500	0.729	1.628	0	2.266	0.921	0	4.698	0	2.282	2.059	0.880
5	1.342	0.577	1.427	0	2.089	0.940	0	4.918	0	2.114	2.326	0.864
6	1.225	0.483	1.287	0.030	1.970	0.952	0	5.078	0	2.004	2.534	0.848
7	1.134	0.419	1.182	0.118	1.882	0.959	0.205	5.204	0.076	1.924	2.704	0.833
8	1.061	0.373	1.099	0.185	1.815	0.965	0.387	5.306	0.136	1.864	2.847	0.820
9	1.000	0.337	1.032	0.239	1.761	0.969	0.546	5.393	0.184	1.816	2.970	0.808
10	0.949	0.308	0.975	0.284	1.716	0.973	0.687	5.469	0.223	1.777	3.078	0.797

from standard tables, such as that shown in Table 4.2, which is an abridged version of Table A.4 in the Appendix. The above formulas give "3-sigma" limits for the statistics $\overline{X}$ and R, in the sense that the limits are located at a distance of three standard deviations of the respective statistics from their average values. A detailed discussion on how these formulas are derived to provide the limits of natural variability for the statistic plotted is given in Chapter 5. The example below shows how the formulas are used to calculate the limits and how the charts are used in process control.

Example 4.1

The process to be controlled is a filling operation in a packaging shop where a powder chemical is filled in bags. The net weight of the chemical in the filled bags is to be controlled using control charts so that the average and variability of the net weight of all the filled bags remain consistent.

Solution

We use the $\overline{X}$- and R-charts to control the process average and variability. The data are shown in Figure 4.4a on a standard control chart form. In this example, 24 samples of five bags each were collected at approximately one-hour intervals. (In this case, only 24 samples were available and we proceeded to use them to calculate limits. Please see the discussion on the issue of the number of samples required to calculate limits later in this section.) The $\overline{X}$ and R values were calculated for each sample and then plotted using a suitable scale. The control limits were calculated as follows: from calculations made from the data, $\overline{\overline{X}} = 21.37$, and $\overline{R} = 3.02$. From Table 4.2, $A_2 = 0.577$, $D_4 = 2.114$, and $D_3 = 0$ for $n = 5$. Therefore,

$$\text{UCL}(R) = 2.114(3.02) = 6.38 \qquad \text{UCL}(\overline{X}) = 21.37 + 0.577(3.02) = 23.11$$

$$\text{CL}(R) = 3.02 \qquad \text{CL}(\overline{X}) = 21.37$$

$$\text{LCL}(R) = 0.0(3.02) = 0.0 \qquad \text{LCL}(\overline{X}) = 21.37 - 0.577(3.02) = 19.63$$

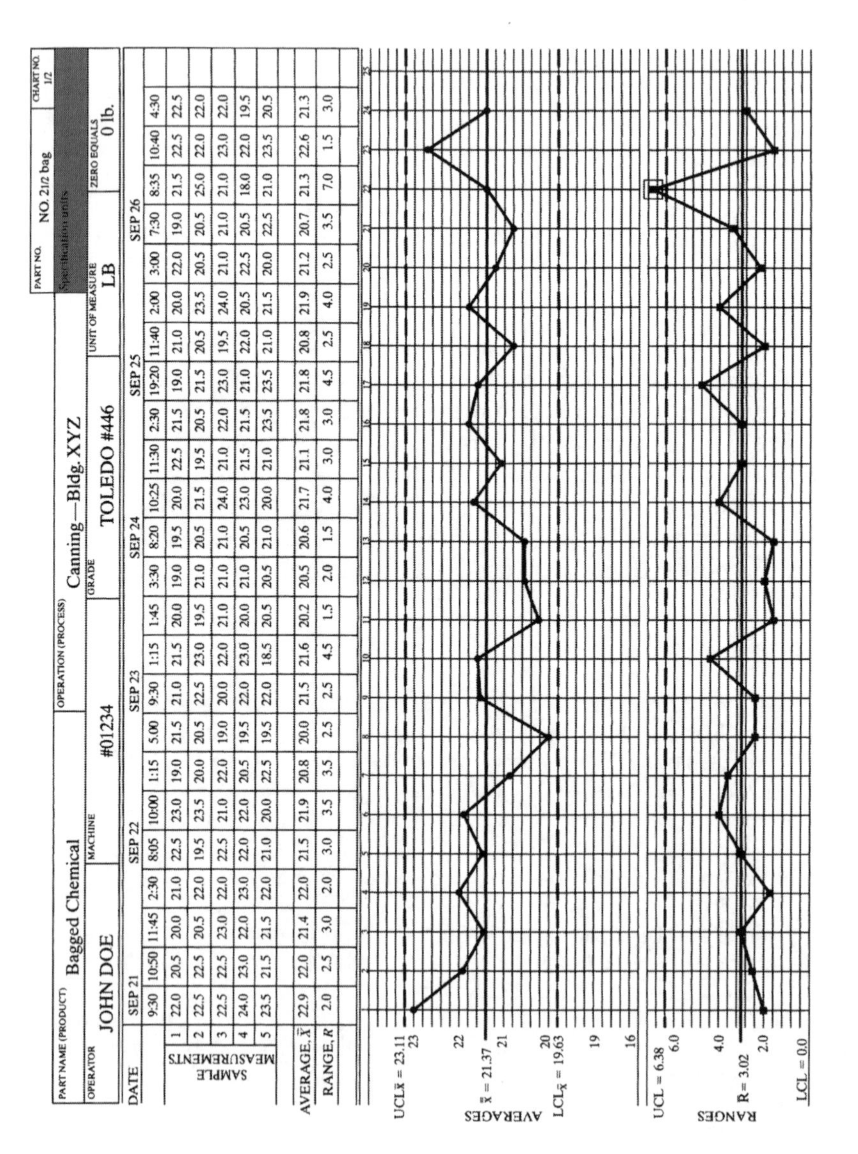

PART NAME (PRODUCT)	OPERATION (PROCESS)	PART NO.	CHART NO.
Bagged Chemical	Canning—Bldg. XYZ	NO. 21/2 bag	1/2

OPERATOR	MACHINE	GRADE	UNIT OF MEASURE	ZERO EQUALS
JOHN DOE	#01234	TOLEDO #446	LB	0 lb.

Specification units

DATE	SEP 21				SEP 22				SEP 23			
	9:30	10:50	11:45	2:30	8:05	10:00	1:15	5.00	9:30	1:15	1:45	3:30
SAMPLE MEASUREMENTS 1	22.0	20.5	20.0	21.0	22.5	23.0	19.0	21.5	21.0	21.5	20.0	19.0
2	22.5	22.5	20.5	22.0	19.5	23.5	20.0	20.5	22.5	23.0	19.5	21.0
3	22.5	22.5	23.0	22.0	22.5	21.0	22.0	19.0	20.0	22.0	21.0	21.0
4	24.0	23.0	22.0	23.0	22.0	22.0	20.5	19.5	22.0	23.0	20.0	21.0
5	23.5	21.5	21.5	22.0	21.0	20.0	22.5	19.5	22.0	18.5	20.5	20.5
AVERAGE, $\bar{X}$	22.9	22.0	21.4	22.0	21.5	21.9	20.8	20.0	21.5	21.6	20.2	20.5
RANGE, R	2.0	2.5	3.0	2.0	3.0	3.5	3.5	2.5	2.5	4.5	1.5	2.0

DATE	SEP 24				SEP 25				SEP 26			
	8:20	10:25	11:30	2:30	19:20	11:40	2:00	3:00	7:30	8:35	10:40	4:30
SAMPLE MEASUREMENTS 1	19.5	20.0	22.5	21.5	19.0	21.0	20.0	22.0	19.0	21.5	22.5	22.5
2	20.5	21.5	19.5	20.5	21.5	20.5	23.5	20.5	20.5	25.0	22.0	22.0
3	21.0	24.0	21.0	22.0	23.0	19.5	24.0	21.0	21.0	21.0	23.0	22.0
4	20.5	23.0	21.5	21.5	21.0	22.0	20.5	22.5	20.5	18.0	22.0	19.5
5	21.0	20.0	21.0	23.5	23.5	21.0	21.5	20.0	22.5	21.0	23.5	20.5
AVERAGE, $\bar{X}$	20.6	21.7	21.1	21.8	21.8	20.8	21.9	21.2	20.7	21.3	22.6	21.3
RANGE, R	1.5	4.0	3.0	3.0	4.5	2.5	4.0	2.5	3.5	7.0	1.5	3.0

FIGURE 4.4a
An example of $\overline{X}$- and R-charts.

The limits calculated as above are drawn on the control chart form in Figure 4.4a. Because one of the plotted R values is outside the limit in the R-chart, the process must be declared not-in-control. If an R value falls outside its limits, this means that the process variability does not remain consistent. If an $\bar{X}$ value falls outside its limits, however, it would indicate that the process mean does not remain consistent.

If we relate what happens on the process to what the control chart shows, it is often possible to identify the assignable cause (or causes) that generates an out-of-limit plot. In situations where process interruptions, such as a broken tool or falling air pressure, are likely to occur, the time of occurrence of the events will be related to the approximate time of the indication of signals on the control chart. For this reason, it is advisable to maintain a log of things that are happening in the process, such as change of operator, change of tool, change of raw material, and so on. This will facilitate the discovery of the assignable causes when they are indicated on the charts. Control charts can tell a lot more about what is happening in the process if we know how to read them correctly. This will be discussed in more detail later.

If all $\bar{X}$ and R values are within limits on both charts, the limits computed from such an in-control process can be used for future control. If not, those $\bar{X}$ and/or R values that are outside the limits can be removed from the data after making sure that the causes responsible for those values have been eliminated from the process. New limits can be calculated from the remainder of the data after removing those outside the limits. Such recalculation of limits from the "remaining" data saves times and money, and it is an accepted procedure for calculating limits for future control. When making these recalculations with the remaining data, it is advisable to fix the R-chart first, because calculation of limits for the $\bar{X}$-chart requires the value of $\bar{R}$. We may as well fix the R-chart first and use a "good" $\bar{R}$ from the in-control R-chart to calculate the limits for the $\bar{X}$-chart. Otherwise, if the $\bar{X}$-chart is fixed first, it may require rework when we find values in the R-chart that are outside the limits. Such swinging back and forth can be avoided if the R-chart is fixed first.

For Example 4.1, the one R value outside the upper limit is removed, assuming that the reason for the value being outside the limit was found and rectified. We then obtain a new $\bar{R}$ of 2.85 from the remaining 23 observations of R. This results in new limits for the R-chart:

$$\mathrm{UCL}(R) = 2.114(2.85) = 6.02$$

$$\mathrm{CL}(R) = 2.85$$

$$\mathrm{LCL}(R) = 0.0$$

All of the R values are now seen to be inside the new limits. The new limits for $\overline{X}$ are calculated using the new values for $\overline{R}$ and $\overline{\overline{X}}$ obtained from the remaining 23 values of $\overline{X}$. This results in new limits for the $\overline{X}$-chart:

$$UCL(\overline{X}) = 21.37 + 0.577(2.85) = 23.02$$

$$CL(\overline{X}) = 21.37$$

$$LCL(\overline{X}) = 21.37 - 0.577(2.85) = 19.72$$

All of the R and $\overline{X}$ values are now found to be inside their respective new limits, and these limits can be used for controlling this process in the future. These represent the limits for the intrinsic, natural variability in the process that is not affected by any assignable causes. The control charts prepared using the Minitab software, with limits calculated from the remaining 23 samples, are shown in Figure 4.4b. This chart can now be used for future control of the process. The control limits calculated from the initial set of data, before removing any out-of-limits values, are called "trial control limits."

A few questions arise while eliminating the values of $\overline{X}$ and R from the data for recalculating the limits as described above. First of all, no data should be eliminated unless there is a reasonable belief that the causes responsible for the out-of-limit values have been found and eliminated, and that those causes will not reoccur. The objective of using a control chart is to bring a process in-control. However, making a chart look "in-control" is not the same as *bringing* the process in-control.

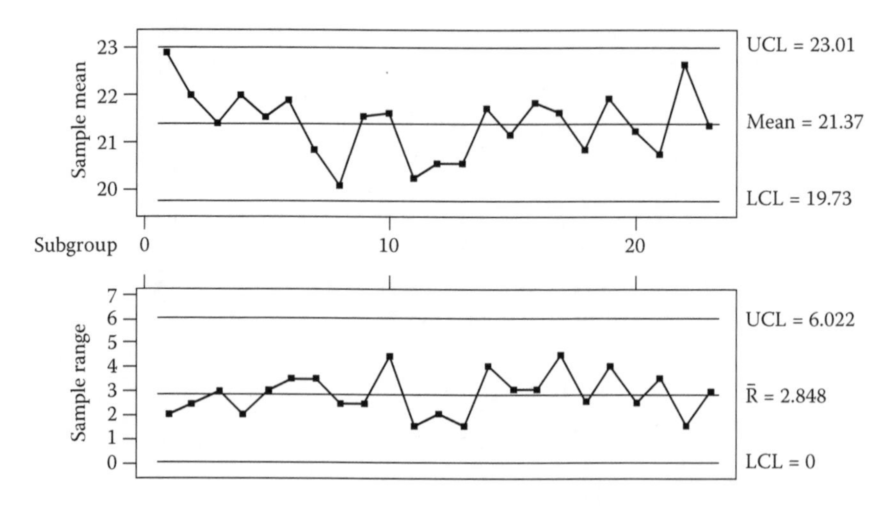

FIGURE 4.4b
(b) $\overline{X}$- and R-charts with revised limits calculated using "remaining" data for Example 4.1.

If an R value is outside its limits and its causes are eliminated, should the corresponding $\overline{X}$ value also be eliminated from the data along with the R value? The answer is yes; it makes sense to eliminate the entire sample from the data because it came from the process when it was not in-control. However, it is not always necessary to remove an $\overline{X}$ value (or R value) from its chart when removing the corresponding value from the other chart. Otherwise, when calculations are done with paper and pencil, excessive arithmetic work may result, and we may find ourselves going back and forth between the charts and redoing the limits. This opinion is justified by the known result that for normal populations, $\overline{X}$ and R are independent in the data and so the two statistics can be treated independently. If an $\overline{X}$ value is removed from the data, the corresponding R value can be left in the data, and vice versa.

When using a computer, however, this question becomes moot, because calculation time is not an issue, and if one statistic is outside its limits then the entire sample can be eliminated and new limits can easily be calculated for both charts. Even while using the computer software, the option of treating the charts independently will be valid and may help in conserving data.

How many of the original samples can be thrown out, leaving only samples that will be considered enough for calculating the limits? Here, the issue is part practical and part statistical. Dr. Shewhart, the author of the control charts, recommended that a process be observed over a period of time (for at least 25 samples of size four) before concluding that a process is in-control (Duncan 1974). His advice, which most practical statisticians would agree with, was that conclusions should not be drawn about a process being in-control based on observations taken over a short period of time. That was the genesis for the practice of using at least 25 samples in calculating the control limits. If 25 samples are initially taken, the process is not-in-control, and some of the sample values have to be thrown out, then obviously there will not be enough samples to claim that the limits are calculated from a process that is in-control. This is where the practical steps become necessary.

If sampling is expensive, and testing takes time, we cannot afford to waste data. Waiting to collect a new set of data may also delay control of the process. Therefore, in practice, people use as few as 15 samples to calculate the limits if that is all that is left after discarding the samples that produced plots outside the limits. Those limits, however, will be updated when additional samples become available. Another solution is to take about 30 samples of size four or five to start with, with the expectation that after discarding a few samples, enough will remain for computing the limits.

A statistical reason also exists for using 25 or more samples for calculating limits. The limits for the $\overline{X}$-chart, as well as for the R-chart, are calculated using $\overline{R}/d_2$ as an estimate for process standard deviation σ (see Chapter 5 for the derivation of the limits). The d_2 factor, which makes $\bar{R}$ an unbiased estimate for σ, is not good if the $\bar{R}$ value is obtained from a small number of samples. For example, for samples of size four or five, the tabulated d_2 values are not good if $\bar{R}$ is calculated from less than 15 samples. However, for larger sample sizes (n > 8), the tabled values of d_2 are good even if computed from less than 15 samples (from Table D3 of Duncan 1974, which gives d_2^*, the correction factor to estimate σ from $\bar{R}$ for varying sample sizes and number of samples). This means that if the sample size for the chart is large, we can do with fewer samples, and if the sample size is small, then we need a larger number of samples from which to calculate the limits. These facts should be kept in mind while deciding if there is adequate number of samples for calculating the limits. If using the conventional sample size of four or five, it might be a good idea not to use less than 15 samples to calculate the limits. Even then, when more samples (>25) become available, the limits must be recalculated.

4.3.2 A Few Notes about the $\bar{X}$- and *R*-Charts

4.3.2.1 The Many Uses of the Charts

The $\bar{X}$- and *R*-chart combination is the most popular statistical process control (SPC) method employed in industrial applications. Their simplicity and effectiveness in discovering significant assignable causes have made them very popular. The fact that many quality characteristics and process variables are normally distributed has also contributed to their popularity. The $\bar{X}$- and *R*-charts are used to accomplish several objectives:

1. *To maintain a process at its current level*

   In the case of many product and process variables, such as strength of steel or amount of impurity in a chemical, the variable cannot easily be moved to any given average level. The process is then first controlled at its "current" mean using the $\bar{\bar{X}}$ value derived from data as the centerline of the control chart. Further effort is then made to move the process mean progressively in the desired direction through additional investigation and experimentation. In the meantime, the process variability is controlled at the current level using the *R*-chart. The charts with limits calculated thus, using the formulas $\bar{\bar{X}} \pm A_2\bar{R}$, are said to be for maintaining current control, and are called the limits for "current control" of a process.

2. *To control a process at a given target or nominal value*

   The formulas given above based on $\bar{\bar{X}}$ and $\bar{R}$, obtained from data taken from the process, bear no relation to any standards the process may have to meet for either the average or the standard deviation. It would be desirable if a process could be made to conform to a given standard while also being controlled to be consistent. Many processes, such as cutting sheet metal to a given width or turning a shaft to a given diameter, can easily be controlled around a given target with a simple adjustment of tools or fixtures. Suppose a process is to be controlled at a given target T, the $\bar{\bar{X}}$ can then be replaced with T in the formulas. The spread for the control limits is, however, obtained as $A_2\bar{R}$ from the data. Similarly, there may be situations in which the σ for the process is specified and the process is to be controlled at this specified variability. These situations are referred to as "standards given" cases, and details of how the limits for these cases are calculated are discussed in Chapter 5.

3. *As a troubleshooting tool*

   Many quality problems in industry can be traced to excessive variability in product characteristics and process variables. Implementation of $\bar{X}$- and R-charts would be a good first step in solving these problems, because a good understanding of the process behavior will be gained from a close observation of the process during data collection. First, the charts will signal the existence of the causes; then they will help in discovering the sources of those causes if we relate happenings in the process to signals on the chart. Once the sources of the causes are known, the causes can be eliminated by taking proper remedial actions.

4. *As an acceptance tool*

   Control charts are also used to prove to a customer that the process has been in a stable condition producing good products. In such situations, the customer can reduce, or altogether eliminate, their inspection at receiving, resulting in considerable savings in labor, time, and space. In fact, customers who operate in a just-in-time environment often demand proof of process control and capability to meet specification so that they can eliminate the need to inspect them. Thus, the control chart can be used for accepting products without incoming inspection by the customer.

4.3.2.2 Selecting the Variable for Charting

Although these charts are very useful tools, they are expensive to maintain, especially if measurements have to be made manually. Thus, their use must be limited to where they are absolutely necessary. Even if measurements are made using automatic instruments and charts are made using computers, too many charts may hide the important ones. Thus, they must be used on product characteristics or process variables that are critical to product quality and customer satisfaction. In certain situations, the criticality of a variable to the process may be obvious. In others, a process failure mode and effects analysis (see Chapter 3) will reveal the critical process variables that will have to be controlled. In some situations, a preliminary study, either involving experiments or use of an attribute-type chart, may have to be made first, before selecting the important characteristics to be controlled using $\overline{X}$- and R-charts. An attribute chart—which is explained later in this chapter—is cheaper to maintain, because one chart can be used to monitor several product characteristics, and attribute inspection is generally less expensive. If one characteristic is found to be repeatedly responsible for the rejection of a product, then that characteristic can be monitored using $\overline{X}$- and R-charts.

4.3.2.3 Preparing Instruments

Before starting a control chart on any process, it is first necessary to decide the instrument to be used for measuring the variable and the level of accuracy at which the readings are to be recorded. It is also necessary to verify if the chosen instrument gives true readings—in other words, whether it has been calibrated—and whether the instrument has adequate capability. (The calibration and capability of instruments are explained later in this chapter.) In many situations, the lack of a suitable instrument will be found as a contributing factor to defective output. This will provide an opportunity to repair, or replace the instrument as a first step toward making a better-quality product.

In one example, the author was asked to help a company making paper cubes, an advertising specialty, to reduce the variability in the height of the cubes. The company president felt that the company was losing money by giving away an excessive amount of paper in cubes with heights larger than they should be. On his first visit to the shop, the author picked one cube from the pile and asked several people in the shop, including the plant superintendent, the shop foreman, the operator, and the inspector, to measure its height. Each person gave a different number for the height of the cube, because each person used a different tool and/or squeezed the cube differently before measuring. One used a foot rule; another, a caliper; yet another, a tape measure, and so on. No standard measuring scheme had been prescribed for determining the height of a paper cube. A new measuring scheme had to be installed that would give the "same" reading for the same cube when measured repeatedly, either by the same person or by different persons. Until this was done, no further work could be undertaken to quantify the variability in the cubes—much less, attempt to find the sources of variability and eliminate them.

Making sure that a good measuring instrument is available is an important first step in the control of any product characteristic or process variable.

4.3.2.4 Preparing Check Sheets

Proper check sheets or standard forms must be developed for recording relevant process information, and for recording and analyzing data.

Standard forms, such as that shown in Figure 4.4a, that call for relevant data and provide space for recording and analyzing these data, greatly facilitate the collection of information. It is often advisable to gather more rather than less information. Any information relating to the process, such as ambient temperature, humidity, operator name, type and ID of the instrument used, or information on any factor that may affect the output quality of the process, must be recorded. The preparation of check sheets forces an analyst to plan ahead the details of data collection—such as what exactly is to be measured, which instrument is to be used, and to what level of accuracy.

4.3.2.5 False Alarm in the $\overline{X}$-Chart

When an $\overline{X}$ value falls outside a control limit, we normally understand that an assignable cause has occurred and changed the process mean to a different level than the desired level. Just as in any statistical procedure, however, the control chart is also subject to Type I and Type II errors. The Type I error, which occurs when the control chart declares a process to be not-in-control when, in fact, it is in-control, is called the "false alarm." This arises from the fact that some $\overline{X}$ values in the distribution of $\overline{X}$ are outside the control limits even when the process is in-control. The probability of this happening for a 3-sigma chart is 0.0027. That is, even when a process is in-control, about 3 in 1000 samples will fall outside the limits.

People who are experienced in using $\overline{X}$-charts in practice, including Dr. Deming (Deming 1986), say that the false alarm should not be a concern while using the chart on a process that has not previously been controlled. According to them, if a process has not been already controlled using a procedure such as the control chart, it is more than likely not in-control. Concern over a false alarm should arise only when monitoring a process that has already been brought in-control.

4.3.2.6 Determining Sample Size

The typically recommended sample size for $\overline{X}$- and R-charts is four or five. Dr. Shewhart, when he first proposed the $\overline{X}$-chart, recommended the use of as small a sample as possible, because averaging over large samples would hide an assignable cause that may occur during the taking of the sample. However, samples should be large enough to take advantage of the effect of the central limit theorem, which makes the sample averages follow the normal distribution even if the population being controlled is strictly not normal. These considerations led Dr. Shewhart (Shewhart 1931) to believe that the sample size should be neither too large nor too small. He was also influenced by the fact that the number four provided an advantage in calculating the root mean square, $\left(\sqrt{\Sigma(X_i - \overline{X})^2 / n}\right)$, which he used as the sample measure of variability. The ease in finding the square root of four was an advantage in those days when electronic calculators were not available. He therefore recommended a sample size of four.

Later users saw some advantage in using a sample size of five, because it provided some advantage in calculating the average. (Obviously, they did not have electronic calculators, either.) Therefore, a sample size of four or five became the norm for $\overline{X}$- and R-charts. In some situations, however, the choice of sample size may be dictated by practical considerations, and the analyst is forced to choose other sizes. In such situations, sample sizes anywhere from 2 to 10 can be used. The minimum size of two is needed to calculate R, and the maximum of 10 is dictated by the fact that larger samples make the R value less reliable. That is, the variability in the sample range R becomes large when the sample size is larger than 10.

Theoretical studies have been made by researchers (see Duncan 1956; and Chapter 9 in Montgomery 2001) to determine optimally the sample size, the spread between the CL and the limits, and the frequency of sampling. Such researchers consider the cost of sampling, the cost of not discovering assignable causes soon enough after they have occurred, and the cost of false alarms, and then choose the above parameters of the chart to minimize the average cost of controlling a process. This has to be done for each individual process, taking into account the cost structure for that particular process. Such determination of sample size and other parameters of the control chart fall under the topic of *the economic design* of control charts. This is an advanced topic, and readers interested in it may refer to Montgomery (2001).

4.3.2.7 Why 3-Sigma Limits?

Again, it was Dr. Shewhart who recommended use of the 3-sigma rule for calculating control limits when he originally proposed the control charts. He was particular that the chance for process interruption due to false alarms resulting from the control procedure should be small. He chose the criterion that the chance for looking for an assignable cause when one does not exist should not be more than 3 in 1000 (Shewhart 1939), which translates to the 3-sigma rule when the process distribution is normal. According to this rule, the control limits will be placed at three standard deviations of the statistic being plotted from the centerline.

The limits could be placed at any distance—say, 2, 2.5, 4 or 4.5 sigma from the centerline. If, for example, the limits are at a 2-sigma distance, the control chart will detect changes in the process more quickly, and detect even smaller process changes more effectively, compared to the 3-sigma limits. However, the chart will also have a larger probability of a false alarm—that is, the probability of looking for an assignable cause when none exists. If the limits are placed at, for instance, a 4-sigma distance, then the false alarm probabilities will be smaller but the control chart will take longer to detect assignable causes; even sizeable changes in the process may remain undetected for a longer period of time. False alarms cost money because they may result in an unnecessary stoppage of the process and incur an expense in searching for trouble that does not exist. Therefore, Dr. Shewhart argued that only those assignable causes that can be found "without costing more than it is worth to find" (Shewhart 1939, 30) should be discovered. In his experience, the 3-sigma limits provided that "economic borderline" between detecting those causes worth detecting and those not worth detecting.

We must realize, however, that the economic borderline depends on the cost structure of a given process. In a situation where the false alarm is not expensive, it may be possible to locate the control limits closer to the centerline, with increased power for detecting changes, which would result in savings through detecting assignable causes swiftly. On the other hand, if the false alarm is expensive, the limits should be located farther than the 3-sigma limits, thus saving the expenses from false alarms. Such modifications to the 3-sigma rule are part of the recommendations by researchers who work on the economic design of control charts.

In this connection, the recommendations by Dr. Ott, a pioneer in the use and propagation of statistical methods in industry during the early 1940s, is worth mentioning. Dr. Ott recommended (Ott 1975) that if we know a process is producing defectives and are in a troubleshooting mode to discover assignable causes, we may as well use a narrow set of limits—say, 2-sigma limits—and detect the causes and repair them. The premise is that the larger false alarm probability associated with narrow limits need not be a concern under these conditions, because there are assignable causes in the process to be detected. False alarms should be a concern only when a process is in-control.

4.3.2.8 Frequency of Sampling

Again, economic studies can be made to determine how often the samples should be taken to optimize the control operation. Such studies try to find a balance between the increased cost of sampling from increased frequency, and the benefits of discovering assignable causes soon to avoid damages in the process output. Such studies tend to become complex, however, and beyond the reach of personnel who usually maintain the control charts. A more practical approach is to take samples more frequently during the initial stages of controlling a process and then reduce the frequency once stability is attained. Practical considerations, such as the production rate, the time needed to make the measurements, and the cost of taking the measurements, must be considered when deciding how often samples are taken.

4.3.2.9 Rational Subgrouping

In control chart parlance, the term "subgroup" and the term "sample" mean the same thing. Some prefer the former to the latter, however, because the former clearly implies there is more than one unit in it, which makes communications clearer on the shop floor. We use both the terms here interchangeably.

The term "subgrouping" refers to the method of selecting subgroups, or samples, from the process in order to obtain data for charting. We want to decide the basis of subgrouping in a rational manner. As an example, suppose a process is likely to deteriorate over time, then taking samples based on time, at regular intervals, would be appropriate. If the process changes over time, the samples would then lead to the discovery of such changes. On the other hand, for a process in which the individual skill of the operators makes a difference in the quality of the output, subgrouping should be done based on the operator. That is, if the first subgroup is taken from Operator 1, the next subgroup should be taken from Operator 2, and so on. Then, if a change occurs to the process because of a difference in the performance of one operator compared to others, the charts will show the change, and that change will be traced to the operator causing it. Such subgrouping, carried out to enable the discovery of a cause when it is indicated on the charts, is called "rational subgrouping."

Rational subgrouping is an important idea that should be understood and employed correctly in order to get the most out of control charts. One of the respected textbooks in statistical quality control by Grant and Leavenworth (1996) has a full chapter on rational subgrouping. We would even add that if control charts have not been successful in controlling a process, or if the assignable causes cannot be discovered when the process is still producing defectives, then we can conclude that proper subgrouping is not being used. A change in subgrouping based on some rational hypothesis would help in discovering the causes. In some situations, an entire project may consist of finding the right way of subgrouping by trying one method of subgrouping after another to disclose the assignable causes affecting the process. The following two case studies illustrate the concept of rational subgrouping.

Case Study 4.1

In a department where customer assistance is provided to callers on the company's 1-800 number, the managers were concerned that the time for answering the calls had recently increased, resulting in higher waiting times for—and greater complaints from—the callers. They wanted to find out what was causing the increased time for answering calls so that they could reduce the average answering time and, thus, reduce the waiting time for callers. A control chart was started, plotting the average time per call on a daily basis, on the hypothesis that the calls probably required more time to answer on some days of the week than on others. The process was in-control; the chart did not show any difference from day to day.

Daily averages of individual operators were then plotted, with the basis of subgrouping now being the operator. The process was in-control, showing no significant difference among operators. The basis of subgrouping was next changed to product type, such as calls on doors being separated from calls on openers. The process was in-control; no difference was found among the calls on product types. Then the basis was changed to type of calls, such as emergency calls for help with installation being separated from calls for ordering parts. It was discovered that the calls for installation help were the assignable causes requiring significantly longer answering times than for the other types of services. A separate group of operators was organized who were given additional training on installation questions. All calls relating to installation were directed to this group. The average time for installation calls decreased, as did the average for all calls.

Case Study 4.2

A filler that put liquid detergent in bottles on an automatic filling line had 18 heads. Each head could lose its adjustment, be clogged, or be affected in some way by assignable causes. The amount of liquid filled in bottles was controlled using $\overline{X}$- and R charts. One way to subgroup was to take five bottles every hour at the end of the line, regardless of which head filled the bottles. Such subgrouping would serve the purpose as long as the process was in-control. If the process became not-in-control, however, there would be no clue as to which head needed to be fixed unless further experimentation was done on all 18 heads.

An alternate method of subgrouping was to take five bottles from each head every hour and then plot the average from each head, which would show with certainty when any one head was not-in-control. However, this would be a very expensive way of subgrouping, because so many checks were needed per hour.

A compromise was to take one measurement from each head per hour and then subgroup them by putting bottles from the first six heads in Subgroup 1, bottles from the next six heads in Subgroup 2, and bottles from the last six heads in Subgroup 3. If an $\overline{X}$ value should fall outside the limits, this would at least indicate which group of heads needed further examination. This was a good compromise between the two previous alternatives, and worked well.

4.3.2.10 When the Sample Size Changes for $\overline{X}$- and R-Charts

This refers to the situation in which samples of same size are not available consistently for charting purposes. Such a problem arises less often in the case of measurement control charts than in the case of the attribute charts (discussed in the next section). This is because the sample size used for the measurement charts is small, on the order of four or five, and obtaining such samples consistently from a large production process is not generally a problem. In some situations, however, we suddenly find that there are not enough units in a sample, whether because some of the units were lost or because some readings became corrupted and were removed from the sample. Such situations can be handled by the following approach.

Consider Example 4.1, in which the first set of samples resulted in control limits as shown in Figure 4.4b. Now, suppose another set of 24 samples is taken from this process, and suppose that some of those samples have sizes different from five, as shown in Table 4.3. The limits for those samples with a size different from five are calculated using $\bar{\bar{X}}$ and $\bar{R}$ from the previous data set but with A_2, D_3, and D_4 appropriate to the differing sample size. The $\bar{X}$ and R from the samples with differing sample size are compared with the limits calculated for those samples. (If there are a few samples with different sizes in the original set taken for calculating limits, it is best to discard those samples and replace them with samples of the planned size.)

For example, in the new set of data, if the sample numbers 4, 14, and 23 have sample sizes of four, three, and three, respectively, as shown in the Table 4.3, the limits for those samples are recalculated, as shown in Table 4.3, using A_2, D_3, and D_4 chosen as appropriate to the sample size. The provision available in Minitab to compute limits with a given mean and given standard deviation has been used to calculate the limits and draw the charts shown in Figure 4.5.

4.3.2.11 Improving the Sensitivity of the $\bar{X}$-Chart

In the next chapter, while discussing its operating characteristics, we will see that the $\bar{X}$-chart with 3-sigma limits and a sample size of four or five—referred to as the "conventional charts"—is capable of detecting only large changes in the process mean. If small changes in the process mean (<1.5σ distance from center) occur, the $\bar{X}$ -chart has a very small chance of detecting such changes. The R-chart also has only a very small chance of discovering changes in process variability unless the changes make the process standard deviation at least twice as large as the original.

Sometimes, this insensitivity of $\bar{X}$- and R-charts in discovering small changes is an advantage, because those small changes in the process may not have to be detected anyway. In some situations, however, small changes in the process mean may cause large damage if the production rate is high or the cost of rejects is high, or both. In such situations, it is desirable to have better power for the chart to discover small changes. Several approaches are available.

4.3.2.12 Increasing the Sample Size

One way to increase the power of the $\bar{X}$- and R-charts is to use larger sample sizes. As will be shown in the next chapter, the power of the $\bar{X}$-chart

TABLE 4.3

Data for $\overline{X}$- and R-charts with Varying Sample Sizes

	1	2	3	4	5	6	7	8	9	10	11	12	13	14	15	16	17	18	19	20	21	22	23	24
	22.1	20.3	18.6	20.7	22.2	20.5	21.7	21.4	21.4	22.7	22.4	22.2	22.7	23.4	20.2	21.0	22.6	22.8	19.6	20.1	20.9	19.8	19.5	21.5
	19.6	20.1	21.1	*	23.1	22.7	21.7	21.7	21.7	21.0	21.4	19.2	21.6	*	19.3	23.4	20.2	23.0	22.0	23.8	21.8	19.3	22.3	21.2
	21.8	21.2	24.4	20.5	20.8	20.9	20.8	21.5	19.2	21.3	19.4	19.3	19.5	22.3	22.9	21.1	19.5	21.7	20.3	21.9	20.7	23.0	*	19.3
	20.0	20.8	24.0	18.9	19.4	19.9	23.5	23.1	21.7	22.2	22.2	23.4	20.4	*	21.1	22.6	22.0	20.7	21.2	22.6	22.8	21.7	*	23.4
	19.8	21.7	21.0	21.8	22.8	19.6	22.2	21.6	20.4	21.9	22.6	20.0	19.6	21.6	20.8	22.8	21.8	21.0	20.4	20.8	20.7	22.7	22.3	19.7
$\overline{X}$	20.4			20.3										22.4									21.4	
R	1.0			2.9										1.8									2.8	
A_2	0.58			0.73										1.02									1.02	
D_4	2.11			2.28										2.57									2.57	
UCL($\overline{X}$)	23.11			23.6										24.4									24.4	
LCL($\overline{X}$)	19.63			19.2										18.3									18.3	
UCL (R)	6.384			6.9										7.8									7.8	

Note: Asterisks indicate missing data or empty cell.

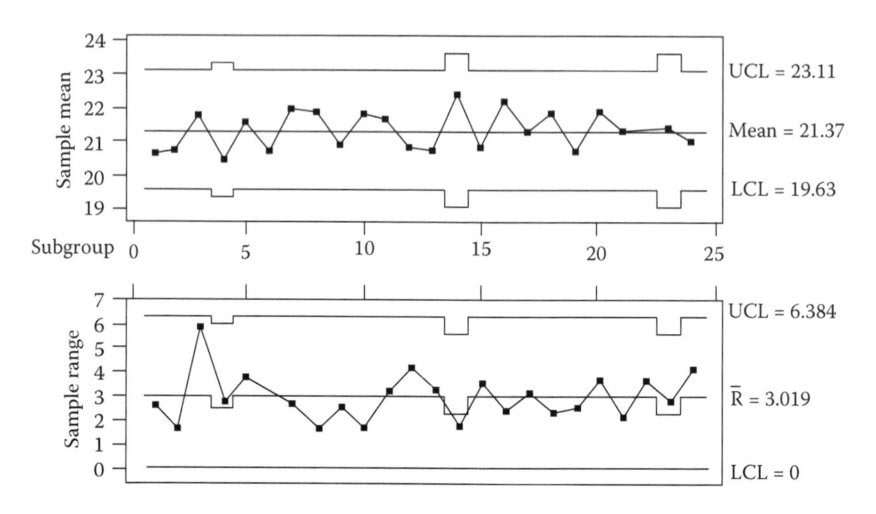

FIGURE 4.5
$\overline{X}$- and R-charts with varying sample sizes.

increases with an increase in sample size, as does the power of the *R*-chart. An increase in sample size, however, also involves an increase in the cost of sampling. When the cost of sampling is not very high, the increase in sensitivity of the charts can be accomplished by increasing the sample size. Of course, we should remember that if a sample size larger than 10 is chosen, then we cannot use the *R*-chart. In this situation we should use the *S*-chart as discussed in a later section.

4.3.2.13 Use of Warning Limits

Another way to improve the sensitivity of the $\overline{X}$-chart is to use warning limits drawn at 1-sigma and 2-sigma distances between the centerline and the (3-sigma) control limits as shown in Figure 4.6a. Rules are then made to help discover assignable causes using the warning limits drawn as above. One such rule: an assignable cause is indicated if two out of three consecutive values fall outside the 2-sigma warning limits on the same side of the centerline. Another rule: an assignable cause is indicated if four out of five consecutive values fall outside of the 1-sigma warning limits on the same side. Examples are shown in Figure 4.6a to illustrate how these rules disclose the occurrence of assignable causes.

4.3.2.14 Use of Runs

Yet another method to enhance the sensitivity of the $\overline{X}$-chart is to use runs. A run is a string of consecutive plots with some common properties. For example, if a sequence of consecutive $\overline{X}$ values occurs below the centerline, this sequence will constitute a "run below the CL," as shown in Figure 4.6b.

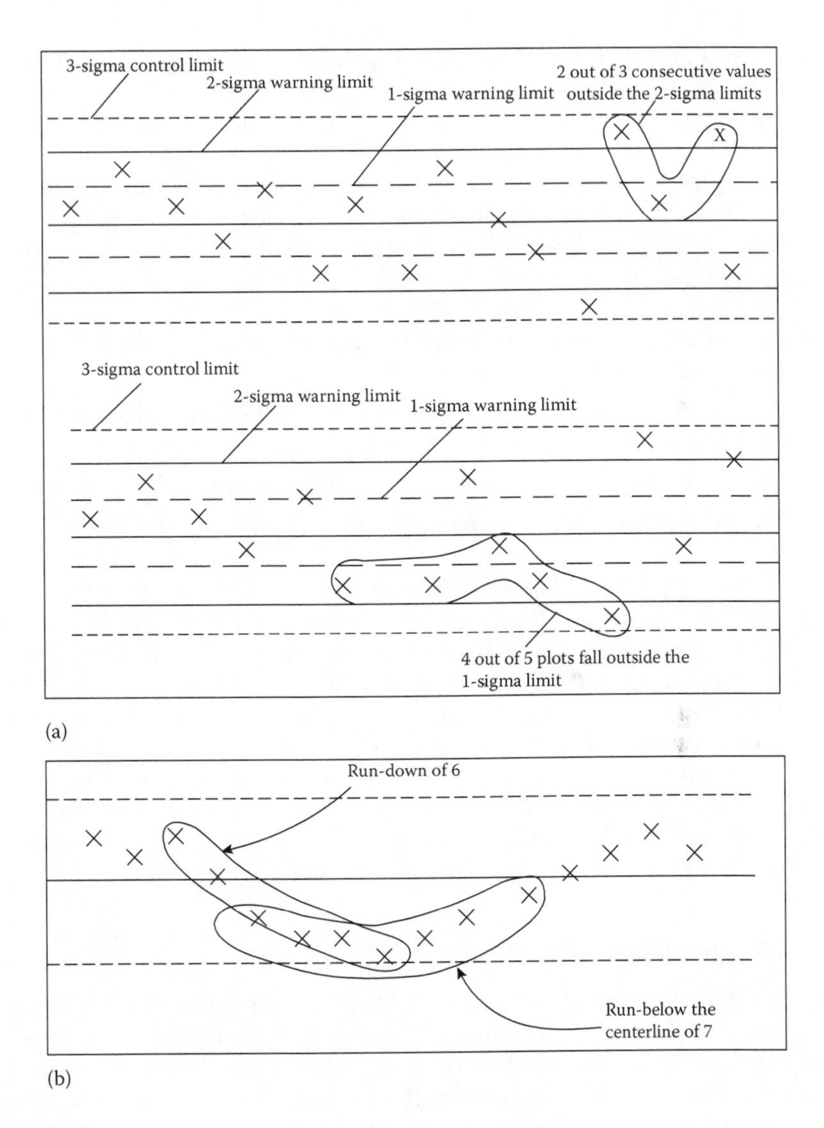

FIGURE 4.6
(a) Use of warning limits on a $\overline{X}$-chart. (b) Examples of runs in a control chart.

Similarly, there could be a "run above the CL." A "run up," would occur if there is a string of plots in which each plot is at a higher level than the previous one and a "run down" would occur if each plot is at a lower level than the previous one. An example of a run down is shown in Figure 4.6b. These runs signify that the process is behaving in a non-normal way because of being affected by an assignable cause. The run above or below the CL would indicate that the process mean has jumped, and the run up or run down would indicate that a drift, or trend, in the process mean is occurring. To help decide when a change has occurred on the process, the following rules are employed. A run longer than seven above or below the centerline is considered to indicate an assignable cause. Similarly, a run longer than seven, up or down, would indicate that an assignable cause is in action (Duncan 1974; Montgomery 2001).

The rules to increase the sensitivity of the $\overline{X}$-chart can be summarized as follows. These rules are used in addition to the rule that any one plot outside of the three sigma limit will indicate an assignable cause. An assignable cause is indicated if:

1. Two of three consecutive plots fall outside of a 2-sigma warning limit on the same side of the centerline.
2. Four of five consecutive plots fall outside of a 1-sigma warning limit on the same side.
3. More than seven consecutive plots fall above or below the centerline.
4. More than seven consecutive plots are in a run up or run down.

These rules are referred to as the "Western Electric rules," because they were originally recommended in the handbook published by the Western Electric Co., which was republished as the *Statistical Quality Control Handbook* (AT&T 1958).

A caution may be in order here. Each rule that is employed to increase the sensitivity of a chart also carries a certain probability of false alarm. When several of the rules are used simultaneously, the overall false alarm probability adds up (according to the addition theorem of probability). Therefore, all the rules should not be used simultaneously. We recommend, based on experience, use of only the rules with runs (Rules 3 and 4 above) to supplement the rule of a single value falling outside the 3-sigma limits.

4.3.2.15 Patterns in Control Charts

Besides runs, other telltale signs also show up in the charts that can be used by an analyst to understand what is happening in the process and possibly discover assignable causes. Some of these patterns are shown in Figure 4.7. The comments accompanying each pattern suggest how an analyst can interpret the patterns. The cycles and trends are the most common patterns that help in discovering causes.

4.3.2.16 Control vs. Capability

A process is in-control if it is operating consistently within its natural variability. A process is capable if it produces products "entirely" within specification. When we find a process in-control, however, it does not mean that the process is also capable. There could be a situation in which a process is producing a characteristic that is uniformly the same but uniformly outside the specification. The process will then be in-control but totally not

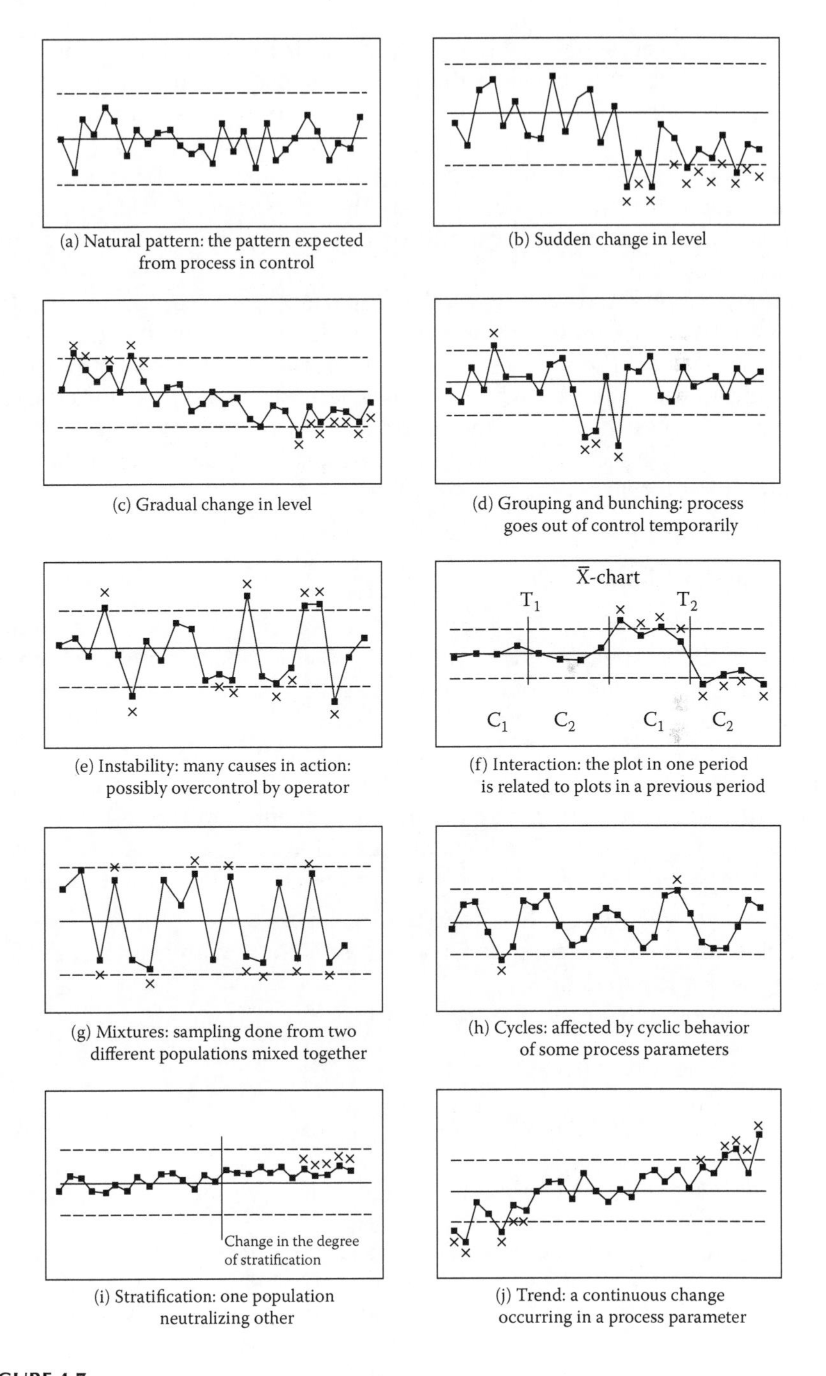

(a) Natural pattern: the pattern expected from process in control

(b) Sudden change in level

(c) Gradual change in level

(d) Grouping and bunching: process goes out of control temporarily

(e) Instability: many causes in action: possibly overcontrol by operator

(f) Interaction: the plot in one period is related to plots in a previous period

(g) Mixtures: sampling done from two different populations mixed together

(h) Cycles: affected by cyclic behavior of some process parameters

(i) Stratification: one population neutralizing other

(j) Trend: a continuous change occurring in a process parameter

FIGURE 4.7
Patterns in control charts. (Compiled from AT&T., *Statistical Quality Control Handbook*. 2nd ed. Indianapolis, IN: AT&T Technologies, 1958.)

capable. Therefore, when we know a process is in-control, a further verification is necessary to check if the process is also capable. If the process is not capable, then further work is necessary to achieve capability. Such work is referred to as a "capability study." There are measures to evaluate quantitatively the extent to which a process is capable; that is, the extent to which the output of a process meets the specification. Such measures, which are known as "process capability indices," are discussed later in this chapter.

For now, it should be emphasized that when a process is brought in-control using a control chart, only part of the work is done. A capability study should be done to verify if the process also meets specifications. However, evaluating a process for capability when it is not-in-control would not make sense. Because a process that is not-in-control has no predictable behavior and, therefore, has no predictable capability. A capability study should be done only after the process is brought in-control.

4.3.3 $\bar{X}$ and S-Charts

Although the *R*-chart is commonly used to control the process variability because of the simplicity in computing *R*, in several situations the *S*-chart, or the standard deviation chart, is preferred. For example, when the sample size must be larger than 10 because that extra sensitivity is needed for the $\bar{X}$-chart, the *R*-chart cannot be used because of the poor efficiency (i.e., larger variability) of the statistic *R* with large sample sizes. Furthermore, with the availability of the modern calculator and implementation of SPC tools on the computer, the simplicity advantage of the *R*-chart may not be a real advantage any longer. Therefore, the *S*-chart can be used to take advantage of its efficiency. For the *S*-chart, the standard deviation $S = \sqrt{\Sigma(X_i - \bar{X})^2 / n-1}$ is calculated for each subgroup and plotted on a chart with limits calculated and drawn for the statistic *S*.

The control limits for the $\bar{X}$- and *S*-charts are:

$$\text{UCL}(S) = B_4\bar{S} \qquad \text{UCL}(\bar{X}) = \bar{\bar{X}} + A_3\bar{S}$$

$$\text{CL}(S) = \bar{S} \qquad \text{CL}(\bar{X}) = \bar{\bar{X}}$$

$$\text{LCL}(S) = B_3\bar{S} \qquad \text{LCL}(\bar{X}) = \bar{\bar{X}} - A_3\bar{S}$$

where A_3, B_3, and B_4 are factors that give 3-sigma limits for $\bar{X}$ and *S* and can be found in standard tables (Table 4.2, or Table A.4 in the Appendix). The difference between using the *S*-chart instead of the *R*-chart in conjunction with the $\bar{X}$-chart is in calculating the value of *S* instead of *R* for each sample, and using the values of $\bar{S}$ for calculating the limits. The methods of charting and interpreting these charts are the same as those for the $\bar{X}$- and *R*-chart combination.

TABLE 4.4

Calculation of $\overline{X}$ and S for Data from Example 4.1

	1	2	3	4	5	6	7	8	9	10	11	12	13	14	15	16	17	18	19	20	21	22	23	24	Average
	22	20.5	20	21	22.5	23	19	21.5	21	21.5	20	19	19.5	20	22.5	21.5	19	21	20	22	19	21.5	22.5	22.5	
	22.5	22.5	20.5	22	19.5	23.5	20	20.5	22.5	23	19.5	21	20.5	21.5	19.5	20.5	21.5	20.5	23.5	20.5	20.5	25	22	22	
	22.5	22.5	23	22	22.5	21	22	19	20	22	21	21	21	24	21	22	23	19.5	24	21	21	21	23	22	
	24	23	22	23	22	22	20.5	19.5	22	23	20	21	20.5	23	21.5	21.5	21	22	20.5	22.5	20.5	18	22	19.5	
	23.5	21.5	21.5	22	21	20	22.5	19.5	22	18.5	20.5	20.5	21	20	21	23.5	23.5	21	21.5	20	22.5	21	23.5	20.5	
$\overline{X}$	22.9	22	21.4	22	21.5	21.9	20.8	20	21.5	21.6	20.2	20.5	20.5	21.7	21.1	21.8	21.6	20.8	21.9	21.2	20.7	21.3	22.6	21.3	21.37
S	0.82	1.00	1.19	0.71	1.27	1.43	1.44	1.00	1.00	1.85	0.57	0.87	0.61	1.79	1.08	1.10	1.78	0.91	1.78	1.04	1.25	2.49	0.65	1.25	1.20

Example 4.2

The data from Example 4.1 is reproduced in Table 4.4 and the calculation of $\overline{X}$ and S is shown in the table. Calculate the limits for the charts based on $\overline{S}$.

Solution

Control limits for S-chart

$$\text{UCL}(S) = B_4\overline{S} = 2.09(1.2) = 2.51$$

$$\text{CL}(S) = \overline{S} = 1.2$$

$$\text{LCL}(S) = B_3\overline{S} = 0$$

Control limits for $\overline{X}$-chart

$$\text{UCL}(\overline{X}) = \overline{\overline{X}} + A_3\overline{S}$$

$$= 21.37 + 1.43(1.2) = 23.09$$

$$\text{CL}(\overline{X}) = \overline{\overline{X}} = 21.37$$

$$\text{LCL}(\overline{X}) = \overline{\overline{X}} - A_3\overline{S}$$

$$= 21.37 - 1.43(1.2) = 19.65$$

The charts made using Minitab are shown in Figure 4.8.

For these same set of data we had prepared the $\overline{X}$-and R-charts in Example 4.1. In comparing the R-chart in Figure 4.4a and the S-chart in Figure 4.8, we notice that Sample 22, which had its R value outside the limits in the R-chart, has its S value inside the limits in the S-chart. The S-chart shows that the process is in-control even if the S value of the one sample is very close to the limits in the S-chart. We have to believe that the S-chart represents the true state of affairs better than the R-chart in this case, because the statistic S is a better estimator of the process variability than the statistic R.

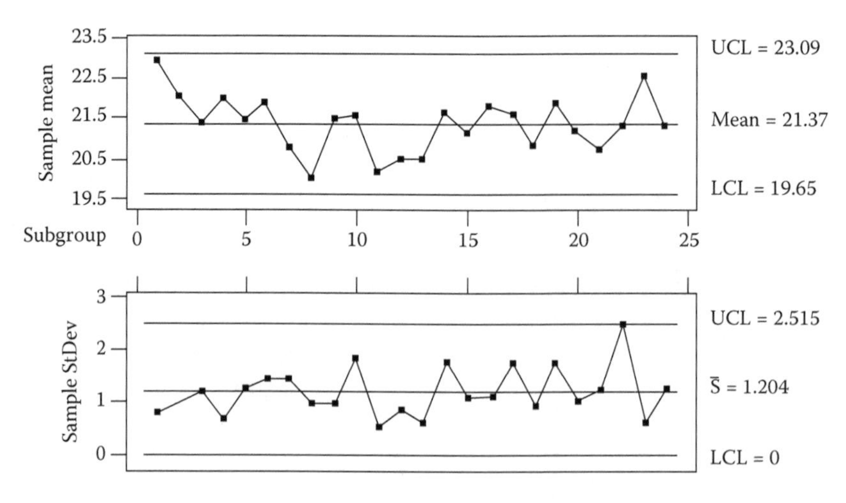

FIGURE 4.8
$\overline{X}$- and S-charts for the data in Example 4.1.

4.4 Attribute Control Charts

We will discuss here two attribute charts: the *P*-chart, and the *C*-chart. The *P*-chart is used to control and minimize the proportion defectives in a process, and the *C*-chart is used to control and minimize the number of defects per unit produced in a process. The details on where they are used and how the limits are calculated are explained below. Some minor variations of the *P*-chart and *C*-chart are also included here.

4.4.1 The *P*-Chart

The *P*-chart is also known as the "fraction defectives" or "fraction nonconforming chart," because it is used to monitor and control the fraction produced in a process that is defective or nonconforming. The terms "defective" and "nonconforming" need to be explained first. The term defective was the original name used in SPC literature for a unit of a product that did not meet the specified requirements, and so the *P*-chart was known as the fraction defectives chart. The term nonconforming has become more commonly used in the literature in place of the term defective, however, since around 1980, when liability suits and awards became more common. The term defective was thought to convey a negative connotation, as if such a unit was unsafe or dangerous to use. Therefore, the term nonconforming, because it simply means that the unit in question is not meeting the chosen specification, became the accepted term. Here, we use the term defective for its easy readability but with the same meaning as the term nonconforming.

The *P*-chart is typically used with a large, continuous production that has some defectives produced and where the proportion of defectives should be monitored and reduced to a minimum—even to zero levels. The *P*-chart usually requires a large sample size ($n > 20$). The method consists of taking a sample of size n at a regular time interval, inspecting the sample, and counting the number of defective units D in the sample. Then, the proportion defectives in the i-th sample is computed as: $p_i = D_i/n$ and plotted on the chart. The control limits are calculated using the formulas:

$$\mathrm{UCL}(P) = \bar{p} + 3\sqrt{\frac{\bar{p}(1-\bar{p})}{n}}$$

$$\mathrm{CL}(P) = \bar{p}$$

$$\mathrm{LCL}(P) = \bar{p} - 3\sqrt{\frac{\bar{p}(1-\bar{p})}{n}}$$

where $\bar{p}$ is the average proportion defectives in about 25 samples. Thus, about 25 samples from the process are necessary to start a P-chart to control a process.

To make the meaning of the notations clear:

- p is the unknown proportion defectives in the population that is being controlled;
- P is the statistic that represents the proportion defectives in a sample and is used to estimate p;
- p_i is the proportion defectives or the value of the statistic P in the i-th sample; and
- $\bar{p} = \sum_{i=1}^{k} p_i/k$ is the average of the p_i values, where k is the number of samples taken.

The chart takes the name of the statistic being plotted.

Derivation of the above formulas for the limits is given in Chapter 5. For now, however, we just want to recognize that P is related to D, which is a binomial random variable, and that the limits for P are calculated as 3-sigma limits where "sigma" stands for $\sigma_{P,}$ the standard deviation of the statistic P.

Example 4.3

The process was an automatic lathe producing wood screws. Samples of 50 screws were taken every half-hour, and the screws were gaged for length, diameter of head, and slot position, and were also visually checked for finish. The number rejected in each sample was recorded, as shown in Figure 4.9a. Prepare a control chart to monitor the fraction defectives in the process.

Solution

The fraction defectives in the samples are computed as p_is. For example, the proportion defectives in Sample 5 is 3/50 = 0.06 and that in Sample 14 is 1/50 = 0.02. The average of the p_is, $\bar{P}$, is computed as 0.027. The limits are:

$$\text{UCL}(P) = 0.027 + 3\sqrt{\frac{0.027 \times 0.973}{50}} = 0.096$$

$$\text{CL}(P) = 0.027$$

$$\text{LCL}(P) = 0.027 - 3\sqrt{\frac{0.027 \times 0.973}{50}} = -0.042 \simeq 0$$

Note that the denominator within the radical sign is entered as 50, being the sample size or number of units in each sample. Note also that when the LCL is obtained as a negative number, it is rounded off to zero. The p_i values are shown

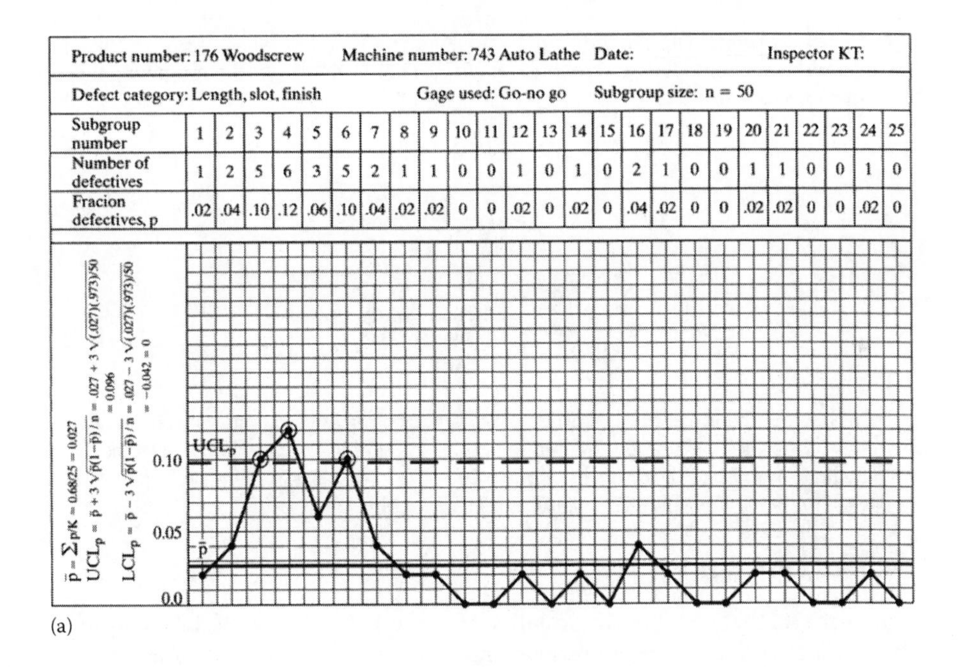

Product number: 176 Woodscrew	Machine number: 743 Auto Lathe	Date:	Inspector KT:
Defect category: Length, slot, finish	Gage used: Go-no go	Subgroup size: n = 50	

Subgroup number	1	2	3	4	5	6	7	8	9	10	11	12	13	14	15	16	17	18	19	20	21	22	23	24	25
Number of defectives	1	2	5	6	3	5	2	1	1	0	0	1	0	1	0	2	1	0	0	1	1	0	0	1	0
Fracion defectives, p	.02	.04	.10	.12	.06	.10	.04	.02	.02	0	0	.02	0	.02	0	.04	.02	0	0	.02	.02	0	0	.02	0

(a)

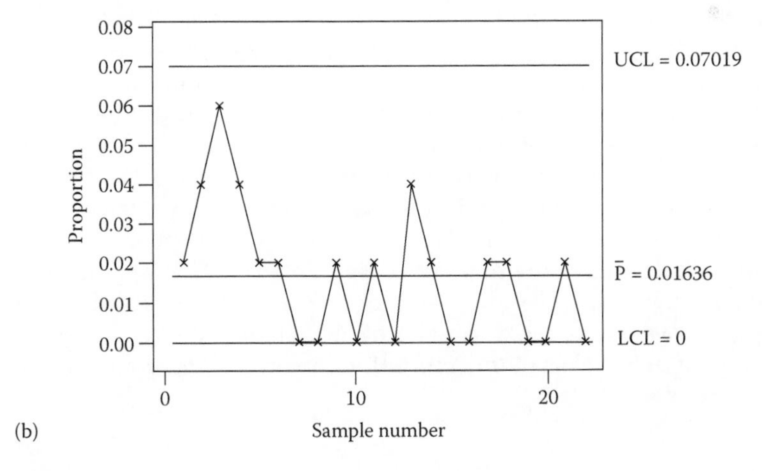

(b)

FIGURE 4.9
(a) An example of a *P*-chart. (b) *P*-chart after removing p values outside the limit.

plotted in Figure 4.9a in relation to the limits. The chart shows that the process is not in-control. Action is necessary to find out why there are p_i values outside the limits. The reason was that several large p_i values occurred at the start of the process because the process was going through a set up. When the set up was complete, the p_i values settled down at a lower level.

When the causes for the values outside the limits are known and there is assurance that these causes would not reoccur, the p_i values outside the limits are removed from the data and new limits are calculated. The p_i values from Samples 3, 4, and 6 are removed, and the new limits are calculated from the remaining 22 subgroups as follows:

$$\bar{p} = 0.36/22 = 0.016$$

$$\text{UCL}(P) = 0.016 + 3\sqrt{\frac{0.016 \times 0.984}{50}} = 0.069$$

$$\text{CL}(P) = 0.016$$

$$\text{LCL}(P) = 0.016 - 3\sqrt{\frac{0.016 \times 0.984}{50}} = -0.037 = 0$$

Again, when the lower control limit is obtained as a negative quantity it is rounded up as 0. We see that all the remaining p_i values fall within the new limits, and these new limits can be used for further control of the process. The *P*-chart with the new limits, drawn using Minitab, is shown in Figure 4.9b.

4.4.2 The *C*-Chart

The *C*-chart is used when the quality of a product is evaluated by counting the number of blemishes, defects, or nonconformities on units of a product. For example, the number of pinholes may be counted to determine the quality of glass sheets, or the number of gas holes may be counted to determine the quality of castings. In these cases, a certain number of nonconformities may be tolerable, but the number must be monitored, controlled, and minimized. The *C*-chart would be the appropriate tool for this purpose. The *C*-chart is also known as the "defects-per-unit" chart or "control chart for nonconformities." (The term nonconformity is just a politically acceptable term for a defect.)

The procedure consists of selecting a sample unit from the process at regular intervals and counting the number of defects on it. These counts per (sample) unit are the observed values of *C*, or c_is. After inspecting about 25 sample units, the average of the c_i values are calculated as $\bar{c} = \sum c_i/k$, where k is the number of units selected for computing the limits. The limits for the control chart are calculated as:

$$\text{UCL}(C) = \bar{c} + 3\sqrt{\bar{c}}$$

$$\text{CL}(C) = \bar{c}$$

$$\text{LCL}(C) = \bar{c} - 3\sqrt{\bar{c}}$$

To explain the notations:

- c is the average number of defects per unit in the population, which is to be controlled;
- C is the statistic representing the number of defects in a sample unit, which is used as the estimator for c;

- c_i is an observed value of the statistic C in any sample unit; and
- $\bar{c}$ is the average of the observed values.

The chart takes the name of the statistic that is plotted. The reader can recognize C as the Poisson variable, and the limits are calculated as 3-sigma limits, where "sigma" stands for the standard deviation of the statistic C, σ_C. The details of the derivation are given in Chapter 5.

Example 4.4

The process is a laminating press that puts plastic lamination on printed art sheets, which are later cut and finished into a credit card-type product. The number of chicken scratches on the laminated sheets is to be controlled by counting the number per sample sheet.

Solution

A C-chart is used to control the defects per sheet. The data obtained from 25 sample sheets is shown in Figure 4.10a. The figure also shows the calculation of control limits and the plot of the observed values of C. The limits are calculated as follows:

$$\bar{c} = 14.1$$

$$UCL(C) = 14.1 + 3\sqrt{14.1} = 25.4$$

$$CL(C) = 14.1$$

$$LCL(C) = 14.1 - 3\sqrt{14.1} = 2.8$$

[Note: As in the case of the P-chart, if the lower control limit came out to be a negative value, it must be rounded up to 0.]

The chart shows the process not-in-control.

It is possible, as in other charts, to remove the points outside the limits from the data and recalculate the limits for future use, assuming that the causes have been rectified. Sample 20 is removed, and new limits are calculated as follows: the new $\bar{c}$ is 13.5, and the new limits are:

$$UCL(C) = 13.5 + 3\sqrt{13.5} = 24.52$$

$$CL(C) = 13.5$$

$$LCL(C) = 13.5 - 3\sqrt{13.5} = 9.82$$

Unit 15 is now seen outside these limits. If this value is removed, the new $\bar{c}$ would be 13.0, and the new limits would be as follows:

$$UCL(C) = 13.0 + 3\sqrt{13.0} = 23.81$$

$$CL(C) = 13.0$$

$$LCL(C) = 13.0 - 3\sqrt{13.0} = 2.18$$

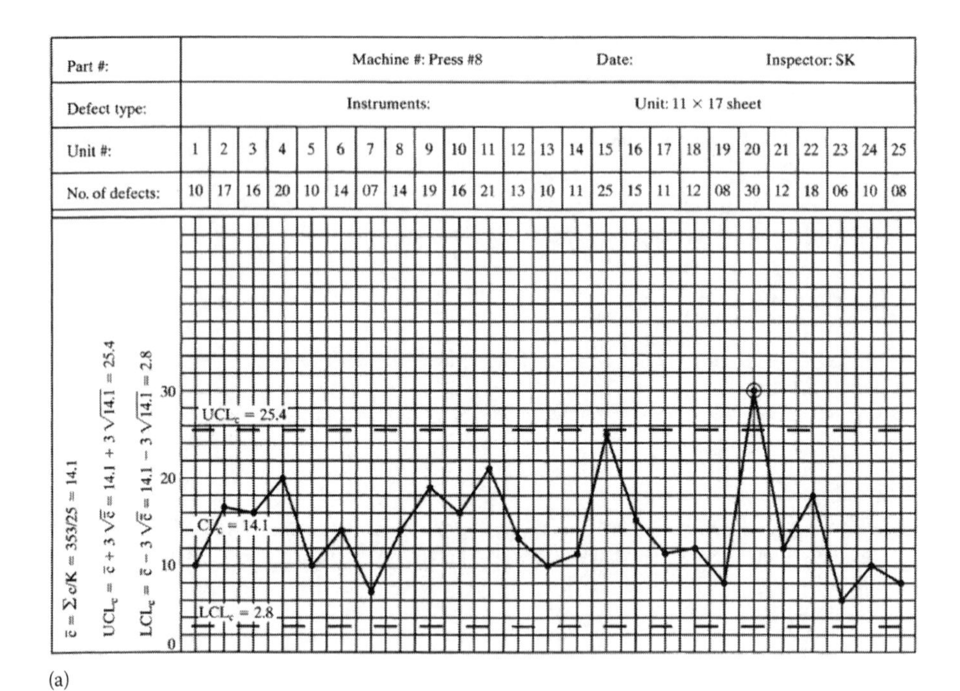

(a)

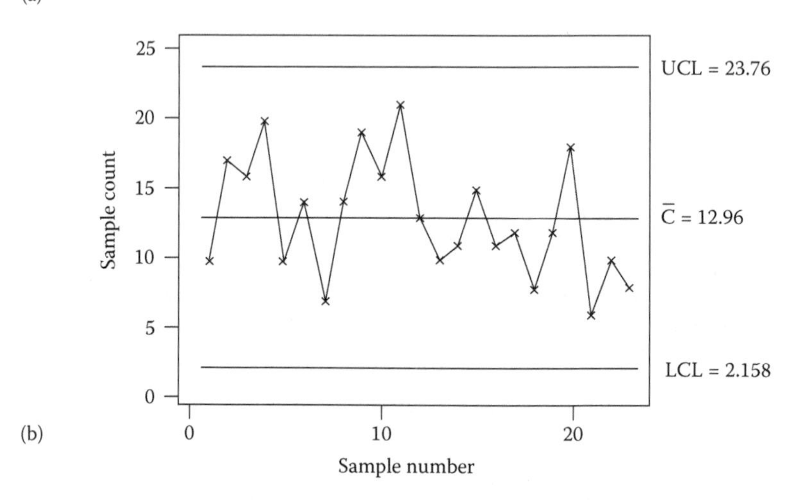

(b)

FIGURE 4.10
(a) Example of a C-chart. (b) C-chart with revised limits.

> The remaining c_i values are all within these limits, and these limits can be used for the next stage of control. The chart with the revised limits, drawn using Minitab, is shown in Figure 4.10b.

This method of calculating control limits for the future by using the remaining data is recommended, because often taking a whole new set of samples is expensive or time consuming. When the process is badly not-in-control and several assignable causes are discovered and removed, the first set of data may have no relevance to the future process. In that case, a new set of data should be taken from the "new" process, and new limits should be computed from them for future use. Experienced observers say that if more than 50% of original data has to be discarded because they came from assignable causes, then it is advisable to take a new set of data from the new process.

The chicken scratches on the laminated sheets were caused by several factors—temperature and surface condition of the platen used in the press, pressure applied, thickness of the plastic sheet, and the atmospheric humidity were the most prominent. Use of the *C*-chart enabled the discovery of a few abnormal conditions that resulted in plots outside the limits. It also brought about the need for further investigation, because the average number of scratches per sheet was too high even after the process was in-control.

An experiment was conducted to discover the optimal pressure and temperature of the press. A specialist in plastic technology was brought in to help with the investigation. Several remedies were applied, and the process improved considerably. This is an example where all the assignable causes were not obvious and some engineering investigation was required to find and rectify them.

4.4.3 Some Special Attribute Control Charts

4.4.3.1 The P-Chart with Varying Sample Sizes

The discussion of the *P*-chart earlier assumed that samples of size n could be drawn repeatedly from the process to be controlled and that the number of defective units in each of the equal-size samples could be counted for control purposes. In many situations, samples of the same size may not be available each time we want to take a sample. For example, in one situation in the manufacture of tool boxes, all boxes produced in a day constituted a sample for a *P*-chart used for controlling defectives due to poor workmanship. The number of boxes produced varied from day to day; hence, a *P*-chart that can accommodate varying sample sizes was needed.

The sample size n goes into the calculation of control limits. When the sample size is different from sample to sample, the question arises as to which sample size should be used in the formula to calculate limits.

One way to handle the varying sample sizes that is recommended in the literature is to use an average value for n—say, $\bar{n}$—if the sample size does not vary too much (the largest and the smallest do not vary by more than 25% from the average). This rule is not based on any science but is considered adequate for practical purposes. A more exact approach would be to calculate limits for each sample, $\mathrm{UCL}_i(P)$ and $\mathrm{LCL}_i(P)$, based on the sample size n_i for that sample, and compare the observed p_i from that sample against the limits

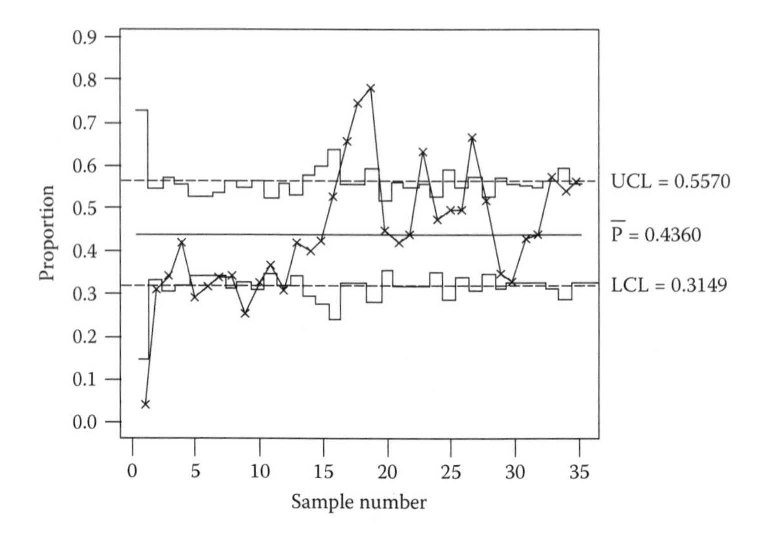

FIGURE 4.11
Example of a *P*-chart with varying sample sizes.

thus calculated. The *P*-chart then looks like the one shown in Figure 4.11. These limits are called "stair-step" limits because of the way they look.

Yet another approach is a compromise between the two approaches described above. In this, the limits for the chart is calculated using an $\bar{n}$, which results in constant limits. If a p_i value falls close to the limit, then the limits for that sample are calculated using the exact size of that sample, and the p_i value is compared with those limits. This contributes to speedy charting while also allowing clarification in doubtful cases.

Example 4.5

Table 4.5 shows data on the daily production of castings in a foundry and the number found each day with the "burn-in" defect for a period of 35 days. Draw a *P*-chart for the proportion of castings with burn-in defects, and make any observation that is discernible from the chart.

Solution

The control limits are calculated for each sample based on the size of that sample using the following formulas:

$$\text{UCL}(P) = \bar{p} + 3\sqrt{\frac{\bar{p}(1-\bar{p})}{n_i}}$$

$$\text{CL}(P) = \bar{p}$$

$$\text{LCL}(P) = \bar{p} - 3\sqrt{\frac{\bar{p}(1-\bar{p})}{n_i}}$$

where n_i is the size of the i-th sample and $\bar{p}$ is the average proportion defectives in 35 samples.

For example, for the 17th sample,

$$\mathrm{UCL}(P) = 0.436 + 3\sqrt{\frac{0.436(0.564)}{158}} = 0.554$$

$$\mathrm{LCL}(P) = 0.436 - 3\sqrt{\frac{0.436(0.564)}{158}} = 0.318$$

Therefore, the observed value of $p_i = 0.658$, for this sample is outside the limits. As another example, for the 35th sample,

$$\mathrm{UCL}(P) = 0.436 + 3\sqrt{\frac{0.436(0.564)}{151}} = 0.557$$

$$\mathrm{LCL}(P) = 0.436 - 3\sqrt{\frac{0.436(0.564)}{151}} = 0.315$$

Therefore, the observed value of $p_i = 0.556$ is inside the limits.

TABLE 4.5

Data on Sample Size and Number of Defective Castings

Sample No.	n	No. Burn-in	p_i	Sample No.	n	No. Burn-in	p_i
1	26	1	0.038	18	150	112	0.747
2	177	54	0.305	19	88	69	0.784
3	128	43	0.336	20	305	137	0.449
4	150	63	0.420	21	147	61	0.415
5	230	67	0.291	22	167	73	0.437
6	236	73	0.309	23	159	101	0.635
7	204	69	0.338	24	266	126	0.474
8	143	49	0.343	25	93	46	0.495
9	171	42	0.246	26	190	94	0.495
10	136	44	0.324	27	128	85	0.664
11	276	100	0.362	28	259	135	0.521
12	156	47	0.301	29	127	43	0.339
13	234	98	0.419	30	158	51	0.323
14	108	43	0.398	31	164	70	0.427
15	85	36	0.424	32	170	74	0.435
16	55	29	0.527	33	140	80	0.571
17	158	104	0.658	34	94	51	0.543
				35	151	84	0.556
						$\bar{p} =$	0.436

Suppose we want to calculate the constant control limits based on an average sample size, $\bar{n} = 161$, and the limits would be:

$$\mathrm{UCL}(P) = 0.436 + 3\sqrt{\frac{0.436(0.564)}{161}} = 0.553$$

$$\mathrm{LCL}(P) = 0.436 - 3\sqrt{\frac{0.436(0.564)}{161}} = 0.319$$

The P-chart with varying sample sizes plotted using Minitab software is shown in Figure 4.11. The constant control limits based on an average sample size are also shown on the figure. In this case, we see that the constant control limits show the same number of plots outside the limits as the varying limits do. The process is not in-control; there is a sudden increase in the defective rate after the 15th sample. The defective rate also shows an upward trend even before the 15th sample. Further investigation is therefore needed.

4.4.3.2 The nP-Chart

For the P-chart, the statistic P, proportion defectives in a sample, is calculated as $P = D/n$, where D is the number of defectives in a sample of size n. If $P = D/n$, then $nP = D$. Instead of plotting P, suppose we plot nP, which is simply the number of defectives found in a sample, we get the nP-chart. The control limits must then be calculated for the statistic nP. The easiest way to calculate the limits for nP is to calculate the limits for the corresponding P-chart first and then multiply the limits by the sample size n. The following formulas can also be used for calculating the limits, which have been obtained by multiplying by n the formulas for the limits of the P-chart:

$$\mathrm{UCL}(nP) = n\bar{p} + 3\sqrt{n\bar{p}(1-\bar{p})}$$

$$\mathrm{CL}(nP) = n\bar{p}$$

$$\mathrm{LCL}(nP) = n\bar{p} - 3\sqrt{n\bar{p}(1-\bar{p})}.$$

The chart with limits calculated as shown above is called the nP-chart. (It could also have been called the D-chart.) The results of using a P-chart and an nP-chart will be the same. For example, the two charts with the following two sets of limits will produce the same result. With the one on the left, the fraction defectives, p_i, will be plotted, and with the one on the right, the number of defectives, np_i, will be plotted.

***P*-chart (*n* = 50)**	***nP*-chart (*n* = 50)**
UCL(P) = 0.07	UCL(nP) = 3.5
CL(P) = 0.016	CL(nP) = 0.80
LCL(P) = 0.0.	LCL(nP) = 0.0.

The advantage in using the nP-chart can easily be seen. First, it avoids calculation of p_i for each sample. Second, most people prefer to deal with the number of defectives, which is in whole number, rather than fraction defectives, which is in decimal fraction. Of course, this chart cannot be used if n does not remain constant.

4.4.3.3 The Percent Defectives Chart (100P-Chart)

If, instead of multiplying the limits of the P-chart by n, they are multiplied by 100, then the limits for percent defectives are obtained. For example, the charts with the following two sets of limits are equivalent. Whereas p_i values will be plotted on the chart with the limits on the left, $100p_i$ (percent defectives) will be plotted on the chart with the limits on the right.

***P*-chart**	**100*P*-chart**
$UCL(P) = 0.07$	$UCL(100P) = 7.0$
$CL(P) = 0.016$	$CL(100P) = 1.6$
$LCL(P) = 0.0.$	$LCL(100P) = 0.0.$

The advantage of using the 100P-chart, or percent defectives chart, is that the plotted numbers will be large compared to the small fractions that are encountered with the use of the P-chart. People like to handle large numbers, and they tend to grasp the meaning of percent defectives more easily than fraction defectives.

4.4.3.4 The U-Chart

The U-chart is a variation of the C-chart. The C-chart described previously can only be used if all the units inspected for the chart are identical—that is, if the opportunity for a defect is the same from unit to unit. Several situations exist in which an inspection station receives units that are similar but not identical, such as television sets of different sizes, cars of different models, or printed circuits of different configurations. When units of different sizes are being inspected and one chart is used to monitor all the units arriving at the inspection station, the U-chart is needed, in which the statistic U represents the average number of defects per (standard) unit.

When the size of units varies from unit to unit, we define one size as a standard unit and then find out the number of standard units in each of the other sizes. This is done by comparing the opportunities for a defect in any unit with that in the standard unit. We then calculate the average number of defects per standard unit in each sample unit. This quantity is designated as U. The limits for U are calculated using the following formulas:

$$\text{UCL}(U) = \bar{u} + 3\sqrt{\frac{\bar{u}}{n_i}}$$

$$\text{CL}(U) = \bar{u}$$

$$\text{LCL}(U) = \bar{u} - 3\sqrt{\frac{\bar{u}}{n_i}}$$

where $\bar{u}$ is the average of the u values from the sample units and n_i is the number of standard units in each of the sample units. Because n varies from sample to sample, the limits need to be calculated for each sample unit and the u_i value from each sample compared with its corresponding limits. We then have the same situation as in the case of the *P*-chart with varying sample sizes (i.e., the chart with stair-step limits). As an alternative, an average of the n values can be calculated as $\bar{n}$ and used in the denominator inside the radical sign in the above formulas, resulting in a constant set of limits. The example below illustrates the use of the *U*-chart.

Example 4.6

In a foundry that produces engine blocks, a salvage welder fills gas holes on castings if the holes do not affect the structural strength of the castings. The number of holes filled per casting, however, must be kept track of, and any unusual occurrence (i.e., when the process is not-in-control) must be conveyed to the mold line. A control chart is needed to control the number of holes in the castings.

Solution

The castings are of different sizes and the block I-4 (inline 4-cylinder) is chosen as the standard unit. The numbers of standard units in other blocks have been determined based on their surface areas (opportunities for defects). The data on the holes per casting and the number of standard units in each casting are given in Table 4.6 for 20 castings. Because the casting size varies, a *U*-chart will be used. The calculation of u_i for each sample unit is shown in Table 4.6.

$$\bar{u} = \frac{\sum c_i}{\sum n_i} = \frac{143}{32.7} = 4.373$$

$$\bar{n} = \frac{32.7}{20} = 1.635$$

The chart drawn using Minitab is shown in Figure 4.12. The chart shows the limits calculated for each sample individually. For example, the limits are calculated for the eighth sample as:

TABLE 4.6

Data on Salvaged Holes in Castings

Sample No.	Casting Type	No. of Standard Units (n_i)	No. of Holes (c_i)	No. of Holes/Standard Unit ($u_i = c_i/n_i$)
1	V6	1.5	4	2.7
2	I–4	1.0	6	6.0
3	I–4	1.0	3	3.0
4	I–6	1.4	7	5.0
5	V6	1.5	9	6.0
6	V8	2.0	13	6.5
7	V6	1.5	10	6.7
8	I–4	1.0	11	11.0
9	I–6	1.4	10	7.1
10	V8	2.0	9	4.5
11	V12	3.0	7	2.3
12	V6	1.5	4	2.7
13	I–4	1.0	5	5.0
14	V8	2.0	7	3.5
15	I–6	1.4	3	2.1
16	I–4	1.0	5	5.0
17	V8	2.0	8	4.0
18	V12	3.0	9	3.0
19	V6	1.5	4	2.7
20	V8	2.0	9	4.5
		$\Sigma n_i = 32.7$	$\Sigma c_i = 143$	

$$\text{UCL}(U) = 4.373 + 3\sqrt{\frac{4.373}{1.0}} = 10.646$$

$$\text{CL}(U) = 4.373$$

$$\text{LCL}(U) = 4.373 - 3\sqrt{\frac{4.373}{1.0}} = -1.9 \simeq 0$$

The negative value for the LCL is rounded off to zero. The observed value $u_8 = 11$ is outside the upper limit. As another example, the limits for the 20th sample are

$$\text{UCL}(U) = 4.373 + 3\sqrt{\frac{4.373}{2}} = 8.809$$

$$\text{LCL}(U) = 4.373 - 3\sqrt{\frac{4.373}{2}} = -0.06 \simeq 0$$

The observed $u_{20} = 4.5$ is inside the limits.

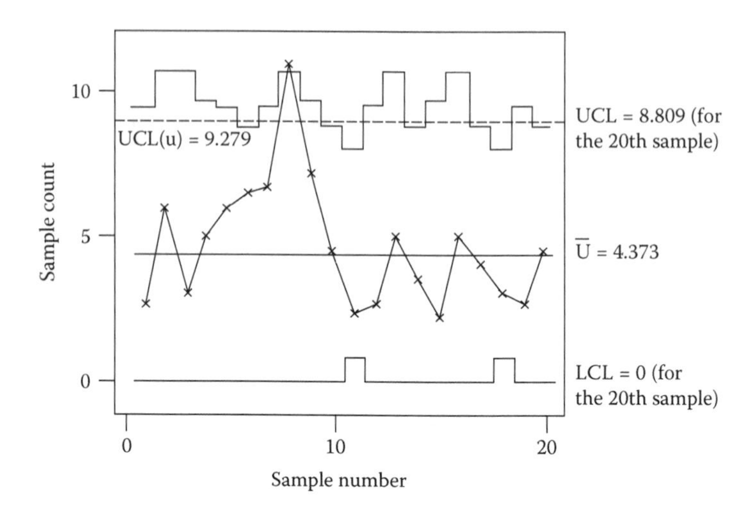

FIGURE 4.12
Example of a *U*-chart for number of holes per casting.

The limits using $\overline{n}$ are calculated as:

$$\text{UCL}(U) = 4.373 + 3\sqrt{\frac{4.373}{1.635}} = 9.279$$

$$\text{LCL}(U) = 4.373 - 3\sqrt{\frac{4.373}{1.635}} = -0.53 = 0$$

The control limit calculated using $\overline{n}$ has been added to the graph in Figure 4.12. In this case, we see that the constant control limits and the varying control limits produce the same result in the sense that if a u value is inside/outside the varying limit, it is also inside/outside the constant limit.

4.4.4 A Few Notes about the Attribute Control Charts

4.4.4.1 Meaning of the LCL on the P- or C-Chart

The lower control limit does not have the same significance with the *P*-chart, or the *C*-chart, as it has with the $\overline{X}$-chart. When a p_i value on a *P*-chart or a c_i value on a *C*-chart falls below the LCL, it indicates that the process has changed for the better. For this reason, when we recalculate limits using remaining data, we do not remove values below the lower limit in either the *P*-chart or the *C*-chart. Sometimes, the values below the LCL will come inside the limit when those limits are recalculated after removing those outside the UCL. A caution, however, seems to be in order: care must be taken to make certain that the values below the lower limit are not caused by a faulty instrument or an erring inspector.

4.4.4.2 P-Chart for Many Characteristics

One of the advantages of the *P*-chart is that one chart can be used for several product characteristics. The product will be classified as defective if any one of the characteristics is outside the acceptable limits. It is often a good idea to start using one *P*-chart for several characteristics in order to identify the characteristic(s) that causes the most problems and then use a *P*- or $\overline{X}$-chart for those characteristics that need a closer watch.

4.4.4.3 Use of Runs

The rules pertaining to runs above or below the CL and to runs up or run down can also be used with the *P*-chart and the *C*-chart. These rules are especially useful when the average of *P*, or *C*, is decreasing and there is no LCL (i.e., LCL = 0). In such circumstances, it is only through the runs that changes in the average *P* or *C* can be noticed.

4.4.4.4 Rational Subgrouping

As in $\overline{X}$- and *R*-charts, proper subgrouping is key to getting the most out of the *P*-chart and the *C*-chart. The subgrouping must be done so as to provide leads to discovering assignable causes when they are present. The following case study shows the value of creative subgrouping when a process is being controlled and improved using a *P*-chart.

Case Study 4.3

This example (from Krishnamoorthi 1989) relates to the assembly of certain specialty cable harnesses. The assembly operation included the soldering of connectors at cable ends, which required considerable eye focus. Therefore, the final quality of a harness depended on the individual assembler. The assembly line was experiencing a large amount of rejects at final inspection, and a quality consultant was asked to help.

There were 14 assemblers who worked in one shift. The first thing the consultant did was to consider each day's production as a sample and plot the *P*-chart with the daily data gathered over a month. Figure 4.13a shows the data plotted on the *P*-chart based on the daily samples. The process was in-control, with an average reject rate of 11.5%—an example of a process in-control but with a totally unacceptable performance. That the process was in-control only meant that the assembly process

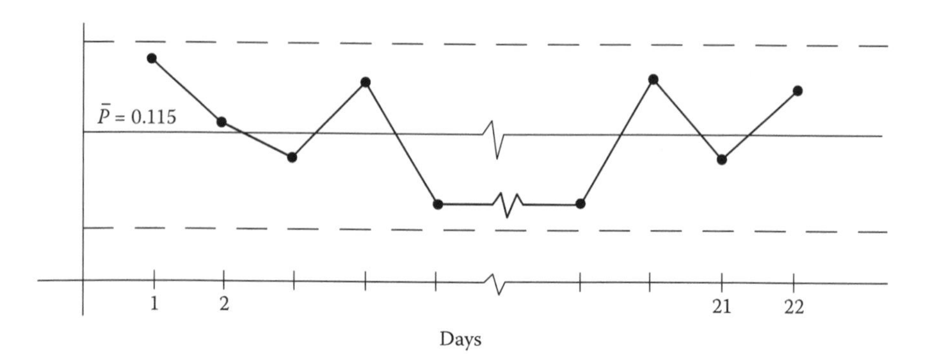

FIGURE 4.13a
Control chart for fraction defectives in cable assembly (subgroups based on days).

was consistently producing the same proportion of defectives day after day. This is the typical situation when the basis of subgrouping should be changed to check if subgrouping by another basis would disclose the assignable causes.

Data were then collected such that each sample represented a one-month production of each assembler, and a *P*-chart was drawn with "assembler" as the basis of subgrouping. Figure 4.13b shows this *P*-chart.

The cause of the problem became obvious. Assembler 4 and Assembler 12 were different from the rest. On further investigation, it was found that both assemblers had very poor eyesight and needed corrective glasses. The company provided free testing and free glasses. In further pursuance of the assignable cause, the company offered free eye examinations and glasses for all assemblers, and several obtained new prescriptions and new glasses.

The results of these actions were astonishing. The *P*-chart for the period during which the improvements were made is shown in Figure 4.13c. The CL and limits were revised when the process average was becoming markedly lower. The average of *P* for the third month after the investigation had started became 0.85%. The percentage of defectives dropped from 11.5% to 0.85%, a tremendous accomplishment with the use of the control chart.

The basis of subgrouping, however, was changed back to daily production after finding that the chart with assembler subgroups was in-control. When the assembler subgroups showed the process in-control, it meant that there were no more significant differences among assemblers. The daily chart involved less work in data collection.

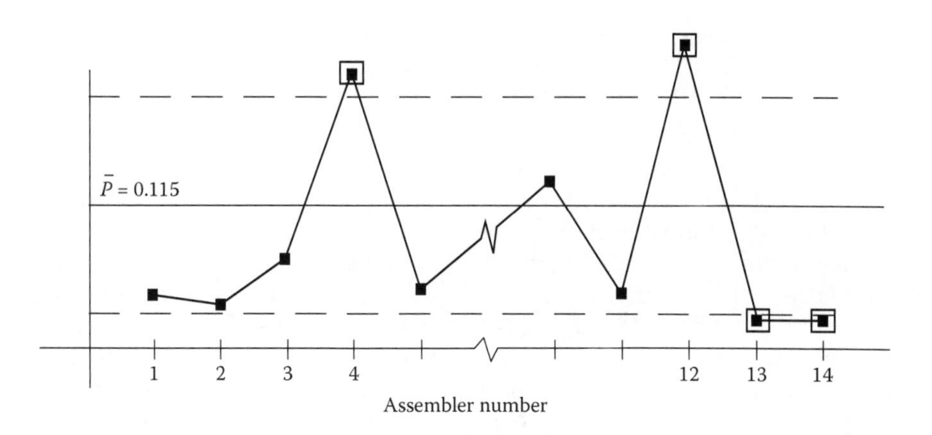

FIGURE 4.13b
Control chart for fraction defectives in cable assembly (subgroups based on assemblers).

Two lessons must be noted from the above case study. The first is about how changing the basis of subgrouping helped in discovering the assignable cause. The second is about how an assignable cause was pursued until it was completely eliminated. The company did not stop after improving the eyesight of the two assemblers; they helped in improving the eyesight of every assembler until eyesight would no longer be the reason for production of a defective cable assembly.

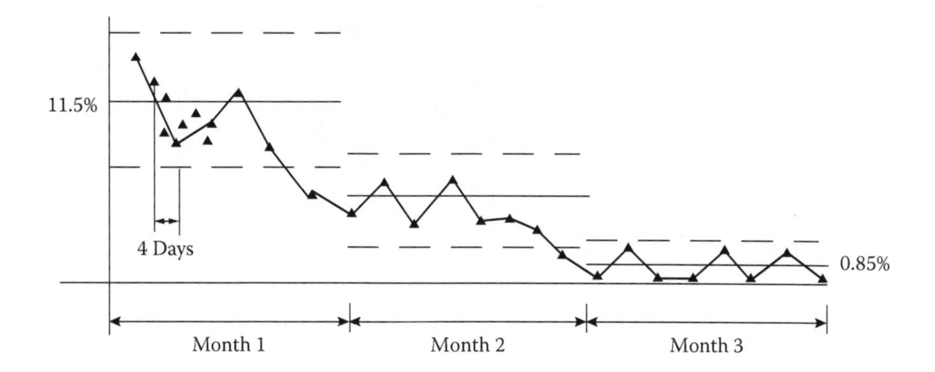

FIGURE 4.13c
(c) Control chart for fraction defectives (subgroups based on days).

4.5 Summary on Control Charts

The discussion so a has been limited to the $\bar{X}$- and R-, P-, and C-charts, and some minor variations of these. These are the basic tools of process control that are most useful in industry because of their applicability to many situations. Table 4.7 summarizes the information on these three basic types.

A few more sophisticated methods are available in the literature that are useful under special circumstances. The cumulative sum chart is used when higher sensitivity is needed to discover small changes. The exponentially weighted moving average (EWMA) chart is also useful when small changes must be detected. In addition, the EWMA chart and the moving average chart are preferred when multiple units are not available in quick succession and, therefore, $\bar{X}$- and R-charts cannot be used. The median chart is known to perform well with populations that are not normally distributed. Details

TABLE 4.7

Summary of the Three Basic Control Charts

	$\bar{X}$- and R-chart	P-chart	C-chart
Where applicable	For controlling the average and variability of a measurement	For controlling fraction defectives in a process	For controlling defects per unit, unit length, unit area, and so on
Also known as	Measurement control charts	Fraction defectives control chart	Defects per unit control chart
Based on the assumption	Normal distribution for the measurement being controlled	Binomial distribution for the occurrence of defectives in a sample	Poisson distribution for the occurrence of defects in a unit
Usual sample size	$n = 4$ or 5	Use $n \geq 20$	Use unit size to obtain LCL > 0.
Formulas for limits	$\text{UCL}(\bar{X}) = \bar{\bar{X}} + A_2\bar{R}$ $\text{CL}(\bar{X}) = \bar{\bar{X}}$ $\text{LCL}(\bar{X}) = \bar{\bar{X}} - A_2\bar{R}$ $\text{UCL}(R) = D_4\bar{R}$ $\text{CL}(R) = \bar{R}$ $\text{LCL}(R) = D_3\bar{R}$	$\text{UCL}(P) = \bar{p} + 3\sqrt{\frac{\bar{p}(1-\bar{p})}{n}}$ $\text{CL}(P) = \bar{p}$ $\text{LCL}(P) = \bar{p} - 3\sqrt{\frac{\bar{p}(1-\bar{p})}{n}}$	$\text{UCL}(C) = \bar{c} + 3\sqrt{\bar{c}}$ $\text{CL}(C) = \bar{c}$ $\text{LCL}(C) = \bar{c} - 3\sqrt{\bar{c}}$
Variations of the chart	$\bar{X}$- and S-chart when $n > 10$	np-chart or $100P$-chart to obtain large numbers to plot. P-chart with varying n when sample size changes	U-chart when unit size changes

of these charts can be found in textbooks in statistical quality control such as Montgomery (2001) and Grant and Leavenworth (1996).

A few of these special charts, with details on how and when they are used, along with their strengths and weaknesses, are covered in the next chapter. Chapter 5 also includes a discussion on the theoretical basis of the formulas used for calculating the control limits for the various control charts we discussed earlier in this chapter. The methods for evaluating the performance of the charts using operating characteristic curves are also included in Chapter 5. The remaining parts of this chapter deal with three important topics related to process control: implementing S PC on processes, process capability, and measurement system analysis.

4.5.1 Implementing SPC on Processes

We discuss here how to begin implementing an SPC program in a plant or an organization. If an engineer finds him- or herself in an organization with no current use of SPC methods, the following discussion will help that engineer in getting started. If the engineer is in an organization that has these methods already in use, the discussion should enable him or her make more effective use of the methods.

The first question to be answered is where does one start? The implementation effort should begin with the organization of a team to be called the quality team, quality improvement team, process control team, or some similar name. The team should be drawn from all the functions capable of contributing to the improvement of the processes in an organization, such as manufacturing, product design, quality, maintenance, and customer relations. The team should also include representatives of operators working on the process once the process to be improved is identified. If any complaints have been received from a customer, an excessive internal rejection rate exists in a process, or considerable field failure of a product is causing excessive warranty payments, then those would be the obvious places to start using the SPC methods. If there are several such opportunities, some prioritization is needed.

The criterion for prioritization depends on the individual circumstances. A product line that is causing some serious damage to a valuable customer may be the first one to be addressed, or it could be a product line that is causing the most economic damage to the producer in terms of in-house rejections. The first project for a team, or for an organization, should be one that is doable, can be completed in a timely manner, and has a significant financial return. Such a choice, when successful, would provide some confidence to the improvement team and help establish the credibility for the improvement process within the organization.

It is also true that collecting data on a product characteristic or a process variable and then plotting a suitable control chart is the best way to begin gaining knowledge of how a process works. Such knowledge is needed for developing theories on how a product characteristic is related to process variables. Experienced analysts would say that the first set of control charts for any process should be made manually, using paper and pencil, on a clipboard—not on a computer screen. There is much to be learned when the analyst observes the process firsthand and sees how the observations are generated and how the instruments respond to process conditions. There is also valuable information to be gained from interacting with operating personnel and working with the raw data by hand.

The next question would be which variable to track? If a product characteristic is giving trouble to a customer, then that characteristic should be monitored. The process variables that are responsible for the characteristic in question must then be identified. A cause-and-effect diagram, prepared by the team in a brainstorming session, may be all that is needed to identify the cause variable(s). However, if the relationship between the characteristic and the process variables is not obvious or needs to be confirmed, an experiment may be necessary. Once the process variables are identified as being responsible for the product quality, they must be controlled using the appropriate control charts.

The next question would be what type of chart to use? If the product characteristic or the process variable is a measurement, and is expected to follow a normal distribution, the $\overline{X}$- and R-chart combination would be the most suitable choice, provided that samples of size between 2 and 10 are available. If the sample size has to be large (>10), either because of a practical necessity or because the extra power from a large sample size is needed, then the $\overline{X}$- and S-chart combination should be used. If multiple units are not available for forming samples because the observations are slow in coming, then a chart, such as the chart for individuals (X-chart), the moving average chart, or the EWMA chart, should be used. (These charts for slow processes are discussed in Chapter 5.)

Use of a control chart should reveal whether the process is in-control. If the process is not in-control, efforts should be made to hunt out the assignable causes and eliminate them. Relating what happens in and around the process—whether through use of event logs, consultation with operators, or close observation of the process—to the signals appearing on the control charts should usually lead to the discovery of assignable causes. An intimate knowledge of the process, including the physics, chemistry, and technology of the process, is helpful and is often necessary in the search of assignable causes. This is when the knowledge of the operators or process engineers becomes most helpful. Patterns in the plots on the control charts, such as trends and cycles, would give clues to the sources of the assignable causes.

When an assignable cause is identified, steps should be taken to *eliminate* it completely; in other words, the remedies applied should be such that the assignable causes will not affect the process again. Furthermore, we should keep in mind that the assignable causes come and go, and lack of indication of them on the control chart for a short period of time should not lull the analyst into believing that the process is in-control. A process should not be declared in-control until the control chart indicates the existence of no assignable cause for a *sustained* period of time. A rough rule of thumb used by many practitioners (originally given by Dr. Shewhart) is that a control chart should show an absence of assignable causes for at least 25 consecutive samples before the process is considered to be in-control.

Eliminating the first set of assignable causes should result in reduced variability. This calls for revised control limits, which, in turn, may reveal the existence of more assignable causes. The iterations of eliminating assignable causes, setting new limits, and discovering and eliminating more assignable causes, should continue until the process shows stability with respect to a set of control limits and there are no obvious assignable causes that can be economically eliminated. Most likely, when the process has been brought in-control thus, the variability is small enough to meet the customer's specifications entirely. The capability of the process is now assessed using the capability indices C_p and/or C_{pk}. (See the next section for the definition of these indices.)

If the capability is not adequate according to the customer's requirements or the internal standards of the organization, experiments must be conducted to explore the process variables and their levels to yield the product characteristic in question at the desired level and the desired limits of variability. The experiments may have to be repeated and remedies applied until the desired reduction in variability is achieved. When the capability goal is reached, the team should then turn their attention to the next project on the prioritized list. The team should make sure, however, that the process that was improved is monitored periodically by operations people to make certain that the process maintains the gains and does not slip back to previous ways of operating. This becomes easier if the team had been in communication with the process owners and process operators and enlisted their cooperation in the discovery and change process. Unless the operating personnel buy into the solutions for removing the assignable causes, the solutions would have little chance of remaining implemented on the process.

If the product characteristic to be controlled is of the attribute type, one of the attribute charts—the *P*-chart, *C*-chart, or one of their variations—will be appropriate to use. The best way to differentiate between the situation when a *P*-chart is needed and when a *C*-chart is needed is to ask the question: is the opportunity for a defect or a defective occurring infinity (very large)? If the opportunity is infinity, then the attribute in question follows the Poisson law, so a *C*-chart is appropriate. If the opportunity for occurrence of the defect or defective is not infinity (i.e., not really large), then occurrence of the attribute is governed by the binomial distribution, so a *P*-chart should be used. Again, if the process is not in-control, the assignable causes must be discovered and eliminated. As mentioned, the discovery of an assignable cause requires a good understanding of the process and its surroundings, as well as a relational study of the control chart signals and the happenings on the process. This requires intimate knowledge about the working of the process combined with an ability to interpret the statistical signals from the charts.

It is quite possible that when the assignable causes are found and eliminated, the process achieves defective levels that are satisfactory to the customer.

We want to emphasize a point here: the value of these attribute charts lies not so much in bringing a process in-control (although it is a necessary first step) as it is in using them to continuously identify and eliminate assignable causes until the proportion defectives or average number of defects per unit is reduced to very low (near zero) levels.

The capabilities of processes that produce attribute outputs are measured in terms of defectives per thousand units or defects per thousand opportunities. When a very high level of quality is required, the capabilities are measured in terms of defectives per million units or defects per million opportunities. The capability can also be measured in the number of sigmas according to Motorola's scale with six sigma as a benchmark (see Chapter 5, Table 5.10). If the capability is not adequate, experiments should be conducted to investigate the factors that are responsible for the attribute and to discover the important factors and their optimal levels to bring the attribute to the acceptable level. When the capability has reached the desired level, the team should move on to the next project on the priority list. Figure 4.14 summarizes the procedure for implementing the SPC methods in a production environment.

After most of the urgent projects have been completed, an organization-wide survey should be made to identify the key processes—and the key variable in those processes—that determine the quality and customer satisfaction of products produced by the organization. Through data collection on the key variables and computation of the capability indices C_p and C_{pk} (discussed in the next section), the variability in the key variables, and thus the capability of the key processes in the organization, can be established and monitored. Of course, when a key variable is found to be not-in-control, effort must first be made to bring the process in-control with respect to that variable.

The computerization of data collection, data analysis, and presentation of summary information would be of great help in this effort in capturing the capabilities of key processes. The summary information on the important process capabilities will present a perspective on the quality capability of the whole organization. Sometimes, an average C_p or C_{pk}, averaged over all key processes, or the proportion of the key processes above 1.33 capability, is used as an overall measure of the capability of an organization, and is used to monitor progress in this regard. Processes with inadequate capabilities must be identified and prioritized. Variability in the processes should

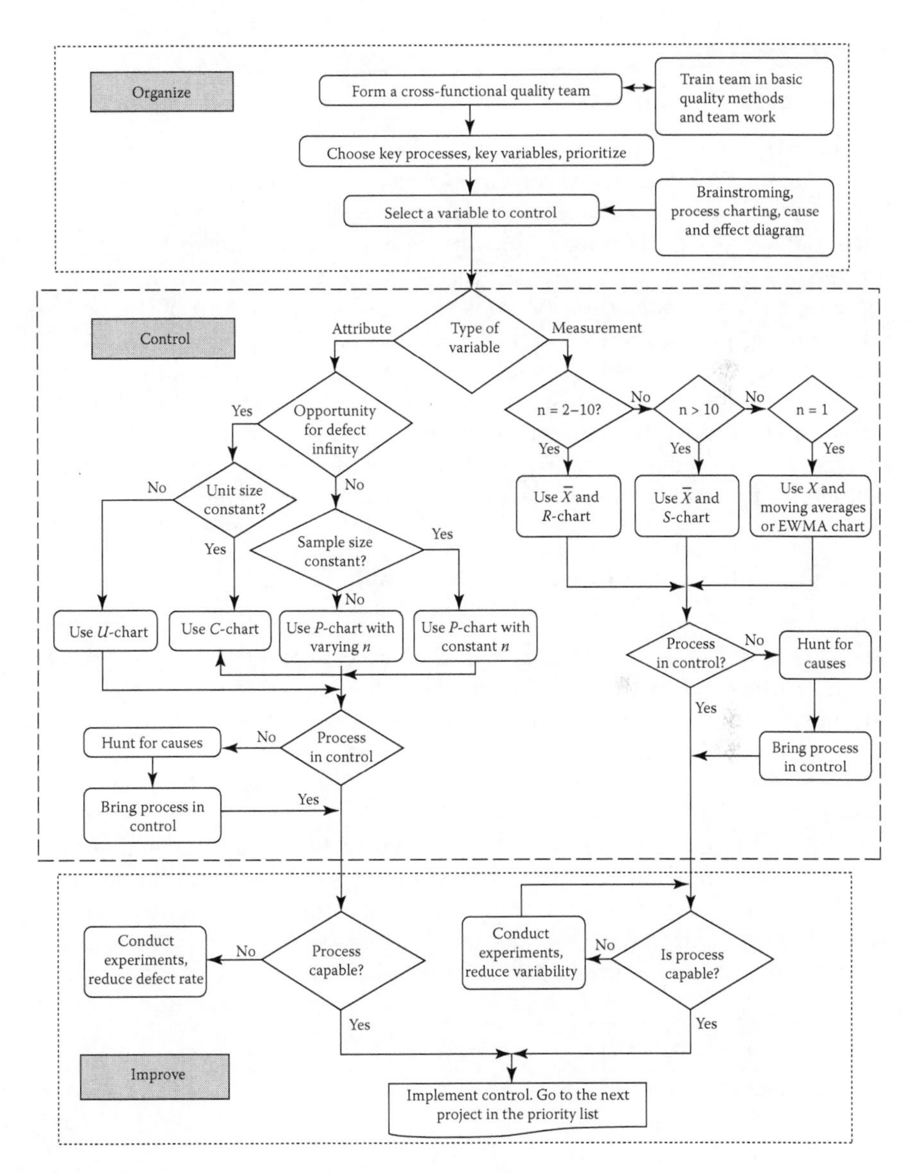

FIGURE 4.14
A scheme for implementing the SPC.

be continuously reduced, and the capabilities continuously improved, using repeated experiments, which will improve product quality and customer satisfaction. [Note: When we mention that experiments are needed, this does not always mean that a multifactor, factorial, or fractional factorial experiment is needed. Simple experiments with two levels of a single factor can often reveal much-needed information.]

4.6 Process Capability

As pointed out earlier, a process that is in-control may not be fully capable of producing products that meet the specifications. An analysis is necessary to verify if the in-control process is also in-specification. A process that produces "all" units of its output within specification is said to be a capable process. To assess its capability, a process, represented by a distribution, is compared with the specifications it is expected to meet. If all of the process is not within specification, adjustments to the process must be made in order to bring the process to full capability. Such an analysis is called the "process capability analysis." It is necessary here to re-emphasize a point made earlier that a process must be brought in-control before its capability is assessed, because a process that is not-in-control cannot be expected to behave in any predictable manner, and so its capability cannot be predicted.

4.6.1 Capability of a Process with Measurable Output

When a process produces a measurable output that can be assumed to have a normal distribution, the condition of the process can be fully described by two measures—its mean μ and its standard deviation σ. The μ and σ can be estimated from data obtained from a random sample (≥50) or from the data collected for control charts. If from the former source, the average $\overline{X}$ and the standard deviation S will estimate μ and σ, respectively. If from the latter source, the $\overline{\overline{X}}$, the CL of the $\overline{X}$-chart will estimate μ, and $\overline{R}/d_2$ will estimate σ, where $\overline{R}$ is the CL of the R-chart and d_2 is the correction factor that makes the $\overline{R}$ an unbiased estimator for σ. If the S-chart is used instead of the R-chart, then $\overline{S}/c_4$ will provide an estimate for σ. Values of d_2 and c_4 for various sample sizes are available in standard tables such as Table A.4 in the Appendix. The information on how process parameters μ and σ are estimated is summarized below.

Source of Data	Estimate for μ	Estimate for σ
Random sample ($n \geq 50$)	$\overline{X}$	S
$\overline{X}$ and R-charts	$\overline{\overline{X}}$	$\overline{R}/d_2$
$\overline{X}$ and S-charts	$\overline{\overline{X}}$	$\overline{S}/c_4$

When a process is compared to a given set of specifications, there are several possible situations that could arise, as shown in Figure 4.15. From such a figure, a qualitative assessment of whether or not the process meets the specifications can be seen. If the process does not fully meet the specifications, we can see whether it is the excess variability or the lack of centering of the process that is causing values outside of specification. Often, however, a quantitative assessment of the extent to which a process meets the specification

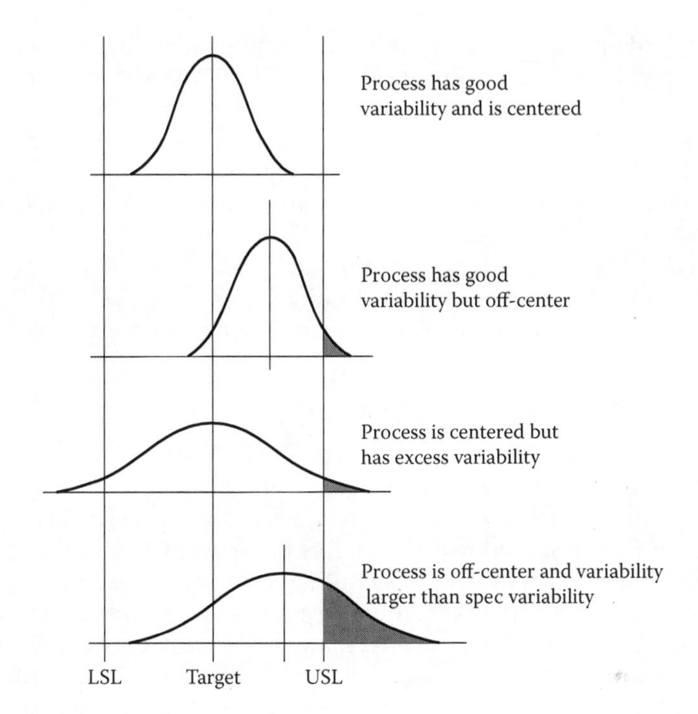

FIGURE 4.15
Qualitative evaluation of process capability.

is desired. For example, we may want to compare the capabilities of two machines, or two vendors, or we may want to set goals and monitor progress with regards to capability while making improvements to processes. The capability indices, which provide numerical measures quantifying the capability of a process, then become useful. The two most common process capability indices, C_p and C_{pk}, are described next.

4.6.2 Capability Indices C_p and C_{pk}

The index C_p is defined as:

$$C_p = \frac{\text{Variability allowed in spec}}{\text{Variability present in process}} = \frac{\text{USL} - \text{LSL}}{6\sigma}$$

where USL and LSL are the upper and lower specification limits, respectively, and σ is the standard deviation of the process.

The C_p index simply compares the natural variability in the process, which is given by 6σ based on normal distribution for the process, with the variability allowed in the specification, which is given by (USL – LSL). If the value of C_p is 1.0, the process is just within the specifications. To be exact, 99.73% of the process output is within specification. If the value is less than 1.0, the process variability is larger than the variability allowed by specifications, and so is producing rejects; if it is larger than 1.0, the process variability is smaller than the variability allowed in the specifications, and so nearly all of the products will be within specification. In general, the larger the value of C_p, the better is the process, as shown in Figure 4.16.

As defined above, C_p is a population parameter, the exact value of which is never known. It can, however, be estimated from sample data. The value of C_p can be estimated once an estimate for the value of σ is available. Thus, we can write

$$\hat{C}_p = \frac{USL - LSL}{6\hat{\sigma}}$$

This means that by using an estimate of σ, we can get an estimate, or a point estimate, for C_p. We should realize that this point estimate is just an observation of a statistic and is subject to sampling variability. That is, if we obtain estimates of C_p from different samples from the same population, the estimates will all be different. However, if the sample size is large (≥ 100), then the variability can be considered negligible (see Krishnamoorthi, Koritala, and Jurs 2009), and the estimate can then be used as the capability of the process.

Customers usually stipulate that the (estimated) value of C_p should be at least 1.33 in order to guarantee that the process variability is well within specification. This allows for the fact that most processes do not stay at one central location all the time and have a tendency to drift around the target. The requirement that $C_p = 1.33$ is meant to ensure that the product units will remain within specification even when some (small) change occurs in the process mean. The 1.33 requirement will also provide for possible error caused by sampling variability in the value of C_p computed from sample data.

We should realize when estimating C_p that, if the sample size is small, the estimate is not reliable. Therefore, it is advisable to estimate the capability

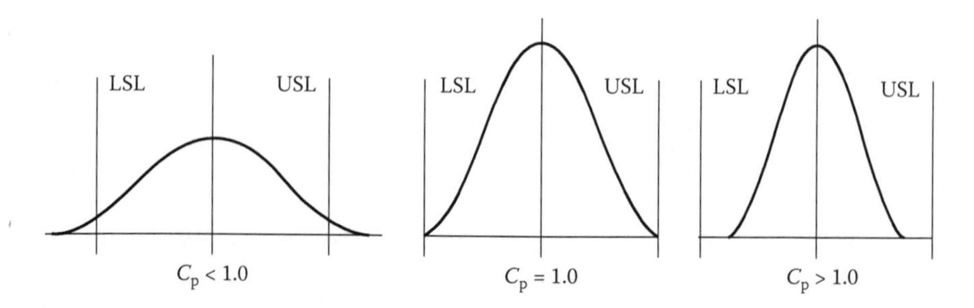

FIGURE 4.16
Process conditions and their C_p values.

index from a sample size of at least 100, which is equivalent to 25 subgroups of four units each if a control chart is used to monitor the process. If such a large sample is not available, we have to resort to a confidence interval for the index, which is discussed in Chapter 5.

Example 4.7

Calculate the C_p for the two processes in Figure 4.17.

Solution

Using the formulas given above:

C_p for process in Figure 4.17a: 24/18 = 1.33
C_p for process in Figure 4.17b: 24/18 = 1.33

Both processes have the same C_p in spite of the fact they are different in terms of their capabilities. One produces almost all values inside specification, whereas the other produces a sizeable proportion of values outside specification.

From the above example, we see that the C_p index is not able to recognize the lack of centering in the process shown in Figure 4.17b; it only evaluates the process variability in comparison to the specification variability. This drawback of the C_p index is rectified in the next capability index, C_{pk}, which is defined as:

$$C_{pk} = \frac{\text{Distance between process center and the nearest specification}}{3\sigma}$$

$$= \frac{\text{Min}\left[(\text{USL}-\mu),(\mu-\text{LSL})\right]}{3\sigma}$$

where μ is the process mean.

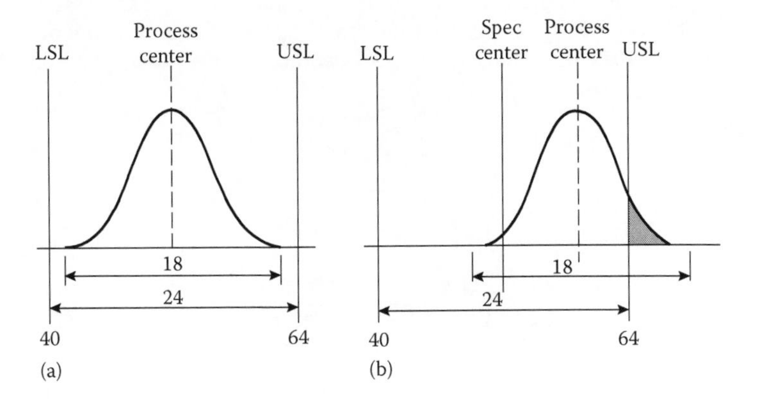

FIGURE 4.17
Two processes with the same C_p.

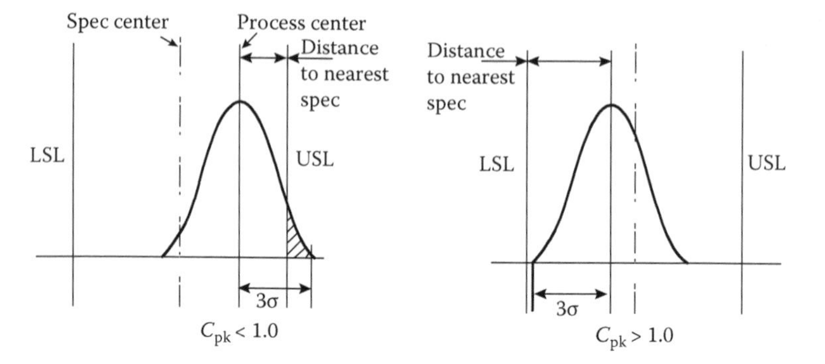

FIGURE 4.18
How the C_{pk} index measures the capability of a process.

Figure 4.18 has been drawn to illustrate the meaning of this index. Notice how, by comparing the distance of the nearest specification from the process center with 3σ, the index evaluates whether the part of the process which is most likely to exceed the specification does in fact exceed it. If the value of C_{pk} is less than 1.0, the process is producing values outside specification. The larger the value of C_{pk}, the better the process is in meeting specification. Again, when we use estimates for μ and σ in the above formula, we only get an estimate for C_{pk}. We have to use large samples ($n \geq 100$) to minimize the sampling variability. Furthermore, customers usually require a value of at least 1.33 for C_{pk} to cover possible drift in the process center and provide for possible sampling error.

Example 4.8

A process that has been brought in-control using control charts has process average $\bar{\bar{X}} = 41.5$ and $\bar{R}/d_2 = 0.92$. If the specification for the process calls for values between 39 and 47, calculate the capability indices C_p and C_{pk} for this process in its present condition.

Solution

$$C_p = \frac{47 - 39}{6 \times 0.92} = 1.45$$

$$C_{pk} = \frac{\text{Min}[(47 - 41.5),(41.5 - 39)]}{3 \times 0.92} = \frac{\text{Min}[5.5, 2.5]}{3 \times 0.92} = \frac{2.5}{3 \times 0.92} = 0.91$$

The process passed the C_p test (≥1.33) because of small variability, but it failed the C_{pk} test (<1.33) because of lack of centering, as shown in Figure 4.19.

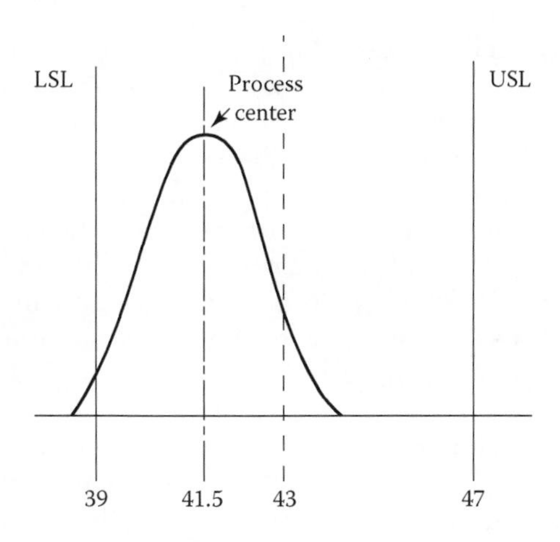

FIGURE 4.19
Example of a process with good C_p but poor C_{pk}.

The C_{pk} is a superior index for measuring the capability of a process, because it checks process centering as well as process variability. Often, however, it helps to compute the value of both C_p and C_{pk} and to make the comparison, as done in the above example. Such a comparison will reveal the condition of the process with respect to its variability and centering and help in determining what needs to be done to improve the process capability. These indices are also used to:

- Track the capability of processes over time
- Prioritize the processes for improvement projects
- Certify the supplier's processes

In the above discussion of capability indices, we covered the two basic indices, C_p and C_{pk}. These are also the most popular indices among industrial users. It is important at this point to take note of some of the shortcomings of these indices. One of them relates to the assumption of normality for the process. The 6σ, for example, in the denominator of the formula for C_p, to quantify the variability in the process, is a characteristic of the normal distribution. When the normal distribution assumption does not hold well, researchers have pointed out that the evaluation of a process's capability by these indices may lead to serious errors. Other methods are then recommended for this situation. Another issue with these indices is that they can be used only where there are two-sided specifications. Although two-sided specifications are the most common type, there are processes subject to one-sided specifications with requirements such as "the larger, the better" or "the smaller, the better." The C_p and C_{pk} are not suitable to use for these situations.

Yet another drawback of these indices, according to some, is the implied assumption in their use that the loss arising from a process is zero as long as the process remains within the specification and the losses jump to a very high and unacceptable level as soon the process starts falling outside the specification. Critics contend that the loss from a process is zero only when the process stays on target, and as the process center deviates from the target the losses increase parabolically, as suggested by Taguchi's loss function (see Figure 3.11). Another index, C_{pm}, has been proposed as an improvement on the above two indices as a response to this criticism. Some of these issues and solutions for them are discussed in a section on process capability in Chapter 5.

When a measurable characteristic is being controlled using $\overline{X}$- and R-charts, the process capability, or the ability of the process to meet the specifications, cannot be read directly from the charts because the control limits do not have any relationship to the specifications. The control chart center does not represent the specification center if the chart is used to maintain the "current level." Furthermore, the spread between the upper and lower control limits in the $\overline{X}$- chart does not reflect the spread in the specification. It will be smaller because the former represents the variability of the averages whereas the latter represents the variability of the individual values X. Comparing $\overline{X}$ values with specification limits makes no sense and will only cause confusion. This is also the reason why we never put specification limits on $\overline{X}$-charts.

In the case of processes producing attribute outputs, however, the capabilities can be read directly off of control charts without the need for a separate capability study, as discussed below.

4.6.3 Capability of a Process with Attribute Output

When the output of a process is an attribute, the capability requirements are specified in terms of the maximum number of defectives that will be tolerated in the output of the process. Specifications such as "not more than 1% defective," "not more than 10 errors per 1000 opportunities," or "not more than 50 parts per million (ppm) defective" are commonly used. If attribute control charts are in use for controlling a process, then the capability of the process can be read directly from the charts. The CL of the P-chart gives the average proportion of defectives in the process, and the CL of the C-chart gives the average number of defects per unit in the process. These quantities can be compared with the capabilities required by the customer or the capability goals established by the producer.

We see that the capability of a process with attribute output is measured differently from the capability of a process with measurable output. Such a lack of uniform measures of capability applicable to both measurable and attribute outputs gave rise to the creation of capability measures in "number of sigmas" by the Motorola statisticians (Motorola Corporation 1992). They proposed to measure the capability of a process by the distance at which specifications are located from the process center in terms of the number of standard deviations (σs) of the process. For processes with measurable outputs, which can be assumed to be normal, it can be calculated if the mean and standard deviation are known or estimated. For processes that produce attribute output, the capability is expressed as the number of sigmas of a normally distributed process that produces the same proportion of defectives as the process in question. Suppose that a process is producing 2% defectives. Then, its capability will be expressed as 2.327σ, because a normal process with specifications at 2.327σ on either side of the center will have 2% of the production outside specifications. Because the quality of many service processes is measured by attribute characteristics, this method of evaluating process capability is useful in both service and manufacturing environments. A more detailed description of how process capability is designated using the number of sigmas, with six sigma as the benchmark, is given in Chapter 5.

4.7 Measurement System Analysis

The measurement system analysis is done to make sure that measuring instruments of adequate capability are available for taking measurements on the product characteristics and process variables of a given process, and that the measuring instruments will measure the characteristics and variables truthfully. Here, the term "measurement system" is used to imply that a scheme, or a system, including several instruments, appliances, operators, and a procedure, is sometimes required to produce a measurement. The main reference for this section on measurement system analysis is the *Measurement System Analysis (MSA)—Reference Manual* published by the big three automakers (Chrysler LLC, Ford Motor Company, and General Motors Corporation 2010).

4.7.1 Properties of Instruments

Every instrument, be it a caliper, micrometer, thermometer, or a measuring scheme to find the percentage of solids in a solution, produces the measurement with certain variability. In other words, if an instrument is used to measure the same characteristic on the same unit of a product repeatedly, then the results from all the trials will not be the same. This variability in the readings can be described by a distribution—usually the normal distribution. The average of the distribution can then be used as the measurement provided by the measuring scheme. The difference between this average and the true value of the measurement is called "bias." The "accuracy" of an instrument is the opposite of bias; the smaller the bias, the better is the accuracy. An instrument that provides measurements with a smaller bias is more accurate.

The variability in the readings is quantified using the standard deviation of the distribution of the readings, which is referred to as the "error standard deviation," or simply the "error." The "precision" of an instrument is the opposite of this variability; the smaller this variability in the readings, the higher the precision of the instrument. Figure 4.20 shows the definitions of bias and precision.

In addition, two other properties of instruments, stability and linearity, are also used. "Stability" refers to how the precision of an instrument remains consistent over time. "Linearity" refers to how the bias remains the same over the range of possible values of the measurement. Figure 4.21 shows graphically the definition of stability and linearity. Finally, "resolution" refers to the smallest division of the unit that the instrument is designed to measure. For example, if the smallest division on a foot rule is 1/16 of an inch, then this is its resolution.

The following discussion is meant to explain how these characteristics of a measurement system are evaluated and controlled in order to provide a good set of measurements for a production process. Here, the term measuring

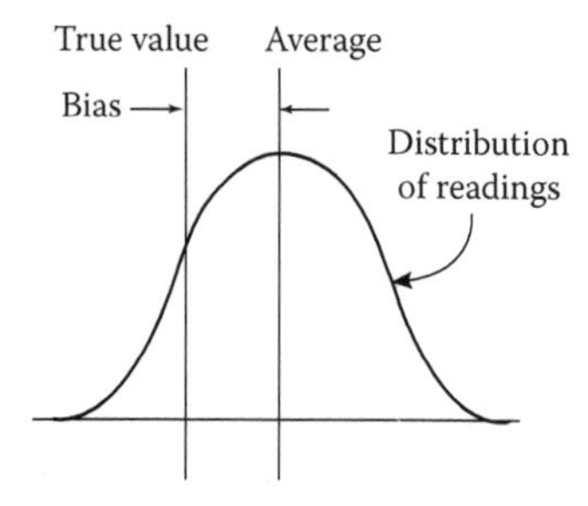

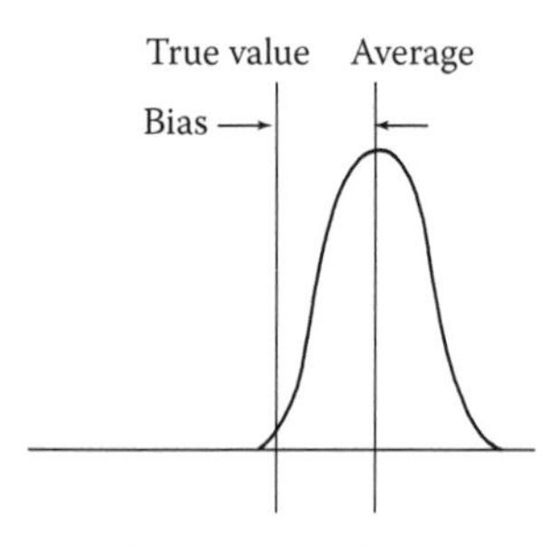

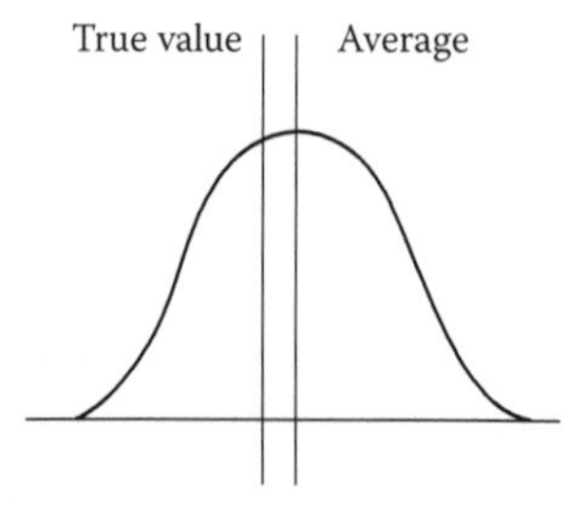

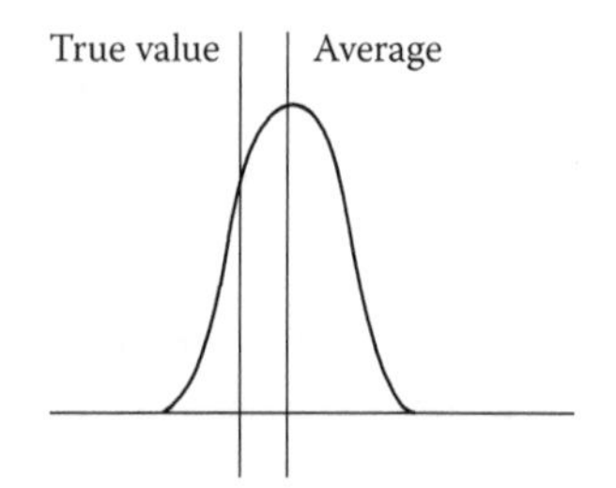

FIGURE 4.20
Definition of bias and precision of instruments.

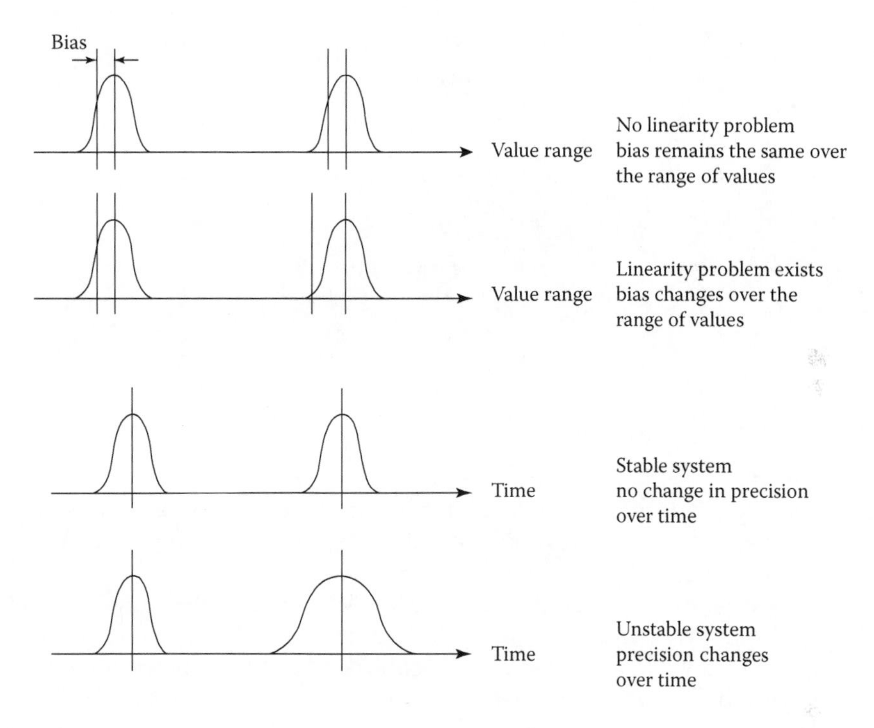

FIGURE 4.21
Linearity and stability of instruments.

system is used in a generic sense to refer to an instrument, such as a caliper or micrometer, or a system, such as that needed to measure the percentage of carbon in steel. We use the terms instrument, gage, and measurement system interchangeably. A brief discussion on measurement standards is provided before the methods of evaluating measurement systems are described.

4.7.2 Measurement Standards

The standards for characteristics, such as length, weight, hardness, color, and so on, are maintained by standards organizations of individual nations and are called "national standards." For the United States, for example, the national standards are maintained by the National Institute of Standards and Technology (NIST), headquartered at Gaithersburg, Maryland. Similarly, every industrialized nation has a national standards organization of its own that maintains the national standards for that country. Hence, the standard for every unit such as ft., kg, or °C is defined by that maintained at the standards organizations, and all the instruments that measure in these units must be compared with the standards for evaluating their accuracy and precision. This is done through a hierarchy of intermediate standards that

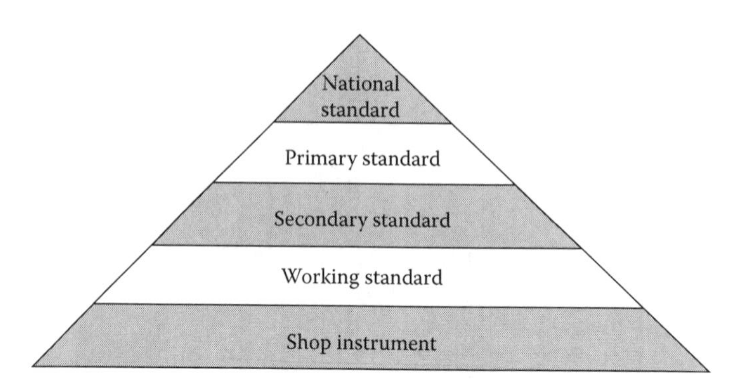

FIGURE 4.22
Hierarchy of instrument standards.

intervene between the national standards and the instruments on the shop floor. The Figure 4.22 shows the hierarchical structure through which this comparison takes place.

The process of comparing a lower-level standard with a higher-level standard and implementing corrective adjustments to make it conform to the higher-level standard is called "calibration." Instruments and standards should be kept calibrated against next higher-level standards to maintain their correctness or integrity. The ability of an instrument to establish its correctness, or integrity, by remaining in calibration with the national standard through the sequence of intermediate standards is called "traceability."

Referring to Figure 4.22, the primary standard is one that is directly compared and calibrated against the national standard. The primary standards are owned by organizations, which in the United States can be another government agency or private laboratory. Primary standards are expensive, and they are too delicate for regular, routine calibration purposes. Therefore, another set of intermediate standards—namely, secondary standards—are created by the labs and used for routine calibration. Primary and secondary standards owned by government or private labs are maintained in environmentally controlled premises under the supervision of specialists in metrology. Working standards are owned by manufacturing companies and are periodically calibrated against the secondary standards. These standards, which are also maintained in environmentally controlled labs in manufacturing facilities, are used to calibrate the shop-floor instruments. In a well-maintained quality system, every instrument used in the production processes must be calibrated so as to be traceable to national standards. This is the traceability requirement specified in quality management system standards such as ISO 9000.

4.7.3 Evaluating an Instrument

4.7.3.1 Properties of a Good Instrument

A good instrument should have:

1. Good resolution suitable to the purpose for which the instrument is used
2. Near-zero bias
3. Measurement variability much smaller than the variability in the product it is used to measure
4. Variability that is small compared to the tolerance allowed in the product
5. Stability in precision
6. No linearity problem

If the same instrument is used to measure different measurements, the above requirements should be met with respect to the most demanding measurement.

4.7.3.2 Evaluation Methods

The methods (or testing procedures) for assessing the properties of instruments and their acceptability are described below. Many of the recommendations made here on instrument capability are those given by the *Measurement Systems Analysis (MSA)—Reference Manual* (Chrysler LLC et al. 2010).

This manual recommends that before starting to test the acceptability of an instrument, the question of whether the measurement in question is needed at all must be considered. There are many measurements being taken in production shops that serve no useful purpose. There is no use for elaborate checking of the instrument if the measurement it provides is not needed. An examination of how the results from the measurement are used—and their usefulness in determining the condition of a process or the acceptability of a product—would help in deciding the necessity of the measurement. Also, a preliminary evaluation should be made of whether the instrument in question is affected by the environment in which the measurement is taken. Ambient temperature and humidity, lighting, as well as gas and froth accumulation, are some of the factors known to affect measuring processes. If any such factor is seen to influence the measurement, then experiments to evaluate the instrument should be designed taking those factors into consideration.

The procedures described below are generic methods, and they may have to be modified to suit an instrument being evaluated or a given set of special circumstances that may prevail around its use. These procedures should be written up as part of an organization's instrument calibration program. These procedures are collectively called the "gage R&R" study, referring to the repeatability and reproducibility study of a gage for evaluating its precision. Since evaluation of its precision is the major part of the study of an instrument, the whole evaluation of an instrument is called the gage R&R study.

The properties of an instrument to be evaluated are:

1. Resolution
2. Bias (accuracy)
3. Variability (precision):
   a. From repeatability
   b. From reproducibility
4. Stability
5. Linearity

4.7.3.3 Resolution

As stated earlier, the resolution is the smallest division of a measurement unit that an instrument is capable of distinguishing. This is also known as "discrimination." Good resolution is necessary to discover changes in the measurements when they occur, so the requirement on resolution depends on what changes need to be detected.

The best way to verify adequacy of resolution in an instrument is to make the $\overline{X}$- and R-charts of repeated measurements made by the instrument. As we know, the $\overline{X}$-chart tracks the mean, the R-chart tracks the variability in the measurements made, and the control limits represent the limits for natural variability in the respective statistics being plotted.

Figures 4.23a and 4.23b show two sets of control charts made from two sets of readings taken on the same set of sample units. The charts have been made from the data gathered on the diameter of 15 aluminum shafts measured by a dial caliper recorded in Table 4.8, under Operator A. The control charts in Figure 4.23a are made from the same 15 subgroups of three readings each, made by Operator A, recorded with a resolution of 0.001 in. Figure 4.23b is made from the 15 same subgroups but with the last digit dropped. When the last digit is dropped from each of the readings, it amounts to taking these readings with an instrument with a larger resolution—a resolution of 0.01 in. These two control charts offer the contrast between readings taken by instruments with two different resolutions.

The difference in the resolution between the two sets of data can be seen in the R-charts. The lack of resolution in the second set of readings causes many repeated readings to become identical, thus causing many zero values for R (see Figure 4.23b). Lack of adequate resolution also produces fewer possible

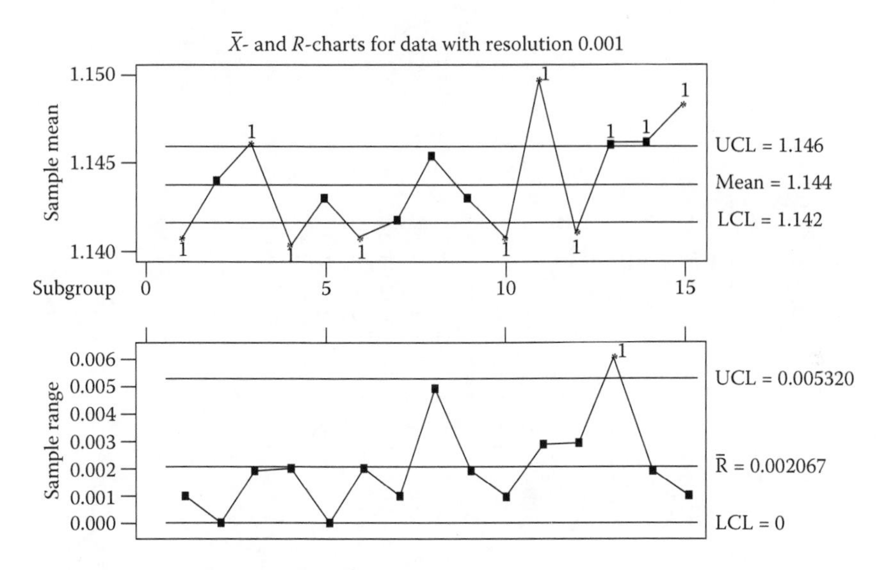

FIGURE 4.23a
Checking for the resolution of an instrument with resolution 0.001.

values for R. There are six possible values for R when the resolution is 0.001 in., whereas there are only two possible values when the resolution is 0.01 in.

Lack of resolution can also be seen in the $\bar{X}$-chart. If the resolution is not adequate, then there are many identical values in the readings resulting in many zeros for R, making the $\bar{R}$ small. Since the spread of the $\bar{X}$ limits is

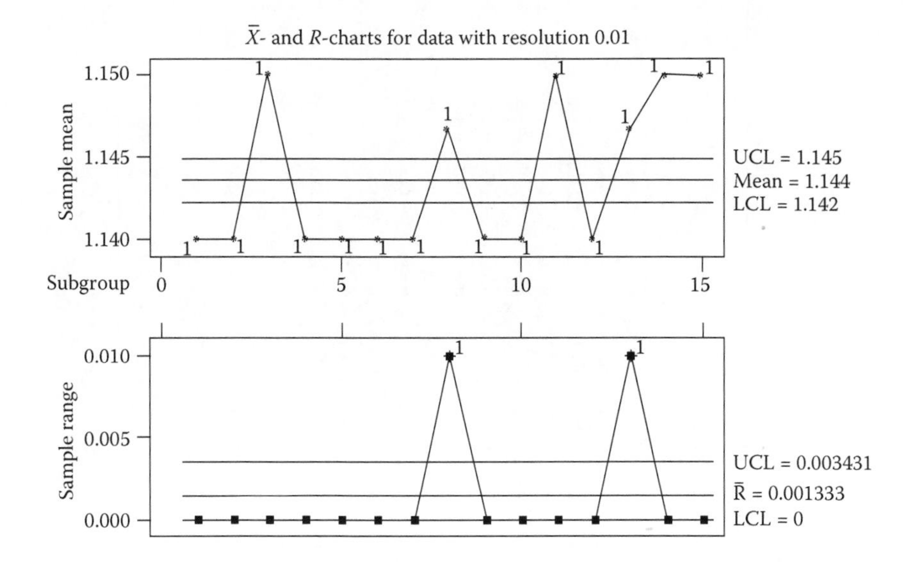

FIGURE 4.23b
Checking for the resolution of an instrument with resolution 0.01.

TABLE 4.8

Example of a Gage R&R Study

	Operator A						Operator B						Operator C						
No.	Trial 1	Trial 2	Trial 3	S	R	$\overline{X}$	Trial 1	Trial 2	Trial 3	S	R	$\overline{X}$	Trial 1	Trial 2	Trial 3	S	R	$\overline{X}$	$R(\overline{X})$
1	1.141	1.141	1.14	0.0006	0.001	1.1407	1.14	1.143	1.14	0.00158	0.003	1.141	1.137	1.136	1.136	0.00058	0.001	1.13633	0.0047
2	1.144	1.144	1.144	0.0000	0.000	1.1440	1.142	1.144	1.142	0.00115	0.002	1.14267	1.14	1.139	1.139	0.00058	0.001	1.13933	0.0047
3	1.146	1.145	1.147	0.0010	0.002	1.1460	1.146	1.152	1.146	0.00346	0.006	1.148	1.144	1.143	1.142	0.001	0.002	1.143	0.0050
4	1.139	1.141	1.141	0.0012	0.002	1.1403	1.14	1.141	1.139	0.00051	0.002	1.14	1.137	1.136	1.135	0.001	0.002	1.136	0.0043
5	1.143	1.143	1.143	0.0000	0.000	1.1430	1.143	1.143	1.143	0.000	0.000	1.143	1.139	1.137	1.138	0.001	0.002	1.138	0.0050
6	1.14	1.142	1.14	0.0012	0.002	1.1407	1.139	1.14	1.139	0.00084	0.001	1.13933	1.136	1.135	1.135	0.00058	0.001	1.13533	0.0053
7	1.142	1.141	1.142	0.0006	0.001	1.1417	1.142	1.145	1.141	0.00184	0.004	1.14267	1.139	1.138	1.137	0.001	0.002	1.138	0.0047
8	1.142	1.147	1.147	0.0029	0.005	1.1453	1.146	1.145	1.146	0.00051	0.001	1.14567	1.143	1.141	1.142	0.001	0.002	1.142	0.0037
9	1.143	1.144	1.142	0.0010	0.002	1.1430	1.143	1.146	1.143	0.00173	0.003	1.144	1.147	1.139	1.139	0.00462	0.008	1.14167	0.0023
10	1.14	1.141	1.141	0.0006	0.001	1.1407	1.14	1.141	1.141	0.00051	0.001	1.14067	1.136	1.136	1.135	0.00058	0.001	1.13567	0.0050
11	1.15	1.151	1.148	0.0015	0.003	1.1497	1.15	1.153	1.149	0.00184	0.004	1.15067	1.139	1.146	1.146	0.00404	0.007	1.14367	0.0070
12	1.14	1.14	1.143	0.0017	0.003	1.1410	1.148	1.141	1.141	0.00404	0.007	1.14333	1.136	1.137	1.136	0.00058	0.001	1.13633	0.0070
13	1.148	1.142	1.148	0.0035	0.006	1.1460	1.148	1.149	1.147	0.00153	0.002	1.148	1.144	1.144	1.143	0.00058	0.001	1.14367	0.0043
14	1.146	1.147	1.145	0.0010	0.002	1.1460	1.144	1.149	1.145	0.00252	0.005	1.146	1.142	1.142	1.142	0.000	0.000	1.142	0.0040
15	1.148	1.149	1.148	0.0006	0.001	1.1483	1.147	1.149	1.148	0.00102	0.002	1.148	1.144	1.143	1.143	0.00058	0.001	1.14333	0.0050
				Average: 0.0011	Average: 0.00207					Average: 0.00154	Average: 0.00287					Average: 0.00118	Average: 0.00213		Average: 0.0048

Overall $\overline{X}$: 1.143 Overall s: 0.004

equal to $A_2\overline{R}$, when $\overline{R}$ is small, the limits become narrow. In turn, when the limits are narrow, many $\overline{X}$ values fall outside the limits. Hence, an inadequate resolution can also be recognized by too many $\overline{X}$ values outside the limits on the $\overline{X}$-chart. The following rules are recommended to recognize a lack of sufficient resolution in an instrument when using $\overline{X}$- and R-charts:

1. If the R-chart shows very few possible values (<3) falling inside the control limits, conclude that the instrument has inadequate resolution.
2. If the R-chart shows more than three possible values inside the control limits but more than one-fourth of the values are zero, then conclude that the instrument has inadequate resolution.

In the example shown in Figure 4.23, the resolution of 0.01 seems to be inadequate on both counts, whereas the resolution of 0.001 seems to be satisfactory on both counts. Five possible values for R fall within the limits of the R-chart when the resolution is 0.001, whereas no R value falls inside the limits when the resolution is 0.01. Also, with 0.01 resolution, there are too many zeros.

Problems relative to inadequate resolution are easily resolved. Either an instrument with finer divisions is sought or, if the measurements are taken by rounding off the readings, the measurements are rounded to a larger number of decimal places. A recommendation made by the *Measurement Systems Analysis (MSA)—Reference Manual* (Chrysler LLC et al. 2010) is to choose an instrument with a resolution smaller than 1/10 of the spread in process variability represented by $6\sigma_p$, where σ_p represents the actual variability in the product.

4.7.3.4 Bias

The bias of an instrument is measured by taking repeated measurements on a product for which the "true" value is known. The true value can be obtained by measuring the product using a "tool-room" instrument (i.e., an instrument known to have high precision and small bias). The difference between the average of the readings by the instrument being evaluated and the true value gives the bias of the instrument. The bias is also sometimes expressed as a percentage of the spread in process variation ($6\sigma_p$) for the measurement in question.

Example 4.9

The following 10 readings were obtained by measuring a 1-in. gage block (a working standard) by a caliper. What is the bias of the caliper in this range?

No.	1	2	3	4	5	6	7	8	9	10
Readings (in.)	1.032	1.045	1.035	1.030	1.030	1.035	1.040	1.035	1.040	1.035

Solution

The true value corresponding to these readings is 1.00, since the gage block is a tool-room instrument and its designated dimension is 1.00 in.

$$\bar{X} = 1.0357$$

$$s = 0.0048$$

$$\text{Bias} = 1.0357 - 1.0000 = 0.0357 \text{ in.}$$

This means that the caliper gives, on average, values that are larger than the true value by 0.0357.

When the measurement destroys the unit being measured, it will not be possible to make repeated measurements on the same unit; a different approach is necessary, as shown in the following example.

Example 4.10

A sand tester for testing the tensile strength of sand mixes is being evaluated. Twenty dog-bones were made from the same batch of (well-mixed) sand and were divided randomly into two groups of 10 each. One group was tested by a standard tester (tool-room instrument) and the other group was tested by the tester under evaluation. The following readings were recorded. Calculate the bias in the tester under evaluation.

No.	1	2	3	4	5	6	7	8	9	10	$\bar{X}$
Standard tester (lb.)	262.2	261.5	267.7	260.9	262.2	261.7	269.3	272.4	263.2	265.2	264.6
Shop tester (lb.)	279.9	267.1	287.1	256.9	259.9	260.6	263.2	270.2	275.2	261.9	268.2

Solution

$$\text{Bias} = 268.2 - 264.6 = 3.6 \text{ lbs.}$$

Therefore, the shop tester has a bias of 3.6 lbs.

Such information on bias will help in making an adjustment to the instrument to correct for bias, or adjusting the reading by applying a compensation for bias.

4.7.3.5 Variability (Precision)

Instrument variability, or error, could come from two sources:

1. Repeatability
2. Reproducibility

Repeatability error comes from the instrument hardware that prevents the instrument from giving identical readings when measuring the same unit of a product repeatedly using the same operator. Reproducibility error arises from the instrument allowing different operators to do the measurement differently, thus preventing the instrument from giving identical readings when measuring the same unit using different operators.

Repeatability error is quantified by the standard deviation of the readings from the instrument taken repeatedly on the same unit by the same operator. Reproducibility error is quantified by the standard deviation of the readings taken by different operators on the same unit. The example below shows how we design and conduct an experiment to evaluate the repeatability and reproducibility errors of an instrument.

Example 4.11

Table 4.8 shows readings taken on 15 units of a product (diameters of aluminum shafts measured by a dial caliper) by three operators, with each operator repeating the measurement three times. Estimate the repeatability and reproducibility errors of the caliper.

Solution

Repeatability Error

The difference among readings on the same unit by any one operator is caused by the repeatability error. The standard deviation of the three readings on any one unit taken by any one operator provides an estimate of the repeatability error. For example, from the three readings of Unit 1 by Operator A, the repeatability error can be estimated as follows. Calculate the standard deviation s of the three readings (already done in the table) and divide it by the correction factor c_4, for $n = 3$, taken from Table A.4 in the Appendix. We get:

$$\hat{\sigma}_e = \frac{s}{c_4} = \frac{0.0006}{0.8862} = 0.00068$$

where the subscript e in σ_e stands for equipment or hardware.

The repeatability error can also be estimated from the range R of the three readings as:

$$\hat{\sigma}_e = \frac{R}{d_2} = \frac{0.001}{1.693} = 0.0006$$

The value of the correction factor d_2 is obtained from Table A.4 in the Appendix for $n = 3$.

Forty-five such estimates, however, are possible from the 45 sets of readings made by three operators on 15 units of the product. These can be averaged to obtain a better estimate for repeatability error. Using $\bar{s}$, the average value of s from the 45 sample readings, we have $\bar{s} = (0.0011 + 0.00154 + 0.00118)/3 = 0.00126$. Therefore,

$$\hat{\sigma}_e = \frac{\bar{s}}{C_4} = \frac{0.00126}{0.8862} = 0.00142$$

Using $\bar{R} = (0.00287 + 0.00287 + 0.00213)/3 = 0.00236$, we get:

$$\hat{\sigma}_e = \frac{\bar{R}}{d_2} = \frac{0.00236}{1.693} = 0.00139$$

It is interesting to see that the estimates obtained from the two methods, one using S and the other using R, agree so closely. We will therefore take the repeatability error as $\hat{\sigma}_e = 0.0014$.

Reproducibility Error

The reproducibility error of the instrument causes the difference in $\bar{X}$ values obtained by two different operators for the same unit. We will assume that the repeatability error has been neutralized in each $\bar{X}$ as a result of averaging of the three readings. Thus, the range of the three $\bar{X}$ values from the three operators, for any one unit of the product, will provide an estimate for the reproducibility error. For example, $R(\bar{X})$ from Unit 1 = (1.141 − 1.13633) = 0.00467, and the estimate of the reproducibility error is:

$$\hat{\sigma}_o = \frac{0.0047}{1.693} = 0.0028, \text{ (d_2 for $n = 3$ was chosen from Table A.4)}$$

where the subscript o in σ_o is for operator.

There are 15 values of $R(\bar{X})$, however, that can be averaged to provide $\overline{R(\bar{X})} = 0.0048$, and the estimate for σ_o is:

$$\hat{\sigma}_o = \frac{0.0048}{1.693} = 0.0028.$$

It is just a coincidence that the estimate from Unit 1 is the same as from the average of all 15 units.

Gage Error

Having obtained estimates for repeatability and reproducibility standard deviations, the total variability from the measuring system can be obtained as:

$$\sigma_g = \sqrt{\sigma_e^2 + \sigma_o^2}$$

where the subscript g is for gage.

For the example, $\hat{\sigma}_g = \sqrt{0.0014^2 + 0.0028^2} = 0.0031.$

Variability in the Parts

There are a total of 135 observations, and the variability among them is made up of the actual variability in the parts and the variability arising from the instrument. We have already estimated the variability from the instrument, and will therefore be able to isolate the variability in the parts. The overall variability, estimated by the computed standard deviation s of all 135 observations, is $\hat{\sigma}_{all} = 0.004$. If we designate the variability in the product as σ_p, then $\sigma_{all} = \sqrt{\sigma_p^2 + \sigma_g^2}$. Solving for σ_p, $\sigma_p = \sqrt{0.004^2 - .0031^2} = 0.0025$.

The ratio σ_g/σ_p represents the variability from the instrument (gage) compared to the actual variability in the parts. In the above example, $\sigma_g/\sigma_p = 0.0031/0.0025 = 1.24$, so the instrument variability is larger than the real variability in the product—an unacceptable situation. If we consider only the repeatability error, however, then the ratio $\sigma_e/\sigma_p = 0.0014/0.0025 = 0.56$, and the instrument variability is smaller than the product variability, indicating that it was the reproducibility error (caused by the differences in the use by the operators) that gave rise to the inflated variability in the instrument. A desirable situation is when the ratio σ_g/σ_p is less than 0.10—that is, when the variability from the instrument is no larger than 10% of the variability in the product. The reproducibility error can be reduced by making the measurement procedure, which is used by different operators for making the measurement, more uniform through proper instructions and training.

In the above example, Operator C was a veteran machine operator, and Operators A and B were professors who taught statistics but could claim no experience in using the caliper; therefore, the instrument was used differently by the different "operators," which explains the high level of reproducibility error. The repeatability error comes from the instrument hardware. If it is a caliper, then it could be due to the friction in the sliding surfaces or unevenness of the knife-edges. If it is a hydraulic tensile tester, it could be from the temperature rise in the fluid or the lack of a positive grip on the specimen. Awareness of the existence of the error and understanding its source would contribute to minimizing such errors and improving the capability of the instrument. Generally speaking, smaller repeatability error comes with more expensive equipment because protection needs to be built into the instrument in order to prevent the various sources from contributing to this error. The precision required in the measurements in a particular situation determine the choice of measuring equipment.

4.7.3.6 A Quick Check of Instrument Adequacy

A quick way of comparing the variability in the instrument and the variability in the product, and thus of assessing the adequacy of the instrument, is as follows. Take repeated measurements using the instrument on several units of a product using one operator, as shown in Table 4.8 for Operator A. Make $\overline{X}$- and R-control charts, as shown in Figure 4.24. The R-chart in

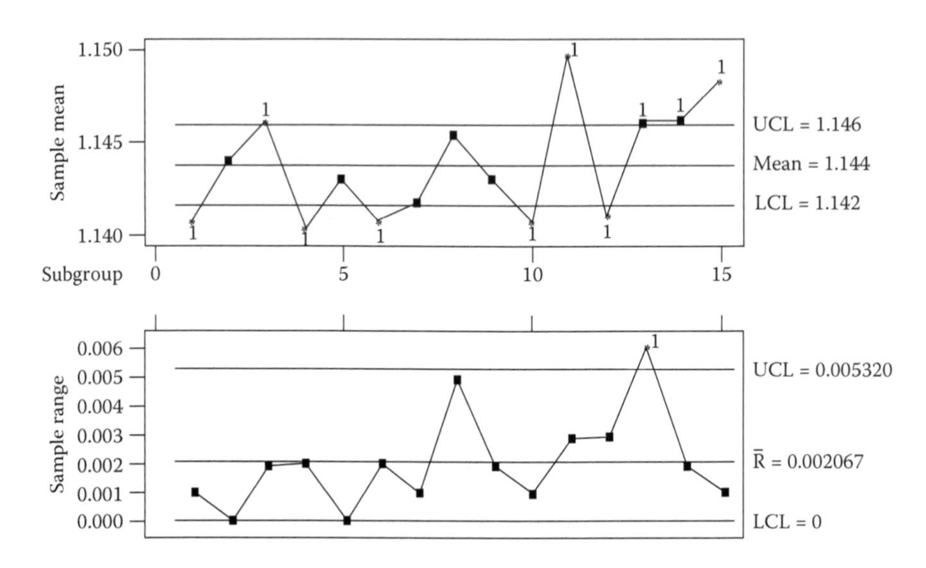

FIGURE 4.24
A quick check of instrument adequacy.

this figure tracks the variability in the instrument, not the variability in the product, since the three measurements from which a value of R is calculated come from the same unit; hence, the R-chart will show if the variability in the instrument is consistent, which is also called the "stability." If R values fall outside the limits, it would mean that the variability is not stable. If R values show a trend over time, or show wide variability, it should lead to an investigation as to why the variability changes over time.

The $\overline{X}$-chart in this figure is an interesting chart. The limits for the $\overline{X}$-chart have been calculated from the variability of the instrument—not the natural variability of the product. The differences in the values of $\overline{X}$, however, represent only the variability in the product. Therefore, this $\overline{X}$-chart presents a comparison of product variability and instrument variability. If the variability in the instrument is small compared to the variability in the parts, then the band width of the control limits will be small, causing a large number of $\overline{X}$ values to fall outside the limits. Therefore, if there is a large number of $\overline{X}$ values outside the limits—that is, the $\overline{X}$-chart shows that the "process" is not-in-control—then it indicates a good situation: that the instrument variability is much smaller compared to product variability, which is desirable. According to the rules recommended by the *Measurement Systems Analysis* (*MSA*)—*Reference Manual* (Chrysler LLC et al. 2010) at least 50% of $\overline{X}$ values should fall outside the control limits in the $\overline{X}$-chart constructed as above before the instrument can be considered to have adequate capability.

We should remember that lack of adequate resolution of the instrument could also cause a large number of $\overline{X}$ values to fall outside the limits in the $\overline{X}$-chart constructed above. The resolution question must be first resolved before the experiment is done to compare the product and instrument variability.

4.8 Exercise

4.8.1 Practice Problems

4.1 The following data were collected on the amount of deodorant in cans, in grams, filled in the filling line of a packaging company. Prepare $\overline{X}$- and R-charts for the data, and comment on the condition of the process.

Time	8 AM	9 AM	10 AM	11 AM	12 PM	1 PM	2 PM	3 PM	4 PM	5 PM	6 PM	7 PM	8 PM	9 PM	10 PM	8 AM	9 AM	10 AM	11 AM	12 PM
1	25.0	24.3	25.2	24.3	26.2	24.1	25.1	24.2	24.1	25.1	24.6	23.9	25.1	24.8	24.4	23.9	24.5	24.4	25.2	23.1
2	24.5	25.1	24.0	24.0	25.1	24.0	25.2	24.1	24.3	25.4	24.9	23.7	27.3	24.4	24.3	23.4	24.8	24.5	25.6	23.2
3	24.5	24.2	25.1	24.3	25.2	24.4	25.0	24.3	24.1	25.2	24.1	24.2	27.3	24.1	24.9	23.8	24.8	24.8	25.3	23.4
4	25.1	24.1	24.2	24.0	24.2	25.0	25.4	24.2	25.0	24.4	24.0	24.1	27.1	24.3	25.0	24.0	24.9	24.6	25.2	23.4

4.2 For the data in Problem 4.1, prepare $\overline{X}$- and S-charts, and compare the results with those from the $\overline{X}$- and R-charts.

4.3 To control the weight of jumbo ravioli packaged in a food processing plant, the following data were collected on the weight of four packages of ravioli, approximately every 30 minutes. Prepare $\overline{X}$- and R-control charts to monitor the weights, and comment on the results. Only 15 subgroups were available. The weights are in ounces. (Data from Schmillan and Johnson 2001.)

Trial	**1**	**2**	**3**	**4**	**5**	**6**	**7**	**8**	**9**	**10**	**11**	**12**	**13**	**14**	**15**
Time	7:30	8:15	8:45	9:15	9:45	10:15	10:45	11:15	11:45	12:15	12:45	1:30	1:55	2:15	2:45
Package 1	35.1	34.4	34.7	39.0	35.1	36.1	34.8	35.7	35.8	36.5	35.2	35.3	35.3	36.9	35.7
Package 2	36.5	33.3	35.7	35.5	37.1	35.7	34.5	34.6	36.4	37.5	37.1	34.6	36.3	35.3	36.9
Package 3	34.3	33.9	35.6	34.3	34.8	36.2	35.8	36.7	34.2	36.3	36.4	33.8	37.4	35.8	35.3
Package 4	34.3	34.6	34.9	36.4	35.4	35.6	36.6	36.6	34.9	37.6	36.1	36.3	36.5	36.3	37.7

4.4 For the data in Problem 4.3, prepare $\overline{X}$- and S-charts, and compare the results with those from the $\overline{X}$- and R-charts.

4.5 A sample of 30 bottles was taken at the end of a filling line every 30 minutes; inspected for cleanliness, proper labeling, location of traceability code, and so on; and were classified as either good or defective. The table below shows the data on 25 such samples. Prepare a P-chart, and comment on the results.

Subgroup	1	2	3	4	5	6	7	8	9	10	11	12	13	14	15	16	17	18	19	20	21	22	23	24	25
No. of defectives	0	0	4	3	2	0	1	2	0	3	0	4	1	1	1	2	1	0	0	1	1	0	2	1	0

4.6 In a fabrication shop where cabs are built for different models of bulldozers, the cab shells are inspected for proper welds before painting. The table below gives the total number of welds in each cab and the number found to be defective. Recommend an appropriate control chart for monitoring the welds, and compute the control limits.

Cab no.	1	2	3	4	5	6	7	8	9	10	11	12	13	14	15	16	17	18	19	20	21	22
Total no. of welds	46	62	80	75	32	69	45	48	46	92	121	110	55	64	32	95	68	42	99	101	66	75
No. of defectives	7	3	6	9	5	11	10	5	6	6	8	11	2	2	3	4	12	9	14	16	8	6

4.7 The number of printing errors per page in a newspaper was recorded from a randomly selected page, each day, for 20 days. The data are to be used to prepare a control chart to monitor and control printing mistakes in the newspaper. Recommend a suitable control chart, and compute the control limits.

Days	1	2	3	4	5	6	7	8	9	10	11	12	13	14	15	16	17	18	19	20
No. of errors	7	6	1	2	4	1	8	14	2	9	10	2	4	9	21	3	5	4	2	5

4.8 The inspection crew of a state department of transportation counts the number of potholes per mile on a major highway as part of their quality assurance program. The data below show the number of holes on 32 randomly selected miles. Is the pothole producing process in-control? In other words, does the incidence of potholes seem to be consistent from mile to mile, or do some miles have more potholes than the others?

Mile no.	1	2	3	4	5	6	7	8	9	10	11	12	13	14	15	16
No of potholes	12	8	6	4	2	21	3	2	8	13	25	0	1	5	4	8
Mile no.	17	18	19	20	21	22	23	24	25	26	27	28	29	30	31	32
No. of potholes	11	31	3	0	1	7	0	7	5	2	9	13	0	2	12	15

4.9 The data in the table below represent the production quantities of paper bags, and the number rejected each hour for 23 hours in a paper mill. What type of control chart would be used for monitoring and controlling the production of defective paper bags? Calculate the control limits for the chart(s).

Hour	1	2	3	4	5	6	7	8	9	10	11	12	13	14	15	16	17	18	19	20	21	22	23
Production (×100)	12	14	21	18	16	20	13	12	11	10	9	13	14	23	21	22	15	17	23	18	17	16	14
No. of defectives	34	56	43	44	32	67	43	56	87	55	50	120	87	90	32	44	23	21	87	34	30	23	55

4.10 A group of IE majors was intent on helping a friend improve the quality of service she provides to customers in her house-cleaning business. The IE group visited randomly selected houses after cleaning crews had completed their jobs and obtained the following data. The data show the number of rooms in each of the houses visited and the number of "defects" that were found. The data also indicate the ID of the crews (A, B, or C) that cleaned the houses. Is the cleaning process in-control? Is the cleaning process capable if the cleaning business assures its customers that there will be "zero" defects in the houses after the crews have done the cleaning?

Home	1	2	3	4	5	6	7	8	9	10	11	12	13	14	15	16	17	18	19	20	21	22	23	24	25
No. of rooms	8	6	10	7	6	6	8	9	7	10	11	7	8	9	9	7	7	6	6	8	7	11	12	9	8
No. of defects	1	0	1	0	1	0	0	0	1	1	5	0	1	4	0	0	0	0	1	0	1	1	2	1	1
Crew ID	B	B	C	A	B	A	C	C	B	C	C	B	A	C	A	A	A	C	B	C	A	C	C	A	B

(Data from Van Sandt and Britton 2001.)

4.11 Calculate the capability indices C_p and C_{pk} for the process in Problem 4.1. The LSL and USL are 24 and 26 g, respectively; the target being at the center of the specification. Use $\overline{R}/d_2$ to estimate the process standard deviation. Make sure that the estimates for the process mean and process standard deviation are made from a process that is in-control.

4.12 If the specifications for the weight of ravioli packages in Problem 4.3 are 31.5 and 32.3 oz, and the target is at the center of the specification, find the values of C_p and C_{pk}. Make sure that the estimates for the process parameters are made from a process that is in-control.

4.13 The data below are the thicknesses of rubber sheets measured in millimeters by a thickness gage. Two measurements have been taken of each specimen by each of the two operators. The specifications are at 0.50 and 1.00 mm.

a. Calculate the reproducibility and repeatability errors (standard deviations), and separate the gage variability and part variability.

b. Verify the adequacy of the gage in resolution and precision.

	Operator 1		Operator 2	
Part	**Trial 1**	**Trial 2**	**Trial 1**	**Trial 2**
1	1.00	1.00	1.04	0.96
2	0.83	0.77	0.80	0.76
3	0.98	0.98	1.00	1.04
4	0.96	0.96	0.94	0.90
5	0.86	0.83	0.72	0.74
6	0.97	0.97	0.98	0.94
7	0.63	0.59	0.56	0.56
8	0.86	0.94	0.82	0.78
9	0.64	0.72	0.56	0.52
10	0.59	0.51	0.43	0.43

4.14 The table below shows part of the data from Table 4.8. Complete a gage R&R study using only this part of the data, and compare the results with those of Example 4.11.

No.	A-Trial 1	A-Trial 2	B-Trial 1	B-Trial 2
1	1.141	1.141	1.140	1.143
2	1.144	1.144	1.142	1.144
3	1.146	1.145	1.146	1.152
4	1.139	1.141	1.140	1.141
5	1.143	1.143	1.143	1.143
6	1.140	1.142	1.139	1.140
7	1.142	1.141	1.142	1.145
8	1.142	1.147	1.146	1.145
9	1.143	1.144	1.143	1.146
10	1.140	1.141	1.140	1.141
11	1.150	1.151	1.150	1.153
12	1.140	1.140	1.148	1.141
13	1.148	1.142	1.146	1.148
14	1.146	1.147	1.146	1.144
15	1.148	1.149	1.148	1.147

4.8.2 Mini-Projects

Mini-Project 4.1

The table below shows data on the breaking strength of water-jacket cores used in making cylinder-head castings in a foundry. The jacket core is in three sections, identified as the B, C, and R sections. In other words, the three sections together make one core for one cylinder head. These cores, which are made of sand mixed with bonding material, should have sufficient strength to withstand handling stresses and the stress generated when hot iron is poured around them in the mold. Baking them in a hotbox improves their strength, but overbaking will make them brittle. The three sections of the core are baked at different locations in a hotbox, leaving room for suspicion that the three sections may not be baked uniformly and may not be of the same strength. It is essential that these three parts of the core are equally strong, because the strength of the total core is equal to that of the weakest section. The data were collected to verify if the cores have consistently "good" strength.

Two sets (each set contains a B, C, and R core) of cores were taken every half-hour (in the night shift) and were broken in a transverse testing machine, which gave the beam strength of the cores in kilograms. After 3:00 A.M., however, only one set of cores was tested because the production people were running short of cores and did not want to spare cores for breaking.

Analyze the data and see if the process is producing consistently good cores. The strength should be consistent both among the three sections and from time to time. What should the CL and limits of the control chart be for future control? What would your suggestion be to improve the quality of the process?

What is the current process capability? Note that no specifications are given and you may have to come up with an appropriate set of specifications for the strength. The strength of the cores is a process parameter. The foundry does not sell cores; it sells the iron that is produced using the cores. Strength specifications are created to assist in maintaining the consistency of the process. Such specifications are usually created as natural tolerance limits (NTLs) of the process, which are obtained as $(\bar{\bar{X}} \pm 3\hat{\sigma})$. The capability can be described by the NTLs.

Time	11:15 PM			11:45 PM			12:15 AM			12:45 AM			1:15 AM			1:45 AM		
Core	B	C	R	B	C	R	B	C	R	B	C	R	B	C	R	B	C	R
Set 1	22	8.9	15.3	17.9	11.6	18.2	18.9	12.7	19.6	22.9	19.6	20.1	19.4	14.8	16.5	21.8	16	16.8
Set 2	15.5	8.9	12	22.5	15.8	9.0	15.1	17.1	19.6	20.5	11.5	18.8	19.3	18.9	17.1	18.8	9.7	16.4

Time	2:15 AM			3:00 AM			3:30 AM			4:00 AM			4:30 AM			5:00 AM		
Core	B	C	R	B	C	R	B	C	R	B	C	R	B	C	R	B	C	R
Set 1	22.5	13.9	18.3	19.5	18.5	13	5.6	13.8	12.1	11.8	14.1	13	18	12.8	17	21.1	14.3	15.6
Set 2	17.8	13.5	19.0	—	—	—	—	—	—	—	—	—	—	—	—	—	—	—

Mini-Project 4.2

The data below represent the proportion of rejected castings for various defects on a particular production line. Make a *P*-chart using the data on the daily proportion of defectives. Draw any useful information that can be gleaned from these data based on the *P*-chart.

Date	5/1	5/2	5/3	5/6	5/7	5/8	5/9	5/10	5/13	5/28	5/29	5/30	5/31	6/3	6/12
No. made	115	192	146	350	284	236	353	193	212	174	329	289	238	157	162
No. inspected	26	177	128	150	230	236	204	143	171	136	276	156	234	108	85
No. defective	1	54	43	63	67	73	69	49	42	44	100	47	98	43	36

Date	6/13	6/14	6/15	6/17	6/18	6/19	6/20	6/21	6/24	6/25	6/26	6/27	6/28	7/1	7/2
No. made	151	158	180	135	305	191	169	170	309	104	190	162	313	138	295
No. inspected	55	158	150	88	305	147	167	159	266	93	190	128	259	127	158
No. defective	29	104	112	69	137	61	73	101	126	46	94	85	135	43	51

Date	7/3	7/5	7/8	7/9	7/10	7/11	7/29	7/30	7/31	8/1	8/2	8/5	8/6	8/7	8/8
No. made	164	174	188	149	164	134	319	135	184	180	141	152	142	182	159
No. inspected	164	170	140	94	151	83	240	124	173	166	141	103	128	161	140
No. defective	70	74	80	51	84	43	144	57	63	78	69	38	44	42	39

Date	8/9	8/12	8/13	8/14	8/15	8/16	8/28	8/29	8/30	8/31	9/3	9/4	9/5	9/6	9/7
No. made	266	225	346	355	308	173	101	352	165	333	322	188	179	125	335
No. inspected	142	192	302	329	273	147	23	255	165	296	236	172	144	80	270
No. defective	51	68	128	161	125	66	7	49	35	79	77	36	28	25	65

Date	9/9	9/10	9/12	9/13	9/24	9/25	9/26	9/27
No. made	170	157	239	85	130	188	188	150
No. inspected	147	142	212	76	99	129	14	6
No. defective	35	36	60	21	26	62	2	4

Mini-project 4.3

I have a suspicion that the speedometer in my 1984 VW van is not showing the correct speed of the vehicle, so I decided to evaluate it using the speedometer of my wife's 2005 Toyota Prius. We set out on a road trip (in May 2008) including city streets and highways, and recorded the speedometer readings from the two vehicles as we both cruised at the same speed. We used the cell phone to communicate the readings, which were taken at the same time. The following table shows the readings recorded from the two vehicles. Note that the speed readings, in mph, are at two levels: one at the city street speed level and another at the highway speed level.

Comment on the bias, linearity, precision and stability of the VW speedometer considering the newer Toyota's speedometer as the "tool-room" standard.

This is just the first experiment; I would like to do a revised experiment with better data collection. What improvement would you suggest?

1984 VW	2005 Prius	1984 VW	2005 Prius
28	17	33	26
36	32	42	39
15	10	23	19
42	39	96	81
59	49	91	84
65	56	85	72
66	49	65	55
85	73	45	32
88	72	48	35
32	26	96	89
39	33	42	36
98	81	55	42
86	80	63	52
45	37	41	36
49	36	22	15
77	62	21	16
78	66	39	31
18	12	38	32
19	16	74	68
68	56	75	66
56	52	79	69
54	48	88	76
64	55	87	77
24	19	85	72
33	25	28	15
36	25	21	13
44	39	22	15
76	69	11	6
74	65	9	5
56	50	10	4
52	43	12	5
50	42	26	15
55	46	28	16

References

AT&T. 1958. *Statistical Quality Control Handbook.* 2nd ed. Indianapolis, IN: AT&T Technologies.

Chrysler LLC, Ford Motor Company, and General Motors Corporation. 2010. *Measurement Systems Analysis (MSA)—Reference Manual.* 4th ed. Southfield, MI: AIAG.

Deming, W. E. 1986. *Out of the Crisis.* Cambridge, MA: MIT—Center for Advanced Engineering Study.

Duncan, A. J. 1956. "The Economic Design of $\overline{X}$ Charts Used to Maintain Current Control of a Process." *Journal of the American Statistical Association* 51: 228–242.

Duncan, A. J. 1974. *Quality Control and Industrial Statistics.* 4th ed. Homewood, IL: Irwin.

Grant, E. L., and R. S. Leavenworth. 1996. *Statistical Quality Control.* 7th ed. New York, NY: McGraw-Hill.

Krishanmoorthi, K. S. 1989. *Quality Control for Operators and Foremen.* Milwaukee, WI: ASQC Quality Press.

Krishnamoorthi, K. S., V. P. Koritala, and C. Jurs. June 2009. "Sampling Variability in Capability Indices." *Proceedings of IE Research Conference—Abstract.* Miami, FL.

Montgomery, D. C. 2001. *Introduction to Statistical Quality Control.* 4th ed. New York, NY: John Wiley.

Motorola Corporation. 1992. *Six Steps to Six Sigma.* Schaumburg, IL: SSG 102, Motorola University.

Ott, E. R. 1975. *Process Quality Control—Trouble Shooting and Interpretation of Data.* New York: McGraw-Hill.

Schmillan, P., and C. Johnson. 2001. "Evaluation of Process Control of Jumbo Ravioli." Unpublished project report, IME 522, IMET Department, Bradley University, Peoria, IL.

Shewhart, W. A. 1931. *Economic Control of Manufactured Product.* Princeton, NJ: D. Van Nostrand Co., Inc. Reprinted 1980. Milwaulkee, WI: American Society for Quality.

Shewhart, W. A. 1939. *Statistical Method from the Viewpoint of Quality Control.* Washington, DC: Graduate School of Department of Agriculture. Reprinted 1986. Mineola, NY: Dover Publications, Inc.

Van Sandt, T., and P. Britton. 2001. "A Quality Inspection of A+ Cleaning." Unpublished project report, IME 522, IMET Department, Bradley University, Peoria, IL.

5

Quality in Production—Process Control II

This chapter is a continuation of Chapter 4 and includes some additional material on process control and related topics. The three types of control charts—$\overline{X}$- and R-charts, the P-chart, and the C-chart—that were discussed in Chapter 4 would cover the majority of situations requiring process control in the real world, and the few minor variations of these charts included in the last chapter would cover a few additional, special situations. Many other situations, however, call for even more variations of the basic charts because of practical necessities. In this chapter, we will address some of these special situations and see how special charts are constructed to handle them.

Before we look at these special control charts, we want to discuss how the formulas for the control limits we used for the three major control charts are derived. We also want to study the operating characteristics of these charts, which reveal their strengths and weaknesses in discovering changes in the processes. These topics in the theory of control charts enable a user to understand how the control method works—which in turn helps a user to obtain the maximum benefit out of them—when using them on real processes. Furthermore, many real-world problems do not resemble the simplified textbook versions that can be solved by direct application of the methods. Some do not satisfy the assumptions needed, and some do not yield data in the format needed. It then becomes necessary to modify the basic methods to fit the situation at hand. Such modification of the methods is possible only by a user with a good understanding of the principles behind, and the assumptions that are made, in the derivation of the methods. An understanding of the fundamental theory of the charts is also necessary if one wants to read more advanced technical literature on these topics.

This chapter also includes some additional topics in process capability and experimental design beyond their coverage in earlier chapters. Specifically, in this chapter we will discuss:

- Derivation of the formulas for control limits of $\overline{X}$- and R-charts
- Derivation of limits for P- and C-charts
- Operating characteristic curves of $\overline{X}$- and R-charts
- Operating characteristic curves of P- and C-charts
- Control charts when the standards for μ and/or σ are given

- Control charts for slow processes
    - The chart for individuals
    - The moving average and moving range charts
- The exponentially weighted moving average chart
- Control charts for short runs
- Additional topics in process capability
- Additional topics in design of experiments

5.1 Derivation of Limits

5.1.1 Limits for the $\overline{X}$-Chart

Derivation of the limits for the $\overline{X}$-chart is based on the assumption that the process being controlled follows a normal distribution. We saw in Chapter 2 that if a process is normally distributed, the sample average from that process is also normally distributed according to the following law:

$$\text{If } X \sim N(\mu, \sigma^2), \quad \text{then } \overline{X}_n \sim N\left(\mu, \frac{\sigma^2}{n}\right)$$

Furthermore, according to the central limit theorem, even if the process distribution is not normal, the sample averages will tend to be normal as sample sizes become large. That is, if X has any distribution $f(x)$ with mean μ and variance σ^2,

$$\text{then } \overline{X}_n \xrightarrow[n\to\infty]{} N\left(\mu, \frac{\sigma^2}{n}\right)$$

Suppose we have a process that is normally distributed with a certain mean and standard deviation as shown in Figure 5.1. When we say that we want to

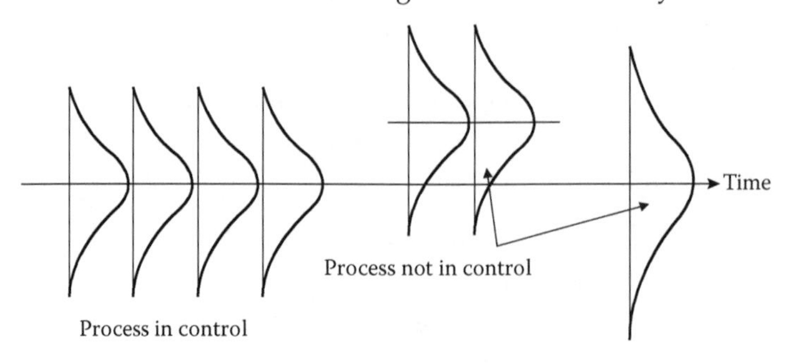

FIGURE 5.1
Example of a process when in-control and when not-in-control.

control this process, it means that we want to make certain the distribution of this process remains the same, with the same mean and the same standard deviation, throughout the time that is of interest to us. To verify this is so, we monitor the $\overline{X}$ values from samples taken periodically from this process.

If the process distribution remains the same, then the distribution of $\overline{X}$ will also remain the same and the observed values of $\overline{X}$ will "all" (99.73% of them) fall within three standard deviations ($3\sigma_{\overline{X}}$) from the mean μ. Conversely, if "all" the observed values of $\overline{X}$ fall within three standard deviations from the mean, we can conclude the process distribution remains the same. Because we know the $\overline{X}$ values have the normal distribution, with $\mu_{\overline{X}} = \mu$ and $\sigma_{\overline{X}} = \sigma/\sqrt{n}$, the limits at three standard deviations, or the 3-sigma limits for $\overline{X}$, are given by (see Figure 5.2):

$$\text{UCL}(\overline{X}) = \mu + 3\sigma_{\overline{X}} = \mu + 3\sigma/\sqrt{n}$$

$$\text{CL}(\overline{X}) = \mu$$

$$\text{LCL}(\overline{X}) = \mu - 3\sigma_{\overline{X}} = \mu - 3\sigma/\sqrt{n}$$

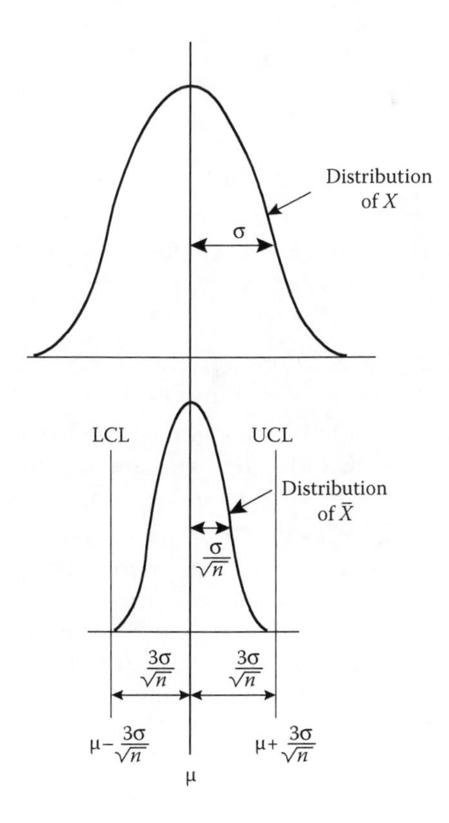

FIGURE 5.2
Deriving the control limits for the $\overline{X}$-chart.

Because the values of μ and σ are not usually known, we can estimate them by using $\overline{\overline{X}}$ and $\overline{R}/d_2$, respectively, where $\overline{\overline{X}}$ is the average of about 25 sample averages and $\overline{R}$ is the average of about 25 sample ranges from an in-control process. The factor d_2 is a correction factor that makes R, or $\overline{R}$, an unbiased estimator of σ. In other words, the factor d_2 is such the average of $(\overline{R}/d_2)$ denoted as $E(\overline{R}/d_2) = \sigma$.

Thus, substituting for μ and σ with estimates, we get:

$$\text{UCL}(\overline{X}) = \overline{\overline{X}} + 3\frac{\overline{R}}{d_2\sqrt{n}}$$

$$\text{CL}(\overline{X}) = \overline{\overline{X}}$$

$$\text{LCL}(\overline{X}) = \overline{\overline{X}} - 3\frac{\overline{R}}{d_2\sqrt{n}}$$

If we make $A_2 = 3/(d_2\sqrt{n})$, then:

$$\text{UCL}(\overline{X}) = \overline{\overline{X}} + A_2\overline{R}$$

$$\text{CL}(\overline{X}) = \overline{\overline{X}}$$

$$\text{LCL}(\overline{X}) = \overline{\overline{X}} - A_2\overline{R}$$

Values of A_2 are computed for various values of n and provided in standard tables, such as Table A.4 in the Appendix.

From the above derivation, we learn the following:

1. The control limit calculations for the $\overline{X}$-chart are based on the assumption that the process is normally distributed.
2. The control limit calculations are good, even for processes that are not normally distributed provided the sample sizes are large. (Even $n = 4$ or 5 is known to be large enough for this purpose).
3. A_2 provides 3-sigma limits for the $\overline{X}$-chart. As mentioned in Chapter 4, Dr. Shewhart, the author of the control chart method, recommended the 3-sigma limits because they provide the economic trade-off between too-tight limits, which would produce excessive false alarms, and too-loose limits, which would allow assignable causes to go undetected. If 2-sigma limits are needed in a particular situation, as when someone wants to detect and eliminate assignable causes quickly, the factor to use would be $2A_2/3$.
4. Because the control limits are at a 3-sigma distance from the centerline (CL), there is a probability of a false alarm (Type I error) of 0.0027. In other words, the control chart could find a process to be

not-in-control when, in fact, it is in-control. This could happen 27 times out of 10,000 samples, or roughly 3 out of 1000 samples.

5. The control limits have been calculated using estimates for μ and σ computed from the data obtained from the process.

5.1.2 Limits for the *R*-Chart

The rule on computing the limits for any 3-sigma control chart can be generalized as follows.

If Θ is the statistic that is plotted to control a process parameter θ, then the CL should be at the average of the statistic, or expected value $E(\Theta)$, and the UCL and LCL should be at $E(\Theta) + 3\sigma(\Theta)$ and $E(\Theta) - 3\sigma(\Theta)$, respectively, where $\sigma(\Theta)$ represents the standard deviation of the statistic Θ. In other words, for any control chart, the centerline will be at the average value of the statistic being plotted, and upper and lower control limits will be placed at a distance of 3 × (standard deviation of the statistic) on either side of the centerline. Therefore, for the *R*-chart, the CL should be at $E(R)$, and the two control limits should be at $E(R) \pm 3\sigma(R)$.

The statistic $w = R/\sigma$, where σ is the process standard deviation, has been studied, and its distribution—along with its mean and standard deviation—are known for samples taken from normal populations (Duncan 1974). The statistic w is called the "relative range." Specifically, $E(w) = E(R/\sigma) = d_2$, a constant for a given sample size n. Therefore, $E(R) = \sigma d_2$. Similarly, st. dev. (w) = st. dev. $(R/\sigma) = d_3$, a constant for a given n, which means that st. dev. $(R) = \sigma d_3$.

Values of d_2 and d_3 have been tabulated for various sample sizes drawn from normal populations, as in Table A.4 in the Appendix. Using these, the limits for the *R*-chart can be computed as:

$$\text{UCL}(R) = \sigma d_2 + 3\sigma d_3 = \sigma(d_2 + 3d_3)$$

$$\text{CL}(R) = \sigma d_2$$

$$\text{LCL}(R) = \sigma d_2 - 3\sigma d_3 = \sigma(d_2 - 3d_3)$$

If σ is estimated by $\overline{R}/d_2$, then:

$$\text{UCL}(R) = \overline{R} + 3\frac{\overline{R}d_3}{d_2} = \overline{R}\left(1 + \frac{3d_3}{d_2}\right)$$

$$\text{CL}(R) = \overline{R}$$

$$\text{LCL}(R) = \overline{R} - 3\frac{\overline{R}d_3}{d_2} = \overline{R}\left(1 - \frac{3d_3}{d_2}\right)$$

Setting $D_4 = \left(1 + \frac{3d_3}{d_2}\right)$ and $D_3 = \left(1 - \frac{3d_3}{d_2}\right)$, we get:

$$\text{UCL}(R) = D_4\overline{R}$$

$$\text{CL}(R) = \overline{R}$$

$$\text{LCL}(R) = D_3\overline{R}$$

Values of the constants D_3 and D_4 have been computed and tabulated for various values of n and are available in tables such as Table A.4 in the Appendix.

From this derivation, we note the following:

1. The control limits for the R-chart have been calculated on the assumption that the samples are drawn from normal populations.
2. The control limits are 3-sigma limits. If, for example, 2-sigma limits are needed, then the limits can be calculated as $\overline{R}(1 \pm 2(d_3/d_2))$, using d_3 and d_2 from Table A.4.
3. The control limits are not equidistant on both sides of the CL, because D_3 will be smaller than D_4 for all sample sizes.
4. These control limits are based on estimates for σ obtained from process data.

5.1.3 Limits for the *P*-Chart

To determine the control limits for the P-chart, we should first determine the expected value and the standard deviation of the statistic P that is being plotted.

$P = D/n$, where D is the number of defective units in a sample of size n drawn from a population with p fraction defectives. From Chapter 2, D is a binomial variable; that is, $D \sim Bi(n, p)$ and $E(D) = np$ and $V(D) = np(1-p)$. Therefore,

$$E(P) = E\left(\frac{D}{n}\right) = \frac{1}{n}E(D) = \frac{1}{n}np = p.$$

$$V(P) = V\left(\frac{D}{n}\right) = \frac{1}{n^2}V(D) = \frac{1}{n^2}np(1-p) = \frac{p(1-p)}{n}$$

Thus, the standard deviation $(P) = \sqrt{\frac{p(1-p)}{n}}$.

Therefore, the 3-sigma limits for the *P*-chart are:

$$\mathrm{UCL}(P) = p + 3\sqrt{\frac{p(1-p)}{n}}$$

$$\mathrm{CL}(P) = p$$

$$\mathrm{LCL}(P) = p - 3\sqrt{\frac{p(1-p)}{n}}$$

Because the value of p, the proportion defectives in the population, is generally not known, it is estimated using the average $\bar{p}$ of the observed values of P from about 25 samples. Hence, the limits are:

$$\mathrm{UCL}(P) = \bar{p} + 3\sqrt{\frac{\bar{p}(1-\bar{p})}{n}}$$

$$\mathrm{CL}(P) = \bar{p}$$

$$\mathrm{LCL}(P) = \bar{p} - 3\sqrt{\frac{\bar{p}(1-\bar{p})}{n}}$$

In the above derivation we have used the following notations:

- p is the proportion of defectives in the population to be controlled;
- P is the statistic used as an estimator for p;
- p_i is an observed value of P in the i-th sample; and
- $\bar{p}$ is the average of the p_i values.

5.1.4 Limits for the C-Chart

The parameter that we want to control is c, the average number of defects per unit in the population. We plot the statistic C, which represents the number of defects in any sample unit.

The random variable C has a Poisson distribution, and from Chapter 2, we know that: $E(C) = c$, $V(C) = c$, and sta. dev.$(C) = \sqrt{c}$.

Therefore, the 3-sigma limits for the statistic C are:

$$\mathrm{UCL}(C) = c + 3\sqrt{c}$$

$$\mathrm{CL}(C) = c$$

$$\mathrm{LCL}(C) = c - 3\sqrt{c}$$

Because the value of c, the population parameter, is usually not known, we can estimate it by $\bar{c}$, or the average of the observed values of C from about 25 sample units. The limits, then, are:

$$\mathrm{UCL}(C) = \bar{c} + 3\sqrt{\bar{c}}$$

$$\mathrm{CL}(C) = \bar{c}$$

$$\mathrm{LCL}(C) = \bar{c} - 3\sqrt{\bar{c}}$$

In the above derivation we have used the following notations:

- c is the average number of defects per unit in the population, which is to be controlled;
- C is the statistic used as an estimator for c;
- c_i is an observed value of C from the i-th sample unit; and
- $\bar{c}$ is the average of the c_i values.

5.2 Operating Characteristics of Control Charts

5.2.1 Operating Characteristics of an $\bar{X}$-Chart

The operating characteristics (OC) curve of a control chart describes how well the control chart discovers assignable causes of various magnitudes. When they occur on a process, assignable causes could disturb the process

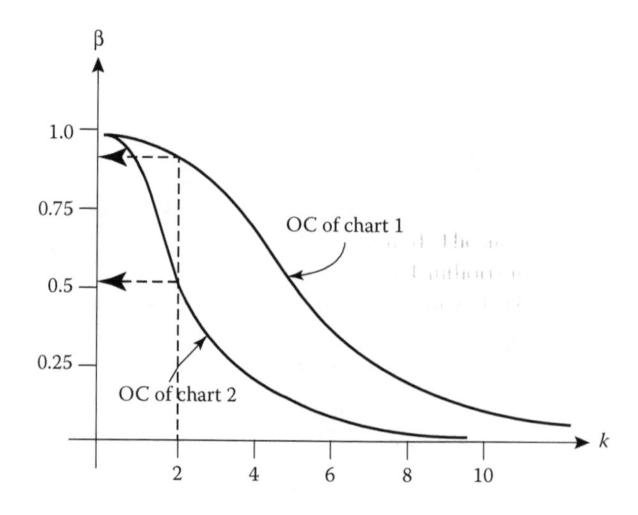

FIGURE 5.3
Examples of OC curves.

mean as well as the process standard deviation. In the following discussion, we will assume, for the sake of simplicity, that an assignable cause changes only the process mean and not the standard deviation. Therefore, the magnitude of an assignable cause is designated by the amount of shift it causes in the process mean, measured in number of standard deviations of the process.

The change in the process mean is represented on the x-axis by k, where k is the distance, in number of standard deviations, that the process mean is moved by an assignable cause. The ability of the control chart to detect a change is represented on the y-axis by β, the probability that the control chart will accept the changed process. In Figure 5.3, two examples of OC curves are shown. The OC curve of Chart 1 indicates that the chart will accept a process when its mean has been moved by two standard deviations with a probability of about 0.9; whereas according to the OC curve of Chart 2, this chart will accept such a change with a probability of only about 0.5. Chart 2 is the stricter, more discriminating chart.

5.2.1.1 Computing the OC Curve of an $\overline{X}$-Chart

Figure 5.4 shows a process with a standard mean μ being moved by an assignable cause to a new location $\mu + k\sigma$, where σ is the standard deviation of the process, which remains unchanged. The process is controlled by an $\overline{X}$-chart with control

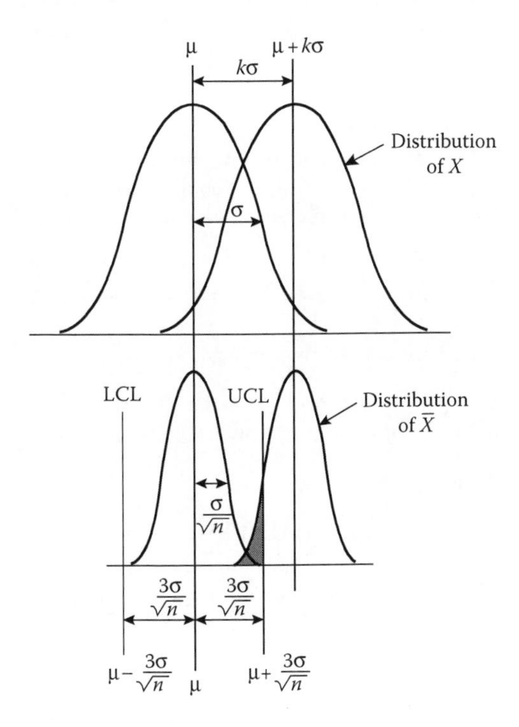

FIGURE 5.4
Computing the OC curve of an $\overline{X}$-chart.

limits at $\mu \pm 3\sigma/\sqrt{n}$. When the process is moved to the new location, the distribution of $\overline{X}$ is also moved to the new location. The shaded area in the new distribution of $\overline{X}$ represents the probability that the $\overline{X}$ values will fall within the original control limits, and thus the probability that the control chart will still accept the process now located at the new mean. This area represents β. Therefore:

$$\beta = P\left(\text{LCL} \le \overline{X} \le \text{UCL} \mid \mu = \mu + k\sigma\right)$$

$$= P\left(\mu - 3\sigma/\sqrt{n} \le \overline{X} \le \mu + 3\sigma/\sqrt{n} \mid \mu = \mu + k\sigma\right)$$

$$= P\left(\frac{\mu - 3\sigma/\sqrt{n} - \mu - k\sigma}{\sigma/\sqrt{n}} \le Z \le \frac{\mu + 3\sigma/\sqrt{n} - \mu - k\sigma}{\sigma/\sqrt{n}}\right)$$

$$= P\left(-3 - k\sqrt{n} \le Z \le 3 - k\sqrt{n}\right) = \Phi\left(3 - k\sqrt{n}\right) - \Phi\left(-3 - k\sqrt{n}\right)$$

We notice that β is a function of k, the size of the assignable cause; and n, the sample size used for the control chart.

Example 5.1

Calculate the OC curve for a 3-sigma control chart for sample sizes of $n = 4$ and $n = 16$.

Solution

Calculating the OC curve involves calculating the values of β for several selected values of k. The calculated values of the OC function are shown in Tables 5.1a and 5.1b for $n = 4$ and $n = 16$, respectively. As an example of the calculations, suppose that $n = 4$ and $k = 1$. Then,

$$\beta = \Phi\left(3 - \sqrt{4}\right) - \Phi\left(-3 - \sqrt{4}\right) = \Phi(1) - \Phi(-5) = 0.8413.$$

The graphs of the OC curves are shown in Figure 5.5.

Notice that in Example 5.1 the $\overline{X}$-chart with $n = 4$—a commonly used chart, called the conventional chart—has a probability of about 0.8 of accepting

TABLE 5.1a

OC Function of an $\overline{X}$-chart with 3-sigma Limits and $n = 4$

k	0.0	0.5	1.0	1.5	2.0	2.5
β	0.9973	0.9772	0.8413	0.5	0.1587	0.0228

TABLE 5.1b

OC Function of an $\overline{X}$-chart with 3-sigma Limits and $n = 16$

k	0.0	0.5	1.0	1.5	2.0	2.5
β	0.9973	0.8413	0.1587	0.0014	0.0	0.0

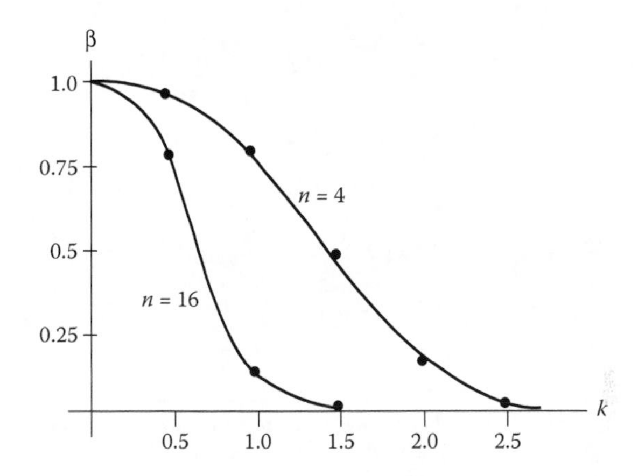

FIGURE 5.5
OC curves of an $\overline{X}$-chart with 3-sigma limits.

(or 0.2 of rejecting) a process that has moved through a 1σ distance. Only when the process shift is greater than 1.5σ does this $\overline{X}$-chart have any reasonable probability (>0.50) of rejection. Also, notice that when the sample size increases, the power of the control chart, (1 − β), increases for detecting a given amount of change in the process mean. When $n = 16$, the power to detect a change of lσ distance in the process mean becomes 0.84.

We can see from the above OC curves that the conventional $\overline{X}$-chart with $n = 4$ is not quite sensitive to changes in the process mean unless the change in the mean is more than 1.5σ distance. This may not be a bad feature for the control chart when such small changes need not be discovered. When small changes in the process mean are important, however, an increase in power can be obtained by increasing the sample size.

5.2.2 OC Curve of an *R*-Chart

The OC curve of an *R*-chart will show the probability of acceptance by the chart when the process standard deviation changes from the original value. The change in the process standard deviation is denoted by $\lambda = \sigma_1/\sigma$, where σ and σ_1 are the original and the changed (new) standard deviations, respectively. The probability that the process will be accepted by the chart is denoted as β and is calculated as follows. This calculation of β makes use of the distribution of the relative range R/σ:

$$\begin{aligned}\beta &= P\left(\text{LCL} \le R \le \text{UCL} \mid \sigma = \sigma_1\right)\\ &= P\left(\frac{\text{LCL}}{\sigma_1} \le \frac{R}{\sigma_1} \le \frac{\text{UCL}}{\sigma_1}\right)\\ &= P\left(\frac{\text{LCL}}{\sigma_1} \le w \le \frac{\text{UCL}}{\sigma_1}\right)\end{aligned}$$

$$= P\left(\frac{\sigma(d_2 - 3d_3)}{\sigma_1} \le w \le \frac{\sigma(d_2 + 3d_3)}{\sigma_1}\right)$$

$$= P\left(\frac{D_3 d_2}{\lambda} \le w \le \frac{D_4 d_2}{\lambda}\right)$$

because $D_3 = (d_2 - 3d_3)/d_2$, $D_4 = (d_2 + 3d_3)/d_2$, and $\lambda = \sigma_1/\sigma$.

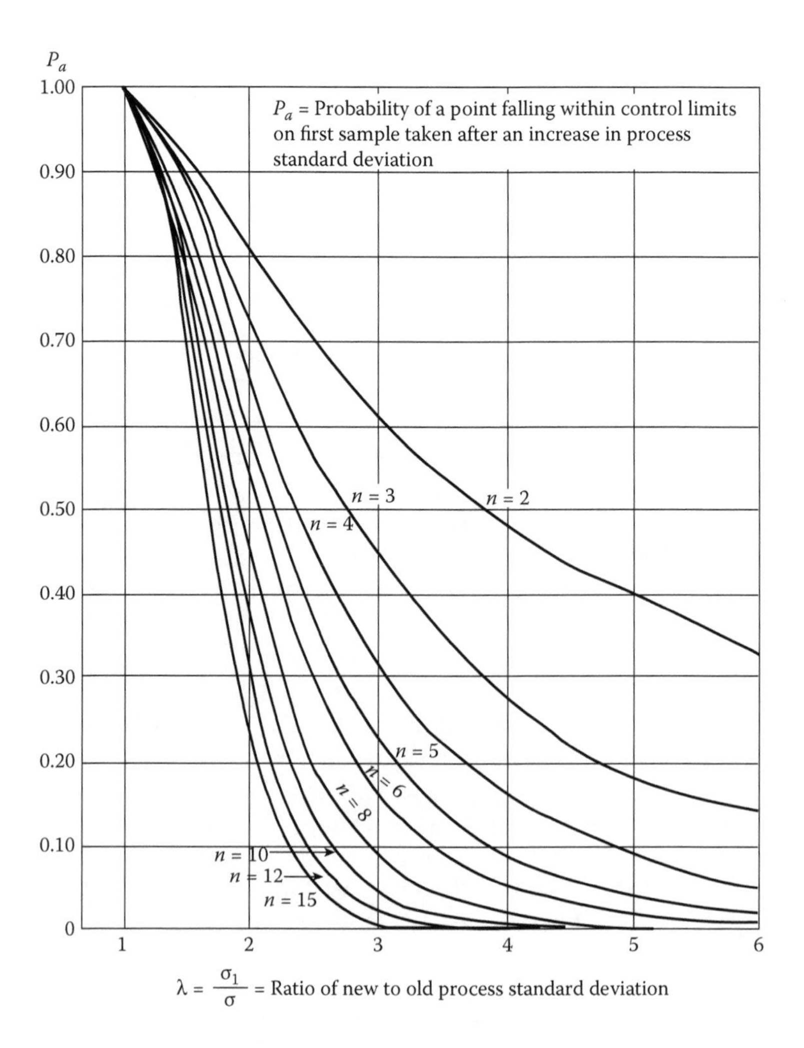

FIGURE 5.6
OC curve of an R-chart for various sample sizes. (Reproduced from Duncan, A.J. "Operating Characteristics of R-Charts." *Industrial Quality Control,* pp. 40–41. Milwaukee: American Society for Quality, 1951. With permission.)

For a given n, β can be calculated for different values of λ using tables of distribution of w, which are available, for example, in Duncan (1974). Figure 5.6, which is reproduced from Duncan (1951), shows the OC curve of the *R*-chart with 3-sigma limits for several values of n.

We notice in Figure 5.6, for the *R*-chart with $n = 4$, that when the standard deviation becomes twice that of the original value (i.e., $\lambda = 2.0$), the probability of acceptance is about 0.66. This means that the probability of rejection when such a large change occurs is only 0.34, which is not much. The probability of rejection becomes greater than 0.5 only when $\lambda > 2.4$. This again shows that the *R*-chart with the conventional sample size of $n = 4$ is not very sensitive to changes in the process standard deviation.

5.2.3 Average Run Length

Average run length (*ARL*) is the average number of samples needed for a control chart to signal a change when a change has occurred on the process being controlled. For example, suppose that a process is currently being controlled by an $\overline{X}$-chart at an average level μ_0, and suppose it is disturbed by an assignable cause that moves the process to a different average level μ_1, as shown in Figure 5.7. Will the $\overline{X}$ value from the sample taken immediately after the change occurred fall outside the limits on the chart? There is a probability that this $\overline{X}$-value will fall outside the limits equal to the area shown shaded in the distribution of $\overline{X}$ in Figure 5.7. There is also a probability that the $\overline{X}$ value from this sample will not fall outside the limits. The question then arises: how many samples will it take for the chart to indicate that the change has occurred? The number of samples needed to discover the change is a random variable.

If Y is the random variable that represents the number of samples needed before an $\overline{X}$ value falls outside a limit after the change has occurred, then Y has the geometric distribution with parameter p, where p is the probability that an $\overline{X}$ value falls outside the limits, given that the change has occurred. It

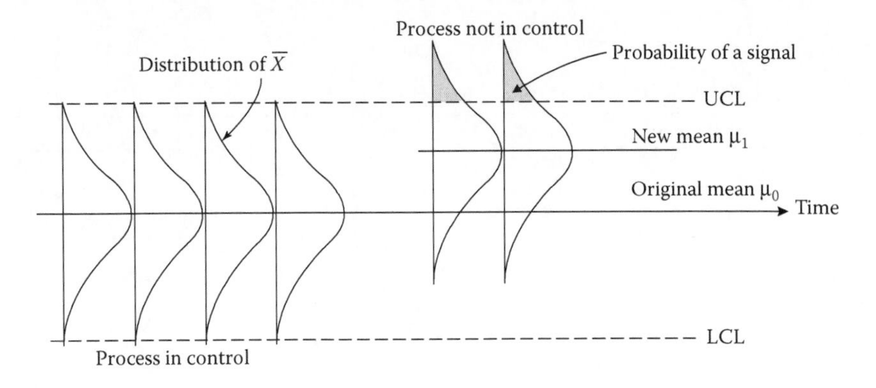

FIGURE 5.7
Probability of detection by an $\overline{X}$-chart when a process is not-in-control.

can be seen that $p = 1 - \beta$, where β is the probability of acceptance calculated for the OC curve. The distribution of the random variable Y, the geometric distribution, is represented by the probability mass function:

$$p(y) = (1-p)^{y-1}\, p, \text{ for } y = 1, 2, 3, \ldots$$

The average of Y can be shown to be:

$$E(Y) = \sum_{y=1}^{\infty} y(1-p)^{y-1}\, p = \frac{1}{p}$$

See Hines and Montgomery (1990) for proof.

This average value of Y is called the *ARL*. Thus, it is the number of samples needed, on average, for the control chart to discover an assignable cause when it occurs. The *ARL* is a function of p, which in turn depends on the size of the assignable cause and the parameters of the control chart—such as sample size and spread for the limits. It is used as a measure of how well a control chart is able to discover changes in a process.

Example 5.2

Calculate the *ARL* values for the $\bar{X}$-chart with 3-sigma limits when $n = 4$ and when $n = 16$ for different assignable causes (i.e., for different values of k). Draw the *ARL* curves.

Solution

The *ARL* values for the two $\bar{X}$-charts are shown in Tables 5.2a and 5.2b. These tables use the data on OC curves computed in Example 5.1. The graphs of the *ARL* as a function of k are shown in Figure 5.8a.

TABLE 5.2a

ARL Calculations for the 3-sigma $\bar{X}$-chart when $n = 4$

k	0.0	0.5	1.0	1.5	2.0	2.5
β	0.9973	0.9772	0.8413	0.5	0.1587	0.0228
p	0.0027	0.0228	0.1587	0.5	0.8413	0.9772
$ARL = 1/p$	370	43.9	6.3	2.0	1.2	1.02

TABLE 5.2b

ARL Calculations for the 3-sigma $\bar{X}$-chart when $n = 16$

k	0.0	0.5	1.0	1.5	2.0	2.5
β	0.9973	0.8413	0.1587	0.0014	0.0	0.0
p	0.0027	0.1587	0.8413	0.9986	1.00	1.00
$ARL = 1/p$	370	6.3	1.2	1.0	1.0	1.0

Figure 5.8b shows the *ARL* curves for the 3-sigma $\overline{X}$-chart for several other sample sizes. Note that the *ARL* curve for $n = 4$ shows that if the change in the process mean is less than 1σ—for instance, 0.5σ—then the *ARL* is very large, meaning that it will take many samples to discover such a change.

The *ARL* curves have the same information in them as the OC curves; however, the *ARL* is a more meaningful measure of performance than the OC function. Hence, it is used more often to evaluate and compare the performance of control charts.

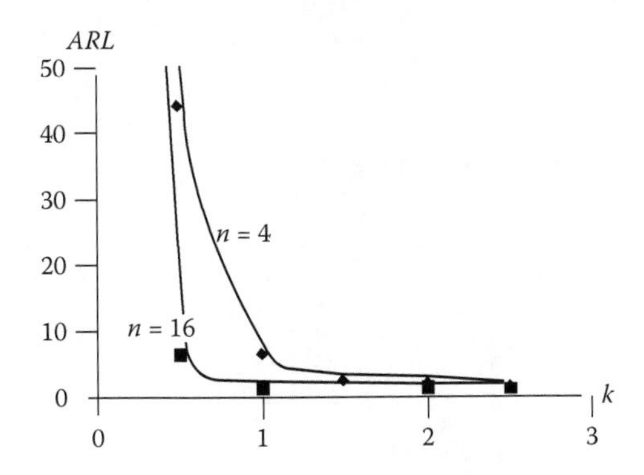

FIGURE 5.8a
ARL curves for a 3-sigma $\overline{X}$-chart.

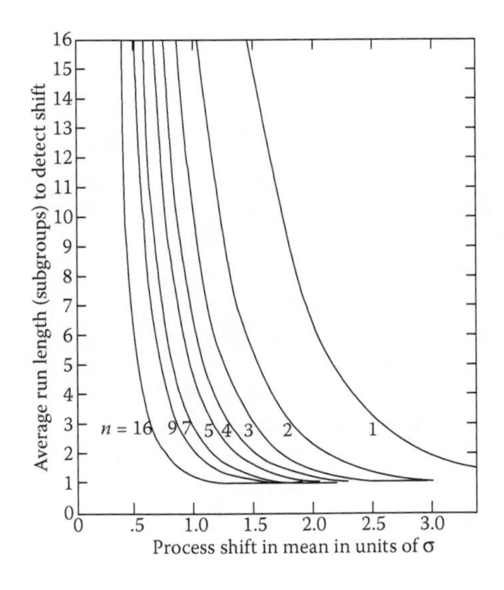

FIGURE 5.8b
ARL curves for an $\overline{X}$-chart for various sample sizes. (Wadsworth, H. M., K. S. Stephens, and A. B. Godfrey. *Modern Methods for Quality Control and Improvement*. 1986. Copyright Wiley-VCH Verlag GmbH & Co. KGaA. Reproduced with permission.)

5.2.4 OC Curve of a *P*-Chart

The OC curve of a *P*-chart is made by plotting the p values, proportion defectives in the process, on the x-axis and the probability of acceptance corresponding to the p values on the y-axis. The OC curve of a chart designed for controlling a process at a particular value—say, p_0—will give the probabilities of acceptance by the chart that the proportion defectives produced becomes a p other than p_0. The method of calculating the OC is illustrated by Example 5.3.

Example 5.3

Suppose that a process is currently producing 10% defectives—that is, $p_0 = 0.10$—and we use a *P*-chart to control the process at this level of p, with $n = 100$. We want to find the OC curve of the *P*-chart.

Solution

The limits for the chart would be:

$$\mathrm{UCL}(P) = 0.1 + 3\sqrt{\frac{(0.1)(0.9)}{100}} = 0.19$$

$$\mathrm{CL}(P) = 0.1$$

$$\mathrm{LCL}(P) = 0.1 - 3\sqrt{\frac{(0.1)(0.9)}{100}} = 0.01$$

Suppose that the process fraction defective becomes p_1. The probability of acceptance $Pa(p_1)$ is then given by:

$$\begin{aligned}
\beta &= P(\text{an observed value of } P \text{ falls in between the two control limits given } p = p_1) \\
&= P(0.01 < P < 0.19 \mid p = p_1) \\
&= P(1 < 100P < 19 \mid p = p_1) \\
&= P(1 < D < 19 \mid p = p_1)
\end{aligned}$$

where D is the number of defectives in the sample of $n = 100$. (We have assumed that on-the-line is out.)

Therefore, $\beta = P(D \le 18 \mid p = p_1) - P(D \le 1 \mid p = p_1)$, where $D \sim Bi(100, p_1)$.

We can assign different values for p_1 and obtain the values for β using the Poisson approximation to binomial and the cumulative Poisson tables (see Table A.5 in the Appendix). The OC function is shown in Table 5.3, and the OC curve

TABLE 5.3

OC Function of a *P*-chart Designed to Control a Process at $p = 0.10$

p_1	0.005	0.01	0.02	0.05	0.1	0.15	0.2	0.25
$np = (100p)$	0.5	1.0	2.0	5.0	10	15	20	25
$Pa(p_1) = \beta$	0.091	0.265	0.594	0.960	0.992	0.819	0.381	0.092

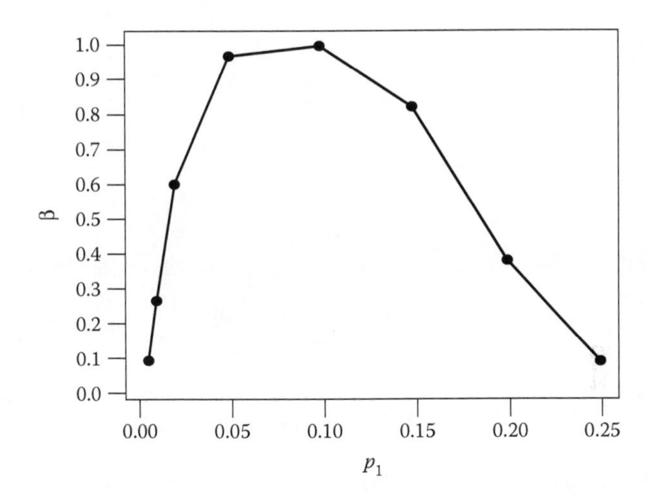

FIGURE 5.9
OC curve of a *P*-chart designed to control a process at $p = 0.10$.

is shown in Figure 5.9. For example, when $p_1 = 0.01$, $\beta = P(D \leq 18 \mid p = 0.01) - P(D \leq 1 \mid p = 0.01)$, where D is a Poisson variable with $np = 100 \times 0.01 = 1.0$. Then, using the Poisson table, $\beta = 1.0 - 0.735 = 0.265$.

In Figure 5.9, the probability of acceptance is very high—nearly 1.0—for p values near the value of $p = 0.1$, for which the chart has been designed. The probability of acceptance falls off when the p value moves farther from 0.1 on both sides. We also observe that the *P*-chart is very tolerant of the changes in p. When p in the process decreases by about 50% of the original value, or when p increases by 50%, the control chart still has a high probability of acceptance (>0.8). That is, the power is still very low (<0.2) to detect the 50% change. The change has to be more than 50% of the original value of p for the control chart to detect it with any significant power. The use of runs and warning limits could improve the sensitivity of *P*-charts.

5.2.5 OC Curve of a C-Chart

The OC curve of a C-chart designed for controlling a process at, say, $c = c_0$, will show the probabilities of the chart accepting the process when the value of c changes to values other than c_0. Drawing the OC curve of the C-chart is illustrated using an example.

Example 5.4

Suppose that a process produces an average number of defects per unit $c_0 = 9$ and we use a *C*-chart to control the process at this c value. We want to find the OC curve of this control chart.

TABLE 5.4

OC Function of a C-chart Designed to Maintain a Process In-control at $c = 9$

c	1	2	3	5	7	9	10	15	20
$Pa = \beta$	0.633	0.865	0.941	0.994	1.000	0.994	0.985	0.748	0.381

Solution

The control limits for the C-chart to control this process at its current level would be:

$$\text{UCL}(C) = 9 + 3\sqrt{9} = 18$$

$$\text{CL}(C) = 9$$

$$\text{UCL}(C) = 9 - 3\sqrt{9} = 0$$

The OC curve of the chart should show the probability of acceptance β of process conditions with various values for c besides c_0. The β is the probability that an observed value of C falls within the control limits given that c is, say, c_1:

$$\beta = P(0 < C < 18 \,|c = c_1)$$

$$= P(C < 18 \,|c = c_1) - P(C \leq 0 \,|c = c_1)$$

$$= P(C \leq 17 \,|c = c_1) - P(C \leq 0 \,|c = c_1)$$

The OC function is calculated in Table 5.4, and the OC curve is shown in Figure 5.10. The probability of acceptance, *Pa*, in Table 5.4 is obtained using the cumulative Poisson table. (We have again assumed that on-the-line is out.) As an example, when $c = 2$, the *Pa* is given by:

$$\beta = P(C \leq 17 \mid c = 2) - P(C \leq 0 \mid c = 2) = 1.00 - 0.135 = 0.865.$$

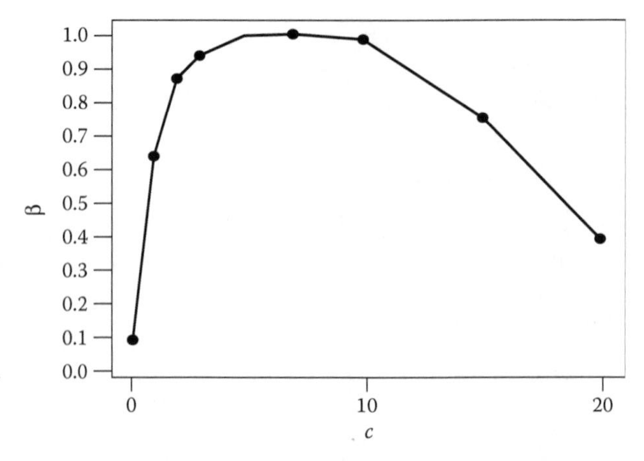

FIGURE 5.10
OC curve of a C-chart designed to control a process at $c = 9$.

We see in the above example that the probability of acceptance is high—almost equal to 1.0—when $c = 9$, the design c value. We also see that the C-chart is not very sensitive to changes in the value of c. For example, even with a 75% increase or decrease in the value of c, the probability of rejection is still less than 0.5. This may not be an undesirable feature for the control chart when the discovery of a change less than 75% of the process average is not important. When smaller changes are important, however, the conventional C-chart is not very powerful. The use of warning limits and runs can improve the sensitivity of the C-chart.

The study of the control charts using their operating characteristics reveals some of their important characteristics. It tells us when the charts are effective in discovering changes in the processes and when they are not. Such studies have led researchers to create newer and better control charts with features that overcome some of the drawbacks inherent in the basic charts. These topics are covered in more advanced literature on the subject.

5.3 Measurement Control Charts for Special Situations

5.3.1 $\overline{X}$- and R-Charts When Standards for μ and/or σ are Given

When we discussed the $\overline{X}$ - and R-charts, we used the following formulas to calculate the CL and control limits:

Control limits for the R-chart:

$$\text{UCL}(R) = D_4\overline{R}$$

$$\text{CL}(R) = \overline{R}$$

$$\text{LCL}(R) = D_3\overline{R}$$

Control limits for the $\overline{X}$-chart:

$$\text{UCL}(\overline{X}) = \overline{\overline{X}} + A_2\overline{R}$$

$$\text{CL}(\overline{X}) = \overline{\overline{X}}$$

$$\text{LCL}(\overline{X}) = \overline{\overline{X}} - A_2\overline{R}$$

where A_2, D_3, and D_4 are tabled factors that giwve 3-sigma limits for the charts.

These formulas use $\overline{\overline{X}}$ and $\overline{R}$ obtained from the data collected from the process to estimate the process mean and process standard deviation, respectively. When we use these limits, we should realize that we are controlling the process to behave consistently at the current level of the process mean and within the current level of process variability. In other words, we are not imposing any outside standards on the process. Therefore, we say that we use these limits to maintain "current control." In many process situations, such maintaining of current control is necessary or desirable. In many others, however, we may be given a standard for the mean (i.e., a target), or a standard for the standard deviation, or both, and required to control the process to conform to the given standards. For example, there may be a process producing 3/4-in. bolts, and the bolt diameter is being controlled using $\overline{X}$ - and R-charts. In this context, it may make sense to control the process average at 3/4 in., in effect forcing the process to a mean of 3/4 in. When we have to control a process against given standards, we have to use different sets of formulas. Two cases arise.

5.3.1.1 Case I: μ Given, σ Not Given

This case arises when it is known the process mean should be controlled at a standard average level but the variability at which the process must be controlled is not given. In this case, data will be collected as for the regular $\overline{X}$- and R-charts, and the $\overline{R}$ value (from at least 25 subgroups of an in-control process) will be used to estimate the process variability. The following formulas will be used, which employ the given standard for the mean, T, as the center of the $\overline{X}$-chart and an estimate of σ using $\overline{R}$ for calculating the spread of the limits. Similarly, the CL and limits for the R-chart are computed from the estimates of σ obtained using $\overline{R}$:

Control limits for the R-chart:

$$\text{UCL}(R) = D_4\overline{R}$$

$$\text{CL}(R) = \overline{R}$$

$$\text{LCL}(R) = D_3\overline{R}$$

Control limits for the $\overline{X}$-chart:

$$\text{UCL}(\overline{X}) = T + A_2\overline{R}$$

$$\text{CL}(\overline{X}) = T$$

$$\text{LCL}(\overline{X}) = T - A_2\overline{R}$$

These formulas are no different from those given for current control, except for T in place of $\overline{\overline{X}}$.

5.3.1.2 Case II: μ and σ Given

In this case, we do not use process data to estimate the process mean or the process standard deviation. The formulas reflect the limits of variability that we should expect in the statistics, $\overline{X}$ and R, if the values of the mean and standard deviation of the process are as given. Suppose that the standard given for the mean is T and that for the standard deviation is σ_0. The formulas will then be as follows:

Control limits for the R-chart:

$$\text{UCL}(R) = D_2\sigma_0$$

$$\text{CL}(R) = d_2\sigma_0$$

$$\text{LCL}(R) = D_1\sigma_0$$

Control limits for the $\overline{X}$-chart:

$$\text{UCL}(\overline{X}) = T + A\sigma_0$$

$$\text{CL}(\overline{X}) = T$$

$$\text{LCL}(\overline{X}) = T - A\sigma_0$$

The values for A, D_1, D_2, and d_2 are available from standard tables, such as Table A.4 in the Appendix. The reader should be able to see why these formulas are good for the case when σ is given. The CL of R is the $E(R)$ in terms of the given standard deviation σ_0, and D_2 and D_1 give the 3-sigma limits for R in terms of σ_0. In fact, $D_2 = (d_2 + 3d_3)$, and $D_1 = (d_2 - 3d_3)$. To obtain the limits for $\overline{X}$ we use A, which is simply equal to $3/\sqrt{n}$ because $\sigma_{\overline{X}} = \left(\sigma_0/\sqrt{n}\right)$.

Example 5.5

Prepare $\overline{X}$- and R-charts for the data in Table 5.5 if the process is to be controlled at a target $T = 10$. No standard value for σ is given.

Solution

The limit calculations for the case when μ is given but the value of σ is not are shown below. From Table 5.5, the value of $\overline{R}$ is obtained as 3.307. The control limits for the R-chart are calculated using values for D_3 and D_4 for $n = 5$.

TABLE 5.5

Data for $\overline{X}$- and R-charts when μ is Given but σ Is Not

No.	X_1	X_2	X_3	X_4	X_5	$\overline{X}$	Range
1	10.09	11.32	10.81	9.19	14.93	11.269	5.739
2	10.16	8.88	9.35	11.16	7.07	9.323	4.092
3	10.93	9.70	9.63	10.89	9.99	10.227	1.296
4	8.36	10.06	9.42	11.06	1.0.25	9.828	2.702
5	8.89	9.62	11.27	9.03	11.49	10.060	2.605
6	10.66	9.91	10.03	10.45	8.51	9.913	2.143
7	7.68	8.38	9.13	11.29	10.04	9.301	3.610
8	13.21	8.94	10.95	12.48	11.96	11.509	4.273
9	10.07	10.85	11.25	11.06	11.78	11.002	1.719
10	6.89	8.12	11.09	12.67	13.80	10.514	6.919
11	8.01	10.80	10.65	9.20	9.42	9.615	2.780
12	11.84	10.32	10.01	10.22	10.09	10.498	1.836
13	11.11	10.31	8.88	12.52	10.70	10.704	3.644
14	9.94	9.61	8.76	12.18	10.42	10.184	3.419
15	9.21	12.02	10.05	9.32	8.53	9.829	3.488
16	1161	12 22	1116	12 05	10.52	11.514	1700
17	11.68	9.52	10.43	9.03	12.92	10.715	3.891
18	8.87	11.78	7.41	13.14	10.17	10.274	5.724
19	11.36	8.13	6.78	11.73	10.74	9.747	4.951
20	6.35	9.67	9.94	11.42	10.31	9.537	5.075
21	8.67	10.28	9.55	10.12	10.17	9.758	1.614
22	10.24	9.59	11.19	11.30	10.32	10.528	1.710
23	10.30	9.93	10.86	12.09	9.99	10.631	2.160
24	10.32	8.31	8.13	10.73	10.44	9.586	2.598
25	9.14	10.55	10.27	7.96	7.58	9.100	2.975
						$\overline{R} =$	3.307

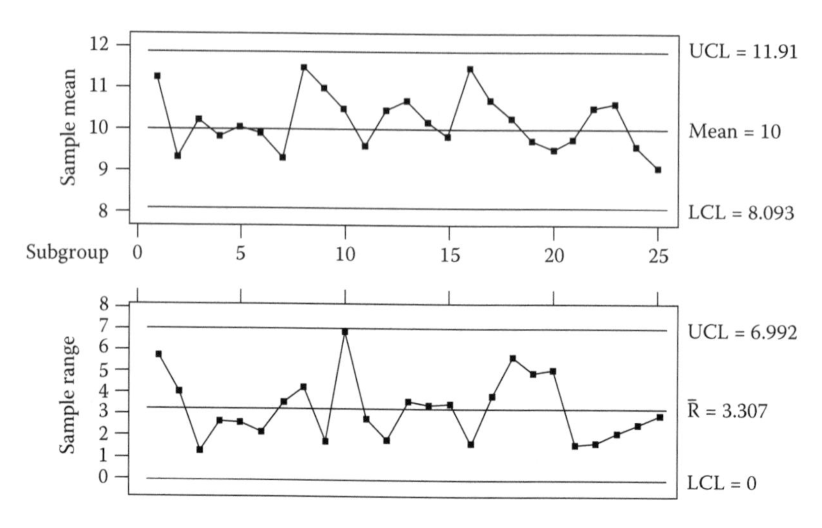

FIGURE 5.11
Example of $\bar{X}$- and R-charts when μ is given but σ is not.

$$\text{UCL}(R) = D_4\bar{R} = 2.114 \times 3.307 = 6.991$$

$$\text{CL}(R) = \bar{R} = 3.307$$

$$\text{LCL}(R) = D_3\bar{R} = 0$$

All R values are within limits, and the $\bar{R}$ value from the in-control process is used to calculate the limits for the $\bar{X}$-chart. The control limits for the $\bar{X}$-chart are:

$$\text{UCL}(\bar{X}) = T + A_2\bar{R} = 10 + 0.577 \times 3.307 = 11.91$$

$$\text{CL}(\bar{X}) = T = 10$$

$$\text{LCL}(\bar{X}) = T - A_2\bar{R} = 10 - 0.577 \times 3.307 = 8.09$$

Figure 5.11 shows the graph of the control charts drawn using Minitab. Incidentally, notice the close agreement between the calculated values for the limits and those computed by Minitab.

Example 5.6

Make $\bar{X}$- and R-charts for the data in Table 5.5 for the process to be controlled at a target $T = 10$ and $\sigma_0 = 1.0$.

Solution

The limits for the control chart are calculated as follows:

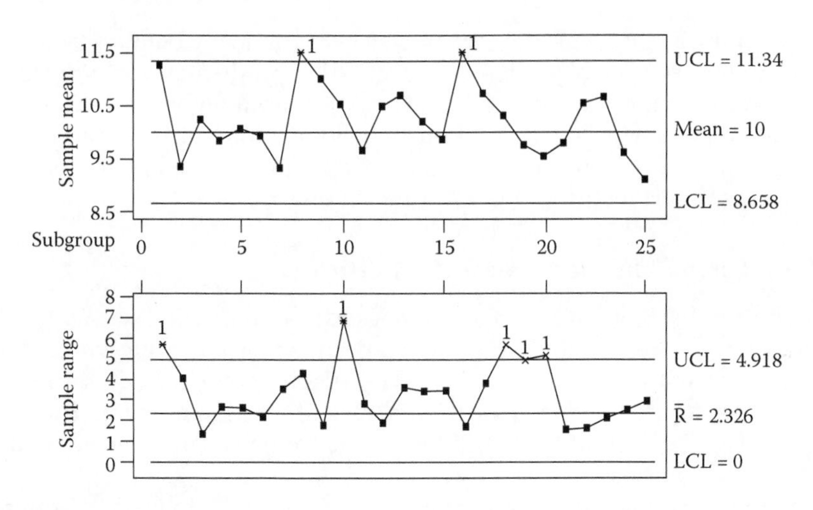

FIGURE 5.12
Example of $\overline{X}$- and R-charts when μ and σ are given.

Control limits for the R-chart:

$$\text{UCL}(R) = D_2\sigma_0 = 4.918 \times 1.0 = 4.918$$

$$\text{CL}(R) = d_2\sigma_0 = 2.326 \times 1.0 = 2.326$$

$$\text{LCL}(R) = D_1\sigma_0 = 0$$

Control limits for the $\overline{X}$-chart:

$$\text{UCL}(\overline{X}) = T + A\sigma_0 = 10 + 1.342 \times 1.0 = 11.342$$

$$\text{CL}(\overline{X}) = T = 10$$

$$\text{LCL}(\overline{X}) = T - A\sigma_0 = 10 - 1.342 \times 1.0 = 8.658$$

Figure 5.12 shows the control charts drawn for Example 5.6 using Minitab. We again see a close agreement between the limits calculated above and those computed by Minitab. We also see several R values and $\overline{X}$ values outside the control limits in the new chart because an external standard on variability has been imposed on the process.

5.3.2 Control Charts for Slow Processes

Recall that in order to use the $\overline{X}$- and R-charts, we need four or five sample units from the process so that the measurements from them can be used to calculate $\overline{X}$ and R. These sample units must be taken (almost) at the same time to represent the condition of the process at the time that the sample is taken. In many situations, it may not be possible to take such multiple sample units at the same time, or in quick succession. Chemical processes are examples of such slow processes, in which production cycle times are long and the analysis and reporting of sample measurements take time. The simplest approach in such a situation is to plot the individual values against the control limits calculated for the individual value X. Such a chart is called the "control chart for individuals," or the "X-chart."

5.3.2.1 *Control Chart for Individuals (X-Chart)*

The control chart for individuals, or the X-chart, is normally used along with a chart for successive differences, which is known as a moving range chart, or MR chart, with subgroup size $n = 2$. The MR chart will track the variability in the process, and the X-chart will keep track of the process mean. The moving range is the absolute value of the difference between the current value and the previous value.

The limits for the two charts are calculated using the following formulas if no standards are given:

Limits for the MR chart:

$$UCL(R) = D_4\bar{R}$$

$$CL(R) = \bar{R}$$

$$LCL(R) = D_3\bar{R}$$

Limits for the X-chart:

$$UCL(X) = \bar{X} + 3\frac{\bar{R}}{d_2}$$

$$CL(X) = \bar{X}$$

$$LCL(X) = \bar{X} - 3\frac{\bar{R}}{d_2}$$

If standards are given—say, T for the mean and σ_0 for the standard deviation—then the limits for the X- and MR charts are calculated as follows:

Limits for the MR chart:

$$UCL(R) = D_2\sigma_0$$

$$CL(R) = d_2\sigma_0$$

$$LCL(R) = D_1\sigma_0$$

Limits for the X- chart:

$$UCL(X) = T + 3\sigma_0$$

$$CL(X) = T$$

$$LCL(X) = T - 3\sigma_0$$

The values for d_2, D_1 D_2, D_3, and D_4 are chosen for $n = 2$ from Table A.4 in the Appendix.

Example 5.7

The first 20 observations of data shown in Table 5.6 come from a process in which the mean is expected to be at 10 and the standard deviation at 1.0. The process mean changes to 11.5, a distance of 1.5σ, after the 20th observation. The data after the 20th observation come from the changed process. Prepare a control chart for

TABLE 5.6

Data for Control Chart for Individuals (X-chart) and MR chart

i	X_i	MR	i	X_i	MR
1	10.09		21	12.16	1.32
2	10.62	0.53	22	11.11	1.05
3	10.52	0.1	23	9.91	1.2
4	10.16	0.36	24	11.77	1.86
5	9.94	0.22	25	11.58	0.19
6	10.72	0.78	26	11.34	0.24
7	8.77	1.95	27	12.63	1.29
8	8.91	0.14	28	11.37	1.26
9	10.31	1.4	29	12.29	0.92
10	10.84	0.53	30	9.44	2.85
11	10.8	0.04	31	11.58	2.14
12	10.53	0.27	32	9.85	1.73
13	8.61	1.92	33	11.1	1.25
14	9.81	1.2	34	10.68	0.42
15	10	0.19	35	11.9	1.22
16	12.39	2.39	36	13.01	1.11
17	10.19	2.2	37	12.33	0.68
18	9.31	0.88	38	12.69	0.36
19	10.57	1.26	39	10.9	1.79
20	10.84	0.27	40	10.62	0.28

individuals, and see how the control chart responds to the change that we know has occurred in the process mean.

Solution

We prepare the control limits using the first 20 observations. The standards are given as $T = 10$ and $\sigma_0 = 1.0$. The values for D_2, d_2, and D_1 for the MR chart are chosen for $n = 2$ from the Table A.4 in the Appendix.

Limits for the MR chart:

$$UCL(R) = 3.686 \times 1.0 = 3.686$$

$$CL(R) = 1.128 \times 1.0 = 1.128$$

$$LCL(R) = 0$$

Limits for the X- chart:

$$UCL(X) = 10 + 3(1.0) = 13$$

$$CL(X) = 10$$

$$LCL(X) = 10 - 3(1.0) = 7$$

The charts prepared using Minitab are shown in Figure 5.13. The chart shows the process to be in-control when the first 20 observations were taken. The change in the process mean occurring after the 20th observation has been detected by the X-chart, but only after 16 samples from the time that the change really occurred. The X-chart is not a powerful chart and, because the individual values have a lot more variability compared to the averages, it is difficult to see the signals in the presence of noise. Also, the performance of the X-chart is very

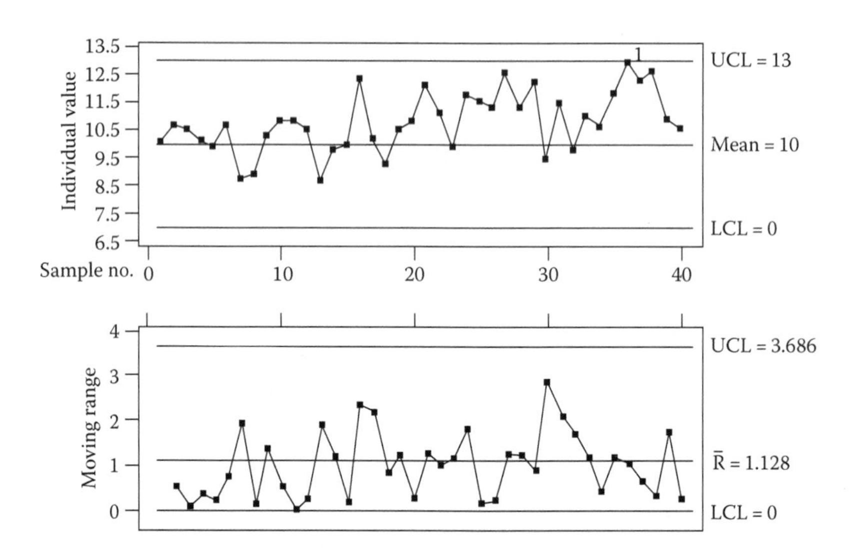

FIGURE 5.13
Example of a control chart for individuals and chart for successive differences.

sensitive to deviation of the process distribution from normal as the X values do not have the robustness present in the $\overline{X}$ values (central limit theorem). Therefore, the moving average and MR charts, to be discussed next, are preferred in these circumstances. The X-chart, however, has some advantages; one of them is that it is simple to use and easy to understand. Another is that specification lines can be drawn on the chart and the characteristic can be monitored against the given specifications. Such specification lines, as we have mentioned before, should not be drawn on charts for averages, because specifications are limits for individuals and averages should not be compared with limits for individuals.

5.3.2.2 Moving Average and MR Charts

The moving average (MA) and MR charts use sample averages and sample ranges, as do the regular $\overline{X}$- and R-charts, but the method of forming the samples or subgroups is different. Suppose the sample size is n. When starting the control chart, the first subgroup is made up of the first n number of consecutive observations. A new subgroup is formed when a new observation arrives by adding it to the current subgroup and discarding the earliest observation from it. The following is the mathematical expression to obtain the i-th moving average with a subgroup of size n:

$$M\overline{X}_i = \frac{X_i + X_{i-1} + \cdots + X_{i-n+1}}{n}, \quad i \geq n$$

The limits for the two charts are calculated using the following formulas, which are the same as the ones used for the regular $\overline{X}$- and R-charts with no standards given.

Limits for the MR chart:

$$UCL(R) = D_4\overline{R}$$

$$CL(R) = \overline{R}$$

$$LCL(R) = D_3\overline{R}$$

Limits for the MA chart:

$$UCL(\overline{X}) = \overline{\overline{X}} + A_2\overline{R}$$

$$CL(\overline{X}) = \overline{\overline{X}}$$

$$LCL(\overline{X}) = \overline{\overline{X}} - A_2\overline{R}$$

If standards are given for the process average—say, T—and for the process standard deviation—for say, σ_0—then the limits will be calculated using the following formulas:

Limits for the MR chart:

$$UCL(R) = D_2\sigma_0$$

$$CL(R) = d_2\sigma_0$$

$$LCL(R) = D_1\sigma_0$$

Limits for the MA chart:

$$UCL(\overline{X}) = T + A\sigma_0$$

$$CL(\overline{X}) = T$$

$$LCL(\overline{X}) = T - A\sigma_0$$

The factors d_2, D_1, D_2, D_3, D_4, A, and A_2 are the same as used for regular $\overline{X}$ and R-charts for the standards given case. The following example illustrates the use of the MA and MR charts.

TABLE 5.7

Data for the MA and MR Charts

i	X_i	*MA*	*MR*	*i*	X_i	*MA*	*MR*
1	10.09			21	12.16	11.190	1.59
2	10.62			22	11.11	11.370	1.32
3	10.52	10.41	0.53	23	9.91	11.060	2.25
4	10.16	10.433	0.46	24	11.77	10.930	1.86
5	9.94	10.207	0.58	25	11.58	11.087	1.86
6	10.72	10.273	0.78	26	11.34	11.563	0.43
7	8.77	9.810	1.95	27	12.63	11.850	1.29
8	8.91	9.467	1.95	28	11.37	11.780	1.29
9	10.31	9.330	1.54	29	12.29	12.097	1.26
10	10.84	10.020	1.93	30	9.44	11.033	2.85
11	10.8	10.650	0.53	31	11.58	11.103	2.85
12	10.53	10.723	0.31	32	9.85	10.290	2.14
13	8.61	9.980	2.19	33	11.1	10.843	1.73
14	9.81	9.650	1.92	34	10.68	10.543	1.25
15	10	9.473	1.39	35	11.9	11.227	1.22
16	12.39	10.733	2.58	36	13.01	11.863	2.33
17	10.19	10.860	2.39	37	12.33	12.413	1.11
18	9.31	10.630	3.08	38	12.69	12.677	0.68
19	10.57	10.023	1.26	39	10.9	11.973	1.79
20	10.84	10.240	1.53	40	10.62	11.403	2.07

Example 5.8

In this example, the first half of the data, which are shown in Table 5.7, come from a process in which the average is 10 and the standard deviation is 1.0. After the 20th sample, the process average increases to 11.5, a distance of 1.5σ. The standard deviation of the process remains the same at 1.0. We want to see how the MA and MR control charts designed for the original 20 observations are able to detect the change in the process average when it occurs.

Solution

The computation of the moving averages and moving ranges for $n = 3$ are shown in Table 5.7. Notice how the samples are formed. The first subgroup is formed with the first three observations. The following subgroups are formed by taking in the newest observation and dropping out the earliest. The charts are drawn using Minitab. The limit calculations are verified using the formulas given above. The MA and MR charts drawn from Minitab are shown in Figures 5.14a and 5.14b.

The limits for the two charts are calculated using the data from the first 20 observations and by making use of the given standards: $T = 10$ and $\sigma_0 = 1.0$.

Limits for MR chart:

$$\text{UCL}(R) = D_2\sigma_0$$
$$= 4.358 \times 1.0 = 4.358$$
$$\text{CL}(R) = d_2\sigma_0$$
$$= 1.693 \times 1.0 = 1.693$$
$$\text{LCL}(R) = D_1\sigma_0 = 0$$

Limits for MA chart:

$$\text{UCL}(\overline{X}) = T + A\sigma_0$$
$$= 10 + 1.732 \times 1.0 = 11.732$$
$$\text{CL}(\overline{X}) = T = 10$$
$$\text{LCL}(\overline{X}) = T - A\sigma_0$$
$$= 10 - 1.732 \times 1.0 = 8.268$$

The reader may have noticed that this is the same data used to prepare the control chart for individuals (X-chart) and the MR chart in the previous example. In that example, the change in the process mean was discovered by the X-chart at the 16th sample after the shift. The same change has been discovered by the MA chart at the seventh sample in the current example. Although we should be cautious in making general conclusions out of this one example, the added power as a result of averaging from larger subgroups should be expected. This example shows the power advantage of a chart using averages over a chart using individuals.

5.3.2.3 Notes on Moving Average and Moving Range Charts

What Is a Good Value for n?

The averaging done in the MA and MR charts reduces the noise (i.e., the variability) and helps in discovering the signals. Larger subgroup sizes tend to smooth out the variations in individual observations and bring out the signals better, but they may also hide changes that may occur during the time that

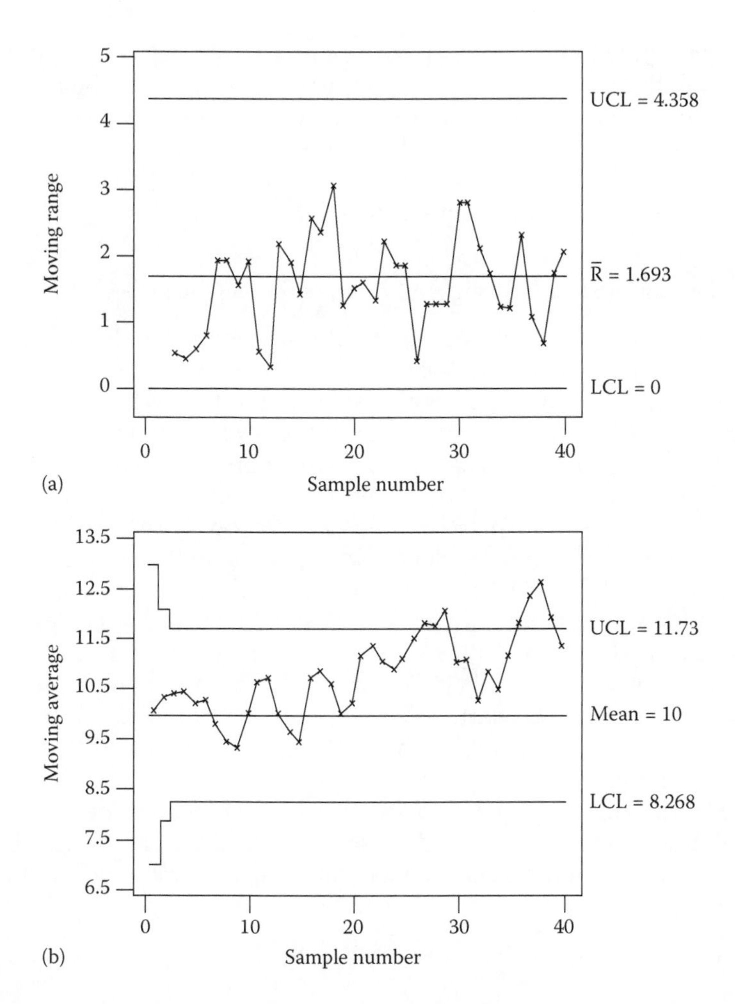

FIGURE 5.14
(a) MR chart for data in Table 5.7. (b) MA chart for data in Table 5.7.

the sample is being taken. Similarly, smaller subgroup sizes may be quick in responding to changes but tend to cause excessive fluctuations in the moving averages and ranges. Subgroup sizes of four or five seem to provide the best trade-off. Subgroup size can also be chosen to reflect what is happening in the process. For example, if approximately three batches of a chemical are produced from one batch of raw material, a subgroup size of three would make sense. If there are differences in the raw material from batch to batch, it will show on the chart.

A Caution

While reacting to plots outside the control limits on the MA and MR charts, caution must be exercised when interpreting the charts. Suppose that an adjustment is made to a process because of a moving average falling outside a limit; the next couple of averages may still be outside the limits even after making the adjustment, because the observations generated when the process was not-in-control may still be in the calculations and may influence some of the subsequent plots. Operators must be warned against overreacting to signals under these circumstances.

5.3.3 The Exponentially Weighted Moving Average Chart

We noted while studying the operating characteristics that the conventional $\bar{X}$-chart with $n=4$ or 5, and 3-sigma limits is not very sensitive to small changes (<1.5σ magnitude) in the process mean. We also noted that their sensitivity could be improved through the use of rules with 1-sigma and 2-sigma warning limits, or with runs. Another approach to obtain improved sensitivity in control charts is through the use of statistics that accumulate information from several past sample observations instead of relying on just one current sample, as is done with the $\bar{X}$- and R-charts. Two important examples of such charts are the exponentially weighted moving average (EWMA) and the cumulative sum (CUSUM) control charts. Researchers have established that the EWMA chart and the CUSUM chart are equally effective (Lucas and Saccucci 1990) in discovering small changes in the process mean. The EWMA chart, however, is easier to understand and construct compared to the CUSUM chart. Therefore, only the EWMA chart is described here.

The EWMA chart is typically used in situations where the sample size $n=1$, although it could be equally useful where $n>1$. If x_i is the observation from the i-th sample, the exponentially weighted moving average w_i is calculated as:

$$w_i = \lambda x_i + (1-\lambda) w_{i-1},$$

where λ, $0<\lambda\le 1$, is called a "smoothing constant." The first value of w_i, w_1, is made equal to x_1; that is, $w_1 = x_1$. If we expand the term on the right-hand side of the above equation for w_i further, we get:

$$\begin{aligned} w_i &= \lambda x_i + (1-\lambda)\left(\lambda x_{i-1} + (1-\lambda) w_{i-2}\right) \\ &= \lambda x_i + \lambda(1-\lambda) x_{i-1} + (1-\lambda)^2 w_{i-2} \end{aligned}$$

Expanding this way further, we get:

$$w_i = \lambda x_i + \lambda(1-\lambda) x_{i-1} + \lambda(1-\lambda)^2 x_{i-2} + \lambda(1-\lambda)^3 x_{i-3} + \cdots + \lambda(1-\lambda)^{i-1} x_1$$

From the above expansion, we can see that the weighted average w_i includes information from the most recent observation along with information from previous observations weighted in an exponentially decreasing manner. By choosing the value of λ appropriately, the relative importance given to the most recent and past values can be changed. Larger values of λ will give more importance to the most recent observation, and smaller values will assign more importance to past values. The example below explains how the EWMA chart is constructed and used.

Example 5.9

Table 5.8 has two sets of data. Column 2 has data from a normal process with mean $\mu = 10$ and $\sigma = 1.0$ up to Sample 20. The data after the 20th sample come from a process with an average of 11.0, a change of one standard deviation ($k = 1$) in the mean of the original process. Column 5 has data from a normal process with $\mu = 10$ and $\sigma = 1.0$ up to the 20th sample, and data after the 20th sample is from a process with a mean of 11.5, a change of 1.5 standard deviations ($k = 1.5$) in the mean of the original process. In both cases, the standard deviation is assumed to remain constant.

Using these two sets of data, compare the performance of the EWMA chart with that of the Shewhart chart by investigating when the charts are able to detect the change occurring in the data sets. We want to see how the charts react when the change in the process is small and when it is large.

Solution

We will use the Shewhart chart for individuals and two EWMA charts, one with $\lambda = 0.6$ and another with $\lambda = 0.2$. The Shewhart and EWMA charts all have 3-sigma limits (calculation of limits for the EWMA chart is explained later) and were produced by Minitab software. Columns 3, 4, 6, and 7 of Table 5.8 show the EWMA calculations. For example, under Column 3,

$$w_1 = x_1 = 9.86,\ w_2 = (0.6)9.3 + (0.4)9.86 = 9.53,\ w_3 = (0.6)10.67 + (0.4)9.53 = 10.22.$$

Figures 5.15a to 5.15f show the charts for the two processes. Figure 5.15a shows the Shewhart chart for the process with a 1σ shift in the process mean, and Figure 5.15b and 5.15c show EWMA charts with $\lambda = 0.6$ and $\lambda = 0.2$, respectively, for the same process.

The Shewhart chart did not detect the change of 1σ distance in the mean. The EWMA chart with $\lambda = 0.6$ detected the change at the 16th sample after the change, and the EWMA chart with $\lambda = 0.2$ detected the change at the 13th sample after the change. In Figures 5.15d, 5.15e, and 5.15f similar charts are shown for the process with the mean changed through a 1.5σ distance. The Shewhart chart barely detects the change at the 16th sample after the change, whereas the EMWA with $\lambda = 0.6$ takes only seven samples to discover the change. The EMWA chart with $\lambda = 0.2$ detects the change even earlier, at the fifth sample.

TABLE 5.8

Data and Calculation of *EMWAs* for Two Different Values of λ

1	2	3	4	5	6	7
Sample no.	X-chart $x_i(k = 1)$	*EWMA* $w_i(\lambda = 0.6)$	*EWMA* $w_i(\lambda = 0.2)$	X-chart $x_i(k = 1.5)$	*EWMA* $w_i(\lambda = 0.6)$	*EWMA* $w_i(\lambda = 0.2)$
1	9.86	9.86	9.86	10.09	10.09	10.09
2	9.30	9.53	9.75	10.62	10.40	10.19
3	10.67	10.22	9.93	10.52	10.47	10.26
4	8.33	9.09	9.61	10.16	10.28	10.24
5	10.93	10.19	9.88	9.94	10.08	10.18
6	11.51	10.98	10.20	10.72	10.46	10.29
7	7.98	9.18	9.76	8.77	9.45	9.98
8	10.18	9.78	9.84	8.91	9.13	9.77
9	9.99	9.91	9.87	10.31	9.84	9.88
10	10.22	10.09	9.94	10.84	10.44	10.07
11	8.59	9.19	9.67	10.80	10.65	10.22
12	10.55	10.00	9.85	10.53	10.58	10.28
13	9.41	9.65	9.76	8.61	9.40	9.94
14	9.25	9.41	9.66	9.81	9.64	9.92
15	9.38	9.39	9.60	10.00	9.86	9.93
16	10.19	9.87	9.72	12.39	11.37	10.42
17	10.88	10.48	9.95	10.19	10.66	10.38
18	10.63	10.57	10.09	9.31	9.85	10.16
19	10.52	10.54	10.17	10.57	10.28	10.24
20	10.18	10.32	10.17	10.84	10.61	10.36
21	11.26	10.88	10.39	12.16	11.54	10.72
22	9.80	10.23	10.27	11.11	11.28	10.80
23	10.44	10.36	10.31	9.91	10.46	10.62
24	11.28	10.91	10.50	11.77	11.24	10.85
25	11.59	11.32	10.72	11.58	11.44	11.00
26	8.53	9.65	10.28	11.34	11.38	11.06
27	10.89	10.39	10.40	12.63	12.13	11.38
28	11.02	10.77	10.53	11.37	11.67	11.38
29	11.62	11.28	10.74	12.29	12.04	11.56
30	11.06	11.15	10.81	9.44	10.48	11.14
31	11.57	11.40	10.96	11.58	11.14	11.22
32	10.79	11.03	10.93	9.85	10.37	10.95
33	11.55	11.34	11.05	11.10	10.81	10.98
34	10.81	11.02	11.00	10.68	10.73	10.92
35	12.18	11.72	11.24	11.90	11.44	11.12
36	12.55	12.22	11.50	13.01	12.38	11.50
37	10.43	11.15	11.29	12.33	12.35	11.66
38	12.00	11.66	11.43	12.69	12.55	11.87
39	10.63	11.04	11.27	10.90	11.56	11.67
40	9.81	10.30	10.98	10.62	10.99	11.46

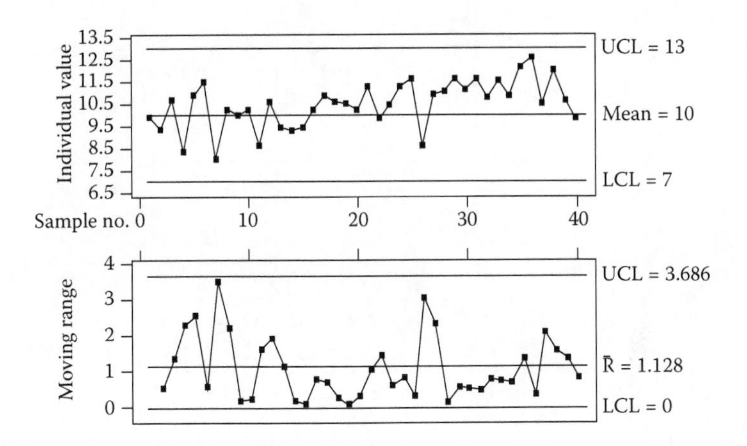

FIGURE 5.15a
Shewhart chart for individuals on a process with 1σ change in the process mean.

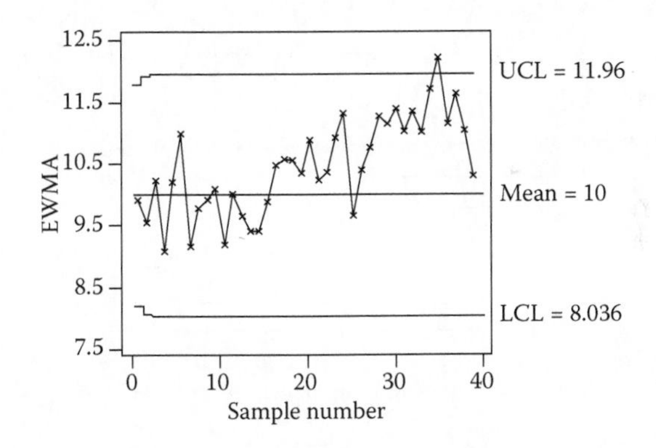

FIGURE 5.15b
EWMA chart ($\lambda = 0.6$) for a process with 1σ change in the process mean.

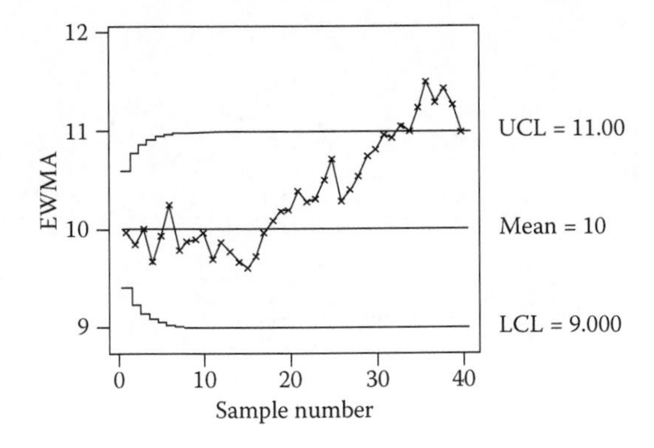

FIGURE 5.15c
EWMA chart ($\lambda = 0.2$) for a process with 1σ change in the process mean.

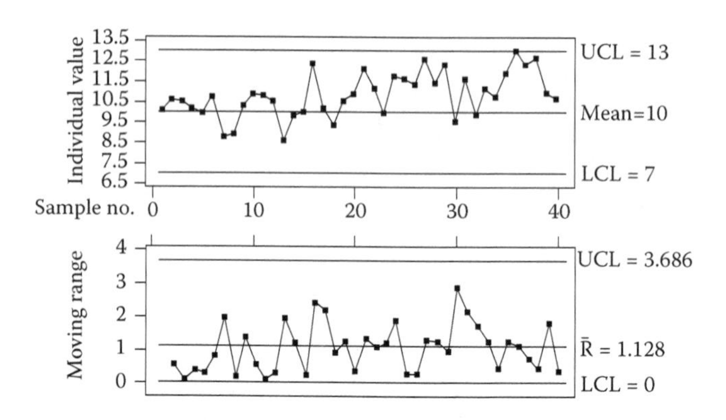

FIGURE 5.15d
Shewhart chart for individuals on a process with 1.5σ change in the process mean.

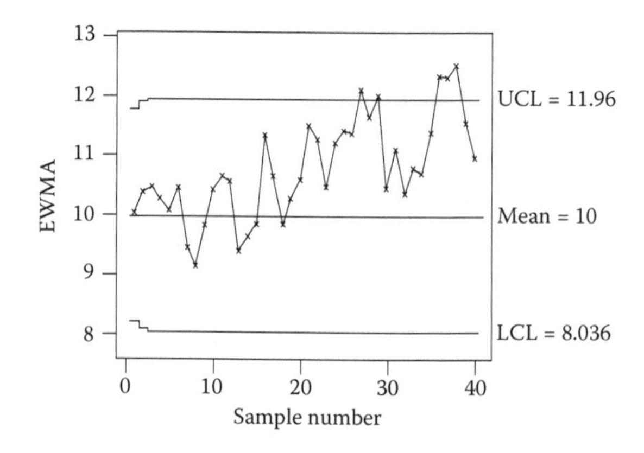

FIGURE 5.15e
EWMA chart (λ = 0.6) for a process with 1.5σ change in the process mean.

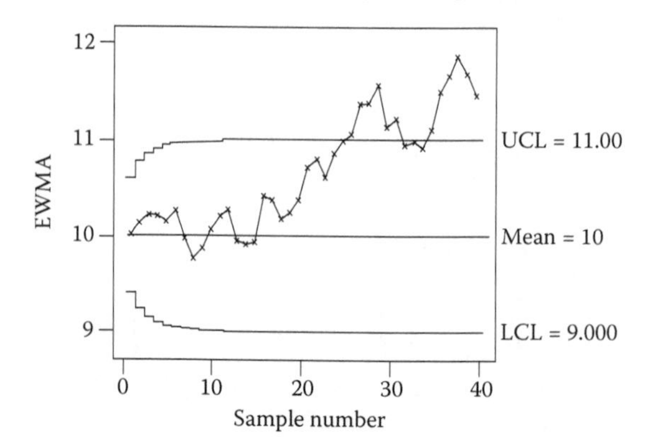

FIGURE 5.15f
EWMA chart (λ = 0.2) for a process with 1.5σ change in the process mean.

We should be careful in making general conclusions out of one example, but it can be safely stated that the EMWA chart has improved sensitivity compared with the Shewhart chart, and that this sensitivity can be adjusted by varying the value of λ. When large changes are significant, large λ values should be used. (An EWMA chart with $\lambda = 1$ is equivalent to the Shewhart chart.) When small changes need to be discovered, small values of λ can be used (Crowder 1989). Roberts (1959), the author of the EWMA control scheme, claimed this as an advantage for the charts he was proposing. Vardeman and Jobe (1999) discuss how to select the value of λ in order to detect the desired magnitude of change in the process average with desired *ARL* values.

In the above illustration, the sample size was taken as $n = 1$ for both the Shewhart chart and the EMWA chart. The EMWA chart is best suited for situations in which the sample size, out of necessity, has to be one. However, the EWMA chart can be used with X from samples of any size. In addition, the improved sensitivity observed in the EMWA chart with $n = 1$ can also be expected with charts with larger n, mainly because the information is accumulated over a larger number of samples and the cumulative measure integrates the information better and discloses the changes sooner.

5.3.3.1 Limits for the EWMA Chart

If $w_i = \lambda x_i + (1-\lambda)w_{i-1}$, it can then be shown (Lucas and Saccucci 1990) that

$$\sigma_{w_i}^2 = \sigma^2\left(\frac{\lambda}{2-\lambda}\right)\left[1-(1-\lambda)^{2i}\right]$$

provided that the x_i values are independent and come from a normal population with variance σ^2. The $K\sigma$ limits for w_i would then be:

$$\text{UCL} = \mu + K\sigma\sqrt{\frac{\lambda}{(2-\lambda)}\left[1-(1-\lambda)^{2i}\right]}$$

$$\text{CL} = \mu$$

$$\text{LCL} = \mu - K\sigma\sqrt{\frac{\lambda}{2-\lambda}\left[1-(1-\lambda)^{2i}\right]}$$

We note that the limits are dependent on the value of i and so have to be calculated for each sample individually. However, when i becomes large, the quantity within the square parentheses inside the radical sign becomes 1.0, and the limits tend to reach the steady-state value:

$$\left.\begin{array}{c}\text{UCL}\\ \text{LCL}\end{array}\right\} = \mu \pm K\sigma\sqrt{\frac{\lambda}{(2-\lambda)}}$$

If the value of K is taken as 3, for simplicity, we will have 3-sigma limits. In the example above, for instance, the 3-sigma control limits for the second sample, when $i = 2$, for the case when $\lambda = 0.2$, are given by:

$$\text{UCL} = 10 + 3(1)\sqrt{\frac{0.2}{2-0.2}\left[1-(1-0.2)^4\right]} = 10.77$$

$$\text{CL} = 10$$

$$\text{LCL} = 10 - 3(1)\sqrt{\frac{0.2}{2-0.2}\left[1-(1-0.2)^4\right]} = 9.23$$

The control limits for the steady-state condition, in a case where $\lambda = 0.2$, are given by:

$$\text{UCL} = 10 + 3(1)\sqrt{\frac{0.2}{2-0.2}} = 11.0$$

$$\text{CL} = 10$$

$$\text{LCL} = 10 - 3(1)\sqrt{\frac{0.2}{2-0.2}} = 9.0$$

Montgomery (2001a) recommends the use of K between 2.6 and 2.8 for $\lambda \leq 0.1$. A practical approach seems to be to use $k = 3$ with a starting value of $\lambda = 0.5$. Based on experience, the value of λ can be increased or decreased depending on how well small changes need to be detected. Crowder (1989) provides a step-by-step procedure for designing a EWMA chart of desired sensitivity through the optimal choice of λ and K. From this study, it appears that a EWMA chart with $K = 3$ and $\lambda = 0.5$ would be a good, middle-of-the-road chart useful for many occasions. It can be fine tuned to whatever sensitivity the situation requires.

As pointed out earlier, the EWMA chart will normally be used where the sample size is one, in situations where the chart for individuals (X-chart) would otherwise be used. The EWMA is an average, so it tends to make the chart robust with respect to the normality of the population, whereas the Shewhart chart, with a sample size of one, cannot claim such robustness. As mentioned, the EWMA chart can also be used with the sample average $\overline{X}$ when $n > 1$. The X_i from sample i will be replaced by $\overline{X}_i$, and σ in the expressions for the limits will be replaced by $\sigma_{\overline{X}} = \sigma/\sqrt{n}$. Charts have also been developed for controlling process standard deviation using exponentially weighted statistics. Further details can be found in Montgomery (2001a) and Vardeman and Jobe (1999).

5.3.4 Control Charts for Short Runs

When a process produces products for short orders—as happens in many job shops, where only a few units of a part number are produced per order—there will not be enough data for computing the control limits from any one part number. Normally, it is recommended to use at least 25 subgroups from an in-control process for calculating the limits for the $\overline{X}$- and R-charts. If there are not enough subgroups, the limits calculated will have considerable error in them. Therefore, assignable causes could escape detection, or additional false alarms could be generated.

Even in a mass-production environment, if control charting is desired in the early stages of production when not enough units of product have been produced, there will not be an adequate number of subgroups from which to compute the limits. Hillier (1969) estimated the changes in the false-alarm probability of the conventional control charts (with 3-sigma limits and $n = 5$) when the limits are computed from a small number of subgroups. The false-alarm probability of the $\overline{X}$-chart that equals 0.004 when the limits are computed from 25 subgroups increases to 0.0067 when the limits are computed from 10 subgroups. (The false-alarm probability equals the theoretical value of 0.0027 only when the limits are computed from an infinite, or a very large, number of subgroups.) Correspondingly, the false-alarm probability for the R-chart increases from 0.0066 to 0.0102. Hillier suggested a remedy, which amounts to computing the limits in two stages. In the first stage, when only a small number of subgroups are available, modified values of the constants A_2, D_3, and D_4, modified to account for the large errors in the estimates of the parameters, are used. In the second stage, the limits are updated progressively as more subgroups become available.

The solution for the job-shop situation, however, lies in combining data from similar products (i.e., products belonging to the same family or of a similar design) and creating a data set that reflects the behavior of the process. The resulting data set should be such that the data can be considered as coming from one homogeneous population. For example, suppose that a machining center produces flanges of different sizes, and suppose that the pitch diameter of the hole centers is being controlled. Flanges in any one size may not produce enough data to calculate the limits, so the data from flanges of different sizes are grouped together. The measurement plotted is not the diameter itself, but rather the deviation of the observed diameter from the target, or the nominal. The parameters of the population of deviations will be controlled. This is the concept employed in the "deviation from the nominal (DNOM)" chart.

5.3.4.1 *The DNOM Chart*

For the DNOM chart, the statistic plotted for the *i*-th sample is the deviation of the sample average from the nominal, or target:

$$(\overline{X}_i - T_i).$$

The units within any sample average $\overline{X}_i$ should be from the same product or part number and should have the same nominal value. Assuming that the X_i values are normally distributed, then under the assumption that the process is centered,

$$\frac{\overline{X}_i - T_i}{\sigma_i/\sqrt{n}} \sim N(0,1)$$

where σ_i, is the standard deviation of the observations in the i-th sample. If we make the assumption that σ_i is a constant $= \sigma$ for all i, then a pooled estimate for σ can be obtained as:

$$S = \sqrt{\frac{\sum S_i^2}{m}}$$

where the values of S_i are the sample standard deviations from m samples. Then, as described by Farnum (1992),

$$\frac{\overline{X}_i - T_i}{S/\sqrt{n}} \approx N(0,1)$$

The 3-sigma limits for $(\overline{X}_i - T_i)$ can then be obtained as follows:

$$P\left(-3 \le \frac{\overline{X}_i - T_i}{S/\sqrt{n}} \le 3\right) = 0.9973$$

$$P\left(-3S/\sqrt{n} \le \overline{X}_i - T_i \le 3S/\sqrt{n}\right) = 0.9973$$

Therefore, the 3-sigma control limits for the "deviation" $(\overline{X}_i - T_i)$ are:

$$\text{UCL} = 3S/\sqrt{n}$$

$$\text{CL} = 0$$

$$\text{LCL} = -3S/\sqrt{n}$$

The control procedure is to calculate the averages of the observations in each subgroup, the deviation of the averages from their respective targets, and to plot these deviations on a chart with the limit lines drawn at values determined by the formulas above. The pooled standard deviation S in the formulas is calculated according to the formula given earlier.

Example 5.10

The following example has been adapted from Farnum (1992). The data, shown in Table 5.9, come from different part numbers, each having a different nominal value. The nominal values, or T_i values, are shown in Column 7. The $\overline{X}_i$ values of the sample averages and the s_i values of the sample standard deviations, calculated for each subgroup, are shown in Columns 8, and 9, respectively. A control chart is to be prepared to verify if the process that generated the data is in-control.

Solution

We have to use a DNOM chart. The deviations of the $\overline{X}$ values from the nominal are shown in Table 5.9 under Column 12.

The pooled standard deviation is calculated as:

$$s = \sqrt{\frac{\sum s_i^2}{m}} = \sqrt{\frac{0.0000923}{20}} = 0.00214$$

The control limits are calculated as:

$$\text{UCL} = \frac{3(0.00214)}{\sqrt{5}} = 0.00287$$

$$\text{LCL} = \frac{3(0.00214)}{\sqrt{5}} = -0.00287$$

The deviations are plotted on the chart in Figure 5.16. There is one value of the DNOM below the lower limit, indicating that the process is not-in-control. The process is not producing the measurements at consistently the same level.

5.3.4.2 The Standardized DNOM Chart

In the above DNOM chart, an assumption was made that the standard deviations of the measurements in each subgroup (i.e., each part number), were the same and equal to σ. This assumption, however, may not always be true. Farnum (1992) suggests it is more reasonable to assume that the coefficient of variation $(CV) = \sigma_i/T_i$ is a constant. This latter assumption means that, for example, if we are measuring the diameters of flanges of different sizes, then the amount of variability is dependent on the size, with the larger flanges having larger variability and the smaller flanges having smaller variability. An estimate for the pooled coefficient of variation CV is obtained as:

TABLE 5.9

Calculations for a DNOM Control Chart

1	2	3	4	5	6	7	8	9	10	11	12	13
i	X_{i1}	X_{i2}	X_{i3}	X_{i4}	X_{i5}	Ti	$\bar{X}_i$	Si	Xi/Ti	Si/Ti	$(\bar{X}_i - T_i)$	$(Si/Ti)^2$
1	0.010	0.011	0.010	0.010	0.011	0.01	0.0104	0.000	1.0400	0.046	0.00039	0.00214
2	0.010	0.010	0.011	0.011	0.010	0.01	0.0101	0.0006	1.0110	0.058	0.00011	0.00333
3	0.021	0.020	0.020	0.020	0.023	0.02	0.0208	0.0010	1.0379	0.051	0.00076	0.00259
4	0.020	0.021	0.019	0.019	0.019	0.02	0.0198	0.0008	0.9890	0.040	−0.00022	0.00158
5	0.021	0.020	0.020	0.021	0.023	0.02	0.0209	0.0010	1.0464	0.049	0.00093	0.00240
6	0.020	0.021	0.020	0.020	0.021	0.02	0.0203	0.0006	1.0146	0.031	0.00029	0.00096
7	0.019	0.021	0.019	0.019	0.022	0.02	0.0200	0.0012	1.0003	0.061	0.00001	0.00374
8	0.029	0.033	0.032	0.031	0.030	0.03	0.0310	0.0015	1.0324	0.049	0.00097	0.00244
9	0.031	0.028	0.029	0.034	0.029	0.03	0.0302	0.0023	1.0075	0.076	0.00023	0.00576
10	0.041	0.042	0.040	0.037	0.037	0.04	0.0393	0.0022	0.9827	0.054	−0.00069	0.00295
11	0.038	0.042	0.040	0.038	0.037	0.04	0.0390	0.0020	0.9762	0.050	−0.00095	0.00246
12	0.043	0.040	0.039	0.042	0.040	0.04	0.0409	0.0017	1.0217	0.042	0.00087	0.00173
13	0.049	0.047	0.052	0.051	0.051	0.05	0.0500	0.0018	1.0006	0.036	0.00003	0.00128
14	0.054	0.050	0.049	0.052	0.053	0.05	0.0516	0.0021	1.0319	0.043	0.00160	0.00184
15	0.056	0.050	0.050	0.051	0.046	0.05	0.0505	0.0035	1.0093	0.071	0.00047	0.00504
16	0.047	0.049	0.045	0.049	0.051	0.05	0.0482	0.0023	0.9642	0.047	−0.00179	0.00220
17	0.061	0.056	0.057	0.057	0.060	0.06	0.0584	0.0023	0.9729	0.039	−0.00163	0.00149
18	0.065	0.070	0.064	0.072	0.061	0.07	0.0665	0.0046	0.9506	0.065	−0.00346	0.00429
19	0.088	0.091	0.092	0.085	0.089	0.09	0.0890	0.0026	0.9884	0.029	−0.00105	0.00086
20	0.098	0.095	0.103	0.098	0.099	0.1	0.0986	0.0031	0.9855	0.031	−0.00145	0.00094
											Total	0.05002

Source: Adapted from Farnum, N. M., *Journal of Quality Technology* 24, 138–144, 1992. With permission.

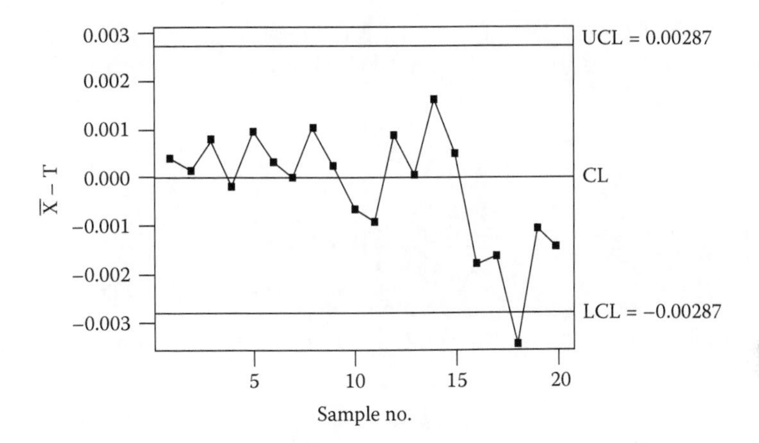

FIGURE 5.16
Example of a DNOM control chart.

$$CV = \sqrt{\frac{1}{m}\sum_i \left(\frac{S_i}{T_i}\right)^2}$$

where S_i/T_i is the estimate of the CV from individual subgroups. Then, according to Farnum (1992):

$$\frac{\overline{X}_i - T_i}{(CV)T_i/\sqrt{n}} \approx N(0,1)$$

and

$$P\left(-3 \le \frac{\overline{X}_i - T_i}{(CV)T_i/\sqrt{n}} \le 3\right) = 0.9973$$

$$P\left(-\frac{3(CV)T_i}{\sqrt{n}} \le \overline{X}_i - T_i \le \frac{3(CV)T_i}{\sqrt{n}}\right) = 0.9973$$

$$P\left(-3(CV)/\sqrt{n} \le \frac{\overline{X}_i - T_i}{T_i} \le 3(CV)/\sqrt{n}\right) = 0.9973$$

$$P\left(-3(CV)/\sqrt{n} \le \frac{\overline{X}_i}{T_i} - 1 \le 3(CV)/\sqrt{n}\right) = 0.9973$$

$$P\left(1 - 3(CV)/\sqrt{n} \le \frac{\overline{X}_i}{T_i} \le 1 + 3(CV)/\sqrt{n}\right) = 0.9973$$

Thus, if we plot the statistic $\overline{X}_i/T_i$, which is the ratio of the individual sample average to the corresponding target, the 3-sigma control limits will be:

$$UCL = 1 + 3(CV)/\sqrt{n}$$

$$CL = 1$$

$$LCL = 1 - 3(CV)/\sqrt{n}$$

where (CV) is the pooled CV as defined above.

For the data in Table 5.9, the pooled CV is calculated from Column 13 of the table as:

$$(CV) = \sqrt{\frac{0.05002}{20}} = \sqrt{0.0025} = 0.05$$

Then,

$$UCL = 1 + \frac{3(0.05)}{\sqrt{5}} = 1.0671$$

$$LCL = 1 - \frac{3(0.05)}{\sqrt{5}} = 0.9329$$

The control chart is shown in Figure 5.17.

None of the values are outside the limits in the standardized DNOM chart, indicating that the process is in-control. A possible explanation for the difference in the results of the two types of DNOM charts for the same data

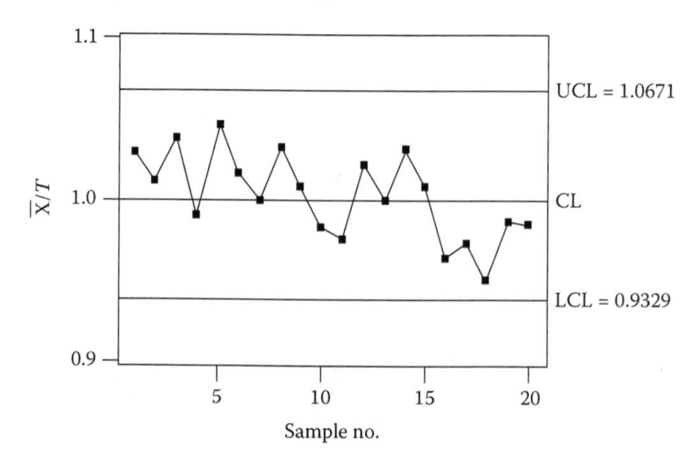

FIGURE 5.17
Standardized DNOM chart for the data in Table 5.9.

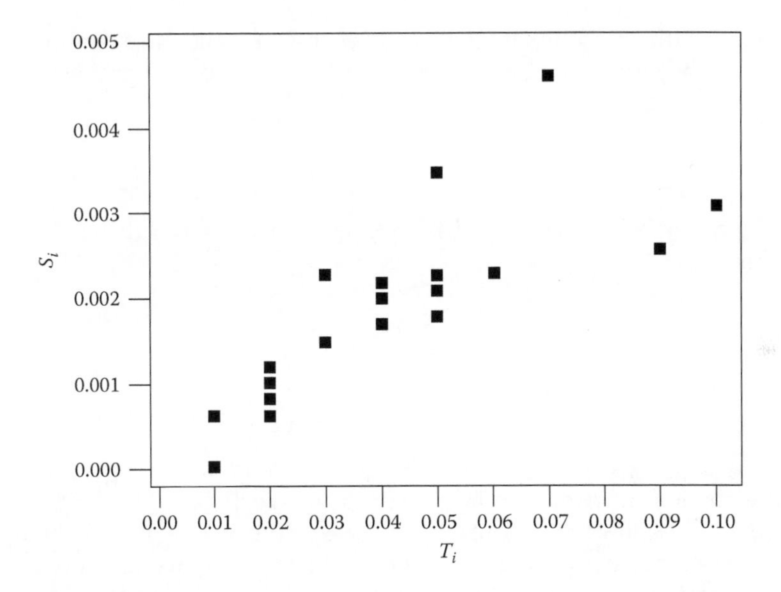

FIGURE 5.18
Plot of the standard deviation of the readings from different part numbers in Table 5.9.

may be that the data in Table 5.9 do not strictly follow the assumption that the standard deviation of the measurement remains constant from one part number to another.

This can be verified by plotting the sample standard deviations of the samples against corresponding target values as in Figure 5.18. We find that the assumption of equal standard deviation is not quite true in this case; the graph shows an increasing trend in the value of S_i with an increase in target value. In this case, it may be more appropriate to use the standardized DNOM chart.

Therefore, it may make sense to verify, before using a DNOM chart, if the assumption of constant standard deviation is true. If the standard deviation is constant over the sizes, a simple DNOM chart can be used; otherwise, the standardized DNOM chart should be used.

5.4 Topics in Process Capability

Process capability was defined in Chapter 4 as the ability of a process to meet specifications, or the ability of the process to produce products that meet specifications. This has to be assessed after the process is brought in-control. The capability indices are used to quantify this capability of a process. The indices C_p and C_{pk}—the basic indices that are popular in industry—were defined and their use explained in Chapter 4. Another such index is C_{pm}, which has been proposed as an index with some superior properties compared with C_p and C_{pk}.

5.4.1 The C_{pm} Index

The C_{pm} index was first introduced by Chan, Cheng, and Spiring (1988) and is defined as:

$$C_{pm} = \frac{\text{USL} - \text{LSL}}{6\sigma'}$$

where $\sigma' = \sqrt{\Sigma(X_i - T)^2/n}$, T is the target for the process and USL and LSL are the upper and lower specification limits, respectively.

Figure 5.19 shows the difference between the usual process standard deviation σ and the σ' as defined above. Whereas the usual process standard deviation σ measures the variability of the individual values from the process mean, σ' measures the variability of the individual values from the target. It can be shown that:

$$C_{pm} = \left(C_p \big/ \sqrt{1+V^2}\right), \text{ where } V = (T - \mu)/\sigma, \text{ and}$$

μ and σ are the process mean and process standard deviation, respectively (see Exercise 5.14 and the solution manual for the proof).

Example 5.11

Calculate C_p, C_{pk} and C_{pm} for the process with the following data: USL = 30, LSL = 10, T = 20, process mean μ = 22, and process standard deviation σ = 3.

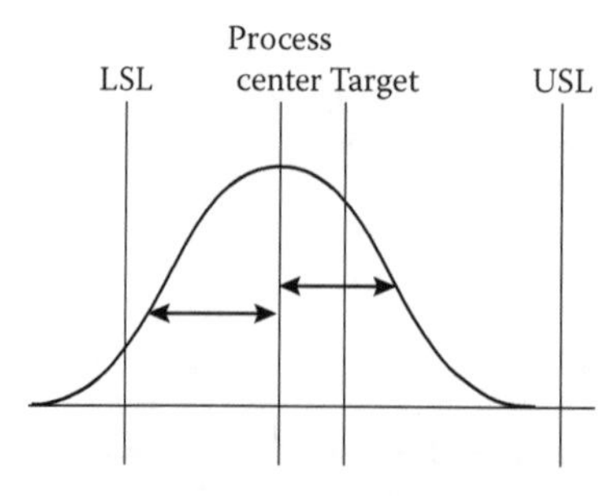

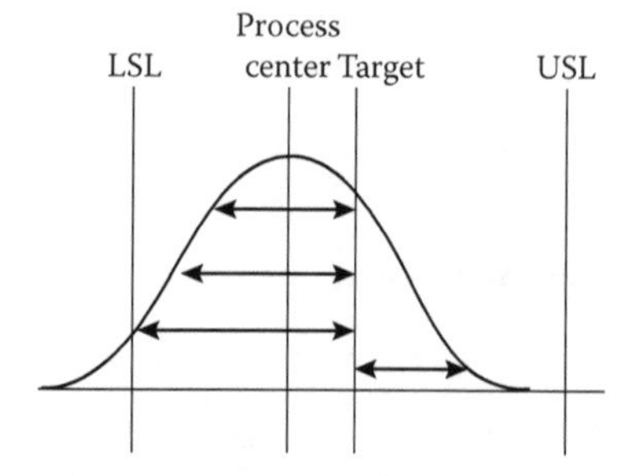

FIGURE 5.19
Difference in calculation of σ and σ' for the C_{pm} index.

Solution

$$V = \frac{20-22}{3} = -\frac{2}{3}$$

$$C_p = \frac{30-10}{6\times 3} = \frac{20}{18} = 1.11$$

$$C_{pk} = \frac{8}{9} = 0.88$$

$$C_{pm} = \frac{1.11}{\sqrt{1+(2/3)^2}} = 0.93$$

In Example 5.11, the value of C_{pm} fell between the values of C_p and C_{pk}, but this need not always be the case. The following comparison of the three indices illustrates the difference between them.

5.4.2 Comparison of C_p, C_{pk}, and C_{pm}

The major advantage claimed for the C_{pm} index is that it does not depend on the availability of the usual two-sided tolerance limits, which are dubbed the "goal-post" limits (Bothe 1997). The goal-post limits imply that no loss is incurred as long as the characteristic falls within these limits, whereas according to the Taguchi loss function the loss is zero only when the characteristic is on target and the loss increases quadratically when the value moves away from the target. Construction of the C_{pm} index follows this loss function philosophy, and maximizing the value of the C_{pm} amounts to minimizing the loss function.

A good comparison of the three indices is provided by Kotz and Johnson (1993). Their comparison is in the context of using the indices where the usual bilateral (goal-post) tolerance is applicable. If the process is centered—that is, the process center μ is equal to the target T, which is at the midpoint of the specification limits—then $C_p = C_{pk} = C_{pm}$. If the process is not centered, then C_{pm} and C_{pk} will be different from C_p, and both will be smaller than C_p. The C_{pk} and the C_{pm} both evaluate the lack of centering of a process, in addition to the variability, although they follow different approaches.

The first thing to note is that C_{pm} will never have a negative value, whereas C_{pk} can assume negative values if the process center is outside the specification limits. Second, C_{pk} is calculated from the specification limits, and there is an implied assumption that the target is at the center of the specifications. On the other hand, the calculation of C_{pm} does not use the specification limits but instead uses a specified value for the target. If the target happens to be at the center of the specifications, as it happens to be in the majority of cases, then the C_{pm} and C_{pk} are comparable. Both indices will decrease in value when the process variability increases or when the process mean gets away from the target. When the target does not fall at the center of the specifications, these two indices behave totally differently.

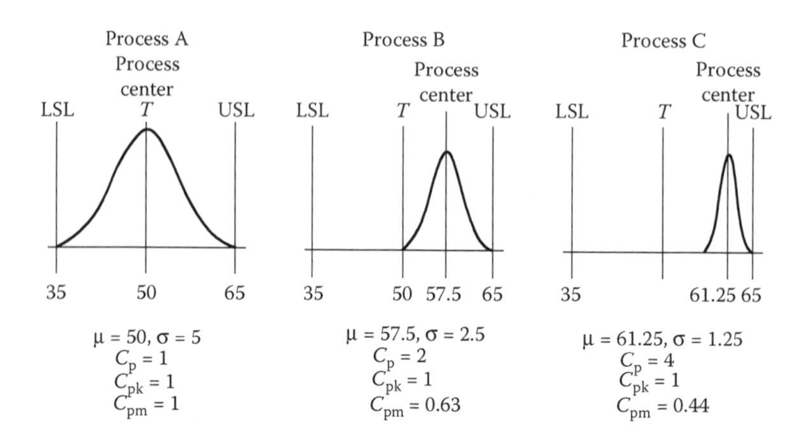

FIGURE 5.20
Comparison of C_p, C_{pk}, and C_{pm} indices.

An example (from Boyles 1992) that assumes the usual two-sided specification with the target at the center of the specifications is shown in Figure 5.20. In this figure, all three processes (A, B, and C) have the same specifications and target but have different process means and standard deviations.

Looking at Process A, when the process mean is at the center of the specification, which is also the target, the values of C_p, C_{pk}, and C_{pm} are all the same. In Processes B and C, when the process standard deviation decreases, the C_p values increase, but the values of C_{pk} and C_{pm} are smaller compared to C_p because of lack of centering of the process. The value of C_{pm}, however, is smaller than that of C_{pk} in both Process B and Process C, reflecting the fact that the C_{pm} reacts much more sharply than the C_{pk} when the process moves away from the target. Notice also that while the C_{pk} value equals 1.0 for Process B and Process C, reflecting that almost "all" the units of the population are still within specification, the values of C_{pm} are exceedingly small. This aspect of the C_{pm} index—its concern more with the location of the process center with respect to the target rather than reflecting how well the units of the population meet the specifications—is what makes it different from the other two indices. If the purpose of using an index is to see how well a process produces a characteristic within the specification stipulated by a customer, the C_{pm} index will give misleading information. The loss-function advocates, on the other hand, would insist that C_{pm} truly reflects the losses suffered by a process when the process center moves away from the target and so would help to minimize losses when it is used to measure the process's performance.

5.4.3 Confidence Interval for Capability Indices

As mentioned in Chapter 4, the value of C_p or C_{pk} calculated from the data using the following formulas is only an estimate, or a point estimate, for the true population value of the index: (The "hats" on $\hat{C}_p, \hat{\mu}$, and $\hat{\sigma}$ indicate that they are estimates.)

$$\hat{C}_p = \frac{USL - LSL}{6\hat{\sigma}}$$

$$\hat{C}_{pk} = \frac{Min\left[(USL - \hat{\mu}),\ (\hat{\mu} - LSL)\right]}{3\hat{\sigma}}$$

The estimate will have a sampling error; that is, for the same population different samples will give different values for the index. If the estimate is made from a large size sample ($n \geq 100$) we can consider the error to be negligible. Then, for all practical purposes, we can use the estimate as the population value. The simulation study reported in Krishnamoorthi, Koritala, and Jurs (2009) supports this statement. If, for some reason, the estimate must be made from a smaller sample, then we need confidence intervals (CIs) to estimate the true value of the index. The following formulas, taken from Kotz and Johnson (1993), provide the confidence intervals and confidence bounds for C_p:

$$100(1 - \alpha)\ \%\ \text{CI for } C_p\text{:} \left[\sqrt{\frac{\chi^2_{n-1,1-\alpha/2}}{n-1}}\hat{C}_p,\ \sqrt{\frac{\chi^2_{n-1,\alpha/2}}{n-1}}\hat{C}_p\right]$$

$$100(1 - \alpha)\ \%\ \text{upper confidence bound for } C_p\text{:} \left[\sqrt{\frac{\chi^2_{n-1,\alpha}}{n-1}}\hat{C}_p\right]$$

$$100(1 - \alpha)\ \%\ \text{lower confidence bound for } C_p\text{:} \left[\sqrt{\frac{\chi^2_{n-1,1-\alpha}}{n-1}}\hat{C}_p\right]$$

In the formulas above, $\hat{C}_p$ is the point estimate for C_p, and $\chi^2_{\nu,\alpha}$ is such that $P(\chi^2_\nu > \chi^2_{\nu,\alpha}) = \alpha$.

Example 5.12

A sample of 15 gear blanks gave an average thickness of 1.245 in. and a standard deviation of 0.006. The specification for the thickness is 1.25 ± 0.015 in.

Calculate an estimate for C_p.
Calculate a 95% CI for C_p.
Calculate a 95% lower bound for C_p.

Solution

$$\hat{C}_p = \frac{1.265 - 1.235}{6 \times 0.006} = 0.83$$

For calculating a 95% CI, $\alpha/2 = 0.025$, and $(n - 1) = 14$. From the χ^2 tables, Table A.3 in the Appendix, $\chi^2_{14,0.025} = 26.12$, and $\chi^2_{14,0.975} = 5.63$. Thus, the 95% CI for C_p is:

$$\left[\sqrt{\frac{5.63}{14}} \times 0.83, \sqrt{\frac{26.12}{14}} \times 0.83\right] = [0.526, 1.134]$$

For calculating a 95% lower bound, $\alpha = 0.05$, $1 - \alpha = 0.95$, and $\chi^2_{14,0.95} = 6.57$. Therefore, the 95% lower bound is:

$$\left[\sqrt{\frac{6.57}{14}} \times 0.83\right] = 0.57$$

We can make the statement that the C_p for this process is not less than 0.57 at 95% confidence.

The CI formula for the C_{pk} index is a long expression and is not provided here. The expression becomes long because, in estimating C_{pk}, two parameters—the mean and the standard deviation of the process—are estimated. These expressions for CIs are also known to be very conservative because of assumptions and approximations made to err on the safe side. For example, a simulation study by Ghandour (2004) estimated that the above formula for the CI for C_p gives intervals that are seven times wider than those that are empirically derived.

At this point, it is important to understand the following. The values of C_p and C_{pk} calculated from sample data are only estimates for the true population values of the indices, and contain sampling error. To minimize this error, we must use large samples—the larger, the better. The simulation study cited above showed that sample sizes of 100 and greater provide good estimates for the indices with a small amount of variability. Therefore, if the estimates are made from a set of data taken from a typical control chart in which at least 25 subgroups of four have been used, the estimates for the indices must be reasonably precise. If large samples are not available, we should at least be aware that the computed values are point estimates and have sampling error. They should therefore be used with caution.

5.4.4 Motorola's 6σ Capability

In the last chapter, and again in this chapter, we have studied how the capability indices C_p, C_{pk}, and C_{pm} are used to quantify the capabilities of processes in their ability to produce products that meet customers' specifications. We will now discuss another way of measuring process capabilities, in number of sigmas, which was originally proposed by the Motorola Corporation, the electronics and communication manufacturer headquartered in Schaumburg,

Illinois. They introduced this method to measure the capability of processes as part of their huge effort to reduce variability and improve the capability of processes. They made a strategic decision in 1986 that they would improve every one of their processes, whether for products or services, to have 6σ capability. According to their definition, a process is said to have 6σ capability if the defects produced in the process are less than 3.4 defects per million opportunities (DPMO). This definition for process capability, or process quality, provides a common yardstick whether quality is checked by measuring a characteristic of a product, counting the number of blemishes on a surface, or tallying the number of mistakes made in delivering a service.

The genesis for the 6σ capability, however, is in the normal distribution used to model many measurable characteristics and the areas lying under different regions of the normal curve. Figure 5.21 shows the percentages of a normal population lying within different regions of the curve. We can see that the percentage under the curve between (μ – 6σ) and (μ + 6σ)—that is, the proportion of a normal population lying between 6σ distance on either side of the mean—is 99.9999998. This means that if a process is normally distributed and the specifications it is expected to meet are located at ± 6σ distance from the center of the process, only 2 of 10^9 (or 0.002 ppm) units in such a population would be outside the specification. Such a process is said to have 6σ (6-sigma) capability.

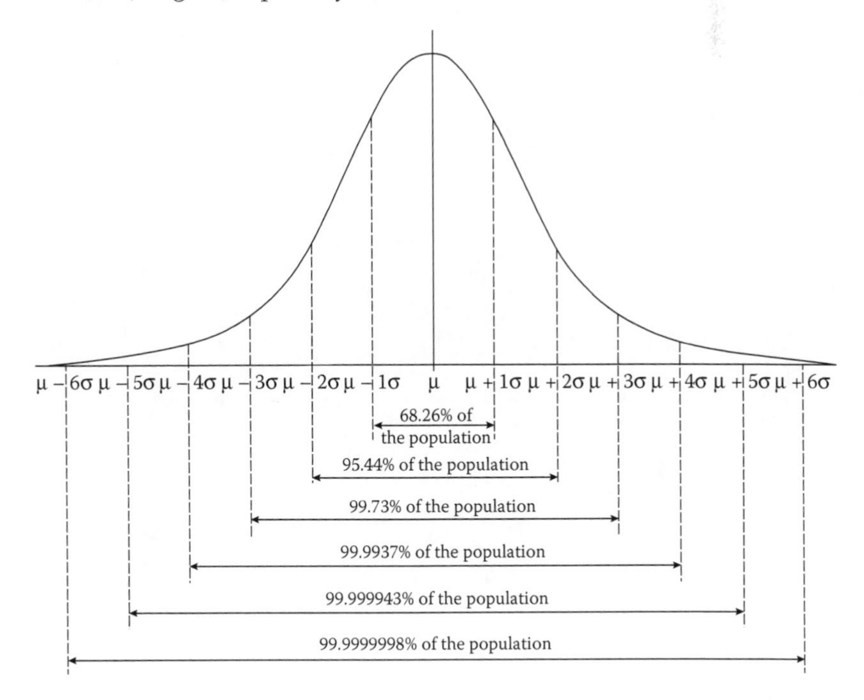

FIGURE 5.21
Proportions of a normal population under different regions of the curve.

Most processes, however, do not have their mean exactly at the center of the specification because the processes tend to drift off center. Therefore, Motorola provides for the fact that processes could be off as far as 1.5σ from the specification center, or target. A process with 6-sigma capability and its mean 1.5σ distance off of the target will produce 99.99966% of the products within specification. That is, the 6-sigma process will produce only 3.4

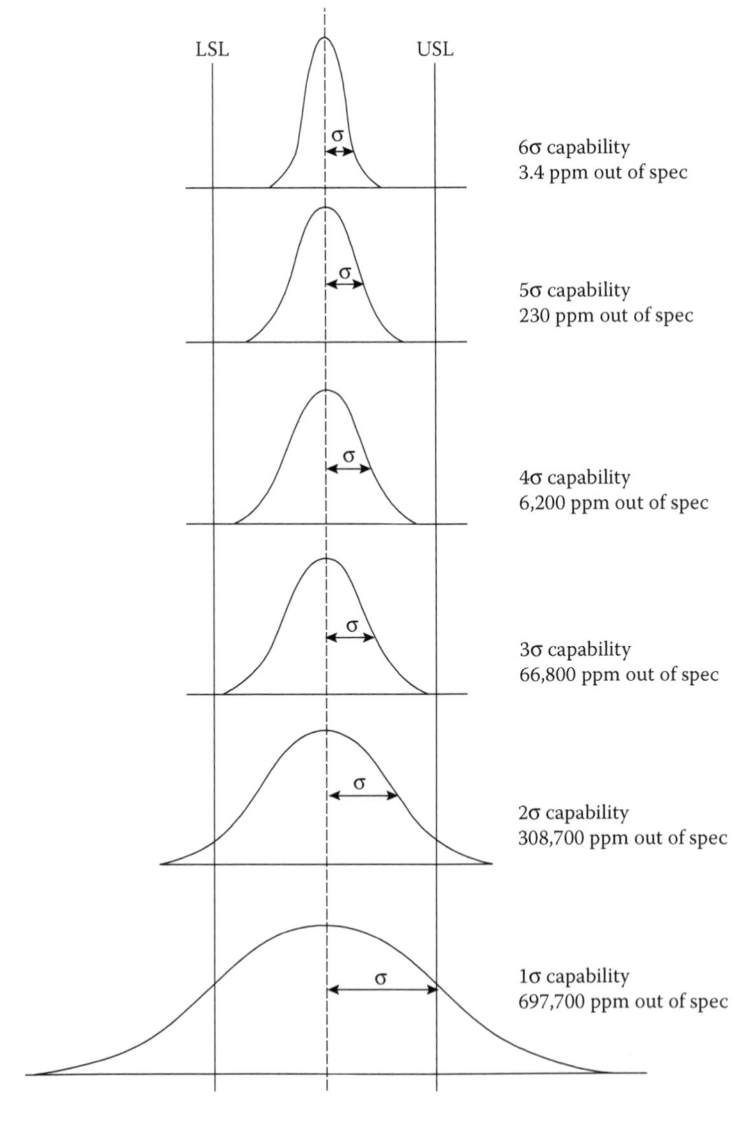

FIGURE 5.22
Processes with capabilities designated by number of sigmas. (Out-of-spec quantities calculated when process center is 1.5σ away from target.)

defects per million (DPM) of total units even in the worst condition when the process center is 1.5σ distance away from the target.

Thus, a 6-sigma process with its center on target produces only 0.002 DPM and will produce, at most, 3.4 DPM at the worst possible condition of the process center being away from target by 1.5s distance. Such a process was designated as the benchmark for minimum variability.

For the sake of uniformity, any process that produces less than 3.4 DPM, whether the process produces a characteristic that can be measured or an attribute characteristic that can be counted, is said to have 6-sigma capability. In the case of processes providing a service, the quality is measured by counting the number of occasions that a "defective" service is provided out of all the opportunities for providing such a service. Processes that are producing higher levels of defects will have smaller sigma capabilities. Figure 5.22 shows process conditions with different sigma capabilities and the proportions of defects produced in each case when the process center is off of the specification center by 1.5σ. Table 5.10 summarizes the relationship between sigma capabilities and the proportion of defectives produced.

The percentage yield and DPMO figures in Table 5.10 can be easily computed by finding the probabilities under the normal curve inside and outside the specifications, respectively, given the process center is 1.5σ away from the specification center. For example, the percentage yield for a 3-sigma process is obtained as follows. If $X \sim N(\mu, \sigma^2)$, then,

TABLE 5.10

Sigma Capability and Proportion Out of Specification

Process Quality in No. of Sigmas	ppm Outside Spec (DPMO)	Yield Percentage
1	697,700	30.23
1.5	501,300	49.87
2	308,700	69.13
2.5	158,700	84.13
3	66,800	93.32
3.5	22,800	97.72
4	6,200	99.38
4.5	1,300	99.87
5	230	99.977
5.5	30	99.997
6	3.4	99.99966

$$
\begin{aligned}
\%\text{ yield} &= 100P(\mu - 3\sigma \le X \le \mu + 3\sigma \mid \mu = \mu + 1.5\sigma) \\
&= 100P\left(\frac{\mu - 3\sigma - \mu - 1.5\sigma}{\sigma} \le Z \le \frac{\mu + 3\alpha - \mu - 1.5\sigma}{\sigma}\right) \\
&= 100P(-4.5 \le Z \le 1.5) \\
&= 100(\Phi(1.5) - \Phi(-4.5)) \\
&= 100(0.9332 - 0.0000) = 93.32\%
\end{aligned}
$$

Therefore, the percentage outside the specifications = 6.68% = 66,800 ppm. Figure 5.23, which is from a Motorola training manual (Motorola Corporation 1992), shows dramatically the comparison between a process with 6-sigma capability (benchmark process) and those "average" processes with 3- and 4-sigma capabilities. The y-axis of the graph shows DPMO, and the x-axis shows the number of sigma capabilities.

The Motorola figure shows the average quality level of many commonly provided services, such as order write-ups, airline luggage handling, and

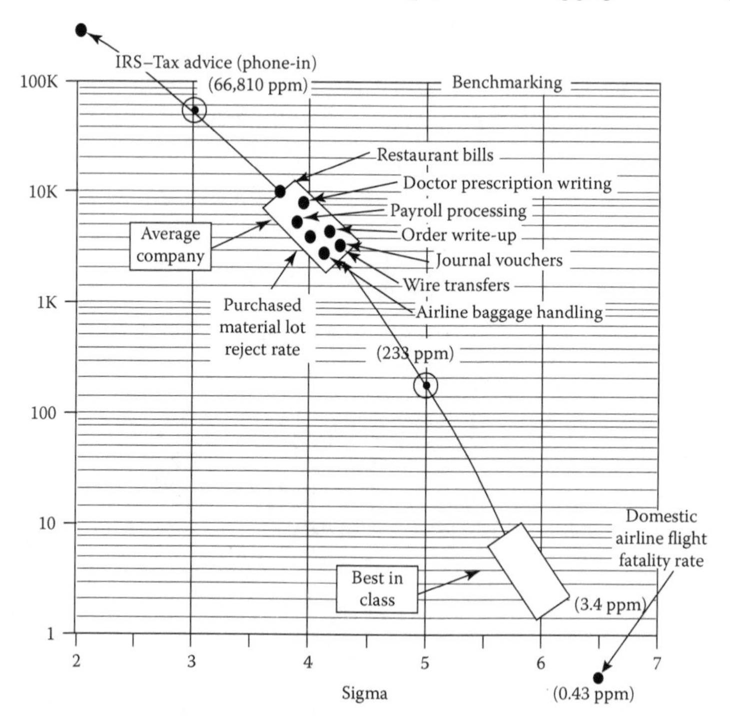

FIGURE 5.23
Examples of processes with different sigma capabilities. (From Motorola Corporation. 1992. *Utilizing the Six Steps to Six Sigma*. Personal Notebook. SSG 102, Motorola University, Schaumburg, IL. Reprinted with permission of Motorola Inc.)

so on, measured in the number of sigmas. The figure also shows that going from the "average" level of 4-sigma capability to the level of 6-sigma capability means a quality improvement, or defect reduction, on the order of better than 1000-fold from the original condition.

The sigma measure is a useful way of measuring the quality produced by processes when we want to create a baseline, set goals, and monitor progress as improvements are achieved. Motorola followed a "six steps to 6-sigma" procedure (popularly known as the "Six Sigma procedure"), which is a step-by-step methodology to improve processes to the 6-sigma level. More detail on the 6-sigma process as a system for continuous improvement is provided in Chapter 9.

Before we leave this section we want to make clear the terminology we use. We have used the "6σ" or "6-sigma" to denote the capability or quality of a process, and use the term "Six Sigma" to refer to the procedure for achieving the 6-sigma capability in processes.

5.5 Topics in the Design of Experiments

5.5.1 Analysis of Variance

When the design of experiments was introduced in Chapter 3 for choosing product and process parameters, the procedure for designing factorial experiments, with each factor considered at two levels, was introduced. The methods for analyzing data from experimental results to calculate factor and interaction effects were explained. A method to test whether the effects were significant using an approximate 95% confidence interval was also included.

Another method, the analysis of variance (ANOVA) procedure, is extensively used for analyzing data from experiments. Many computer programs use ANOVA to interpret data and present results. The ANOVA procedure is used to decide if a *significant* difference exists among the *means* of populations of outcomes generated by different treatment combinations in an experiment.

The ANOVA method is explained here with reference to a factorial experiment with two factors, A and B, with a number of levels in Factor A, and b number of levels in Factor B. There are n replicates of the trials, so there are n observations in each treatment combination. The data from the experimental trials are laid out in Table 5.11. This design includes an equal number of observations in the cells. Such designs with an equal number of observations in each cell are called "balanced" designs.

In Table 5.11, $\bar{y}_{ij.} = (1/n)\sum_{k=1}^{n} y_{ijk}$ are the cell averages; $\bar{y}_{i..} = (1/n)\sum_{j=1}^{b} \bar{y}_{ij.}$ are the row averages; $\bar{y}_{.j.} = (1/a)\sum_{i=1}^{a} \bar{y}_{ij\cdot}$ are the column averages; and $\bar{y}_{...} = ((1/a)\sum_{i=1}^{a} \bar{y}_{i..}) = ((1/b)\sum_{j=1}^{b} \bar{y}_{.j.})$ is the overall average.

TABLE 5.11

Data from an $a \times b$ Factorial Experiment

		Factor B					
		Level 1	...	**Level j**	...	**Level b**	**Average**
Factor A	**Level 1**	$y_{111}, \ldots, y_{11n}$ Average: $\bar{y}_{11\cdot}$	...	$y_{1j1}, \ldots, y_{1jn}$ Average: $y_{1j\cdot}$	...	$y_{1b1}, \ldots, y_{1bn}$ Average: $\bar{y}_{1b\cdot}$	$\bar{y}_{1..}$
	...	...	...	...	...	...	...
	Level i	$y_{i11}, \ldots, y_{i1n}$ Average: $\bar{y}_{i1\cdot}$	...	$y_{ij1}, \ldots, y_{ijn}$ Average: $\bar{y}_{ij\cdot}$	...	$y_{ib1}, \ldots, y_{ibn}$ Average: $\bar{y}_{ib\cdot}$	$\bar{y}_{i..}$
	...	...	...	...	...	...	...
	Level a	$y_{a11}, \ldots, y_{a1n}$ Average: $\bar{y}_{a1\cdot}$	...	$y_{aj1}, \ldots, y_{ajn}$ Average: $\bar{y}_{aj\cdot}$	...	$y_{ab1}, \ldots, y_{abn}$ Average: $\bar{y}_{ab\cdot}$	$\bar{y}_{a..}$
	Average	$\bar{y}_{.1.}$	...	$\bar{y}_{.j.}$	...	$\bar{y}_{.b.}$	$\bar{y}_{...}$

An observation in any cell in Table 5.11 can be considered as generated by the following model:

$$y_{ijk} = \mu + \alpha_i + \beta_j + (\alpha\beta)_{ij} + \varepsilon_{ijk}$$

where μ is the overall mean, α_i is the effect caused by level i of Factor A, β_j is the effect caused by level j of Factor B, $(\alpha\beta)_{ij}$ is the effect caused by the interaction between Factors A and B, and ε_{ijk} is a random error assumed to come from a normal distribution, $N(0, \sigma^2)$, independently at each cell.

This model is known as the fixed-effect model, because the levels in the factors are assumed to be the only levels available in the factors. There could be situations in which the levels chosen in the design are random selections from several possible levels available for the factors. In such situations, the factor and interaction effects become random variables, and the analysis method becomes different. For models of this latter kind, see Montgomery (2001b).

In the above model, we can substitute the overall mean and the factor effects with their estimates from data as:

$$\hat{\mu} = \bar{y}_{...}$$

$$\hat{\alpha}_i = (\bar{y}_{i..} - \bar{y}_{...})$$

$$\hat{\beta}_j = (\bar{y}_{.j.} - \bar{y}_{...})$$

The estimate for the interaction effects $(\alpha\hat{\beta})_{ij}$ is the remainder in the cell average after accounting for the overall mean and the factor effects:

$$\left(\hat{\alpha\beta}\right)_{ij} = \left\{\bar{y}_{ij.} - \left(\bar{y}_{i..} - \bar{y}_{...}\right) - \left(\bar{y}_{.j.} - \bar{y}_{...}\right) - \bar{y}_{...}\right\} = \left(\bar{y}_{ij.} - \bar{y}_{i..} - \bar{y}_{.j.} + \bar{y}_{...}\right)$$

The estimate for the error is the difference between the individual values and the cell average:

$$\hat{\varepsilon}_{ijk} = \left(y_{ijk} - \bar{y}_{ij.}\right)$$

Thus, the observation in a cell can be represented as being comprised of the components as:

$$y_{ijk} = \bar{y}_{...} + \left(\bar{y}_{i..} - \bar{y}_{...}\right) + \left(\bar{y}_{.j.} - \bar{y}_{...}\right) + \left(\bar{y}_{ij.} - \bar{y}_{i..} - \bar{y}_{.j.} + \bar{y}_{...}\right) + \left(\bar{y}_{ijk} - \bar{y}_{ij.}\right)$$

Therefore:

$$\left(y_{ijk} - \bar{y}_{...}\right) = \left(\bar{y}_{i..} - \bar{y}_{...}\right) + \left(\bar{y}_{.j.} - \bar{y}_{...}\right) + \left(\bar{y}_{ij.} - \bar{y}_{i..} - \bar{y}_{.j.} + \bar{y}_{...}\right) + \left(\bar{y}_{ijk} - \bar{y}_{ij.}\right)$$

By squaring both sides and summing over all observations in all cells, it can be shown, because the sum of cross-products such as $\sum_i \sum_j \sum_k \left(\bar{y}_{i..} - \bar{y}_{...}\right)\left(\bar{y}_{.j.} - \bar{y}_{...}\right)$ are all equal to zero, that the following equation involving the sum of squares (SS) is true:

$$\sum_{i=1}^{a}\sum_{j=1}^{b}\sum_{k=1}^{n}\left(y_{ijk} - \bar{y}_{...}\right)^2 = \sum_{i=1}^{a}\sum_{j=1}^{b}\sum_{k=1}^{n}\left(\bar{y}_{i..} - \bar{y}_{...}\right)^2 + \sum_{i=1}^{a}\sum_{j=1}^{b}\sum_{k=1}^{n}\left(\bar{y}_{.j.} - \bar{y}_{...}\right)^2$$
$$+ \sum_{i=1}^{a}\sum_{j=1}^{b}\sum_{k=1}^{n}\left(\bar{y}_{ij.} - \bar{y}_{i..} - \bar{y}_{.j.} + \bar{y}_{...}\right)^2$$
$$+ \sum_{i=1}^{a}\sum_{j=1}^{b}\sum_{k=1}^{n}\left(y_{ijk} - \bar{y}_{ij.}\right)^2$$

Therefore,

$$\sum_{i=1}^{a}\sum_{j=1}^{b}\sum_{k=1}^{n}\left(y_{ijk} - \bar{y}_{...}\right)^2 = nb\sum_{i=1}^{a}\left(\bar{y}_{i..} - \bar{y}_{...}\right)^2 + na\sum_{j=1}^{b}\left(\bar{y}_{.j.} - \bar{y}_{...}\right)^2$$
$$+ n\sum_{i=1}^{a}\sum_{j=1}^{b}\left(\bar{y}_{ij.} - \bar{y}_{i..} - \bar{y}_{.j.} + \bar{y}_{...}\right)^2$$
$$+ \sum_{i=1}^{a}\sum_{j=1}^{b}\sum_{k=1}^{n}\left(y_{ijk} - \bar{y}_{ij.}\right)^2$$

We can write:

SS (Total) = SS (Factor A) + SS (Factor B) + SS (AB interaction) + SS (Error).

Using abbreviations, the equation can be rewritten as:

$$SS_T = SS_A + SS_B + SS_{AB} + SS_E$$

Each of the sums of squares in the above equation has associated with it a certain number of degrees of freedom (*df*) equal to the number of *independent* squared terms that are summed to produce the sum of squares. Take, for example, $SS_E = \sum_{i=1}^{a}\sum_{j=1}^{b}\sum_{k=1}^{n}\left(y_{ijk} - \bar{y}_{ij.}\right)^2$. Although *abn* number of squared terms go into this sum of squares, there are only $ab(n-1)$ independent terms in this sum. In each cell, even if n terms are summed, there are only $(n-1)$ independent terms, because the cell average $\bar{y}_{ij.}$ is calculated out of the cell observations. Therefore, there is only a total of $ab(n-1)$ independent terms in the error sum of squares. We say that the SS_E has $ab(n-1)$ degrees of freedom. Similarly, the SS_A, SS_B, and SS_{AB} each have $(a-1)$, $(b-1)$, and $(a-1)(b-1)$ degrees of freedom, respectively. The sums of squares and the associated degrees of freedom are listed in Table 5.12, which is called an ANOVA table.

The mean square (MS) column in the ANOVA table gives the sum of squares divided by the corresponding degrees of freedom. The ANOVA table, in effect, shows the total variability in the data divided into contributions coming from different sources. The mean squares represent the standardized

TABLE 5.12

ANOVA Table for an $a \times b$ Factorial Design

Source	*Df*	SS	MS	*F*
Factor A	$(a-1)$	$nb\sum_{i=1}^{a}\left(\bar{y}_{i..} - \bar{y}_{...}\right)^2$	$SS_A/(a-1)$	MS_A/MS_E
Factor B	$(b-1)$	$na\sum_{j=1}^{b}\left(\bar{y}_{.j.} - \bar{y}_{...}\right)^2$	$SS_B/(b-1)$	MS_B/MS_E
Interaction AB	$(a-1)(b-1)$	$n\sum_{i=1}^{a}\sum_{j=1}^{b}\left(\bar{y}_{ij.} - \bar{y}_{i..} - \bar{y}_{.j.} + \bar{y}_{...}\right)^2$	$SS_{AB}/(a-1)(b-1)$	MS_{AB}/MS_E
Error	$ab(n-1)$	$\sum_{i=1}^{a}\sum_{j=1}^{b}\sum_{k=1}^{n}\left(y_{ijk} - \bar{y}_{ij.}\right)^2$	$SS_E/ab(n-1)$	
Total	$(abn-1)$	$\sum_{i=1}^{a}\sum_{j=1}^{b}\sum_{k=1}^{n}\left(y_{ijk} - \bar{y}_{...}\right)^2$		

contributions of variability by different sources, standardized for the number of degrees of freedom within each source.

The mean square error (MSE) represents the variability in the data solely due to experimental error, or variability from unexplained noise factors. If the variability from a factor is much larger than the variability from experimental error, we can then conclude that factor to be significant. Thus, the ratio of each of the mean squares to the MSE is taken, and if it is "large" for any source, we then conclude that source to be significant. It can be proved, assuming that the populations of observations generated by the treatment combinations are all normally distributed, that the ratio MS (Source)/MSE has the F_{v_1,v_2} distribution, where v_1 and v_2 are the degrees of freedom for the numerator and denominator, respectively, of the F distribution (Hogg and Ledolter 1987). If the computed value of the F statistic for a source is larger than the critical value F_{α,v_1,v_2}, then that source is significant; otherwise, that source is not significant. The critical values of the F distribution for $\alpha = 0.01$ and 0.05 are available in Table A.6 in the Appendix.

This is the basic principle of the ANOVA method for determining the significance of a source. We will illustrate the use of this method with an example. However, we first want to recognize the existence of expressions to compute the sum of squares, which are equivalent to the expressions given in Table 5.12 but are computationally more efficient. The following equivalent expressions for the sum of squares can be easily proved using algebra:

$$SS_T = \sum_i \sum_j \sum_k (y_{ijk} - \bar{y}_{...})^2 = \sum_i \sum_j \sum_k y_{ijk}^2 - abn(\bar{y}_{...})^2$$

$$SS_A = bn \sum_i (\bar{y}_{i..} - \bar{y}_{...})^2 = bn \sum_i (\bar{y}_{i..})^2 - abn(\bar{y}_{...})^2$$

$$SS_B = an \sum_j (\bar{y}_{.j.} - \bar{y}_{...})^2 = an \sum_j (\bar{y}_{.j.})^2 - abn(\bar{y}_{...})^2$$

$$SS_{AB} = n \sum_i \sum_j (\bar{y}_{ij.} - \bar{y}_{i..} - \bar{y}_{.j.} + \bar{y}_{...})^2$$

$$= n \sum_i \sum_j (\bar{y}_{ij.})^2 - abn(\bar{y}_{...})^2 - SS_A - SS_B.$$

$$SS_E = \sum_i \sum_j \sum_k (y_{ijk} - \bar{y}_{ij.})^2 = SS_T - SS_A - SS_B - SS_{AB}$$

Example 5.13

Analyze the data from the 2^2 experiment for designing the lawnmower product parameters in Example 3.7 of Chapter 3 using the ANOVA method, and draw conclusions.

TABLE 5.13

Calculating Effects of Factors in Lawnmower Design

Treatment Combination Code	Design Columns		Calculation Column	Response (min)		
	Factor A (angle)	Factor B (height)	Interaction (AB)	Replicate 1	Replicate 2	Average
(1)	–	–	+	72	74	73
a	+	–	–	47	49	48
b	–	+	–	59	61	60
ab	+	+	+	83	83	83
Contrasts	–2	22	48	—	—	—
Effects	–1	11	24	—	—	—

Solution

The data from Example 3.7 is reproduced in the Table 5.13, and the main and interaction effects are recalculated using the formula: Effect = (Contrast)/2^{k-1} = (Contrast)/2.

The contrasts and effects are shown in the rows at the bottom of the table of contrast coefficients. The data are rearranged in Table 5.14 to facilitate calculation of the sums of squares, and we get:

$$SS_T = \left(72^2 + 74^2 + \cdots + 83^2\right) - 8\left(66^2\right) = 36{,}250 - 34{,}848 = 1{,}402$$

$$SS_A = 4\left(66.5^2 + 65.5^2\right) - 8\left(66^2\right) = 34{,}850 - 34{,}848 = 2$$

$$SS_B = 4\left(60.5^2 + 71.5^2\right) - 8\left(66^2\right) = 35{,}090 - 34{,}848 = 242$$

$$SS_{AB} = 2\left(73^2 + 60^2 + 48^2 + 83^2\right) - 8\left(66^2\right) - 2 - 242 = 1{,}152$$

$$SS_E = 1{,}402 - 2 - 242 - 1{,}152 = 6$$

The calculated sums of squares are entered in the ANOVA table (Table 5.15). From the completed ANOVA table, we see that Factor A is not significant, whereas

TABLE 5.14

Data from a 2^2 Experiment for Designing Lawnmower Product Parameters

		Factor B (Deck Height)		
		5 in.	7 in.	Average
Factor A (blade angle)	12°	72, 74 $\bar{y}_{11.} = 73$	59, 61 $\bar{y}_{12.} = 60$	$\bar{y}_{1..} = 66.5$
	16°	47, 49 $\bar{y}_{21.} = 48$	83, 83 $\bar{y}_{22.} = 83$	$\bar{y}_{2..} = 66.5$
Average		$\bar{y}_{.1.} = 60.5$	$\bar{y}_{.2.} = 71.5$	$\bar{y}_{...} = 66$

TABLE 5.15

ANOVA Table for the 2^2 Experiment for Lawnmower Design

Source	SS	*df*	*MS*	*F*	$F_{0.05,\nu_1,\nu_2}$	Significant?
Factor A	2	1	2	$2/1.5 = 1.33$	$F_{0.05,1,4} = 7.71$	No
Factor B	242	1	242	$242/1.5 = 161.3$	$F_{0.05,1,4} = 7.71$	Yes
Interaction AB	1,152	1	1152	$1152/1.5 = 768$	$F_{0.05,1,4} = 7.71$	Yes
Error	6	4	1.5			
Total	1,402					

Factor B and Interaction AB are significant. This is the same conclusion reached using the approximate 95% CI for the effects in Example 3.7.

There is also a formula to calculate the sums of squares directly from the table of contrast coefficients Table 5.13:

$$\text{SS from a factor} = \frac{n(\text{Contrast of the factor})^2}{(\text{Number of contrast coefficients})}$$

where n is the number of observations in each cell and the number of contrast coefficients is the number of signs (+ or –) under each factor.

For the lawnmower example, the sums of squares calculated using the above formula are seen to agree with those in the ANOVA table.

$$SS_A = \frac{2(-2)^2}{4} = 2 \quad SS_B = \frac{2(22)^2}{4} = 242 \quad SS_{AB} = \frac{2(48)^2}{4} = 1{,}152$$

5.5.2 The General 2^k Design

When the idea of using designed experiments was introduced in Chapter 3, the two most important designs, the 2^2 and the 2^3, were explained with examples. The objective then was to emphasize the need for experimentation to select the product and process parameters, and to explain the fundamentals of designed experiments and related terminology using simple designs. Some additional topics on designed experiments will now be discussed to make some additional designs available for occasions when the simpler ones are not adequate. The designs discussed below are useful when several (or many) factors are influencing a response and the ones that are most important in terms of their effect on the response have to be culled. These designs are called the "screening designs" because they are used to screen out less influential factors and select the important ones. The important ones will possibly be included in another experiment, following the screening experiment, to determine the best combination of their levels to yield the desired results in the response. Specifically, we will discuss the 2^k designs, involving k factors, $k > 3$, each at two levels. We will start with an example of a 2^4 design.

We want to point out that although we are concentrating here on the designs in which factors are considered at only two levels, on some occasions these two levels may not be adequate. The two-level designs will be appropriate if the response changes between the levels of a factor *linearly*. If the response is *nonlinear*, with respect to the factor levels, then the two-level designs will not be adequate. In these situations, more than two levels must be considered for a factor, and designs such as 2 × 3 factorials (two factors at three levels each) or 3^k (k factors, each at three levels) designs must be considered. Interested readers should refer to books on the design of experiments, such as Montgomery (2001b), listed in the references at the end of this chapter.

5.5.3 The 2^4 Design

In the 2^4 design, there are four factors of interest, each being considered at two levels. For example, in the 2^3 design that was used for process design (Case Study 3.1), if "Iron Temp" is included as the fourth factor, a 2^4 factorial design will result, which can be represented graphically as shown in Figure 5.24. The design has a total of 16 treatment combinations, as indicated by the dots at the corners.

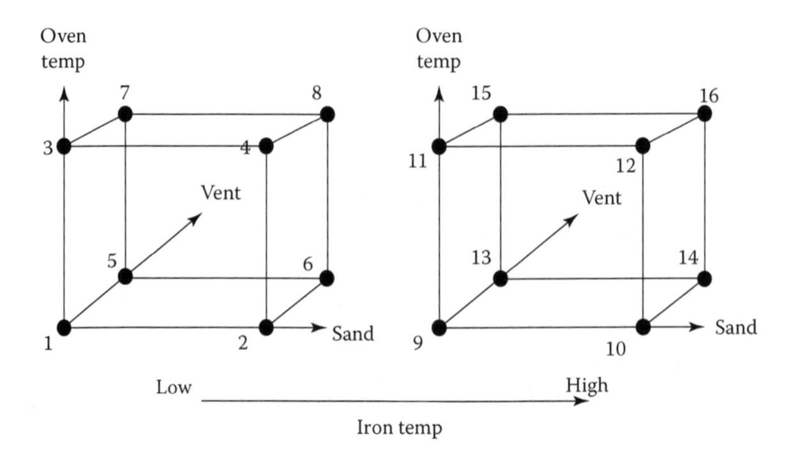

FIGURE 5.24
A 2^4 design with 16 treatment combinations.

The major problem with a 2^k design is that when k increases, the number of trials needed increases exponentially—even for a single replicate of the experiment. If multiple replications are needed to estimate the experimental error, the number of trials becomes impractical. For example, if $k = 5$ and two replicates are needed, the required number of trials is 64, which is an enormous number considering the time and other resource constraints in the context of real processes. Therefore, statisticians have devised methods to eliminate the need for a second replicate to estimate the error. They have also devised *fractional factorial* designs, which call for running only a fraction of the total number of runs in a full factorial, with the fraction being chosen in such a way that useful information from the experimental results can be extracted while only some noncritical information is sacrificed.

5.5.4 2^k Designs with Single Trial

In an effort to avoid too many trials, statisticians try to make do with only one trial at each treatment combination when k is large (>3). When there are no replicates, the SS_E is zero. To test for the significance of effects, the sum of squares from higher-order interactions, higher than two-factor interactions, are pooled together and used as the error sum of squares. This is done assuming that interactions higher than two-factor interactions do not exist, so the sums of squares of higher-order interactions are, in fact, experimental error. This seems to be a somewhat artificial way of obtaining the estimate for experimental error, and if any higher-order interactions do exist, then the results of the significance tests will become questionable. The approach preferred when multiple replicates are not available in a 2^k experiment is to use the normal probability plot (Hogg and Ledolter 1987).

If none of the factor effects is significant, the observations from the treatment combinations should all be from one normal distribution with a mean and a variance. Then, the factor effects, which are differences of the averages of normally distributed observations, should also all be normally distributed. If a normal probability plot (as described in Chapter 2) is made of the factor effects, then all the factor effects should plot on a straight line. If an effect is significant it will plot as an outlier, which can then be identified as such.

Example 5.14

An experimental study was made on a machining process to examine the effect of four factors on the output. The results from the trials (with no replication) were as shown in Table 5.16. Analyze the data and estimate the effects and their significance.

TABLE 5.16

Experimental Results from a Machining Process

Run	Design				Response		Calculations										
	A	B	C	D			AB	AC	AD	BC	BD	CD	ABC	ABD	ACD	BCD	ABCD
1	–	–	–	–	(1)	7.9	+	+	+	+	+	+	–	–	–	–	+
2	+	–	–	–	*a*	9.1	–	–	–	+	+	+	+	+	+	–	–
3	–	+	–	–	*b*	8.6	–	+	+	–	–	+	+	+	–	+	–
4	+	+	–	–	*ab*	10.4	+	–	–	–	–	+	–	–	+	+	+
5	–	–	+	–	*c*	7.1	+	–	+	–	+	–	+	–	+	+	–
6	+	–	+	–	*ac*	11.1	–	+	–	–	+	–	–	+	–	+	+
7	–	+	+	–	*bc*	16.4	–	–	+	+	–	–	–	+	+	–	+
8	+	+	+	–	*abc*	7.1	+	+	–	+	–	–	+	–	–	–	–
9	–	–	–	+	*d*	12.6	+	+	–	+	–	–	–	+	+	+	–
10	+	–	–	+	*ad*	4.7	–	–	+	+	–	–	+	–	–	+	+
11	–	+	–	+	*bd*	7.4	–	+	–	–	+	–	+	–	+	–	+
12	+	+	–	+	*cbd*	21.9	+	–	+	–	+	–	–	+	–	–	–
13	–	–	+	+	*cd*	9.8	+	–	–	–	–	+	+	+	–	–	+
14	+	–	+	+	*acd*	13.8	–	+	+	–	–	+	–	–	+	–	–
15	–	+	+	+	*bcd*	10.2	–	–	–	+	+	+	–	–	–	+	–
16	+	+	+	+	*abcd*	12.8	+	+	+	+	+	+	+	+	+	+	+
Contrast	10.9	18.7	5.7	15.5			8.3	–8.3	15.5	–9.3	4.1	–5.7	–37.7	33.7	8.3	–15.9	–9.9
Effect	1.36	2.34	0.71	1.94	Ave. = 1.68		1.04	–1.04	1.94	–1.16	0.51	–0.71	–4.71	4.21	1.04	–1.99	–1.24

Source: Data from an exercise in Ch 15 of Walpole et al. 2002. *Probability and Statistics for Engineers and Scientists*. 7th ed. Upper Saddle River, NJ: Prentice Hall. Used with permission of Pearson Education Inc.

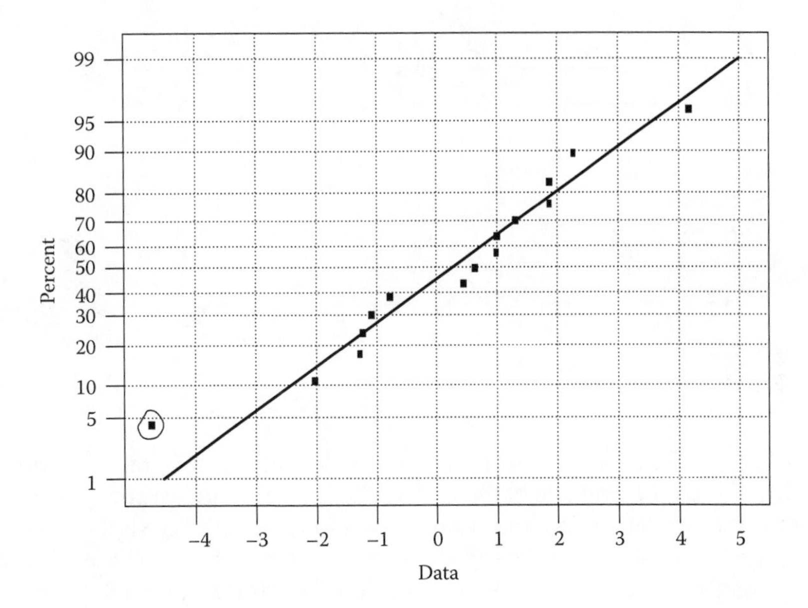

FIGURE 5.25
Normal probability plot of effects of the 2^4 factorial experiment shown in Table 5.16.

Solution:

The effects of the main factors and interactions of the 2^4 experiment are calculated in Table 5.16 and shown at the bottom row. The effects are calculated as:

$$\text{Effect} = \frac{(\text{Contrast})}{2^{k-1}} = \frac{(\text{Contrast})}{8}$$

A normal probability plot of the effects was made using Minitab software, and the resulting graph is shown in Figure 5.25. The normal probability plot shows that the effect due to Interaction ABC is an outlier in the graph and is therefore considered to be significant. No other effect shows as being significant.

5.5.5 Fractional Factorials: One-Half Fractions

As already mentioned, in this approach we perform only a fraction of the number of trials needed for a full factorial. Take, for example, the full 2^3 factorial design, which has eight treatment combinations. Suppose that only four trials (one-half of the number of trials in the full factorial) can be performed in this experiment because of resource constraints. The question then arises: which four of the treatment combinations should be chosen to obtain the best possible information on factor and interaction effects?

One choice is the four treatment combinations represented by the four corners marked with large dots on the cube in Figure 5.26. This intuitively seems to be a good "representative" selection of four treatment combinations

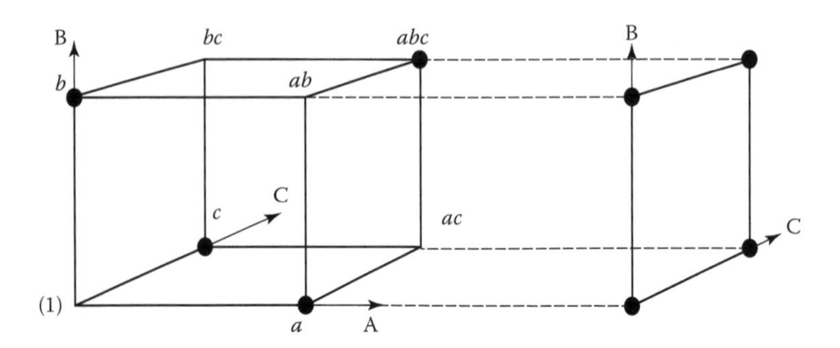

FIGURE 5.26
Choosing a one-half fraction from a 2^3 factorial design.

out of the eight. Also, this selection will result in a full 2^2 factorial design when collapsed on any one of the three factors. This is an advantage, because when one of the factors is found not to be significant, the design can be collapsed on that factor, the data at the corners aggregated, and the resulting design analyzed as a 2^2 design. Figure 5.26 shows the 2^3 factorial design collapsed on one factor, Factor A.

The above choice of one-half fraction of the 2^3 design is designated as 2^{3-1}. Note that there is another half-fraction consisting of the four treatment combinations not included in the first half-fraction. The first half-fraction is sometimes referred to as the "principal fraction" of the 2^3 factorial design, and the second is called the "alternate fraction." The two half-fractions are shown in tabular form in Table 5.17a and 5.17b. Either of the fractional designs can be run when a one-half fraction is desired; in terms of the information generated, the two are equal.

In the above selection of one-half fraction of a 2^3 design, we selected the treatment combinations to be included in the fractional design intuitively. When the number of factors becomes large, we need a formal procedure to select the treatment combinations to make up the fractional design. Described below is a general algorithm for generating fractional designs of 2^k factorials. We will illustrate the procedure using it to generate the half-fraction of the 2^3 design, and we will indicate the extension to generating other fractions—such as one-fourth and one-eighth fractions—at the end.

TABLE 5.17a
One-Half Fractions of the 2^3 Factorial Design (Principal Fraction)

Run	Response	Design			Calculation		
		A	B	C = AB	BC	AC	ABC
1	*c*	−	−	+	−	−	+
2	*a*	+	−	−	+	−	+
3	*b*	−	+	−	−	+	+
4	*abc*	+	+	+	+	+	+

TABLE 5.17b

Other-Half Fractions of the 2^3 Factorial Design (Alternate Fraction)

Run	Response	Design			Calculation		
		A	B	C = – AB	BC	AC	ABC
1	(1)	–	–	–	+	+	–
2	*ac*	+	–	+	–	+	–
3	*bc*	–	+	+	+	–	–
4	*ab*	+	+	–	–	–	–

5.5.5.1 Generating the One-Half Fraction

With reference to Table 5.17a, when we have three factors and want a design with only four runs, we first write down the design for a 2^2 factorial design (which has four runs) using the (–) and (+) signs under A and B of the design columns, as shown. Then, we generate the column for Interaction AB as the product of Columns A and B, and we assign Factor C to this column. In other words, make C = AB. This gives the one-half fraction of the 2^3 factorial design. In this design, for example, Run 1 will be made with Factors A and B at the low level and C at the high level, and this run will be designated with the code *c*, according to the coding convention. The other runs will all be coded accordingly, as shown in the table. The other half of the factorial is generated in a similar fashion, except Factor C is equated to –AB. The correspondence between the fraction indicated in Figure 5.26 with circled treatment combinations and that shown in Table 5.17a can be easily verified.

5.5.5.2 Calculating the Effects

To calculate the effects, we must first complete the calculation columns of the above tables showing the columns of signs for the other interaction terms. The effects of the factors and interactions can be then calculated as:

$$\text{Effect of a factor} = \frac{\text{Contrast of a factor}}{2^{k-1}} = \frac{\text{Contrast of a factor}}{2^{3-1-1}}$$

$$= \frac{\text{Contrast of a factor}}{2}$$

Thus, in the case of the principal fraction, the effect of Factor A is given by $(a - b - c + abc)/2$. Notice, however, that this is also the effect of Interaction BC, because the Columns A and BC are identical. In fact, the effect calculated is the sum of the effect of Factor A and the Interaction effect BC. There is no way that the effect of Factor A can be separated from the effect of Interaction BC in this design, and the main effect of Factor A is said to be *confounded* with the Interaction BC. We write A = BC. Similarly, it can be seen that B = AC and

C = AB. Each of the main factors is confounded with a two-factor interaction. In this one-half fractional design, however, the main factors are not confounded among themselves.

In the above fractional designs, the effects that are confounded with each other are said to be the *alias* of each other. Thus, A is the alias of BC, B the alias of AC, and C is the alias of AB. The basic relationship C = AB, which was used to generate the 2^{3-1} design, is called the "generator" of the fractional design. Another equivalent way of expressing this relationship that defined the one-half fraction is to multiply both sides of the generator by C to obtain CC = ABC. Here, A, B, and C represent the column vectors in the design matrix, and AA = BB = CC = *I*, where *I* is the identity vector with (+) sign for each element. Therefore, C = AB is equivalent to *I* = ABC. The latter expression is in a more generalized form and can be adopted for generating fractional factorials of any 2^k design. Once this generator is identified, it is easy to obtain the confounding pattern as seen below:

$$\text{Generator: } I = \text{ABC}$$

Multiply both sides of the generator by A, and obtain $I\text{A} = \text{A}^2\text{BC}$, which is equivalent to A = BC. Multiply both sides of the generator by B, and obtain $I\text{B} = \text{AB}^2\text{C}$, which is equivalent to B = AC. Multiply both sides of the generator by C, and obtain $I\text{C} = \text{ABC}^2$, which is equivalent to C = AB.

For the alternate fraction:

$$\text{Generator: } I = -\text{ABC}$$

Multiply both sides of the generator by A, and obtain $I\text{A} = -\text{A}^2\text{BC}$, which is equivalent to A = −BC. Multiply both sides of the generator by B, and obtain $I\text{B} = -\text{AB}^2\text{C}$, which is equivalent to B = −AC. Multiply both sides of the generator by C, and obtain $I\text{C} = -\text{ABC}^2$, which is equivalent to C = −AB.

5.5.6 Resolution of a Design

Resolution of a design provides a way of categorizing designs in terms of their ability to gather needed information from an experiment without sacrificing important ones. The resolution number of a design indicates to an experimenter what levels of interactions are lost due to selecting only a fraction of treatment combinations from the full factorial. As explained below, the larger the resolution of a design, the better the design is.

The two one-half fractions of the 2^3 design we dealt with above are both called the 2^{3-1} design of Resolution III and are denoted as 2^{3-1}_{III}. It is designated as Resolution III because one-factor main effects are confounded with two-factor interactions, giving (1 + 2) = 3. Only when higher-level interactions are confounded among themselves or with lower-level interactions or main effects, will the resolution number be large. When lower level interactions are confounded among themselves or with main effects the resolution number will be small. As we would not like to see the main effects or two-level interactions not confounded among themselves, we would not prefer designs with small resolution numbers. When the resolution of a design is large, and a high level interaction is known to be confounded with a two-level interaction or main effect, since the higher-level interactions can be assumed to be negligible, a calculated effect can be taken to be that of the two-level interaction or of the main effect.

In the two 2^{3-1} designs above, we have the main factors confounded with the two-factor interactions. This is not good as we would want to be able to identify the main effects and two-factor interactions separately. For this reason, fractional designs are chosen for only 2^k factorials when $k > 3$. The three-factor experiments are usually run as full factorials unless the experimenter is concerned only about the main effects and not about the two-factor interactions. Because it was convenient to explain the fractioning procedure and the associated terminology, the 2^3 factorial was chosen as an example in the above discussion. In the next example, we will generate the 2^{4-1} design, a one-half fraction of the 2^4 factorial.

Example 5.15

Create a 2^{4-1} design that will have only eight runs, or half of the 16 runs needed in the full 2^4 factorial.

Solution

As in the previous example, we first generate the design for the 2^3 factorial that has eight runs. Then, we assign the fourth factor to the column for the highest-level interaction. The highest level interaction is ABC, so we make D = ABC.

Then, we write out the ABC interaction column as shown in Table 5.18 and make it Column D. Now the design is complete, and the codes for the responses are assigned following the usual convention. For example, the response of the

TABLE 5.18

One-half Fraction of the 2^4 Factorial Design

Run	Response	Design			
		A	B	C	D = ABC
1	(1)	−	−	−	−
2	*ad*	+	−	−	+
3	*bd*	−	+	−	+
4	*ab*	+	+	−	−
5	*cd*	−	−	+	+
6	*ac*	+	−	+	−
7	*bc*	−	+	+	−
8	*abcd*	+	+	+	+

treatment combination where factors A and D are at the higher levels is coded as *ad*, the response from the treatment combination in which factors B and C are at high levels is coded as *bc*, and so on.

The defining relationship of the 2^{4-1} design is D = ABC, and the generator is I = ABCD. The confounding pattern will be found as:

$$\mathrm{A}I = \mathrm{AABCD} \Rightarrow \mathrm{A} = \mathrm{BCD}$$

Similarly,

$$\mathrm{B} = \mathrm{ACD} \quad \mathrm{C} = \mathrm{ABD} \quad \mathrm{D} = \mathrm{ABC}$$

$$\mathrm{AB} = \mathrm{CD} \quad \mathrm{AC} = \mathrm{BD} \quad \mathrm{AD} = \mathrm{BC}$$

The treatment combinations in the 2^{4-1} design are shown marked with a dark dot on the graph of Figure 5.27. This design is of Resolution IV and is designated as 2_{IV}^{4-1}, because the main effects are confounded with three-factor interactions, giving (1 + 3 = 4), and the two-factor interactions are confounded among themselves, giving (2 + 2 = 4). In determining the resolution number, we use the smallest of the sum of the numerals designating the levels of the confounded effects.

Thus, the resolution number helps in understanding the confounding patterns in a design. For example, we can understand that, in a design of Resolution III, the main effects are not confounded among themselves. In fact, the main effects will not be confounded among themselves if the resolution is at least III. Similarly, the main effects will not be confounded with two-factor interactions if the resolution is at least IV. This is how the designation of designs by resolution numbers helps in our understanding of the quality of a design.

There are two alternate one-half fractions available when we want to choose a one-half fraction of a full factorial. There will be four alternate one-fourth fractions when we want a one-fourth fraction of a full factorial.

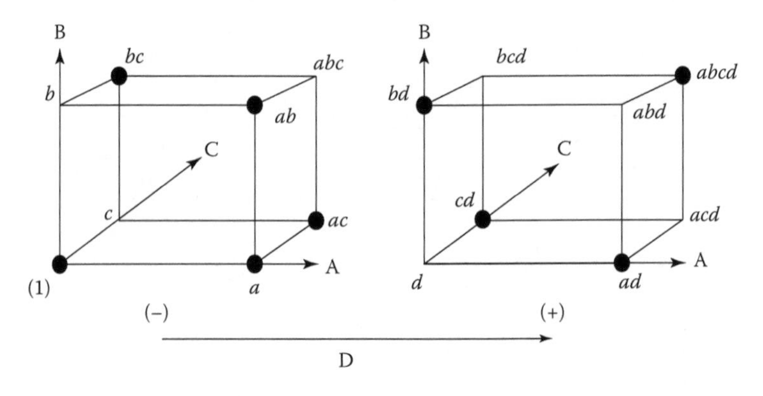

FIGURE 5.27
An example of a 2^{4-1} design.

When the number of factors k increases and even smaller fractions—such as one-eighth and one-sixteenth fractions—are sought, there will be multiple alternate fractions available from which to choose. There will then be design choices with different resolution numbers. The fractional design with the largest resolution will be the most desirable.

Choosing fractions smaller than one-half can be done following the same general logic employed in choosing the one-half fraction. However, when a one-fourth fraction is needed, for example, from a 2^5 factorial design, two main factors will have to be confounded with two higher-order interactions for choosing the generator. The interactions to be confounded have to be chosen judiciously in order to obtain the fraction of maximum resolution. (DeVor, Chang, and Sutherland [1992] provide a good explanation of the procedure for selecting the generator and the defining relationship for selecting fractional factorials.) Fortunately, tables of fractional designs for a given number of factors, and a given number of runs allowed, are available wherein the authors have provided information on the fractional factorials regarding their design generators and resolution numbers. One such example of ready-to-use designs is available in Montgomery (2001b), of which Table 5.19 is an excerpt. Taguchi's orthogonal arrays are also good examples of ready-made fractional factorials. Statistical software packages also present alternate design options once the number of factors and number of runs allowed are specified.

To see how to make use of these tables: suppose an experimenter wants a design for six factors but can afford only eight runs. The experimenter needs one-eighth of the 2^6 factorial (i.e., a 2^{6-3} design). Table 5.19 shows that such a

TABLE 5.19

Examples of Fractional Factorials

Number of Factors	Fraction	Number of Runs	Design Generator
3	2_{III}^{3-1}	4	C = ±AB
4	2_{IV}^{4-1}	8	D = ±ABC
5	2_{V}^{5-1}	16	E = ±ABCD
	2_{III}^{5-2}	8	D = ±AB E = ±AC
6	2_{IV}^{6-1}	32	F = ±ABCDE
	2_{IV}^{6-2}	16	E = ±ABC F = ±BCD
	2_{III}^{6-3}	8	D = ±AB E = ±AC F = ±BC
Etc.			

Source: Montgomery, D. C. *Design and Analysis of Experiments*. 5th ed. 2001b. Copyright Wiley-VCH Verlag GmbH & Co. KGaA. Reproduced with permission.

design can be created using the design generators D = ±AB, E = ±AC, and F = ±BC. Note that there are eight alternative one-eighth fractions available. For example, D = +AB, E = +AC, and F = +BC gives one; D = −AB, E = −AC, and F = −BC gives another; D = +AB, E = +AC, F = −BC gives the third; and so on. All the designs will be of Resolution III.

The objective of the discussion on fractional factorials is to introduce the terminology and to explain the basic method of generating the fractional factorials with desirable properties. Readers interested in the topic should refer to books such as DeVor et al. (1992) and Montgomery (2001b). Readers who understand the concepts explained here, however, have gained enough knowledge to confidently use the designs available in many computer software packages. They should be able to run experiments, analyze data, and interpret results knowing the objectives behind the designs and how the designs are created. We end this discussion on 2^k design with a case study to show how data from a fractional factorial design is analyzed.

Case Study 5.1

This case study comes from an experiment in a chemical plant where the percentage yield of XYZDT, a raw material for plastic products, was to be improved. Four factors were considered at two levels each, and a 2^{4-1} design with eight runs was used. The results of the trials in pounds of chemical are shown in Table 5.20. Factors and levels used were as follows:

	Levels	
Factors	**Low**	**High**
A: RX Temperature	130 °F	150 °F
B: Acid recipe	990 lb.	1050 lb.
C: ORP set point	0 mV	30 mV
D: Slurry concentration	23%	25%

The generator is I = ABCD, and the confounding pattern is

$$A = BCD \quad B = ACD \quad C = ABD \quad D = ABC$$

$$AB = CD \quad AC = BD \quad AD = BC$$

The responses are represented graphically in Figure 5.28

The contrasts and effects are shown calculated at the bottom of the Table 5.21. The effects are calculated as:

$$\text{Effect} = \frac{(\text{Contrast})}{2^{k-1}} = \frac{(\text{Contrast})}{2^{4-1-1}} = \frac{(\text{Contrast})}{4}$$

A normal probability plot of the effects is made using Minitab software, and the plot is shown in Figure 5.29. The dashed line was added to the normal probability plot provided by Minitab, and it seems to be a more plausible representation of the fit than the one drawn by Minitab. With this line we can see that the two effects (B + ACD = 689.75) and (C + ABD = 596.25) are significant. Assuming that the three-factor interactions do not exist, we can conclude that Factors B and C have the most influence on the yield. We also notice that a large effect is caused by the interaction BC + AD, although the normal probability plot did not show this clearly. We should conclude that it is the BC interaction that is significant, because when the main effects A and D are not found to be significant, we usually do not expect their interactions to be significant.

TABLE 5.20

One-Half Fraction of the 2^4 Factorial Design

Run	Response		Design			
			A	B	C	D = ABC
1	(1)	2120	–	–	–	–
2	*ad*	2345	+	–	–	+
3	*bd*	2736	–	+	–	+
4	*ab*	2429	+	+	–	–
5	*cd*	2346	–	–	+	+
6	*ac*	2632	+	–	+	–
7	*bc*	3311	–	+	+	–
8	*abcd*	3706	+	+	+	+

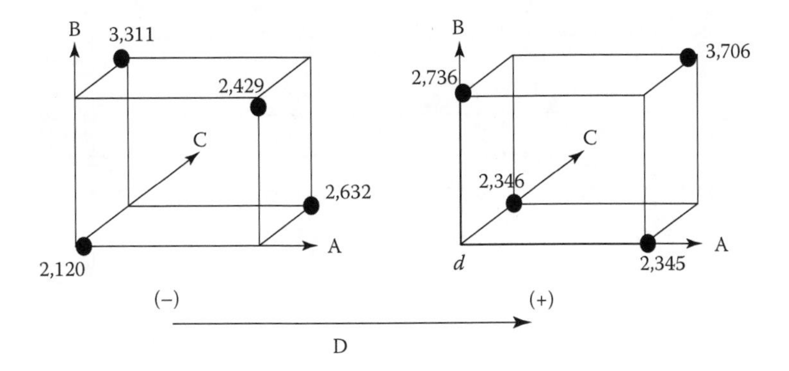

FIGURE 5.28
A 2^{4-1} design for improving the yield from a chemical process.

TABLE 5.21

Calculation of Effects from a 2^{4-1} Design

Run	Response		A(+BCD)	B(+ACD)	C(+ABD)	D(+ABC)	AB + CD	AC + BD	BC + AD
1	(1)	2120	−	−	−	−	+	+	+
2	*ad*	2325	+	−	−	+	−	−	+
3	*bd*	2736	−	+	−	+	−	+	−
4	*ab*	2429	+	+	−	−	+	−	−
5	*cd*	2346	−	−	+	+	+	−	−
6	*ac*	2632	+	−	+	−	−	+	−
7	*bc*	3311	−	+	+	−	−	−	+
8	*abcd*	3706	+	+	+	+	+	+	+
Contrast		2700.6	579	2759	2385	621	−403	783	1319
Effect			144.75	689.75	596.25	155.25	−100.75	195.75	329.75

The analysis of data from the above experiment has provided some useful information. Often this is enough for deciding on process improvement choices for obtaining better quality and/or productivity. Further analysis could be carried out by estimating the error standard deviation and making significant tests, but we did not include them here. Readers are referred to books on experimental design if details on such further analysis are needed.

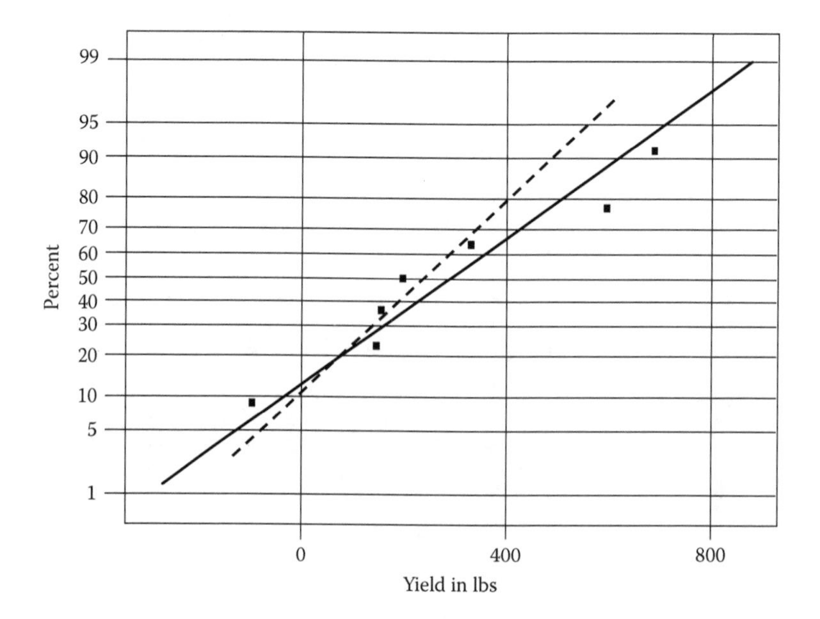

FIGURE 5.29
Normal probability plot of effects from a 2^{4-1} design (Case Study 5.1).

5.6 Exercise

5.6.1 Practice Problems

5.1 The height of 3-in. paper cubes produced in a printing shop is known to be normally distributed with mean $\mu = 3.12$ in. and standard deviation $\sigma = 0.15$. A sample of nine cubes is taken from this process, and the average is calculated.

- **a.** What is the probability that this sample average will exceed 3.42 in.?
- **b.** How would the answer be different if the process distribution is not quite normal but is instead a little skewed to the left, although with the same mean and standard deviation? Why?

5.2 The height of paper cubes is being controlled using $\overline{X}$- and R-control charts with $n = 5$. The current CL of the $\overline{X}$-chart is at 3.1 in., and that of the R-chart is at 0.08 in.

- **a.** Calculate the 3.5-sigma limits for the $\overline{X}$-chart.
- **b.** Calculate the 3.5-sigma limits for the R-chart.
- **c.** What is the probability of a false alarm in the 3.5-sigma $\overline{X}$-chart?

5.3 Draw the OC curve of an $\overline{X}$-chart with a sample size of 9 and 3-sigma control limits. Choose values of $k = 0.5$, 1.0, 1.5, and 2.

5.4 Draw the OC curve of an $\overline{X}$-chart with a sample size of 4- and 2-sigma control limits. Choose values of $k = 0.5$, 1.0, 1.5, and 2, and compare it with the OC curve from Exercise 5.3.

5.5 A process is to be controlled at fraction defectives of 0.05. Calculate the 2.5-sigma control limits for the fraction defectives chart to use with this process. Use $n = 50$.

5.6 Calculate the OC curve of the 2.5-sigma P-chart to control the fraction defectives at 0.05. Use p values of 0.01, 0.03, 0.05, 0.07, and 0.09.

5.7 A control chart is used to control a process by counting the number of defects discovered per engine assembly. Each engine assembly has the same number of opportunities for defects, and the control chart is designed to control the defects per assembly at an average of nine. Calculate the 2.5-sigma control limits for this chart.

5.8 Calculate the OC curve of the 2.5-sigma C-chart to control the number of defects per engine assembly at nine. Use c values of 3, 5, 7, 9, 10, 15, and 20.

5.9 The data in the table below represent the weight in grams of 15 candy bars weighed together as a sample. Two such samples (of 15 bars) were taken at 15-minute intervals and weighed to control the weight of candies produced on a production line.

a. Calculate the control limits for $\overline{X}$- and R-charts to maintain current control of the process at the current levels of average and standard deviation.

b. Calculate the control limits for the process if the process is to be controlled at an average weight of 882 g (for 15 bars). Use the current variability in the process to compute limits. Check if the process is in-control by drawing the graphs of the control charts.

c. Calculate the control limits for the process if it is to be controlled at an average of 882 g and a standard deviation of 3 g. Draw the graphs of control charts, and check if the process is in-control.

Time Period	Weight of Sample 1	Weight of Sample 2	Time Period	Weight of Sample 1	Weight of Sample 2	Time Period	Weight of Sample 1	Weight of Sample 2
1	893.9	894.5	17	880.1	880.6	33	897.0	899.0
2	883.0	893.0	18	891.0	890.0	34	891.0	892.0
3	892.0	893.8	19	892.0	891.0	35	892.0	891.0
4	890.2	891.4	20	890.0	889.0	36	890.0	891.0
5	891.0	893.0	21	890.0	890.0	37	889.0	890.0
6	889.0	881.0	22	889.0	891.0	38	886.0	889.0
7	890.0	891.0	23	888.0	889.0	39	891.8	890.9
8	895.0	893.0	24	890.0	890.0	40	892.4	892.2
9	891.0	890.0	25	889.0	889.0	41	890.0	897.0
10	895.0	892.0	26	889.0	890.0	42	889.0	889.0
11	894.0	890.0	27	891.0	891.0	43	891.0	891.0
12	895.0	896.0	28	891.0	890.0	44	890.0	891.0
13	893.0	891.0	29	891.0	890.0	45	890.0	891.0
14	890.0	898.4	30	900.0	901.0	46	890.0	891.0
15	890.8	896.1	31	900.0	901.0	47	890.0	890.0
16	876.2	878.4	32	891.0	893.0	48	891.0	890.0

5.10 Use only the first sample weight from each time period from the table in Problem 5.9 and use it as the single observation from each time period to prepare:

a. The chart for individuals and the MR chart.

b. The MA and MR charts with $n = 3$.

c. The EWMA chart with $\lambda = 0.2$ and $\lambda = 0.6$

If you are using computer software, verify the control limit calculations in each case.

5.11 The following data on 15 samples come from a job shop with short production runs.

a. Use a DNOM chart, and check if the process is in-control.

b. Use a standardized DNOM control chart and check if the process is in-control.

Sample	Target	Measurements		
		1	2	3
1	50	49	50	51
2	75	76	77	74
3	25	25	27	24
4	25	27	26	23
5	75	79	77	76
6	25	25	24	23
7	50	50	51	52
8	25	24	27	26
9	75	75	76	71
10	50	48	49	52
11	25	24	25	25
12	75	77	74	73
13	50	49	53	51
14	25	26	24	25
15	75	77	77	75

5.12 Calculate the capability index C_{pm} for the process in Problem 4.1. The lower and upper specification limits are 24 and 26 g, respectively, with the target being at the center of the specification. Use $\overline{R}/d_2$ to estimate for process standard deviation. Make sure the estimates for process mean and process standard deviation are made from a process that is in-control.

5.13 If the specifications for the weight of ravioli packages in Problem 4.3 are 31.5 and 32.3 oz, respectively, and the target is at the center of the specification, find the values of C_{pm}. Make sure the estimates for process parameters are made from a process that is in-control.

5.14 Show that $C_{pm} = \dfrac{C_p}{\sqrt{1+V^2}}$, where $V = \dfrac{T-\mu}{\sigma}$.

5.15 In Motorola's scheme of evaluating process capability, if a process is said to have 2.2σ capability, what would be the proportion outside specifications in ppm?

5.16 In Motorola's scheme of evaluating process capability, if a process has 30,000 ppm out of specification, what is the capability of the process in number of sigmas?

5.17 An experiment is conducted so that an engineer can gain insight about the influence of the following factors—Platen temperature, A; time of application, B; thickness of sheet, C; and pressure applied, D—on the chicken scratches produced on laminations of art work. A one-half fraction of a 2^4 factorial experiment is used, with the defining contrast being $I = ABCD$. The data are given in the table below.

Analyze the data, and assume all interactions of three factors and above are negligible. Use $\alpha = 0.05$.

A	B	C	D	Response
−	−	−	−	7
+	−	−	+	7
−	+	−	+	8
+	+	−	−	6
−	−	+	+	9
+	−	+	−	7
−	+	+	−	11
+	+	+	+	7

5.6.2 Mini-Projects

Mini-Project 5.1

A printing shop that produces advertising specialties produces paper cubes of various sizes, of which the 3.5-in. cube is the most popular. The cubes are cut from a stack of paper on cutting presses. The two sides of the cube are determined by the distance of stops on the press from the cutting knife and remain fairly constant; however, the height of the cube varies depending on the number of sheets included in a "lift" by the operator. The lift height varies within an operator and between operators. The difficulty is in judging, without taking much time, what thickness of lift will give the correct height when it comes under the knife and is pressed and cut. The humidity in the atmosphere also contributes to this difficulty because the paper swells with the increase in humidity, thereby making it more difficult to make the correct judgment. The operators tend to err on the safe side by lifting a thicker stack of paper than is necessary.

The company management believes that the cubes are being made much taller than the target, thus giving away excess paper and causing a loss to the company. They have received advice from a consultant that they could install a paper-counting machine, which will give the correct lift containing exactly the same number of sheets each time a lift is made. However, this will entail a huge capital investment. To see if the capital investment would be justifiable, the company management wants to assess the current loss in paper because of the variability of the cube heights from the target.

Data were collected by measuring the heights of 20 subgroups of five cubes and are provided in the table below. Estimate the loss incurred because of the cubes being too tall. A cube that is exactly 3.5 in. in height weighs 1.2 lb. The company produces three million cubes per year, and the cost of paper is $64 per hundred-weight (100 lb.).

Note that the current population of cube heights has a distribution (assume this to be normal) with an average and standard deviation, and the target (ideal) population of the cubes is also a distribution with an average of 3.5 in. and a standard deviation yet to be determined. (Every cube cannot be made to measure exactly 3.5 inches in height.) The target standard deviation should be less than the current standard deviation, especially if the current process is affected by some assignable causes. You must check if the process is under the influence of any assignable causes and decide what would be the best standard deviation of the process if the assignable causes can be found and eliminated.

Estimate the current loss in paper because of the cubes being too tall. You may first have to determine the attainable variability before estimating the loss. If any of the necessary information is missing, make suitable assumptions and state them clearly.

1	2	3	4	5	6	7	8	9	10	11	12	13	14	15	16	17	18	19	20
3.61	3.59	3.53	3.63	3.63	3.57	3.61	3.52	3.47	3.53	3.61	3.65	3.30	3.52	3.60	3.63	3.65	3.60	3.61	3.63
3.59	3.59	3.58	3.58	3.57	3.60	3.64	3.44	3.59	3.61	3.49	3.65	3.52	3.50	3.53	3.66	3.64	3.63	3.62	3.61
3.61	3.58	3.58	3.64	3.57	3.59	3.54	3.52	3.60	3.58	3.44	3.54	3.60	3.51	3.63	3.68	3.65	3.63	3.64	3.63
3.62	3.63	3.52	3.63	3.53	3.55	3.64	3.60	3.59	3.63	3.59	3.54	3.57	3.59	3.63	3.63	3.64	3.61	3.61	3.63
3.60	3.62	3.60	3.62	3.57	3.54	3.58	3.49	3.60	3.55	3.53	3.62	3.51	3.49	3.63	3.64	3.61	3.63	3.61	3.62

Mini-Project 5.2

The data in Columns 2 and 5 of Table 5.8, being the process data from a normal distribution, were generated using the Minitab random number generator. The first 20 observations have $\mu = 10$ and $\sigma = 1.0$, and the second set of 20 observations in Column 2 come from the process that has the mean shifted through a standard deviation of 1.0, and those in Column 5 come from a process that has the mean shifted through a 1.5 standard deviation. In each case, the standard deviation remains unchanged. Create another set of data of this kind, and draw the EWMA charts with $\lambda = 0.2$ and $\lambda = 0.6$. See how these charts react to changes in the process. Also, compare the performance of the EWMA chart with the chart for individuals and the MA and MR charts.

References

Bothe, D. R. 1997. *Measuring Process Capability*. New York, NY: McGraw-Hill.

Boyles, R. A. 1992. "C_{pm} for Asymmetrical Tolerances." *Technical Report*. Portland, OR: Precision Castparts Corporation.

Chan, L. K., S. W. Cheng, and F. A. Spiring. 1988. "A New Measure of Process Capability, C_{pm}." *Journal of Quality Technology* 20: 160–175.

Crowder, S. V. 1989. "Design of Exponentially Weighted Moving Average Schemes." *Journal of Quality Technology* 21 (3): 155–162.

DeVor, R. E., T. Chang, and J. W. Sutherland. 1992. *Statistical Quality Design and Control.* New York, N.Y: McMillan.

Duncan, A. J. March 1951. "Operating Characteristics of *R*-Charts." *Industrial Quality Control.* 40–41. Milwaukee: American Society for Quality.

Duncan, A. J. 1974. *Quality Control and Industrial Statistics.* 4th ed. Homewood, IL: Irwin.

Farnum, N. M. 1992. "Control Charts for Short Runs: Nonconstant Process and Measurement Error." *Journal of Quality Technology* 24: 138–144.

Ghandour, A. 2004. "A Study of Variability in Capability Indices with Varying Sample Size." Unpublished research report (K. S. Krishnamoorthi, advisor). IMET Department, Bradley University, Peoria, IL, 2004.

Hillier, F. S. 1969. "$\overline{X}$- and *R*-Chart Control Limits Based on a Small Number of Subgroups." *Journal of Quality Technology* 1 (1): 17–26.

Hines, W. W., and D. C. Montgomery. 1990. *Probability and Statistics in Engineering and Management Science.* New York, NY: John Wiley.

Hogg, R. V., and J. Ledolter. 1987. *Engineering Statistics.* New York, NY: Macmillan.

Kotz, S., and N. L. Johnson. 1993. *Process Capability Indices.* London: Chapman & Hall.

Krishnamoorthi, K. S., V. P. Koritala, and C. Jurs. June 2009. "Sampling Variability in Capability Indices." *Proceedings of IE Research Conference—Abstract.* Miami, FL.

Lucas, J. M., and M. S. Saccucci. 1990. "Exponentially Weighted Moving Average Control Schemes: Properties and Enhancements." *Technometrics* 32 (1): 1–12.

Montgomery, D. C. 2001a. *Introduction to Quality Control* 4th ed. New York, NY: John Wiley.

Montgomery, D. C. 2001b. *Design and Analysis of Experiments.* 5th ed. New York, NY: John Wiley.

Motorola Corporation. 1992. *Utilizing the Six Steps to Six Sigma.* Personal Notebook. SSG 102, Motorola University, Schaumburg, IL.

Roberts, S. W. 1959. "Control Chart Tests Based on Geometric Moving Averages." *Technometrics* 1 (3): 239–250.

Vardeman, S. B., and J. M. Jobe. 1999. *Statistical Quality Assurance Methods for Engineers.* New York, NY: John Wiley.

Wadsworth, H. M., K. S. Stephens, and A. B. Godfrey. 1986. *Modern Methods for Quality Control and Improvement.* New York, NY: John Wiley.

Walpole, R. E., R. H. Myers, S. L. Myers, and K. Ye. 2002. *Probability and Statistics for Engineers and Scientists.* 7th ed. Upper Saddle River, NJ: Prentice Hall.

6

Managing for Quality

This chapter covers two major topics relating to managing an organization for producing quality results. The first deals with managing an organization's human resources, and the second deals with including quality in the strategic planning of an organization as an equal parameter with financial and marketing metrics.

6.1 Managing Human Resources

6.1.1 Importance of Human Resources

For an organization pursuing quality and customer satisfaction, success greatly depends on how the workforce contributes to this effort. "If total quality does not occur at the workforce level, it will not occur at all" (Evans and Lindsay 1996). The contribution from the workforce depends on how it is recruited, trained, organized, and motivated. This chapter discusses the issues involved and the approaches taken to optimize efforts in developing, organizing, and managing human resources for achieving excellence in product quality, process efficiency, and business success.

Dr. Deming recognized the importance of the contributions that people in an organization make toward producing quality products and achieving customer satisfaction. Nine of the 14 points he recommended to organizations for improving quality, productivity, and competitive position were related to people. The points he made in this regard include (Deming 1986):

- Institute training on the job
- Adopt and institute a new form of leadership
- Drive out fear
- Break down barriers between staff areas
- Eliminate slogans, exhortations, and targets for the workforce
- Eliminate numerical quotas for the workforce and numerical goals for management
- Remove barriers that rob people of pride in workmanship

- Eliminate the annual rating or merit system
- Institute a vigorous program of education and self-improvement for everyone

These points are discussed in detail in Chapter 9. For now, note that Dr. Deming considered people's contributions to be paramount in achieving quality results in an organization. We refer in this regard to a quote by Dr. Deming on how organizations deal with people issues: "In my experience, people can face almost any problem except the problem of people" (Deming 1986). In other words, a typical organization does not recognize the enormous contribution that people can make toward quality and productivity, and fails to exploit this resource to the fullest extent through understanding their problems and finding solutions. The importance of people in the effort for quality improvement cannot be overemphasized. It is necessary to understand the challenges that employees face and to take advantage of their experience, intelligence, and capability to contribute in solving process-related problems. We begin with a discussion on how people are organized and how they work within organizations.

[The material in this chapter has been gathered from several sources, including the authors' observations in organizations they are familiar with. However, the *Certified Quality Manager Handbook* (Oakes and Westcott 2001) and *Juran's Quality Handbook* (Juran and Godfrey 1999) have provided the nucleus of information for this chapter. Readers are referred to these for further information on these topics.]

6.1.2 Organizations

6.1.2.1 Organization Structures

An organization is a collection of hard entities (e.g., facilities, machineries) and soft entities (e.g., methods, people, values) that are gathered together to accomplish a certain goal. Oakes and Westcott (2001) define an organization as "the integration of two major systems:

1. The technical system, which defines how work is to occur (including the equipment, work processes and procedures, and human resources to carry out the processes), and
2. The social system consisting of how people communicate, interrelate and make decisions."

The structure of an organization describes how the different entities are related to one another. The structure is designed based on several factors, such as the nature, size, skill requirements, and geographical dispersion of the activities. The structure for an organization is chosen based on what will best suit the needs of the organization in order for it to accomplish its vision and goals.

A "functional structure," which is found in many organizations, is created by grouping activities based on the functions they perform. Thus, people working different functions such as marketing, engineering, production, quality, maintenance, and customer service are organized in separate departments under individual supervisors. This structure provides flexibility and is cost-effective, because many "products" can share the expertise of the departments, resulting in good utilization of the departmental experts.

A "product structure" is used when activities are grouped by the "product" on which activities are performed. Thus, there will be a product manager for each of the major product categories, with marketing, engineering, maintenance, and other personnel assigned to each product group. This structure allows specialization relative to individual product types and improves accountability for accomplishing goals of the organization relative to the products.

A structure based on the customer being served is used where there are large segments of customers and each segment has its own unique needs. A college of engineering organized to address the needs of engineering majors in a university is an example of a "customer-based structure" for an organization. Similarly, organizations can be structured based on the geographical regions serviced or any special nature of activities performed. Ford Europe is an example of the former type, and Citi Card is an example of the latter.

A "team" is also an organization with no rigid structure. It brings together people with expertise and capabilities in various areas to accomplish a specific project mission, and it is dissolved at the end of the mission. Thus, a team is a temporary organization formed for individual missions. The team has the advantage of being able to assemble the needed expertise for the needed length of time, but it also faces challenges that can arise when a disparate group of people, with no prior social ties, are brought together to accomplish a particular goal. Cross-functional teams that draw expertise from various functional areas specific to the needs of a project are suited for many quality-related activities. They therefore play an important role in enhancing the quality of an organization's output.

A "matrix structure" exists when an organization that is functionally grouped vertically is also grouped horizontally. Specialists from the vertical groups such as product designers, manufacturing engineers and financial managers are assigned to projects or products such as engines, transmissions and body shops, which are grouped horizontally. The specialists are mainly responsible to the project or product managers while reporting to the functional manager for administrative purposes. The functional manager is responsible for coordinating the specialists within the function and for training and evaluating them.

A boundary-less organization is made up of a group of individuals with no structured relationship (e.g., departments and supervisors). Such an organization evolves among groups of people who have equal skills or expertise and do not require regular supervision or guidance. A group of financial advisers or computer system experts are examples of such groups. Without the departmental structure and attendant bureaucracy, they enjoy flexibility, which contributes to creativity and innovation. Boundary-less organizations are also known as "virtual corporations." The availability of modern communication tools, such as the Internet and intranets, has contributed to the growth of such organizations.

It is hard to find an organization, however, that is purely of one type, such as the functional or the product type. Most organizations are a mixture of several types. For example, an organization may be geographically structured at the global level but grouped by product nationally and structured functionally within a regional headquarters.

Traditional organizations were designed following the principle of *unity of command*, by which each employee in an organization reported to a single boss. Such principles ensured that the employee received clear instructions, but they also contributed to the growth of bureaucratic rules and red tape. Modern structures, such as the matrix structure, team structure, and boundary-less organizations, do not follow the unity-of-command principles; they depend on the individual employee's abilities to define their own responsibilities and resolve conflicts in the overall context of achieving the goals and mission of the organization.

"Span of control" is another factor used in the design of organizations. It refers to the number of subordinates that a manager supervises. In traditional organizations, the span of control was once small, on the order of 6 to 10, to maintain close control. Modern organizations are designed with a much larger span of control, which is made possible with improved training and knowledge of the subordinates, and their ability to make decisions for themselves. In other words, as the employees are educated and empowered to make their own decisions, the span of control becomes larger and wider.

6.1.2.2 Organizational Culture

The culture of an organization is a function of the values, norms, and assumptions shared by members of the organization. "Culture is visible in ways such as how power is used or shared; the orientations towards risk or safety; whether mistakes are punished, hidden, or used to guide future learning; and how outsiders are viewed" (Oakes and Westcott 2001).

Thus, the culture of an organization can be seen in the ways in which the members of an organization make decisions, how they treat one another within the organization, and how they treat their contacts outside the organization. The culture of an organization is largely shaped by its leadership—through the values they believe in, their communication style, and how forcefully they influence the other members of the organization. Although a common culture is discernible among the members of a large organization, it is not unusual to observe islands of cultures within an organization that may be different, in degrees, from the overall culture of the larger organization. Such islands are products of individual local leaders or managers who influence their subordinates by the power of their personality and convictions.

When Dr. Deming said: "It is time to adopt a new religion in America" (Deming 1983, 19), he meant that many American organizations needed to change their culture. This strong statement reflects the frustration he felt with the prevailing attitude of people—and especially of management—towards doing quality work. He cited the example of a nametag presented to a friend at the guard gate of a large chemical company when they were visiting that company. Everything on the nametag was correct except for the name and date. Dr. Deming was riled by the indifference of organizations to the errors in the work performed. He was outraged by the "anything goes" attitude. He suggested that in the new culture, everyone should get into the habit of doing everything right the first time—that is the principal hallmark of the quality culture.

Furthermore, in a quality-oriented organization, everyone from the top executive to the operator on the floor is focused on knowing the customer's needs and how those needs should be satisfied. The senior executives are all committed to producing quality, to actively participating in the quality-related activities, and to encouraging their subordinates to do so. Decision making at all levels will be based on data, quantitative evidence on relevant issues, consultations, and consensus. Information about the operation of the organization is open to all, and people are given opportunities to advance their education and skills. Complaints and suggestions for improvement are welcome, and the suggestions are promptly followed and adequately addressed. People making contributions to quality improvement and customer satisfaction are properly rewarded. These are the marks that distinguish an organization that has transformed its culture into a quality culture. The characteristics of the quality culture are further elaborated under the following headings:

1. Quality leadership
2. Customer focus
3. Open communication
4. Empowerment
5. Education and training
6. Teamwork
7. Motivational methods
8. Data-based decision making

The last of these, data-based decision making, is not discussed in detail in this chapter because it is the topic of discussion elsewhere in this book.

6.1.3 Quality Leadership

Behind every quality organization is a quality leader. Robert Galvin of Motorola, David Kearns of Xerox, Jack Welch of General Electric, Don Peterson of Ford Motor Co., and John Young of Hewlett-Packard are all commonly cited examples of leaders who transformed their organizations into quality organizations. Many examples are also known, however, in which lack of commitment to quality from top-level leadership was the reason that organizations could not make any headway in their quality efforts.

"Good leaders come in all shapes, sizes, genders, ages, races, political persuasions, and national origins" (Goetsch and Davis 2003, 288). Authors on leadership (e.g., McLean and Weitzel 1991) agree that leadership qualities can be acquired if not inherited. What makes a leader a quality leader, someone who can transform an organization into a quality organization? The following provide a compilation of the characteristics of good leaders and what they do for accomplishing the quality goals for their organizations.

6.1.3.1 Characteristics of a Good Leader

Good leaders have *vision*: they think and act for the future, not just for the present. They can envision what their organization should be like in order to meet the needs of the customer. They see the big picture, fill out the details, draw the blueprints, and communicate them effectively throughout their organizations. They also anticipate problems and devise solutions to forestall them.

Leaders have good *intuition*, and they follow it. This is the faculty that helps in making decisions when there is not enough information to analyze and arrive at those decisions logically. Good intuition helps leaders to take risks in the face of uncertainty and arrive at good conclusions.

Good leaders are good *learners*. They know their strengths and weaknesses and are not afraid of learning from whatever source that provides them with information they do not already have. They have an appetite for new knowledge, and they know how to process that knowledge into useful skills that they can employ in their work. They are also good *listeners* and *communicators*, and they use their learning to educate their associates.

Good leaders are guided by good sets of *values*. Values are basic beliefs about how one should conduct him- or herself in relation to others, other organizations, or the community. Trust, respect for the individual, openness, honesty, reliability, and commitment to excellence are some of the values that good leaders cherish. They practice these values in their day-to-day work, and they make them the guiding principles for their organizations in dealings with employees, customers, and suppliers.

Good leaders are *democratic* in the sense they believe in consultations and building consensus before making decisions. They willingly delegate their power and responsibility. They are not afraid of empowering their subordinates to assume authority for decision making. They recruit the right people, who share in their vision, and create leadership at all levels of the organization. Empowered employees become motivated to participate in quality-improvement activities and contribute to customer satisfaction.

Good leaders are also *caring* people. They are interested in the welfare of their people, which breeds better morale among the employees; and happy employees make for happy customers.

Good leaders are *ambitious* people and set high expectations for their organizations. They are high achievers, and they expect their organizations to achieve as well. They are risk takers, and they set high goals for themselves and their organizations and then provide the resources necessary for achieving such "stretch goals." Goals such as the reduction of defects per unit of output in every operation by a 100-fold, in four years, were set in Motorola at the initiative of their leader, Robert Galvin, in 1984. This was a very ambitious goal at that time. Such goal setting was one of the factors that propelled Motorola to achieving excellent quality and financial results in the 1990s.

Good leaders show deep *commitment* to quality, and they actively participate in quality activities. They serve as role models by participating as members of quality-improvement teams and being personally involved in quality projects. They are accessible to all in the organization and they listen to customer calls. They become part of the review committees that review reports on quality projects and provide resources and encouragement to project teams. They are present during ceremonies to reward quality achievements and to recognize team accomplishments. They subject themselves to evaluation and implement any resulting suggestions for improvement. In other words, they "walk the talk," create an atmosphere for quality, and remain personally involved in the daily activities related to quality.

"The leader of the future will be persons who can lead and follow, be central and marginal, be hierarchically above and below, be individualistic and a team player, and above all be a perpetual learner" (Schein 1996).

6.1.4 Customer Focus

Customer focus refers to an organization remaining conscious of the needs of its customers and making and delivering products and services to meet those needs.

"Profit in a business comes from repeat customers, customers that boast of your product and service, and that bring friends with them" (Deming 1986). In other words, it is not enough if the customers are merely satisfied; they should be delighted with what they receive. Their needs and expectations should be proactively discovered, and product and services designed and delivered to them in a manner that makes the customers "loyal customers." They will then voluntarily come back for repeat business, and will tell their friends about the product or service with which they are satisfied.

Customer needs and expectations change with time; hence, they must be monitored on a continuous basis. Dr. Deming also said that it is necessary to understand how the customer uses and misuses the product. Rather than making assumptions about how a customer uses a product, it is better to find out how they actually use it. When customer practices and habits are better understood, it is easier to provide a product that meets their needs.

The value a customer places on a product depends on several aspects in addition to the product's quality in terms of possessing the desired attributes. The way in which customers are received and treated when they make their first enquiry is one of those aspects. Greeting customers promptly upon arrival and treating them with courtesy can add to the value that customers place on the product. The prompt and clear explanation and delivery of after-sales and warranty services serves to add further value. A competitive price and competitive lifetime cost (i.e., the cost of running and maintaining the product during its life) is yet another value-adding aspect. Customers place a value on the product that is linked to all these factors, and they develop their loyalty based on such value.

It is therefore necessary to assess how much value the customer places on a product relative to the value placed on a competitor's product. If, in the judgment of the customer, such value is not on a par with a competitor's, then steps must be taken to find opportunities for making improvements to the value and then implementing those improvements.

Dr. Deming conveyed the idea that product quality alone is not enough to create loyal customers through the illustration shown in Figure 6.1. He called the illustration the "three corners of quality." His point was that a producer should address all three corners of the triangle to provide value to the customer. The producer should facilitate proper customer use of the product by providing after-sales services in its use and maintenance. This has to be done in a proactive manner. Waiting for the customer to complain or even to leave, is leaving it too late. In this context he made the famous statement, "[Defective] goods come back, but not the customer" (Deming 1986).

Customer research and receiving customer input *before* the product or service is designed is the best approach. Customer research can be conducted using statistical sampling, which when properly carried out gives reliable results. The survey should elicit information on what the customer wants and does not like in a product, and what it is that they like in a competitor's product—and why.

Moreover, an organization with a customer focus is constantly in communication with its customers and is aligned with their needs at all levels of the organization. The organization delivers what it promises, resolves customer complaints fairly and promptly, and makes it easy for customers to do business with it. The organization invites customer suggestions and treats customer complaints as an opportunity for making things better. It continuously improves the processes and products to keep customers satisfied.

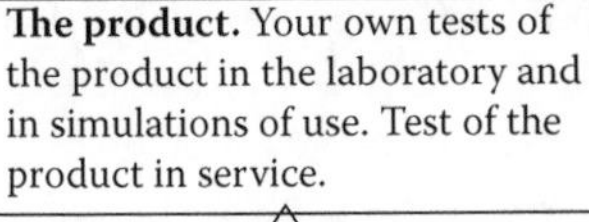

FIGURE 6.1
The three corners of quality. (Reproduced from Deming, W. E. *Out of the Crisis*. Cambridge, MA: MIT—Center for Advanced Engineering Study, 1986. With permission from MIT Press.)

The entire workforce of the organization remains committed to this philosophy, and to practicing it in daily activities.

Most customers who buy and use the products are external to the organization. There also are customers inside the organization who process the products received from the previous station and pass them on to the next customer down the line. Almost every station, operation, and department in an organization is a customer, and it also acts like a supplier. It is important that each supplier gives the same polite and considerate treatment to each customer in the same manner that the organization would treat an external customer. Thus, each person, operation, or department has a dual responsibility: treating well the next person, operator, or department in the line through meeting all their needs, and keeping in focus the needs and expectations of the ultimate end-user—the external customer of the organization. This is the type of customer focus that must be cultivated within an organization.

6.1.5 Open Communications

"If total quality is the engine, communication is the oil that keeps it running" (Goetsch and Davis 2003, 367). Communication refers to the exchange of information among everyone—leaders, managers, workers, suppliers, and customers—in an organization. Effective communication means that the information conveyed is received, understood, and acted upon. Open communication means that nothing is withheld from one person or group, and that the information is transmitted voluntarily and spontaneously. Open communication creates trust, and it motivates people to participate and be involved in the welfare of the organization. Quality organizations have a policy of "open-book management," in which everyone has access to data such as product cost, material and labor cost, the cost of poor quality, the cost of capital, market share, customer satisfaction level, customer complaints, profits, debts, and so on. Only when such information is shared among all employees will they feel responsible and be able to participate effectively in decision making and implementing the decisions that are made.

All employees in an organization should know the vision and mission of the organization. [The mission statement for an organization says what the purpose of the organization is and who the customers are, and the vision of an organization states where the organization wants to be at the end of a planning period.] In fact, they should participate in creating the vision and mission. When employees have been part of the effort to create these, they will become involved in working for and accomplishing them. At quality organizations, senior management meet all of the employees on a regular basis—either monthly or quarterly—and communicate information on organizational performance in accomplishing the vision and mission.

Open communication helps in removing fear among the employees and in furthering the two-way exchange of information and ideas. Unless fear of reprisal is removed, employees will not come forward to point out drawbacks in the system, failures in the production process, or defects in the final product. Many authorities, including Dr. Juran and Dr. Deming, have estimated that more than 80% of problems causing defects in processes can be traced to root causes that only management can correct (Garwood and Hallen 1999). Thus, if the employees are going to point out the existence of poor machinery, poor material, or poor process, it more than likely is the result of some management action or inaction. Unless the possibility of reprisal is removed, the existence of poor conditions in the processes will not be identified, and suggestions for improvement will not be forthcoming.

Quality organizations communicate to their employees not only by word, but by action. They do away with executive dining rooms, reserved parking, punch cards, and time clocks, and they give workers the same freedom to take time off as they do management. They offer the same benefits, such as sick leave, vacation time, and so on, to all levels of payroll. They acknowledge the creativity and collaboration of employees through company newsletters and bulletin boards. They never fail to recognize and reward outstanding contributions, and they publicize them through company channels. They obtain employee surveys, listen to employee needs and suggestions, and are prompt in responding to legitimate concerns.

6.1.6 Empowerment

Empowerment is allowing employees, at all ranks, to make decisions relating to quality, customer satisfaction, employee satisfaction, and waste reduction within an organization. This is the opposite of the situation in which workers do only what management tells them, and all decision making is the responsibility of the supervisors.

Empowered employees have a sense of ownership in the organization, and they feel committed to the goals, mission, and vision of the organization. They make decisions relative to product quality and customer satisfaction without waiting for the approval of supervisors. They will stop a consignment from leaving if they know that that consignment will not satisfy customer needs. They will stop a production line if they observe that the line is producing unacceptable goods. They are authorized to spend money if it will buy certain customer satisfaction. They select projects for quality improvement, and they successfully complete and implement these projects. They even recruit other employees into their teams and departments.

Empowerment does not occur instantly—especially if the organization has operated on a command-and-compliance mode in the past. Figure 6.2, which is taken from Garwood and Hallen (1999), shows the progression of an organization from the least participative to the most participative mode in a certain number of steps. Specifically, the figure depicts the progress made in worker participation at Eastman Chemical, Tennessee, where the system started at the "directive command" stage whereby orders had to be obeyed

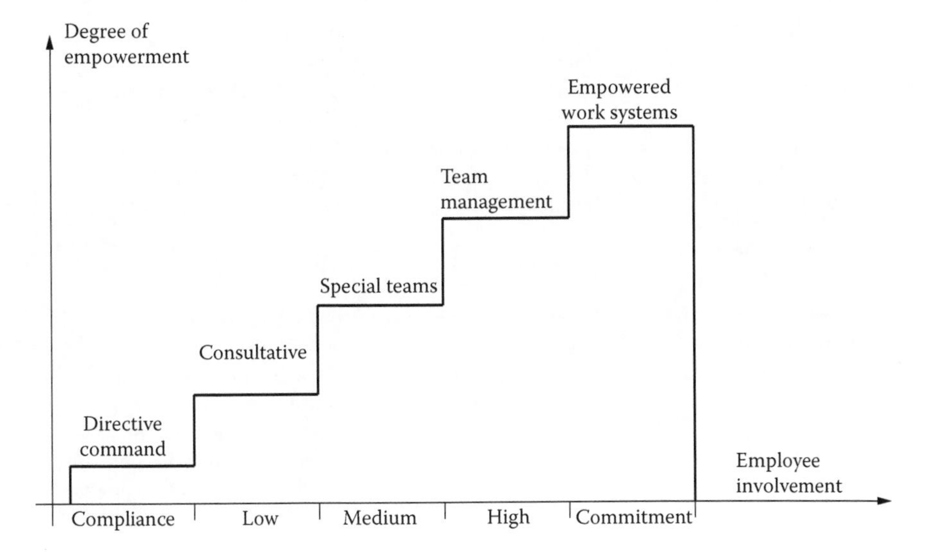

FIGURE 6.2
Relationship between (employee) empowerment and commitment. (Reprinted from Garwood, W. R., and G. L. Hallen, *Juran's Quality Handbook*, 5th ed., McGraw-Hill, New York, NY, 1999. pp. 15.1–15.29. With permission.)

as if in a military organization. At each level of change, worker participation in decision making increased, up to the point where workers could make decisions by themselves. As worker participation in management increased so did their commitment to organizational goals. It took several years for Eastman Chemical to evolve into the final empowered work system.

The ultimate form of employee empowerment occurs in self-managed teams, which are empowered teams that function as if they are departments or divisions without managers. They plan and schedule their own work, prepare budgets, buy material, control quality, improve processes, recruit their own associates, and meet the needs of customers. An organization made up of such empowered teams will become an empowered organization. Because people in empowered organizations are better motivated, have less bureaucracy, enjoy a better mix of expertise, and are willing to take initiatives, they successfully meet and exceed goals in customer satisfaction and business results. They are thus called "high-performance organizations." Table 6.1, which is reproduced from Garwood and Hallen (1999), shows how

TABLE 6.1

Traditional vs. Empowered Organizations

Element	Traditional Organization	Empowered Organization
Guidance	Follow rules/procedure	Actions based on principles
Employee focuses on	Satisfying the supervisor	Satisfying the customer and achieving the business objectives
Operator flexibility	One skill	Multiple skills
Participation	Limited	High
Empowerment	Follows instructions, asks permission	Takes initiative; a can-do attitude; discretionary effort
Employee viewed as	A pair of hands to do defined task	Human resource with head, heart, hands, and spirit
Leadership for work processes	Managers only	Shared by managers and operators
Management communication style	Paternalistic	Adult to adult
Responsibility for continuous improvement	Management	Shared by managers, staff, and operators
Work unit defined by	Function (such as manufacturing or sales)	The work process, which may be cross-functional
Administrative decisions are made by	Management	Shared responsibility of team members and management (if self-directed, the team may be responsible solely for certain administrative decision)
Quality control is the responsibility of	Laboratory	Team capability for process control

Source: Reprinted from Garwood, W. R., and G. L. Hallen, *Juran's Quality Handbook*, 5th ed., McGraw-Hill, New York, NY, 1999. pp. 15.1–15.29. With permission.

empowered organizations differ from traditional organizations in various elements of organizational features.

6.1.7 Education and Training

Education differs from training in that the objectives of education are to build knowledge of the fundamentals and enhance the ability to think, analyze, and communicate. The objectives of training, on the other hand, are to develop the skills needed for a specific job or function. For the sake of convenience, the term "training" in the following discussion includes both education and training.

6.1.7.1 Need for Training

Necessity of Basic Skills

A quality organization needs a quality workforce. The workforce must have capabilities in the basic skills of reading, writing, and arithmetic. In addition, they should be capable of logical thinking, and of analyzing and solving problems. In some industries, knowledge in specialized areas, such as physics and chemistry, is required. People should be able to communicate in the language used within the organization so that they can participate in various teams, where people with different levels of expertise, experience, and capabilities will be participating. Furthermore, they should be able to take leadership roles to plan, organize, and execute projects.

Global Competition

In the twenty-first century marketplace, when competition comes from almost every part of the world, the edge over competitors comes from the knowledge and creativity of the workforce, for which education is a prerequisite. In global competition, where advantages are gained through improvements in quality, customer satisfaction, and reduction in waste, the ability to learn and apply methods of data collection and analysis by the workforce becomes a necessity.

Continuous Change in Technology

Things no longer remain static in the workplace. What is new this year becomes obsolete the next year. This happens not only in the fast-changing communication and computer areas, but also throughout the industrial spectrum, though at a different pace. People must have basic—and sometimes advanced—education in the sciences to catch up with new innovations. Otherwise, people will be left behind and will become lost in the competitive world. Also, because of the quick response needed to meet the changing needs of the market, workers should be able to work in many different workplaces requiring many different skills. An educated workforce will develop multiple skills much more easily than others.

Diversity in the Workplace

The new workforce dynamics in the United States will draw an unusual mix of nontraditional workers, including homemakers returning to work after raising families, minorities, and immigrants, to the workplace. This generates the need for additional resources to train or retrain them in the basic skills—math, science, and communications—to function effectively in the modern workplace.

Need to Improve Continuously

Dr. Deming claimed that: "Competent men in every position, if they are doing their best, know all that there is to know about their work except how to improve it" (Deming 1986). That is a succinct statement of the fact that the ability to analyze and improve processes is a special skill and one that is not part of everyone's education. The new workforce needs to have the ability to continuously learn new ways to think, analyze, solve problems, and be responsive to the changing needs of the customer. Supervisors and managers need training on how to be the coaches and facilitators in the new environment.

We can see from the above discussion that training in the workplace is a necessity, and a major portion of the responsibility for training falls on employers.

6.1.7.2 Benefits from Training

The benefits that come from a trained workforce are many. Some of the obvious benefits from a trained workforce are:

- Increased productivity
- Fewer errors in output and improved quality
- Decreased turnover rate and reduced labor cost
- Improved safety and lower cost of insurance
- Multiskilled employees and better response to change
- Improved communication and better teamwork
- More satisfied employees and improved morale in the workplace

6.1.7.3 Planning for Training

Different organizations may have different needs for training. A manufacturer of heavy machinery may need a different training scheme from that needed by a software developer. Each has to assess their particular training needs, plan accordingly, and execute the necessary plan. There must be a strategic plan for training put together by the executive leadership of the organization, with support from the human resources function and the quality professionals. The plan should have a vision and mission statement, and strategies should be formulated with customer satisfaction in mind. The training plan is usually made by addressing the different segments of an organization using different syllabuses. The following is one such plan for training in quality methodology.

1. Quality awareness training (for all segments)
2. Executive training (for executive leaders)
3. Management training (for supervisors and managers)
4. Technical training:
    a. Basic level (for all technical personnel)
    b. Advanced level (for engineers and scientists)

The typical contents for each of the courses can be outlined as follows:

1. Quality awareness training:
    a. What is quality? What does it mean to us?
    b. How does quality affect customers?
    c. Who is the customer?
    d. What is a process?
    e. Variations in processes
    f. Introduction to the seven magnificent quality tools
    g. Need for quality planning, control, and improvement
    h. Quality management systems (ISO 9000, Six Sigma, or Baldrige criteria)
2. Executive training:
    a. Quality awareness (Module 1 above)
    b. Quality leadership
    c. Strategic planning for quality
    d. Customer satisfaction and loyalty
    e. Benchmarking
    f. Business process quality
    g. Supplier partnering
    h. Teamwork and empowerment
    i. Reward and recognition
3. Management training:
    a. This module will be a mixture of the executive training module shown above and the technical training module given below.
    b. Managers should understand the strategic place of quality in the business, and be able to use some of the technical tools to participate in improvement projects.

   c. They should be trained to be especially sensitive to the human side of the quality system: people's needs, their capabilities, the contributions they can make, and their requirements for education and empowerment.

4. Technical training (basic level):
   a. Brief version of executive training module
   b. Total quality system
   c. Quality costs
   d. Basic statistics and probability
   e. Confidence interval and hypothesis testing
   f. Regression analysis
   g. Correlation analysis
   h. Design of experiments (DOE)
   i. Gauge R&R study
5. Technical training (advanced level):
   a. Technical training (basic level)
   b. DOE
   c. Taguchi designs
   d. Tolerancing
   e. Reliability statistics
   f. Reliability engineering

6.1.7.4 Training Methodology

Training is best done in modules, each no more than two hours' in length. The two-hour segments seem to be the most suitable to fit into the work schedules of managers and executives. The two-hour modules also provide a good break period for the assimilation and absorption of ideas.

Training should be provided in a timely manner, because knowledge stays with people best when they can use what they learn. For example, a module on strategic planning for executives should be given when the executives are making or reviewing the plans. The module on design of experiments (DOE) should be offered to a group of engineers when they are working on an improvement project that needs a designed experiment.

Mentoring by an outside expert will always be helpful, especially if they have practical experience in the skills being taught. For example, an expert with experience in making strategic plans could be of great help to an executive who is making the strategic plans for the first time. Similarly, an expert in DOE who has successfully performed designed experiments could help an engineer in his or her first DOE project. It is always helpful if the trainer is from the peer group (engineers, scientists, or managers), or higher, of the people receiving the training. It is also helpful if the trainer has experience in the line of trade that the majority of the audience belongs to. For example, if a class is made up of chemical engineers, a chemical engineer, or a chemist who can speak the language of the audience and has expertise in the quality methods, would be an ideal choice. However, the choice of a trainer from the same trade as the audience is not an absolute necessity, because many experienced quality professionals can successfully translate their experience in one field to another in short order.

6.1.7.5 Finding Resources

Training programs will usually be budgeted by the training department, with the quality department providing technical consultation in the development of syllabuses and in choosing instructors. Volunteers from within an organization could serve as instructors if they are suitably qualified. We are reminded here of the observations Dr. Deming made regarding the qualification of instructors to teach statistical methods:

> American Management have resorted to mass assemblies for crash courses in statistical methods, employing hacks for teachers, being unable to discriminate between competence and ignorance … No one should teach the theory and use of control charts without knowledge of statistical theory through at least the master's level, supplemented by experience under a master. I make this statement on the basis of experience, seeing every day the devastating effects of incompetent teaching and faulty application. (Deming 1986, 131)

Therefore, it is important to find qualified instructors for training purposes.

Training comes with a price tag. Motorola University, which provides training for all Motorola employees, had a budget of $120 million per year (or about 10% of its payroll) in the year 2000. Intel University spends 5.7% of its total payroll for training each year. General Motor's Saturn division expects all employees to spend about 5% of their time in education and training each year, and it pays a 5% bonus to those who complete the courses. IBM requires all employees to complete at least 40 hours of training each year. When Nissan opened its plants in Smyrna, Tennessee, it spent $63 million in training about 2000 employees at an average cost of $30,000 per employee (information from Garwood and Hallen 1999, and from Goetsch and Davis 2003).

Organizations follow different approaches to maximize return on their training dollars. Using partnership arrangements with local community colleges and other higher education institutions has been one approach. The outside educational institutions are most helpful in imparting basic skills, such as math, statistical thinking, and communications. Trade-specific or work-specific instructions are best prepared and delivered by in-house professionals. The outside institutions generally have well-qualified professional educators with many support facilities, such as libraries and standard course curricula. The in-house experts may have intimate knowledge of the work areas and training needs, but they may lack the qualification, and thus the credibility, that university professors may possess. Furthermore, when an organization spends money to bring in outside instructors, employees will understand that the organization is serious about their training efforts, and so may become serious about learning themselves.

6.1.7.6 Evaluating Training Effectiveness

The test of training is in the learning that the participants acquire. This learning usually becomes apparent in the results of the projects they complete or in the output of the teams in which they participate. Improvements in product and service quality, customer satisfaction, reduction in waste, and increases in productivity are some of the outcomes of effective training. When people have learned improvement tools, they become sensitized to opportunities where improvements can be made. When people are trained, there will be an increased number of suggestions for improvement projects—and an increased number of successfully completed projects. All these can be counted. When employees acquire improved communication and interpersonal skills, there will be a general improvement of the social atmosphere in team meetings and the workplace in general. There will be a decreased number of personal conflicts and turf fights, contributing further to the exchange of information and sharing of ideas for the common goal of the organization. Learning and knowledge provide satisfaction and fulfillment to individuals, and the general morale in the organization will improve, which can be measured by decreased absenteeism, a reduction in complaints, and an increase in volunteers for teams.

At the individual course levels, feedback should be obtained from the participants on the relevancy of the course, the relevancy of the topics, the organization of the material, and the expertise and ability of the instructor to communicate. These will help in making changes to improve the effectiveness of the instruction and the instructors.

6.1.8 Teamwork

Teamwork is an essential component of the effort to produce quality products and services and satisfy customers. Satisfying the customer requires the contributions of many people working in many parts of an organization, who have to participate in teams to accomplish the quality objectives of the organization. The discussion below is about how teams are formed, trained, and are helped to perform at their best, especially relating to quality improvement and customer satisfaction. A good source of reference on teams is *The Team Handbook* by P. R. Scholtes (Scholtes 1988).

"A team is not just any group of people; it is a group with some special characteristics" (Scholtes 1988). The characteristics of people who make up a team include:

- They have a common mission.
- They have agreed on a set of ground rules for working together.
- There is a fair distribution of responsibility and authority.
- They trust and are willing to help one another.
- They have knowledge and expertise in specialized areas and are willing to share this.
- They are willing to learn from one another's education and experience.
- They are willing to subordinate their individual interests to achieve the team's success.

Such teams produce results that are greater than the sum of the individual contributions. The synergy from working together and learning from one another, as well as the enthusiasm arising from mutual encouragement, help the team to achieve such results.

6.1.8.1 Team Building

"It is a folly to assume that a group of people assigned to a task will simply find a way of working cooperatively" (Scholtes 1988). In other words, a certain effort is needed to create good teams, and there is a process for team building. The process includes the following steps.

6.1.8.2 Selecting Team Members

Members who have the most potential for contributing to the mission of a team based on their expertise, experience, and attitude to teamwork should be included. There must be diversity in all respects—education, salary grade, expertise (e.g., engineering, marketing), gender, and race—to take maximum advantage of the contribution that a diverse group can make. The team size should be between six and twelve, with the ideal being eight or nine. The team should choose a leader, and a secretary is elected by the team or appointed by the leader.

6.1.8.3 Defining the Team Mission

A mission statement, written by the team and describing their purpose, should be communicated to all in the organization. It should be broad enough to include all that is to be accomplished, with just enough details to communicate the scope. It should also be simple and understandable to all.

6.1.8.4 Taking Stock of the Team's Strength

A team's strength should be assessed initially, and at regular intervals, based on team members' own perceptions. Strength is assessed in the following areas:

- *Direction of the team*

  (If everyone in the team understands the mission, goals, and time schedule.)
- *Adequacy of expertise and resources*

  (If the team has a good knowledge of the processes they are dealing with, if enough strength exists in their problem-solving skills, and if they have the authority to spend time and money to meet their goals.)
- *Personal characteristics of the members*

  (If all members work as a team with honesty, trust, responsibility, and enthusiasm.)
- *Accountability*

  (If team members understand their responsibilities, how the progress of the team will be measured, and how corrections are to be applied.)

6.1.8.5 Building the Team

Team-building activities must be planned and implemented based on the results of the assessment made in the previous step:

- If the weakness is in terms of lack of direction or how it is communicated, then the team leader should reassess the mission and the goals in consultation with the team and communicate them in clear terms.
- If the weakness is in expertise and other resources, then the team should embark on an education and training program. In general, one of the team members, who is well versed in the methods of problem solving relative to quality—both on the statistical and the managerial side—will be made the quality specialist. This person can provide the necessary training to other members. An outside consultant can be engaged as well. A brief list of basic tools for quality improvement is given later in this chapter, and the list is elaborated in Chapter 8. If there is a shortage of expertise or other resources, the team should approach upper management and secure additional resources
- If the weakness is in personal characteristics or human relations, then the help of a human relations person (either from within or outside

the organization) should be sought. A discussion on the desirable characteristics of team members is given later in this section.

- If the weakness is in the area of accountability, then the leader should identify the goals, divide the responsibilities, and assign those individual responsibilities clearly. Writing minutes of the meetings and making assignments on paper will help avoid weaknesses in the accountability area.

6.1.8.6 Basic Training for Quality Teams

All team members should be familiar with the general problem-solving process and the basic tools for quality improvement. The problem-solving process includes the following steps:

- Defining the problem and the objectives for the project
- Collecting data, and analyzing them for root causes
- Devising solutions to solve the problem
- Selecting the best combination of solutions based on the objectives
- Implementing the solution(s)
- Obtaining feedback and debugging the solutions

There are several versions of this problem-solving methodology, each of which combines or splits some of the above steps. Dr. Deming's "plan-do-check-act" (PDCA) cycle and Dr. Juran's "breakthrough sequence" are examples of problem-solving approaches. These approaches are discussed in detail in Chapter 8. The basic tools for quality improvement are:

1. Flowcharting of processes
2. Pareto analysis
3. Cause-and-effect diagram
4. Histogram
5. Control charts
6. Check sheets
7. Scatter plots

These are known as the "magnificent seven tools," so designated by Dr. Kaoru Ishikawa, the Japanese professor who is recognized as the father of the quality revolution in Japan. These tools are described in detail, using examples, elsewhere in other chapters of this text. Members of a quality-improvement team should be familiar with these tools so that communication among team members becomes easy.

6.1.8.7 Desirable Characteristics among Team Members

Trust

Trust is a very important characteristic among team members because it can transform a group of people into a team. People will not share information openly and assume responsibilities unless they can trust their colleagues to keep their word and pull their weight. Trust comes from honesty and truthfulness, which are part of the culture or values established in a team. The culture established for a team by the team leader and other leading members enable others to follow suit.

Selflessness

Selflessness is the characteristic in individuals that makes them subordinate their individual interests to the interest of the team and the organization. In societies where individuality and personal success are encouraged and celebrated, cultivating this characteristic in team members may be challenging. However, unless members are willing to put their self-interest aside and work for the good of the team, success cannot be guaranteed. The team leader and the senior team members should take the responsibility of reining in those members who may sometimes forget the common interest and start pursuing individual interests.

Responsibility

Responsibility is the characteristic that distinguishes members who complete the tasks assigned to them on time with 100% satisfaction from those who offer excuses for not completing their assigned tasks. Responsible members will also offer a helping hand to those in need rather than blame others for their faults.

Enthusiasm

Enthusiastic individuals are the cheerleaders in a team. They are usually high-energy people with high productivity. They help the team to overcome roadblocks, and they get the team going when the going gets tough. A few such members are always needed for the team to be successful.

Initiative

People with initiative do not wait for tasks to be assigned to them; they offer to take up tasks where they can make contributions. They provide the starting momentum for the team to get moving.

Resourcefulness

Resourceful people are the ones with the ability to find creative ways of resolving difficult issues. They find ways to make the best use of the available resources. They find ways to get to the destination when others feel they have reached a dead end. They usually have a high intellectual capability, with an ability to think on their feet.

Tolerance

There will be differences among team members based on their educational level, gender, age, or race. There will be occasions when people will think or act differently because of the differences arising from cultural, intellectual, or educational differences. Unless the team members respect these differences and tolerate diversity, a team cannot progress toward its goal. People from diverse backgrounds may bring diverse strengths to the team. Team members should learn to take advantage of the positives for the sake of the team's success.

Perseverance

Patience and perseverance are the means of success in any endeavor. There will be team members who are bright, creative, and enthusiastic, but who may get easily disappointed and depressed when the first failure occurs. This is when people with patience and perseverance are needed to keep the team on target and working until the end is reached.

6.1.8.8 Why a Team?

We can easily see from the above discussion of the characteristics needed in team members why a team can succeed where individuals cannot. It is hard to find in one person all of the qualities that team members can bring to bear collectively on a team's work. When several individuals bring qualities that complement one another, and when the team works together as one entity with a common goal, then there are very few problems that cannot be overcome, and very few objectives that cannot be accomplished.

6.1.8.9 Ground Rules for Running a Team Meeting

The following set of rules for conducting team meetings is summarized from *The Team Handbook* (Scholtes 1988), in which further elaboration of these rules can be found:

- Use agendas that have been prepared and approved in the previous meeting, and send them to the participants ahead of time. Identify the people responsible for each item, and the time to be allowed for each item indicated.
- Use a facilitator who will keep the focus on the topics on the agenda; intervene when discussions lack focus, a member dominates, or someone is overlooked; and bring discussions to a close.
- Take minutes to record decisions made, responsibilities assigned, and the agenda agreed on for the next meeting. The minutes will

be recorded by the secretary and distributed among team members within a reasonable time after each meeting.

- Evaluate the meeting to obtain feedback from members on how to improve the team dynamics and make the meetings more productive.
- Seek consensus so that no one in the team has any serious objection to the decisions made. Consensus decisions are better than majority decisions, because majority decisions produce winners and losers, which may not be conducive to the growth of team spirit.
- Avoid interruptions. Scholtes (1988) recommends use of the "100-mile rule," according to which no one in the meeting will be called unless the matter is so important that the disruption would occur even if the meeting was held 100 miles away from the workplace.
- Keep good records so that it is possible to refer to past decisions or understand the reasons behind any decisions made. Data collected, analysis made, and conclusions drawn must be available in the record for ready reference. Minutes of team meetings will provide a good trace of how the team has progressed toward the final goal.
- Keep regular meetings. Activities will fall into a routine once the initial tasks relating to team building, project definition, data collection, and data analysis have been defined and assigned to team members. To keep the momentum of the project going and keep the goals and time frame in focus, the team should still meet on a regular basis even if no new decisions are to be made. A review of the progress made and a clarification of issues that may arise, even if they are minor, can be accomplished in these meetings.

6.1.8.10 Making the Teams Work

A few suggestions are available from experts on how to avoid problems that may develop during the working of a team and, if problems do arise, how to minimize their effects and make progress toward team goals.

Making Team Members Know One Another

Introduce members of a team who have not had prior working relationships through informal introductory chats about their jobs, families, hobbies, and so on. Begin each meeting, especially the early ones, with warm-up exercises in which the team members, through informal small talk, warm to each other before important agenda items are taken up. Help team members visit one another's workplaces to make them all familiar with the working conditions of the other members, and to learn about the details of the projects that the team is working on.

Resolving Conflicts Promptly

Well-defined directions for the team and ground rules for the conduct of business, communicated clearly, will prevent serious conflicts. Disagreements among members are not bad in themselves; in fact, such disagreements may even be healthy for the team's work. However, any overbearing or dominant behavior from members must be discouraged and curbed by the leader and the joint action of the team.

Providing frequent opportunities for members to express their concerns will help to resolve serious conflicts, because it will enable the early resolution of differences. Training team members in managing dissent and expressing it constructively will help avoid flare-ups.

Setting an Example by the Organization

The organization as a whole should set a good example as a team player by working together with the suppliers and customers, and by being a fair, sensitive, and responsible partner in the community.

Rewarding Good Teams

Monetary as well as nonmonetary rewards can be utilized. A wage system that rewards only individual performance may not promote teamwork. A compensation system based on three components—an individual base compensation, an individual incentive compensation, and a team-based incentive compensation—is recommended (Goetsch and Davis 2003).

Nonmonetary awards could be in the form of recognition in company newspapers, travel to conferences, inclusion in important high-powered teams, weekend trips, or tickets to games or other entertainments.

6.1.8.11 Different Types of Teams

Teams acquire their names mainly based on the purpose for which they are constituted. Sometimes, the name reflects the constituents making up the team.

Process Improvement Teams

Process improvement teams are the most relevant type for quality in an organization. They are created in order to address quality improvement and customer satisfaction.

Cross-Functional Teams

Cross-functional teams may be process improvement teams, product design teams, or teams for the improvement of safety in the workplace. These teams are created when there is a need for receiving input from various functions of the organization, such as engineering, manufacturing, marketing, safety and security, or customer relations. The team members are drawn from the functions that are capable of contributing to the team goals.

Self-Managed Teams

Self-managed teams consist of all members of a full department or division of an organization without a department or division head. They manage themselves in that as a team, they have full responsibility for budgeting, staffing, procuring, customer and supplier relationships, product development, and so on, for a product or service that they are supposed to design or produce. Such teams have been employed in manufacturing companies and service companies, and they are credited with achieving improvements in quality and productivity. The synergism of the team, its flexibility and speed in decision making, and the satisfaction that comes from its members being empowered to make their own decisions are all reasons for the success of self-managed teams.

6.1.8.12 Quality Circles

Quality circles are teams of employees mainly working in one area and reporting to the same supervisor, who have come together voluntarily to solve problems relating to the quality of a product or service created in that department. The problems could also be related to safety, environment, or other conditions in the work area that are of concern to the team. The concept of the quality circle originated in Japan as a way of utilizing the knowledge and expertise of workers in solving quality problems, to complement the use of statistical and other technical methods. The Japanese workforce enthusiastically embraced the opportunity to help employers by volunteering their time and knowledge to solve problems relating to quality. The volunteer teams, originally known as "quality control circles" (QCCs), were given training in basic techniques of quality improvement and problem solving in order to be more effective in their activities.

The quality control circles had enormous success in contributing to overall quality improvement in Japanese industry in the 1960s and 1970s. When the concept was transplanted into U.S. organizations, the teams came to be called "quality circles." Although some success was reported with the effectiveness of quality circles in the United States, the overall assessment was that they have had only "mixed" results (Juran and Gryna 1993). However, according to Gryna (1981), there are many beneficial effects from quality circles that can positively contribute to an organization's quality goals. These benefits include helping individuals to gain additional technical and personal skills and improve their self-image; helping workers to better understand the supervisor's and the company's positions when problems cannot be solved immediately; helping to highlight problems relating to the workplace and to resolve them before they become deterrents to worker productivity; and helping to improve the understanding between management and workers and promote the growth of mutual respect.

6.1.9 Motivation Methods

Books on business management expound on theories of how people can be motivated to perform at their best so that they contribute the most toward the goals of an organization. According to one theory, some people enjoy working with others, so their motivation is said to come from *affiliation*. Others are motivated by *achievement* in the sense that they work to accomplish something and to be recognized for it. Still others are motivated by doing good for others, such as a group or a community, and are said to be driven by *altruism*. Then, of course, there are people who are thirsty for "power" and are thus motivated by gaining and wielding this acquired power. Knowledge of what motivates people helps a manager to provide the right motivation to his or her subordinates to get the best out of them.

The famous Maslow hierarchy of needs postulates that people seek satisfaction of their needs in a hierarchical manner (Goetsch and Davis 2003). They first seek to satisfy the basic physiological needs, such as food, clothing, shelter, and rest, and when these needs are satisfied, they seek safety or security. They then want to belong to and be accepted by a group, such as a family, workplace, church, or the community in which they live. After this, they want to earn esteem and be well-regarded by their contemporaries. Ultimately, when all of their other desires are satisfied, they are actuated by the desire to do their best. Thus, if a manager can understand the stage of satisfaction that a person is at, then suitable motivational incentives can be provided to get the best out of that person.

Other theories are based on how people respond to rewards. The *equity theory* says that level of performance depends on the perception of people about how equitably they are treated in being rewarded for their work. The *expectancy theory* tells us that people are motivated by what they expect to receive from the work they perform. According to *reinforcement theory*, people are motivated by their experience of what they received in the past for their different levels of performance—in other words, whether they got a pat on the back or a slap on the hand.

Dr. Deming's theory of motivation is a simple one: if people are allowed to do their job without hindrances from a lack of proper tools or materials, and without being pressured by a supervisor eager to meet daily schedules with no regard to the quality of the product, then these people will do a good job. He believed in the theory that people will want to do a good job if they are allowed to do so. This author has seen this to be true almost everywhere he has had opportunity to work. Almost everyone, whether a supervisor, union worker, nonunion worker, store clerk, maintenance mechanic, or job scheduler, wants to do his or her work in a manner that will satisfy the person who receives that work. With few exceptions, no one likes to turn out bad work if he or she can avoid it.

Incidentally, this author wants to add this piece of advice to future quality analysts. If you establish your credentials as to your ability to help people to do their job better, you will receive cooperation and advice that will help you discover new and better methods for doing the job, and for improving quality and productivity. Some people may be eager to be convinced about your credentials and sincerity, and may offer their help instantly; others will take time to evaluate the situation and wait to join the pursuit of quality. A little bit of perseverance and good communication will almost always enlist the help of people. In short, most workers want to do quality work and be part of the quality effort—if they are presented with the right opportunity.

Maybe it is a fact that doing quality work, or being part of it, provides an inner satisfaction to people. The simple recognition of good work provides encouragement for people to continue this good work. We should, of course, recognize that a totally conducive environment, in which people are treated well, reasonably compensated for their work, and provided with a clean and safe workplace, all contribute to motivating people to do their best.

6.1.10 Principles of Management

The process of management consists of planning, organizing, controlling, and leading the activities of an organization toward achieving its mission. Planning involves determining what the vision and mission of the organization should be in the near and long terms, as well as charting out a scheme of activities and the timeline in which to achieve the planned goals. Organizing involves gathering the resources, capital, people, and facilities, and then allocating them to the different activities to accomplish the goals of the organization. Controlling involves measuring the performance of the organization at the macro as well as the micro process levels, comparing it with the established goals, and then redirecting or reallocating resources to achieve the planned performance within the planned time frame. Leading is the function that formulates the vision and mission of the organization, and takes the guiding role in planning, organizing, and controlling.

The planning function for quality over a long period of time, in the 5–10 year time frame, is called "strategic planning" for quality, and is a principal function of management that can contribute to higher levels of quality accomplishment.

6.2 Strategic Planning for Quality

6.2.1 History of Planning

Godfrey (1999) traces the historical progression of the quality movement in the United States using a figure similar to that shown in Figure 6.3. In that

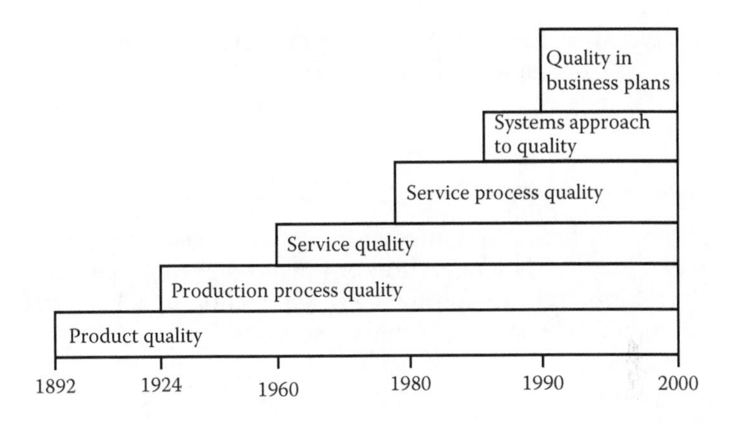

FIGURE 6.3
A perspective on the growth of quality methods in the United States. (Adapted from Godfrey, A. B., *Juran's Quality Handbook*, 5th ed., McGraw-Hill, New York, NY, 1999. pp 14.1–14.35. With permission.)

figure, he marks the year that quality activities began as being 1892, when inspection procedures for telephone equipment in the erstwhile Bell System were created. This practice of inspecting the product at the end of a production line in order to make certain that it was made according to the specifications laid down by designers, still exists today. The idea of monitoring process parameters and controlling them to prevent the production of defective products was introduced by Dr. Walter Shewhart around 1924. The concept of quality assurance through prevention methods was born then and is still practiced today. Around the 1960s, the importance of customer service quality to supplement product quality was recognized, and this idea continues to occupy a place of importance in quality organizations.

Starting in the early 1980s, further emphasis was placed on the quality of services, which was enhanced by identifying service processes and then monitoring them using control methods. This was the time when the luxury carmakers (Acura, Lexus, and Infinity) paid special attention to after-sales services to attract and retain luxury car buyers. Service quality received widespread attention throughout the economic spectrum—from manufacturing to hospitals to hotels and the entertainment industry in the 1980s. The systems approach to quality gained momentum with the introduction of the ISO 9000 standards and the Baldrige Award in 1987. The 1990s saw the introduction of quality at the strategic level, treating it as a parameter for business planning along with finance and marketing measures.

In the words of Godfrey: "In the past few years we have observed many companies starting to integrate quality management into their business planning cycles. This integration of quality goals with the financial goals has been a major thrust of the leading companies" (Godfrey 1999). Such an integration of quality goals with financial and marketing goals at the planning stage—and planning for strategies that would accomplish those goals—is the objective of strategic planning for quality. We will see in some detail below how planning is done and how it is implemented or deployed.

6.2.2 Making the Strategic Plan

Strategic planning for an organization involves developing a set of documented plans for a fairly long period of time (5–10 years). The plan should include the vision and mission for the organization and key strategies to accomplish them. It should also include specific tactics or projects to support the key strategies.

A *mission statement* for an organization is the statement of the purpose for which the organization exists. This is typically a short paragraph that answers the following questions: Why do they exist? What products do they produce? Who are their customers?

A *vision statement* delineates where the organization wants to be at the end of the planning horizon. This is usually a short, pithy statement capturing the aspirations of the organization. The vision will be ambitious yet attainable.

The mission and vision statements are prepared by the executive management, with inputs from all ranks, and are published throughout the organization, including customers and supply partners. Together, they provide the guidelines for decision making regarding the day-to-day issues in the organization.

The executive management will also create a set of values, or guiding principles, to provide the framework within which the organization will pursue its mission and vision. The set of values will typically state what the leadership of the organization believes their obligations are to all the stakeholders of the organization, including investors, customers, employees, and the general public. Statements on their ethical standards, commitment to diversity, responsibility to the environment, and relationship with the community will all be part of the values.

From out of the mission and vision statements are generated key strategies, which are the major action plans needed to accomplish the vision. The key strategies are usually broad in scope, small in number (four to six), and enable the organization to go from where they are to where they want to be.

To understand where it currently is, the organization should first make an extensive study of its present status by collecting data on several organizational parameters, as given below, to understand its internal strengths and weaknesses. Similarly, a study must be conducted to learn about external opportunities and threats. This is known as the "gap analysis," because it explores the gap between where the organization wants to be and where it currently is. The relevant internal parameters to be studied are (Goetsch and Davis 2003):

- Financial strength
- Cost of quality

- Capabilities of processes
- Capabilities of the management team
- Capabilities of the workforce
- Technology leadership
- Flexibility/agility of processes (i.e., the ability to adapt rapidly to changes in customer demands)
- Status of facilities

The relevant external parameters to be studied are:

- Customer satisfaction
- Strength of the distribution system
- Availability of new customers
- Strength of competitors
- Opportunity for diversification
- New technologies in the horizon

The key strategies for the important organizational parameters are selected by the executive management in close consultation with operational managers, based on the gaps discovered between current status and the envisioned goals. Each key strategy should have certain specific, measurable strategic goals. For example, if one of the key strategies is to reduce the cost of poor quality, then a strategic goal would be to reduce the warranty charges by 50% for each year during the planning period. Another example of a strategic goal would be to reduce internal failures by 75% within the planning period.

The key strategies and goals will next be subdivided into subgoals, and short-term tactics or projects will be identified to accomplish the subgoals. For example, if one of the key strategies is to reduce the internal failure costs by 75%, then one of the tactics would be to implement statistical process control on each key characteristic of every production process. The projects will be assigned to individual functional departments or cross-functional teams. These will be short-term (8–12 months) projects with clear goals to be accomplished within the specified time schedule. Figure 6.4 shows how the vision is broken down into key strategies that lead to the identification of subgoals and projects.

6.2.3 Strategic Plan Deployment

This is the implementation phase of the strategic plan, during which the vision is translated into action. The plan should first be distributed to all who would participate in it, and their feedback should be obtained. The plan should then be finalized after suitable modifications have been incorporated

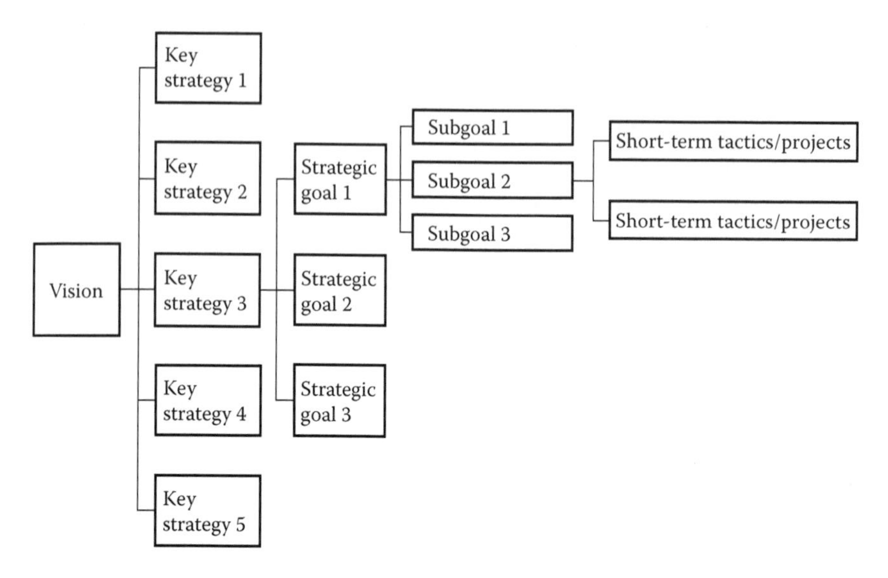

FIGURE 6.4
Breakdown of the vision into key strategies, strategic goals, subgoals, and projects.

to answer concerns and respond to suggestions. The plan is then divided into various tactics, or projects, and these are then assigned to various departments and teams. Project charters should be written and given to individuals or teams along with a time frame for completion. The resources needed should be identified with the help of the departments and teams, and those resources should be made available by the executive management.

Performance measures should be created based on the goals established for each project. Progress of the projects should be monitored and any correction needed for the goals and tactics should be included.

Cascella (2002) points out the three common pitfalls encountered during implementation that lead to poor performance of strategic plans: 1) lack of strategic alignment at all levels; 2) misallocation of resources; and 3) inadequate operational measures to monitor the progress of implementation. He suggests the following remedies to avoid those pitfalls.

The planned strategy must be linked to activities to be performed at the departmental or group level and communicated to them, so that each group knows what is expected of them for the successful implementation of the plan. Next, it is also important to identify the "core" processes in the organization that are critical to achieving the strategic goals and allocate resources to improve them. Otherwise, when the allocation of resources is made randomly, the processes that are strategically important to the business and their customers do not receive the attention they need, and strategic goals are therefore not achieved. Then, measurements need to be put in place to determine the improvements in capabilities of the core processes as the plan implementation progresses. Such measurements would lead to the identification of underperforming processes and to the reallocation of available resources to necessary areas. Another important aspect during implementation is establishing accountability.

When the strategy is linked to individual departmental activities, and is communicated to the departmental leaders to enroll their participation, a certain level of accountability is already established. When they are involved in developing measurement tools and making measurements of the progress made, the process owners and those who work for them derive a sense of participation and responsibility for achieving the goals of the plans. Accountability can also be driven through the performance evaluation of teams and their members, and by linking their financial and nonfinancial incentives to their performance. Yet another way, according to Cascella, is for the business leaders to act as role models themselves. By participating in process improvements and making decisions based on data, they can create a healthy climate for continuous improvement and create confidence in the improvement methods. Thus, they can contribute to achieving planned goals.

Strategic planning for quality has great potential for delivering improved product quality, reduced waste, and improved financial performance for an organization. Standards for quality system management such as the ISO 9000 standards and Balridge Award criteria require that organizations practice strategic planning for quality in order to achieve improvement in quality and excellence in business results.

6.3 Exercise

6.3.1 Practice Problems

6.1 Write the meaning of each of the following terms, in your own words, in a paragraph not exceeding five sentences:

- **a.** Quality leadership
- **b.** Customer focus
- **c.** Open communication
- **d.** Participative management
- **e.** Training and empowerment
- **f.** Teamwork
- **g.** Strategic planning
- **h.** Mission, vision, and values
- **i.** Strategies and tactics
- **j.** Strategic plan deployment

6.3.2 Mini-Project

Mini-Project 6.1

Prepare an essay on any one of the topics, such as empowerment, motivation, strategic planning, and so on, which are covered in this chapter. There are many more references that relate to these topics that are not reviewed here. Limit the essay to between 15 and 20 typed pages.

References

Cascella, V. November 2002. "Effective Strategic Planning." *Quality Progress* 35 (11): 62–67. Milwaukee, WI: American Society for Quality.

Deming, W. E. 1983. *Quality, Productivity, and Competitive Position*. Cambridge, MA: MIT—Center for Advanced Engineering Study.

Deming, W. E. 1986. *Out of the Crisis*. Cambridge, MA: MIT—Center for Advanced Engineering Study.

Evans, R., and W. M. Lindsay. 1996. *The Management and Control of Quality*. 4th ed. St. Paul, MN: South Western Publishing Co.

Garwood, W. R., and G. L. Hallen. 1999. "Human Resources and Quality." In *Juran's Quality Handbook*. 5th ed. Co-edited by J. M. Juran and A. B. Godfrey. New York, NY: McGraw-Hill. 15.1–15.29.

Godfrey, A. B. 1999. "Total Quality Management." In *Juran's Quality Handbook*. 5th ed. Co-edited by J. M. Juran and A. B. Godfrey. New York, NY: McGraw-Hill. 14.1–14.35.

Goetsch, D. L., and S. B. Davis. 2003. *Quality Management*. 4th ed. Upper Saddle River, NJ: Prentice Hall.

Gryna, F. M. 1981. *Quality Circles*. New York, NY: Amacom.

Juran, J. M., and A. B. Godfrey. 1999. *Juran's Quality Handbook*. 5th ed. New York, NY: McGraw-Hill.

Juran, J. M., and F. M. Gryna. 1993. *Quality Planning and Analysis*. 3rd ed. New York, NY: McGraw-Hill.

McLean, J. W., and W. Weitzel. 1991. *Leadership—Magic, Myth, or Method?* New York, NY: American Management Association.

Oakes, D., and R. T. Westcott, eds. 2001. *The Certified Quality Manager Handbook*. 2nd ed. Milwaukee, WI: ASQ Quality Press.

Schein, E. H. 1996. "Leadership and Organizational Culture." In *The Leader of the Future*. Edited by F. Hassebein, M. Goldsmith, and R. Beckherd. San Francisco, CA: Jossey-Bass. 59–69.

Scholtes, P. R. 1988. *The Team Handbook*. Madison, WI: Joiner Associates, Inc.

7

Quality in Procurement

7.1 Importance of Quality in Supplies

In Chapter 3, we discussed the quality methods that are used when products are designed to meet the needs of the customer, and the processes designed to make products according to the design. In Chapters 4 and 5, we discussed the methods employed during production to prevent the manufacture of defective products. This chapter concerns itself with the methods of assuring quality in materials, parts, and subassemblies purchased from outside vendors.

The need for procuring parts and materials of required quality cannot be overemphasized—especially in the context of modern productive organizations, which procure ever larger proportions of the assemblies they build from outside vendors. Statements such as "you cannot make good product from bad material," and "you are as strong as your weakest supplier," which we often hear in production shops, only reinforce the fact that the parts and materials that come into a production facility should be defect free. The modern approach to inventory reduction, which employs a just-in-time production philosophy, makes it even more important to receive defect-free supplies, because in the just-in-time environment there is no cushion in inventory to make up for the part or material found to be defective during assembly.

Several approaches have been adopted by organizations to assure quality in incoming supplies; some involve management methods and some involve statistical tools. We will discuss some of these approaches in this chapter. These include:

- Establishing a good supplier relationship
- Choosing and certifying suppliers
- Specifying the supplies completely
- Auditing the supplier
- Supply chain optimization
- Statistical sampling plans for acceptance

7.2 Establishing a Good Supplier Relationship

7.2.1 Essentials of a Good Supplier Relationship

The Customer-Supplier Division of the American Society for Quality (ASQ), a group of professionals in the procuring business, suggest the following as essentials of a good supplier relationship in Chapter 1 of their *Supplier Management Handbook* (ASQ 2004):

- *Personal behavior*: Professional, personal behavior of parties, with mutual respect for each other.
- *Objectivity*: A moral commitment, by both parties, beyond the legal contract requirements, to attain the goal of quality for the end product.
- *Product definition*: A clear, unambiguous, and complete definition of the product requirements furnished by the customer in writing, with a willingness to provide further clarification if and when needed.
- *Mutual understanding*: Understanding of each other's needs from direct, open communication between the quality functions of the parties, to avoid confusion.
- *Quality evaluation*: Fair, objective evaluation of the quality of supplied goods by the customer.
- *Product quality*: Honest effort by the supplier to provide materials according to the needs of the customer, including disclosure of any weaknesses.
- *Corrective action*: Good faith effort by the supplier in making corrective action when supplies are found to be deficient.
- *Technical aid*: Willingness on the part of the customer to share technical expertise with the supplier whenever such an exchange is needed.
- *Integrity*: Supplier's willingness to provide facilities and services—as needed by the customer—to verify product quality; and customer using those facilities and services only to the extent agreed to in the contract.
- *Rewards*: Customer's use of only qualified suppliers, and offer of reward and encouragement for good performance by the suppliers.
- *Proprietary information*: Protecting each other's privileged information.
- *Reputation safeguard*: Neither party making unsupported or misleading statements about the other; maintaining truthfulness and professionalism in the relationship.

It is easy to see that these are the essentials on which a healthy relationship can be built, and that they will contribute to the exchange of quality information and quality supplies between the supplier and the customer. An organization interested in procuring quality parts and materials would do well to make sure that these elements exist in their relationship with their suppliers.

7.3 Choosing and Certifying Suppliers

7.3.1 Single vs. Multiple Suppliers

During the 1980s, 1970s, and before, organizations in the United States, in general, believed in cultivating and retaining as many suppliers for an item as possible. The belief was that the larger the supplier base, the better the competition; and thus, the better the price to be paid for the procured item. That was also the time when contracts were issued routinely to the lowest bidder and materials and supplies were bought mainly on the basis of price, without much regard for quality. Dr. Deming described the situation in the following words:

> A buyer's job has been, until today, to be on the look out for lower prices, to find a new vendor that will offer a lower price. The other vendors of the same material must meet it.... Economists teach the world that competition in the marketplace gives everyone the best deal.... This may have been so in days gone by.... It is different today. The purchasing department must change its focus from lowest initial cost of material purchased to lowest total cost. (Deming 1986, 32, 33)

Dr. Deming stresses the need for the buyers to understand the quality requirements of purchased material in the context of where that material is used in the production process. Otherwise, when bought on the basis of price alone, the material may make subsequent operations costly or—yet worse—result in defective products being made and delivered to the final customer.

According to Dr. Deming, a single supplier must be chosen and cultivated for each part or material, based on the supplier's ability to provide quality material and their willingness to cooperate with the customer in the design and manufacture of the final product. There are many advantages to choosing a single supplier compared to having multiple suppliers for each item. These include:

- A single supplier with a long-term commitment to supplying a part or material will be able to innovate, realize economy in their production processes, and share the benefits with the customer.

- With a single supplier, it is possible for the supplier and the customer to work together to continuously improve the quality of the supplies and, hence, of the final product, with a resulting reduction in costs.
- Supplies from one supplier can be expected to have more uniformity than supplies received from a number of suppliers.
- The paperwork and accounting system are less complex with a single supplier.
- If customers decide to keep single suppliers for their various purchases, then there will be healthy competition among the potential suppliers to improve in order for them to become the single supplier.
- One supplier per item will result in a smaller overall inventory for that item, because multiple suppliers will create multiple inventories, thus adding to the overall cost of the item.
- Having one supplier reduces the risk involved in searching and experimenting with newer suppliers.
- Single suppliers can be included as part of the team for the concurrent design and development of future products and processes.

Many American businesses have followed Dr. Deming's advice and have benefited from the overall economy and quality improvements that have resulted from this. Juran and Gryna (1993, 317) made the following observation:

> A clear trend has emerged: Organizations are significantly reducing the number of multiple suppliers. Since about 1980, reductions of 50% to 70% in the supplier base have become common. This does not necessarily mean going to single source for all purchases; it does mean a single source for some purchases and fewer multiple suppliers for other purchases.

For organizations in the United States, the main obstacle to realizing the ideal recommended by Dr. Deming—to retain only one source for every procured item—is the possibility of the disruption of supplies due to work stoppages at the supplier or transportation agencies as a result of labor disputes or natural disasters. The long distances between many suppliers and customers in the United States makes this problem more challenging.

7.3.2 Choosing a Supplier

From the earlier discussion, it becomes clear that using a single supplier per item can contribute in many ways toward quality and economy in the manufacture of a product. It then becomes necessary to choose this supplier with care, paying close attention to their relevant qualifications. The qualifications they should have include:

1. Management with a quality philosophy enforced through policies and procedures.
2. Control of design and manufacturing information, with completeness of drawings, specifications, and test procedures, and a positive recall of obsolete information.
3. Adequate procurement control to avoid poor quality in their incoming supplies.
4. Material control to ensure that the materials in storage are protected, verified periodically, and issued only to authorized users.
5. Use of capable machinery and qualified production personnel.
6. Use of prevention-based process control to avoid the production of defective material.
7. Use of final inspection methods to verify that the final product meets the needs of the customer.
8. Use of measuring instruments having sufficient accuracy and precision, and a program to periodically verify and maintain the accuracy and precision of these instruments.
9. Collection and processing of information regarding the performance of processes; use of this information for improving the processes on a continuous basis.
10. Adequate resources, manpower, and facilities to supply goods in the quantities required to meet time schedules.

As will be discussed in a later chapter, these are the features of a good quality system. In essence, suppliers are chosen based on their ability to create and maintain a good quality system within their own production facilities.

7.3.3 Certifying a Supplier

Many customers use a certification procedure as a means of improving the quality of supplies, because certification offers an incentive for suppliers to improve their performance in quality and delivery. In addition, certification offers the customer the advantage of not having to verify the quality of supplies coming from certified suppliers. Certification is also attractive to many suppliers, because certification—especially by well-known industry leaders—is a recognition that they can use to obtain business from other customers. A supplier is designated as a "certified supplier" based on *sustained* good performance over a significant length of time.

Certification takes different names at different organizations. "Quality supplier," "certified supplier," "approved supplier," "excellent supplier," and "preferred supplier" are some of the titles conferred on certified suppliers. Organizations sometimes use some gradation in the level of certification. For example, certification as an "approved supplier" may be given to one who just meets the minimum criteria, and the designation "preferred supplier" is given to one who exceeds the minimum requirements with regards to several certification criteria.

The criteria on which a supplier is evaluated for certification may vary from one organization to the next based on what is important to each. The following is an example of a criteria list used by a manufacturing organization:

- No product-related rejection for the preceding 12 months
- No nonproduct-related (e.g., packaging) rejection for the preceding six months
- No production-related negative incidents (e.g., downtime from late delivery) in the preceding six months
- Must have a fully documented quality system
- Must use statistical process control for critical process parameters, with their capabilities evaluated quantitatively
- Must have a program for continuous improvement
- Must have passed an on-site quality system evaluation by the customer within the past 12 months
- Must have the ability to provide certificates of analysis, test results, and process control and process capability documentation

7.4 Specifying the Supplies Completely

One of the things a customer can do to assure the quality of supplies is to make certain the supplier fully understands the requirements for those supplies. Drawings and specifications are meant to convey information about what the customer wants in the supplies. These must be made part of the contract document—the purchase order (PO).

We should realize that many supplied items, including modern, complex, high-tech hardware and software components, cannot be completely specified on paper. Some consignments of supplies may meet the specifications fully but may offer difficulty in assembly or perform poorly when used by the end-user. It therefore is recommended (Juran and Gryna 1993)—in addition to documentation on specifications, test requirements, and so on—that there is continuous communication between the supplier and buyer in order to ensure that the supplier understands where and how the supplies are used. The objective of both the supplier and the buyer must be to ensure that the supplies meet the needs not only of the manufacturing processes, but also of the end-user of the finished product.

It often is necessary to specify on the initial documentation how the supplies will be checked for conformance to specifications. A customer may accept a consignment on the basis of certification by the supplier that the items meet the specification. Another customer may require control chart documentation, showing that the process was in-control and capable when the goods were produced. Yet another customer may require a sampling inspection of the consignments using plans from a standard sampling tables (e.g., MIL-STD-105E), or a customer may require a 100% inspection of the supplies because of their criticality to the performance of the final product. Whatever the mode for acceptance, it must be made clear in the original PO.

Specifications can range from a mere statement of a dimension of a part on a drawing—along with the allowable tolerance (e.g., 2.000 ± 0.001 in.)—to an elaborate description of the testing procedure and the acceptable range of test results. The use of national and trade association standards, such as those of the American National Standards Institute (ANSI) and American Society for Testing Materials (ASTM), is common practice for specifying material characteristics and acceptable workmanship. There are, however, many materials for which no commonly accepted standards exist. It is not uncommon for the buyer to specify the material in terms of its performance when used, and the supplier then determines its composition or design and how it is made and tested. There are also proprietary products that cannot be specified except by the name given to them by the manufacturer (e.g., Pentium III or COREX coating).

Specifying the reliability of parts and subassemblies is another challenge. Compressors, cooling fans, printed circuit boards, and belts and hoses have requirements for the length of time they will perform without failure so that the final product can meet reliability goals. The part reliability goals should be obtained from the system reliability goals through the apportionment exercise, and they should be specified on the purchase documents in terms of minimum mean-time-to-failure (MTTF) or maximum failure-rate requirements.

Providing complete specifications for the parts and materials is important—but it is also challenging. It requires education and training on the part of buyers who should be able to understand the properties of materials and how they are used in product design and production processes. This is the reason why increasing numbers of engineering organization's purchasing departments are employing technically qualified people in their effort to procure quality supplies.

7.5 Auditing the Supplier

An initial audit and subsequent periodic audit of a supplier's facilities to evaluate their quality system is one of the ways of assuring the quality of supplies. The criteria used in the audit are the same as those listed earlier under the criteria for selecting a supplier, which include management commitment, design control, procurement control, process control, and so on.

In the days preceding the adoption of the ISO 9000 standards, it was not unusual for a supplier to be audited by each of its customers at least once a year. That resulted in many audits per year for the suppliers, which consumed a considerable amount of time and resources. The advent of the ISO standards—which lay down the specifications for a good quality system—and the availability of third-party audit and certification procedures by independent registrars have contributed to a reduction in the number of audits that a supplier must undergo.

The provisions of the ISO 9000 standards and how they ensure creation and maintenance of a quality system at a supplier's facility is discussed in Chapter 9. Choosing a supplier with certification under ISO 9000 serves the same purpose as the customer auditing a supplier themselves to verify that a good quality system exists in the supplier's facilities.

7.6 Supply Chain Optimization

Supply chain optimization is the ultimate approach to organizing relationships among suppliers and customers through the creation of value in the supply chain by joint collaboration. This approach is known not only to improve product quality, but also to lead to reduced cycle time, reduced cost of poor quality, improved technology and innovation, and reduced total cost of ownership of the product.

A supply chain includes all the activities, processes, and interactions among all suppliers and customers (including end-users) in the development, procurement, production, delivery, and consumption of a given product or service (Figure 7.1). The total cost of ownership includes all

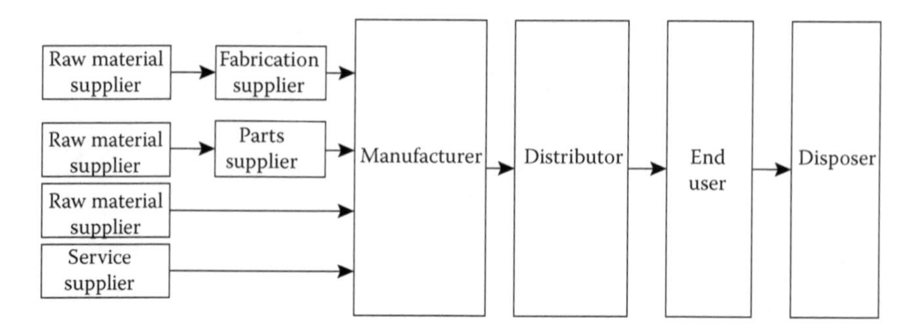

FIGURE 7.1
Components of a supply chain.

costs relating to a product that are incurred by the suppliers and customers through all the stages of design, production, and product use. Typical costs included in the total cost of ownership are the cost of engineering/design; the cost of material, parts, shipping and storage, tools, labor, training, inspection, and testing; the cost of ownership, such as routine testing and maintenance; the cost of downtime; the cost of energy; the cost of environmental control; and the cost of disposal when the product life ends. Under the supply chain management model, the decisions will be made based not on the cost of a particular exchange of goods between a supplier and a customer, but on the total cost of ownership, which includes the costs incurred by all suppliers and customers during preproduction, production, and use of the product.

Donovan and Maresca (1999) enumerated the difference between the traditional purchasing process and the process under the new supply chain model, as shown in Table 7.1. They use the term "strategic view" to signify the approach to purchasing with a cooperative, long-term, continuous partnership between customers and suppliers, which are characteristics of the modern supply chain management philosophy.

7.6.1 The Trilogy of Supplier Relationship

The supplier relationship under the new supply chain optimization is to be managed under a trilogy, patterned after Juran's quality trilogy (see Chapter 9), with the following three components:

- Planning for the supplier relationship
- Control for the supplier relationship
- Improvement for the supplier relationship

All activities under the three phases are to be performed by a cross-functional team made up of a customer's purchasing, operations, quality, and financial representatives. The team will collaborate with the supplier's representatives in "mirror" functions (corresponding functions), as appropriate. This collaboration is especially important during the "improvement" stage.

7.6.2 Planning

During the planning phase for the supplier relationship, a thorough understanding of the needs of the customer is first obtained. Next, available sources of supplies are researched through industry databases, company records, and other such avenues. Data regarding costs of current expenditure as well as the current total cost of ownership are generated. Based on these, recommendations for consolidation of needs, sourcing strategy, and supplier base reduction are made.

TABLE 7.1

Comparison of Traditional vs. Strategic View of the Purchasing Process

Aspect of the Purchasing Process	Traditional Approach	Strategic View
Supplier/buyer relationship	Adversarial, competitive, and distrusting	Cooperative, partnership based on trust
Length of relationship	Short term	Long term, indefinite
Criteria for quality	Conformance to specification	Fitness for use by the immediate and end-users
Quality assurance	Inspection on receipt	No incoming inspection necessary
Communication with suppliers	Infrequent, formal, focus on purchase orders, contracts, and legal issues	Frequent focus on the exchange of plans, ideas, and problem-solving opportunities
Inventory valuation	An asset	A liability
Supplier base	Many suppliers, managed in aggregate	Few suppliers, carefully selected and managed
Interface between suppliers and end-users	Discouraged	Required
Purchasing strategy	Manage transactions, troubleshoot	Manage processes and relationships
Purchasing business plans	Independent of end-user organization business plan	Integrated with end-user organization business plan
Geographical coverage of suppliers	As required to facilitate leverage	As required to facilitate problem solving and continuous improvement
Focus of purchasing decisions	Price	Total cost of ownership
Key for purchasing success	Ability to negotiate	Ability to identify opportunities and collaborate on solutions

Source: Reprinted from Donovan, J. B., and F. P. Maresca, *Juran's Quality Handbook*, 5th ed., McGraw-Hill, New York, NY, 1999. With permission.

7.6.3 Control

A measure for evaluating the performance of the supply chain is defined, which will include performance measures on quality (e.g., percentage rejects), warranty claims, on-time delivery, and financial and environmental effects. The minimum standards for the performance measures are specified based on customer requirements and benchmarking on best of class. The cross-functional team then evaluates current suppliers based on their performance against the set standards. This evaluation will include an assessment of the suppliers' quality system, business plans (e.g., capacity, know-how, facilities, and financial results), and history of product quality and associated service. Suppliers who do not measure up to the standards are eliminated, thus reducing the supplier base.

7.6.4 Improvement

The supply chain performance must be continuously reviewed for additional opportunities to create value. Improvement is expected to occur in a five-step progression as follows:

1. Aligning with customer needs

   During the first phase, the joint team, drawn from all links of the chain, directs its attention toward aligning the goals of each individual link with the ultimate goal of satisfying the end-user of the product. The team then analyzes the business processes of the supply chain, usually through use of flow charts; collects data on their performances; and identifies area of chronic problems, such as excessive cycle time, scrap, or rework. The team then undertakes to remedy the problems. Improved relationships among team members, with enhanced mutual trust and respect, is a healthy by-product during this phase and will be of help during future stages of the team's work.

2. Cost reduction

   During the second phase, the team undertakes a more in-depth analysis of the supply chain processes, searching for further opportunities to reduce waste and improve efficiency. In the first phase, some of the "low-hanging fruits" (i.e., the more obvious opportunities) would have been exploited. In the second phase, the second- and third-tier suppliers, (i.e., supplier-to-supplier links) will be explored. A quality cost study for the entire chain would be an appropriate methodology to use to identify areas of poor performance and opportunities for improvement.

3. Value enhancement

   As the team makes progress in reducing the cost of poor quality, they also subject each step of the processes to the following question: Does this step add any value to the supply chain? If the answer is no, then the step is eliminated. Thus, nonvalue-added activities in the chain are eliminated.

4. Information sharing

   At this stage, there are no secrets among the members of the chain. What used to be confidential in the traditional relationship is now open to all. Data are exchanged through Internet and intranet media, and there is free flow of information among the supply chain participants.

5. Resource sharing

   By this time, the entire supply chain works as one entity, with personnel from suppliers working together in a "borderless" environment. Personnel from different links of the chain are even physically located in the same facility (i.e., co-located) to facilitate better communication and sharing of resources and information.

Impressive results from the use of supply chain management programs have been reported: a reduction in variability from 20% to 70%, a 30% to 90% reduction in cycle time, a 15% to 30% reduction in waste from cost of poor quality, an increase in research and development (R&D) resources by a factor of three or more, and an overall reduction of risk from sharing among the links, have all been observed (Donovan and Maresca 1999).

7.7 Using Statistical Sampling for Acceptance

7.7.1 The Need for Sampling Inspection

A customer performs an inspection using sampling plans when there is a need to verify that the products submitted by a supplier conform to specifications. Such an inspection before acceptance may not be necessary if the supplier has a good quality system, uses effective process control, provides assurance through certification that the supplies are defect-free, and has a proven history of defect-free supplies. However, all suppliers of a given customer may not be able to provide such assurances, and there may be new suppliers for whom quality must be verified during the initial stages before a history is created. Furthermore, situations often arise in productive environments where supplies have to be received and accepted from processes even when they are known to produce less than 100% acceptable products. There are situations in which processes are known not to be in-control, or capable, but the assignable causes are either not apparent or are not fully understood. There also are situations in which the known assignable causes cannot be immediately removed for want of time and capital. The products from these processes may still be needed for subsequent assemblies, so every batch of the product must be inspected and its acceptability determined. Although quality professionals would not want to be in situations like these, the reality is that such situations exist, even if only temporarily. In all these situations, performing a 100% inspection of the supplies may not be economical—or even possible—and sampling inspection needs to be done instead.

In sampling inspections, the supplies are typically received in boxes, totes, or batches, which we call "lots." The number of units in a lot is called the "lot size." When a lot has to be accepted, or rejected, based on what we find in a sample taken from it, questions arise as to the size of the sample and the rule to be used to accept or reject the lot. The answers can be found in the many sampling plans available in tables such as the MIL-STD-105E and Dodge-Romig inspection tables. The discussion below is intended to explain the principles behind the creation of these tables and to provide guidelines on how to choose the right sampling plan for a given occasion.

Generally, a sampling plan is specified by the size of the sample to be chosen from a submitted lot, and the rules for accepting or rejecting the lot are based on the number of defectives found in the sample.

The available sampling plans can be categorized into two major categories:

1. Sampling plans for attribute inspection
2. Sampling plans for measurement (or variable) inspection

Attribute inspection is done based on characteristics such as appearance, color, feel, and taste, and it results in the classification of products into categories such as good/bad, bright/dark, tight/loose, smooth/rough, and so on. The data from attribute inspection will be in counts such as 3 of 20 bad, 12% too tight, and so forth. In measurement (or variable) inspection, however, a characteristic is actually measured using an instrument. Measuring the length of shafts, the weight of sugar in bags, or the strength of bolts will yield measurement data such as 2.2 in., 48.63 lb., 834 psi, and so on. The sampling plans for attribute inspection will specify the number of defectives that can be tolerated in a sample of specified size for the lot to be accepted. They are usually easy to understand and implement. The sampling plans for measurements, however, usually require the calculation of an average, range, or standard deviation (or a function of them) before deciding to accept or reject a lot. Thus, implementation of sampling by measurement is rather complicated and may require specially trained personnel. However, measurement sampling plans are more efficient in the sense that they require less sampling compared with attribute plans.

Under each of the above two categories are single sampling plans, double sampling plans, multiple sampling plans, and continuous sampling plans, based on the number of samples taken per lot submitted (see Figure 7.2).

The continuous sampling plans are suitable when inspecting products at the end of a continuously producing line without the need for grouping products

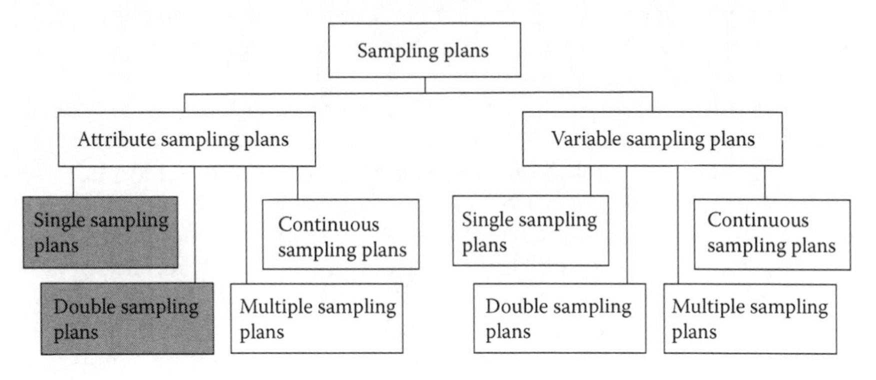

FIGURE 7.2
Categories of sampling plans (shaded plans are covered in this chapter).

into lots. The double or multiple sampling plans, which require more than one sample per lot, are more efficient in that they require less sampling compared to the single sampling plans, which require just one sample per lot. Double and multiple sampling plans will be needed where inspection is expensive, either from units destroyed during inspection or from time-consuming inspection procedures. However, single sampling plans are simple and easy to learn and use. We will restrict our discussions here to single and double sampling plans for attribute inspection. Most sampling needs, however, can be met using these plans, which are the more commonly used plans in industry. It is necessary to understand how these plans are created, the statistics fundamentals behind them, and their strengths and weaknesses, in order to use them correctly. Readers interested in other plans should refer to books such as Bowker and Lieberman (1972), Grant and Leavenworth (1996), or Schilling (1982).

7.7.2 Single Sampling Plans for Attributes

A single sampling plan (SSP) involves selecting one (random) sample per lot submitted for inspection and then determining whether the lot should be accepted or rejected based on the number of defectives found in the sample. An SSP is defined by two numbers, n and c, where n is the sample size and c is the *acceptance number*. The scheme of an SSP is as follows. Suppose a single sampling plan has $n = 12$ and $c = 1$. This means that a sample of 12 items should be taken from each lot submitted and each item in the sample be categorized as acceptable (i.e., conforming to specifications) or defective (i.e., not conforming to specifications). If the number of defectives found in the sample is not more than one, then the lot is accepted; otherwise, the lot is rejected.

7.7.2.1 The Operating Characteristic Curve

When we are faced with choosing a single sampling plan for a given situation, we have numerous alternative plans to choose from. For example, any of the combination of numbers (10, 0), (12, 0), or (24, 2) describes an SSP, with the first number representing the sample size and the second representing the acceptance number. The question then arises: which sampling plan is good for the situation at hand?

The operating characteristic curve (OC curve) is the criterion used to decide which plan is the most suitable for a given situation. Every sampling plan has an OC curve that tells us how the plan will accept or reject lots of different quality. The graph in Figure 7.3 is an example of an OC curve, in which the x-axis represents the quality of the lot, in proportion defectives p, and the y-axis represents the probability of acceptance by the sampling plan Pa. For example, the sampling plan whose OC curve is shown in Figure 7.3 will accept lots with 1% defectives with probability 0.9, and lots with 8% defectives with probability 0.1. The OC curve tells us how strict, or how lenient, a sampling plan is in accepting good lots and rejecting bad lots. We should know how the OC curves are computed.

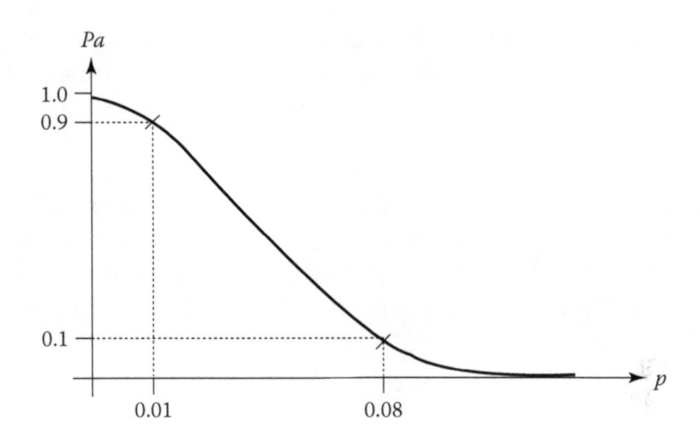

FIGURE 7.3
Example of an OC curve.

For any given sampling plan, the OC curve can be drawn by computing the probabilities of acceptance by the plan of lots of different quality, with the quality of a lot being denoted by the fraction of defectives in that lot. The method of calculating the probability of acceptance by a sampling plan is explained below using an example. For a practical understanding, the probability of acceptance can be interpreted as follows. If a sampling plan has probability 0.9 of accepting a lot with a certain fraction defectives, it means that if 100 such lots are submitted for inspection, about 90 of them will be accepted and about 10 will be rejected. A desirable sampling plan will have high probability (e.g., 0.95 or 0.99) of accepting good lots, and low probability (e.g., 0.10 and 0.05) of accepting bad lots.

7.7.2.2 Calculating the OC Curve of a Single Sampling Plan

Suppose that a lot with p fraction defectives is submitted to a single sampling plan with sample size n and acceptance number c. The probability of acceptance of this lot by the single sampling plan is given by $Pa = P(D \leq c)$, where D is the number of defective units in the sample. The random variable D is binomially distributed, with parameters n and p, and the above probability is given by the binomial sum:

$$Pa = \sum_{x=0}^{c} \binom{n}{x} p^x (1-p)^{n-x}$$

If n is large (>15), this probability can be approximated by the cumulative probabilities of the Poisson distribution with a mean of np. The calculations become easier if we use the cumulative probabilities from the Poisson table, Table A.5 in the Appendix. Hence, we calculate the Pa corresponding to certain chosen p values and plot Pa versus p to obtain the OC curve.

We want to point out here that in quality control literature (e.g., Duncan 1974), the OC curve we defined above is referred to as the "Type-B" OC curve, because the probabilities are calculated using the binomial distribution on the assumption that sampling is done from "large" lots. If the sampling is not from large lots, hyper-geometric distribution should be used to calculate the probabilities of acceptance. The OC curve is then called the "Type-A" OC curve. In our discussion here, we will assume that the sampling is always from large lots and that all the OC curves calculated are Type B. A large lot is usually defined as a lot having 30 or more items. The example below shows how the OC curve is drawn.

Example 7.1

Calculate the OC curve of the three SSPs (20, 2), (20, 1), and (20, 0). Assume that the sampling is from a large lot.

Solution

The calculations are shown in the table below, and the OC curve is shown in Figure 7.4a.

p	0.02	0.04	0.06	0.08	0.10	0.15	0.20	0.25	0.30	0.35
np	0.4	0.8	1.2	1.6	2.0	3.0	4.0	5.0	6.0	7.0
$Pa(20, 2)$	0.992	0.952	0.879	0.783	0.676	0.423	0.238	0.124	0.061	0.029
$Pa(20, 1)$	0.938	0.808	0.662	0.524	0.406	0.199	0.091	0.04	0.017	0.007
$Pa(20, 0)$	0.670	0.449	0.301	0.201	0.135	0.049	0.018	0.006	0.002	0.000

To explain the calculations, take, for example, the calculation of the probability of acceptance of 6% lots ($p = 0.06$) by the plan (20, 2). The probability of acceptance Pa is the probability a Poisson variable with a mean of 1.2 ($np = 20 \times 0.06 = 1.2$) being less than or equal to 2. That is, $Pa = P(Po(1.2) \leq 2)$. This probability can be read off the cumulative Poisson table (Table A.5 in the Appendix) as 0.879.

In this example, we calculated the OC of three different sampling plans with the same sample size but with different acceptance numbers. Figure 7.4a shows that for a given sample size, the OC curve of an SSP with a smaller acceptance number becomes steeper, compared to that of a plan with a larger acceptance number. A steeper curve indicates that the sampling plan will be more discriminating between good and bad lots.

The next example shows how the OC of an SSP changes when the sample size is changed but the acceptance number remains the same.

Example 7.2

Draw the OC curves of the SSPs (30, 2) and (40, 2), and compare them with that of (20, 2).

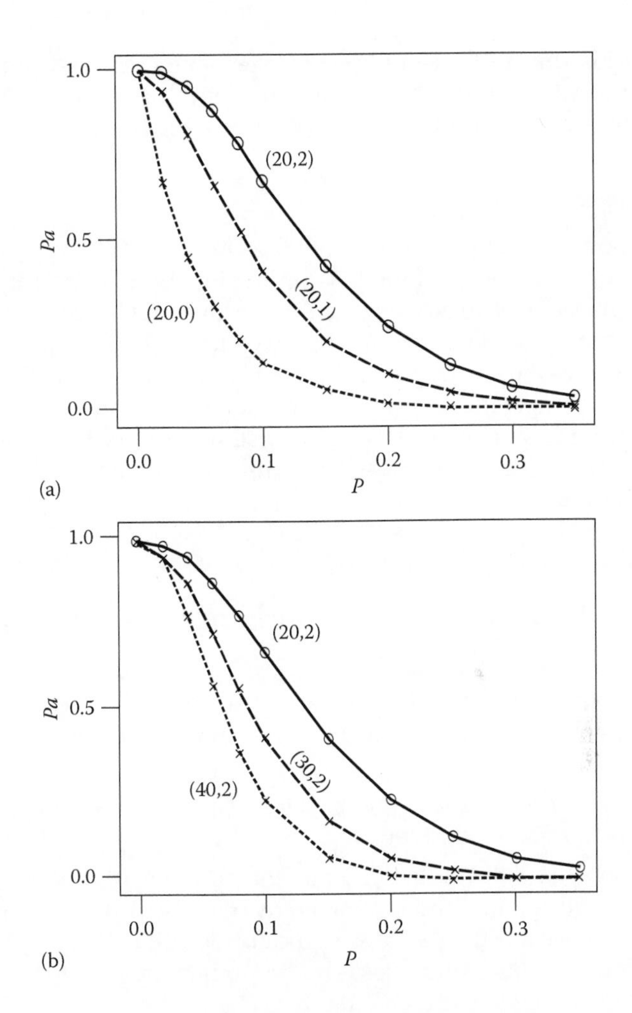

FIGURE 7.4
(a) OC curves of single sampling plans with varying acceptance numbers. (b) OC curves of single sampling plans when sample size is varied.

Solution

p	0.02	0.04	0.06	0.08	0.1	0.15	0.2	0.25	0.3	0.35
$np = 20p$	0.4	0.8	1.2	1.6	2.0	3.0	4.0	5.0	6.0	7.0
$Pa(20, 2)$	0.992	0.952	0.879	0.783	0.676	0.423	0.238	0.124	0.061	0.029
$np = 30p$	0.6	1.2	1.8	2.4	3.0	4.5	6.0	7.5	9.0	10.5
$Pa(30, 2)$	0.976	0.879	0.730	0.569	0.423	0.173	0.061	0.020	0.006	0.02
$np = 40p$	0.8	1.6	2.4	3.2	4.0	6.0	8.0	10.0	12.0	14.0
$Pa(40, 2)$	0.952	0.783	0.569	0.380	0.238	0.061	0.013	0.002	0.001	0.000

The graphs of the OC curves with varying samples sizes for a given acceptance number are shown in Figure 7.4b. They show that increasing the sample size also makes the sampling plan steeper (i.e., more discriminating). We note, however, that increasing the sample size does not produce as dramatic a change in discrimination as does decreasing the acceptance number.

7.7.2.3 Designing an SSP

From the above discussion, we see how the OC curve for a sampling plan is computed and how it shows the discriminating ability of a sampling plan. Next, we will see how to select an SSP for a given OC curve. Often, a purchaser and a supplier agree on an OC curve to guide them in determining how the supplies will be inspected before acceptance. The quality engineer is then asked to select the sampling plan that will have an OC curve "equal" to the OC curve that has been agreed to. First, we will see how a suitable OC curve is specified; then, we will see how to select an SSP that will "fit" the chosen OC curve.

7.7.2.4 Choosing a Suitable OC Curve

An OC curve is specified by selecting two points that lie on it. A few definitions are needed:

- *Acceptable quality level (AQL)* is the largest proportion defectives, or largest value of p that is considered acceptable to the purchaser.
- *Lot tolerance percent defective (LTPD)* is the smallest value of p that the purchaser considers must be rejected. The *LTPD* is also called the *rejectable quality level (RQL).*
- *Producer's risk* is the probability that the sampling plan will reject lots having *AQL* quality, and is denoted by α. It is called the "producer's risk" because it is the risk that a producer is willing to accept of their good quality lots being rejected. The probability that the sampling plan will accept lots of *AQL* quality is $(1 - \alpha)$.
- *Consumer's risk* is the probability that the sampling plan will accept lots having *LTPD* quality, and it is denoted by β. It is referred to as the "consumer's risk" because it is the risk that a consumer is willing to accept of lots of rejectable quality being accepted.

The graph in Figure 7.5a is called the "ideal OC" curve. In this OC curve, the sampling plan will accept with probability 1.0 lots of *AQL* or better quality, and reject with probability 1.0 lots of quality worse than *AQL*. This is the best that we can get out of a sampling plan, but the sample size needed for such a plan is infinity. It is, therefore, just an ideal plan, and we must settle for a more practical plan. An OC curve of a practical plan is shown in Figure 7.5b. In this plan, the *AQL* is a small number, such as 0.005 or 0.02 (0.5% or 2%) and is chosen based on how critical the part, or the characteristic being inspected, is. For a critical characteristic, a much

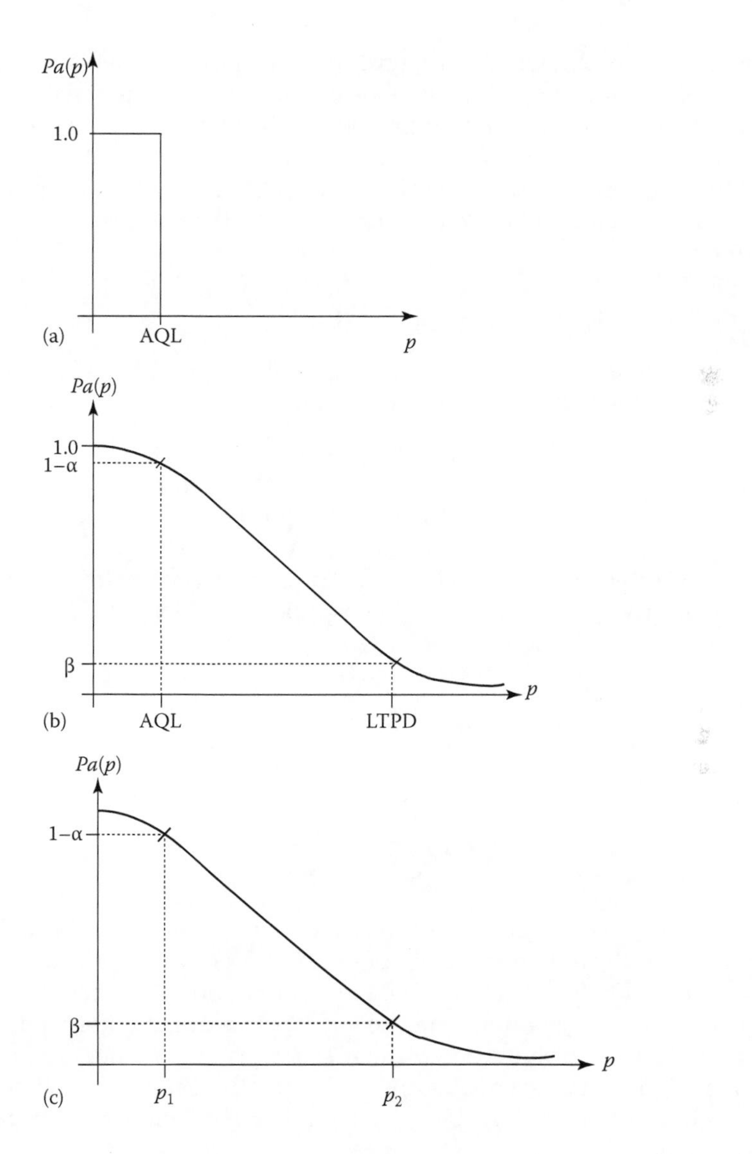

FIGURE 7.5
(a) The ideal OC curve. (b) Specifying a practical OC curve. (c) Designing a single sampling plan for a given OC curve.

smaller *AQL* will be chosen than for a characteristic with only cosmetic value. The producer's risk α is also a small number, such as 0.05 or 0.1, and it represents the risk the producer accepts of their good supplies (supplies of *AQL* quality) being rejected. The *LTPD* is larger than the *AQL*, and it represents the proportion of defectives that the customer can hardly tolerate and would like to reject. The consumer's risk β is also a small number, such as 0.05 or 0.1, which represents the risk the consumer is willing to take in accepting lots they consider to be of poor quality. These risk parameters, $(AQL, 1 - \alpha)$ and $(LTPD, \beta)$, must be agreed to between the buyer and the supplier. These parameters define the two points on the OC curve that is desired. The single sampling plan with an OC curve that will pass through the two points $(AQL, 1 - \alpha)$ and $(LTPD, \beta)$ can then be selected as described below.

7.7.2.5 Choosing a Single Sampling Plan

Let us denote AQL by p_1 and $LTPD$ by p_2 (see Figure 7.5c). The problem is to determine the values of n and c such that:

$$P(D \le c \mid p = p_1) = 1 - \alpha$$

$$P(D \le c \mid p = p_2) = \beta$$

where D is the number of defectives in a sample of size n drawn from the lots of respective quality. Assuming a large lot, D is binomially distributed, and the above probabilities can be written as:

$$\sum_{x=0}^{c} \binom{n}{x} p_1^x (1 - p_1)^{n-x} = 1 - \alpha$$

$$\sum_{x=0}^{c} \binom{n}{x} p_2^x (1 - p_2)^{n-x} = \beta$$

Therefore, we have two equations in two unknowns, n and c, and we have to solve for the unknown values of n and c. These are nonlinear equations in n and c, and they cannot be solved in a closed form. Numerical root-finding methods using a computer may be needed; however, nomographs are available to facilitate solving these equations. One such nomograph is given in Figure 7.6a, which is a nomograph of cumulative binomial probabilities. For any given n and c, the graph gives the following cumulative probability for a given p:

$$Pa = P(D \le c) = \sum_{x=0}^{c} \binom{n}{x} p^x (1 - p)^{n-x}$$

We can recognize this as the probability of acceptance of lots with p fraction defectives by an SSP defined by (n, c). For example, suppose $n = 20$, $c = 1$, and $p = 0.02$. To find the above cumulative probability, find the intersection of the lines representing $n = 20$ and $c = 1$ in the main graph. Draw a line through that intersection and the value of $p = 0.02$ on the scale on the

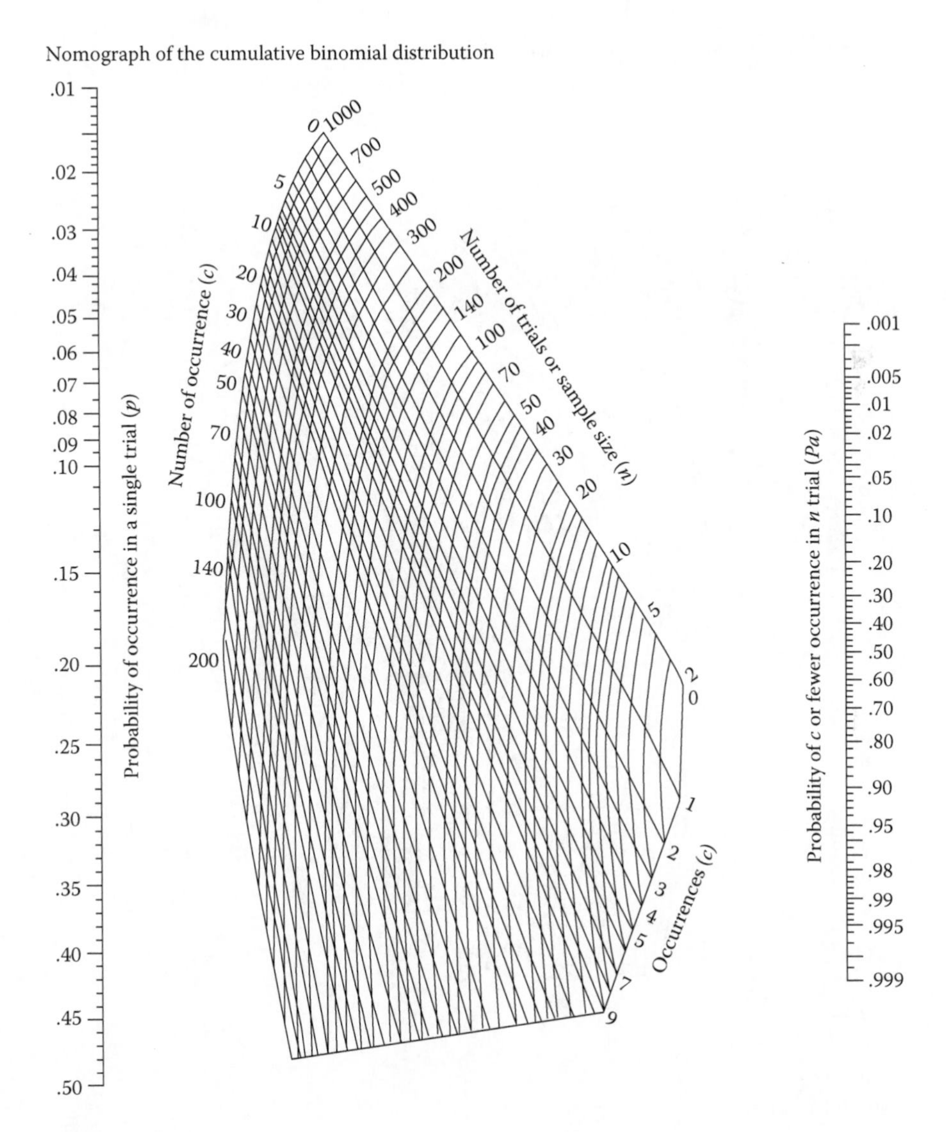

FIGURE 7.6a
Nomograph to select single sampling plans for a given OC curve. Reproduced from Larson, H. R., *Industrial Quality Control* 23 (6), 270–278, 1996, with permission of American Society for Quality, Milwaukee.

left-hand side of the graph. The intersection of this line with the scale on the right-hand side of the graph gives $Pa = 0.94$. Because designing the SSP is equivalent to finding the value of (n, c) that is common to two given values of $(p, Pa(p))$, the graph can be used to design SSPs, as shown in the following example.

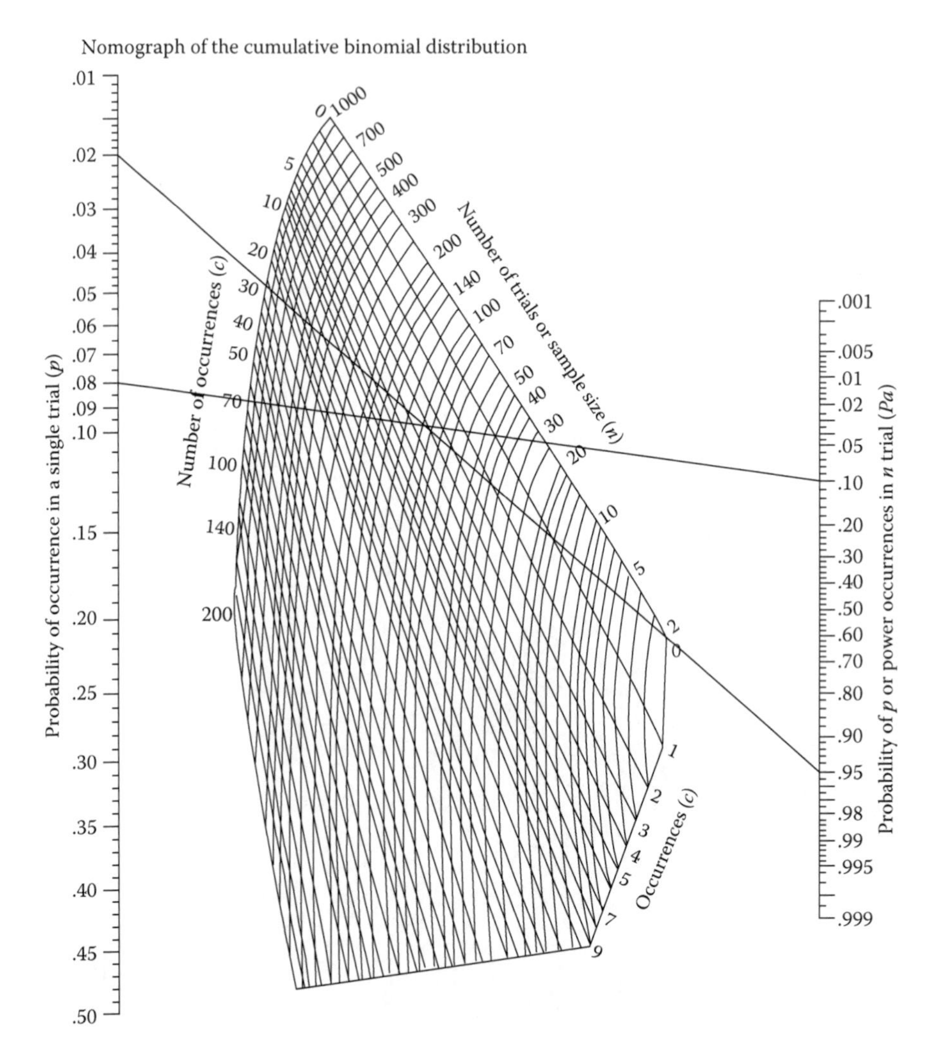

FIGURE 7.6b
Designing a SSP using the nomograph.

Example 7.3

Design an SSP for the following data: $AQL = p_1 = 0.02$, $Pa(p_1) = 0.95$, $LTPD = p_2 = 0.08$, and $Pa(p_2) = 0.1$.

Solution

Draw a line connecting p_1 on the vertical scale on the left-hand side of the graph, and $Pa(p_1)$ on the vertical scale on the right-hand side. Draw another similar line connecting p_2 and $Pa(p_2)$. Read the values of n and c at the intersection of the two lines. For the example, $n = 98$, and $c = 4$ (Figure 7.6b).

The availability of standard tables, such as the MIL-STD-105E, makes it even easier to choose sampling plans without the use of nomographs such as the one presented above. The MIL-STD-105E tables give plans based on *AQL* values and a predetermined (1 – α) value. Similarly, sampling plans known as Dodge-Romig plans will have OC curves passing through given (*LTPD*, β) points. We will discuss here only the MIL-STD-105E plans, after describing the double sampling plan and defining the *average outgoing-quality limit* (*AOQL*). The MIL-STD-105E will provide sampling plans that will meet most of the needs of a quality engineer.

7.7.3 Double Sampling Plans for Attributes

A double sampling plan (DSP) requires taking, at most, two samples per lot for deciding whether to accept or reject it. A DSP is defined by five numbers, compared with the two that are needed to define an SSP. The scheme of a DSP is described in Figure 7.7. "Double and multiple sampling plans reflect the tendency of many experienced inspectors to give a questionable lot an

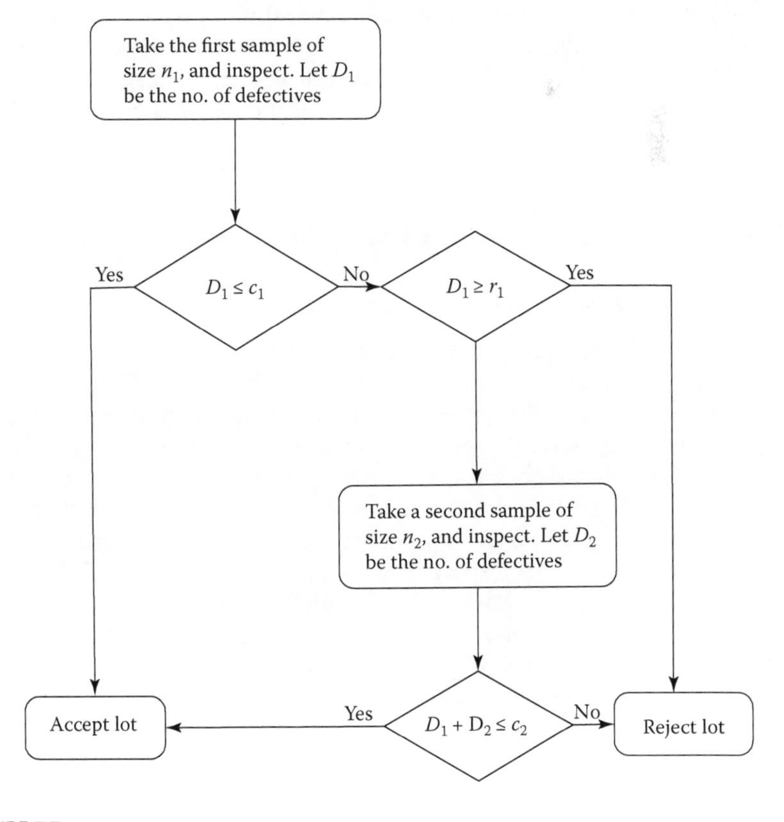

FIGURE 7.7
The scheme of a DSP.

additional chance" (Schilling 1982). The five numbers used to specify a DSP are as follows:

n_1—the size of the first sample

c_1—the acceptance number for the first sample

r_1—the rejection number for the first sample

n_2—the size of the second sample

c_2—the acceptance number for both samples

Some authors (e.g., Montgomery 2001) describe a DSP using only four numbers instead of five, because they assume some relationship between r_1 and c_2, such as $r_1 = c_2$. This assumption may sometimes be true, but not always. The MIL-STD-105E, for example, does not make such an assumption and uses five numbers to define DSPs. Here, we follow the convention used by Schilling (1982) and use five numbers.

7.7.3.1 Why Use a DSP?

The DSP will require, on average, a smaller amount of inspection than a comparable SSP, comparable in the sense that both will have about "equal" OC curves. We will later quantify the amount of inspection needed by a sampling plan using the average sample number, and show how the DSP compares favorably with the SSP in this respect. Where inspection results in the destruction of units or the cost of inspection is high for other reasons, a double sampling plan will be preferred, although administering a DSP is more difficult and may require personnel with additional training.

7.7.3.2 The OC Curve of a DSP

In the case of the DSP, there is more than one OC curve. In fact, a DSP has a primary OC curve and two secondary OC curves. The primary OC curve shows the relationship between the quality of a submitted lot and the probability of it being accepted by the overall plan; that is, accepted either in the first or the second sample. One of the secondary OC curves shows the relationship between lot quality and the probability of *acceptance* in the first sample; the other shows the relationship between lot quality and the probability of *rejection* in the first sample. Figure 7.8 shows an example of the three OC curves of a DSP.

The two secondary OC curves are simply the OC curves of SSPs with sample size n_1, with acceptance number c_1 in one case and $(r_1 - 1)$ in the other case. However, the calculation of the primary OC curve is a bit more involved and is illustrated using an example.

Example 7.4

Calculate the primary OC curve of the DSP with $n_1 = 20$, $c_1 = 0$, $r_1 = 3$, $n_2 = 40$, and $c_2 = 3$.

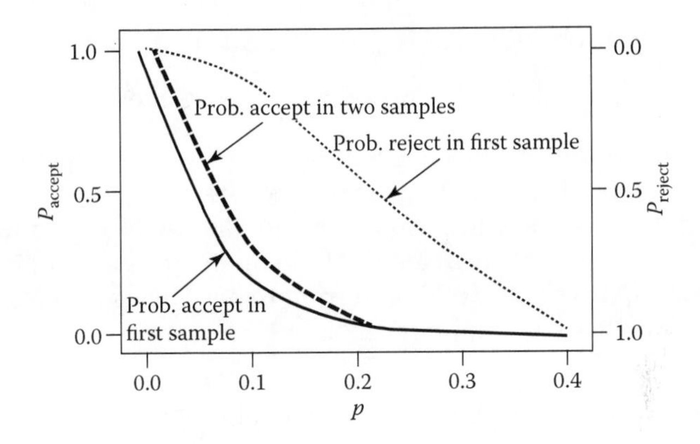

FIGURE 7.8
Primary and secondary OC curves of a DSP.

Solution

The primary OC curve of the DSP is calculated by identifying all possible events in which acceptance of the lot could occur and then finding the probability that acceptance occurs through any one of the events. There are three possible ways in which acceptance of the lot could occur: through event A, B, or C. Event A occurs if the number of defectives in the first sample $D_1 \leq 0$. There is no need for the second sample. Event B occurs when the number of defectives in the first sample $D_1 = 1$, and the number of defective in (the required) second sample $D_2 \leq 2$. Similarly, Event C occurs if the number of defectives in the first sample $D_1 = 2$ and the number of defectives in (the required) second sample $D_2 \leq 1$. The reader can verify that there is no other possible event through which acceptance of the lot can occur.

The calculation of the probabilities of acceptance for a lot with 5% defectives ($p = 0.05$) is carried out in the following table:

Event	Events Leading to Acceptance in First Sample	Events Leading to Acceptance in Second Sample	Prob. of Events using Poisson Approximation
	$n_1p = (20)(0.05) = 1.0$	$n_2p = (40)(0.05) = 2.0$	
A	$D_1 \leq c_1 = 0$	No second sample	0.367
B	$D_1 = 1$ AND	$D_2 \leq 2$	$(0.735 - 0.367) \times (0.676) = 0.249$
C	$D_1 = 2$ AND	$D_2 \leq 1$	$(0.919 - 0.735) \times (0.406) = 0.075$
(*A* or *B* or *C*)			0.692

The calculations in the above table using cumulative Poisson probabilities are fairly straightforward except for the following. To find the probability that the number of defectives in a sample (e.g., the first sample) exactly equals one, we calculate it as the difference of the two cumulative probabilities from the Poisson tables. For example:

$$P(D_1 = 1) = P(D_1 \leq 1) - P(D_1 \leq 0)$$

Furthermore, the probability $P(A$ or B or $C)$ is calculated as:

$P(A \cup B \cup C) = P(A) + P(B) + P(C)$, because A, B, and C are mutually exclusive.

Also, we have assumed independence between events in the two samples, which may be true when sampling from large lots.

This example shows how to calculate the probability of acceptance for one value of $p = 0.05$. Similar calculations can be made for several other values of p and the primary OC curve drawn from them. The OC curves shown in Figure 7.8 belong to this sampling plan.

7.7.4 The Average Sample Number of a Sampling Plan

The average sample number (*ASN*) of a sampling plan is the average number of units that must be inspected per lot to reach an accept/reject decision. Suppose that an SSP is employed in inspecting certain lots. The *ASN* for the plan would simply be the sample size of the plan as long as the inspection is not curtailed (i.e., inspection is not stopped when the number of defectives exceeds the acceptance number). Thus, the *ASN* of an SSP is independent of lot quality. On the other hand, if a DSP is used, some lots will be accepted or rejected in the first sample, and some may require two samples to reach a decision—even if all the lots have the same quality. The *ASN* of a sampling plan reflects how much inspection is required when using a sampling plan. For the DSP, it is a function of lot quality and is calculated using the following formula:

$$ASN(p) = n_1P_1 + (n_1 + n_2)(1 - P_1)$$

where P_1 represents the probability that a decision is made (accept or reject) in the first sample. For a given p, P_1 is calculated as follows:

$$\begin{aligned} P_1 &= P(\text{lot is accepted or rejected in the first sample}) \\ &= P(\text{lot accepted in first sample}) + P(\text{lot rejected in first sample}) \\ &= P(D_1 \le c_1) + P(D_1 \ge r_1) \\ &= P(D_1 \le c_1) + 1 - P(D_1 \le r_1 - 1), \end{aligned}$$

where D_1 is the number of defectives in the first sample.

The formula for *ASN* computes the expected value, or average, of the random variable that represents the number of samples inspected per lot of a given quality p. The random variable takes two possible values, n_1 and $(n_1 + n_2)$, the former with probability P_1 and the latter with probability $(1 - P_1)$. The value of *ASN* lies between n_1 and $(n_1 + n_2)$. The following example illustrates the calculation of *ASN* for a DSP using the formula above.

Example 7.5

Calculate the *ASN* for the following DSP with $n_1 = 50$, $c_1 = 1$, $r_1 = 3$, $n_2 = 100$, and $c_2 = 3$ when used to inspect lots of different quality as shown in the table below.

Solution

The *ASN* calculations are done in the table below. In this table,

$$P_1 = P(D_1 \le c_1) + 1 - P(D_1 \le r_1 - 1) = P(D_1 \le 1) + 1 - P(D_1 \le 2),$$

where D_1 is a Poisson random variable with mean $= n_1p$.

p	0.01	0.02	0.03	0.04	0.05	0.06	0.07	0.08	0.09	0.10	0.12
n_1p	0.50	1.0	1.5	2.0	2.5	3.0	3.5	4.0	4.5	5.0	6.0
$P(D_1 \le 1)$	0.909	0.735	0.557	0.406	0.287	0.199	0.135	0.091	0.061	0.040	0.017
$1 - P(D_1 \le 2)$	0.014	0.080	0.192	0.323	0.457	0.577	0.680	0.762	0.827	0.876	0.938
P_1	0.924	0.816	0.749	0.729	0.744	0.776	0.815	0.854	0.888	0.916	0.955
ASN	57.6	68.4	75.1	77.1	75.6	72.4	68.5	64.6	61.2	58.4	54.5

The next example compares the *ASN* of a DSP with that of an "equivalent" SSP. We design the equivalent SSP by first finding the probabilities of acceptance of the DSP at certain chosen *AQL* and *LTPD* values, and then finding the SSP that will have the same (or approximately the same) probabilities for acceptance at these selected *AQL* and *LTPD* points.

Example 7.6

a. Compute the probability of acceptance corresponding to an *AQL* of 0.01 and an *LTPD* of 0.08 for the DSP defined by $n_1 = 50$, $c_1 = 1$, $r_1 = 3$, $n_2 = 100$, and $c_2 = 3$.
b. Choose an SSP corresponding to the given values of *AQL* and *LTPD* and the calculated values of the probabilities of acceptance. Compare the *ASN* of the DSP and SSP.

Solution

a. To find *Pa*(0.01) of the DSP with $n_1 = 50$, $c_1 = 1$, $r_1 = 3$, $n_2 = 100$, and $c_2 = 3$.

Event	Events Leading to Acceptance in First Sample of Size 50	Events Leading to Acceptance in Second Sample of Size 100	Prob. of Events using Poisson Approximation
	$n_1p = (50)(0.01) = 0.5$	$n_2p = (100)(0.01) = 1.0$	
A	$D_1 \le c_1 = 1$	No second sample	0.909
B	$D_1 = 2$ AND	$D_2 \le 1$	$(0.985 - 0.909) \times (0.735) = 0.056$
(*A* or *B*)			0.966

To find *Pa*(0.08) of the DSP with $n_1 = 50$, $c_1 = 1$, $r_1 = 3$, $n_2 = 100$, and $c_2 = 3$.

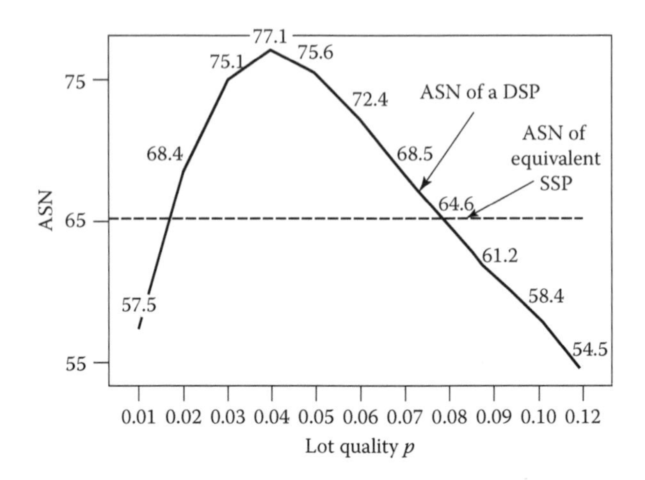

FIGURE 7.9
Graph of the *ASN* for a DSP and an equivalent SSP.

Event	Events Leading to Acceptance in First Sample of Size 50	Events Leading to Acceptance in Second Sample of Size 100	Prob. of Events using Poisson Approximation
	$n_1p = (50)(0.08) = 4.0$	$n_2p = (100)(0.08) = 8.0$	
A	$D_1 \le c_1 = 1$	No second sample	0.091
B	$D_1 = 2$ AND	$D_2 \le 1$	$(0.238 - 0.091) \times (0.003) = 0.0004$
(*A* or *B*)			0.0924

b. To select the single sampling plan:

$$p_1 = 0.01 \quad (1-\alpha) = 0.966$$

$$p_2 = 0.08 \quad \beta = 0.092.$$

Using the nomograph in Figure 7.6a, we find that $n = 65$ and $c = 2$.

The *ASN* for DSP and SSP are shown together in Figure 7.9. We see that the DSP has a smaller *ASN* compared to the SSP for some lot qualities, and a larger *ASN* for other lot qualities. However, note that in the quality range of *AQL*, where we would expect the supplier to submit lots, the *ASN* for the DSP is smaller. This is the advantage of DSP over SSP, and is preferred when inspection is expensive.

7.7.5 MIL-STD-105E (ANSI Z1.5)

The MIL-STD-105E standard provides SSPs, DSPs, and multiple sampling plans, and it is the most popular source of sampling plans among industrial users. [The multiple sampling plans, which are not covered in this book, use more than two samples per lot to arrive at an accept/reject decision. Their design is based on an extension of the logic used in designing DSPs.] The plans in the military standard are called *AQL plans,* because they are all designed such that their OC curves will pass through a chosen (AQL, $1 - \alpha$) point. Also, only an AQL value and the size of the lot to be inspected are needed to choose a sampling plan for a given situation. The value of $(1 - \alpha)$, or the probability of acceptance corresponding to a chosen AQL, is between 0.91 and 0.99 for all plans in the military standard.

These sampling plans were created during World War II to help the U.S. War Department procure quality material for the war effort. The tables, which have appeared under different names since 1942, were published for the first time in 1950 as the MIL-STD-105A. The last revision, the MIL-STD-105E, was published in 1989 (U.S. Department of Defense 1989). Although this revision is still available for use the Department of Defense is not publishing any more revisions. The readers are referred to an equivalent standard ANSI/ASQ Z1.5-2008, published as an American National Standard by the American Society for Quality (ASQ 2008). The discussion of the MIL-STD that follows is applicable to the ANSI/ASQ standard as well since the latter uses the same procedure and tables as the MIL-STD.

The MIL-STD-105E provides for three levels of inspection—normal, tightened, and reduced. The normal level is the level at which the inspection will be started when a supplier begins submitting supplies. The level of inspection will be changed, however, if the performance of the supplier changes. If the quality of the supplies deteriorates, then "tightened inspection," with its stricter acceptance criteria, will be imposed as a way of pressuring the supplier to improve their quality. If the quality performance is good, then the level will be changed to "reduced inspection," which entails a smaller amount of inspection as a way of rewarding good performance. Switching rules are provided by the standard to determine when a supplier's performance is good or bad. Figure 7.10 has been drawn to represent graphically the switching rules described in the standard in words.

It is claimed that the tables with the three levels of inspection and the switching rules constitute a system for obtaining quality in supplies; they are not just isolated sampling plans. When properly used, with the inspection results being fed back to the suppliers, they can be used to improve the processes, prevent the production of defective units, and achieve improved product quality. The sampling inspection tables are a valuable source of sampling plans for those situations in which supplies must be inspected using sampling inspection before they are accepted.

The selection of a sampling plan from the military standard starts with Table 7.2 (Table 7.1 of the military standard), which is called the "sample size code table." This table gives the sample size based on lot or batch size. The examples below illustrate the selection of plans from the tables of the standard.

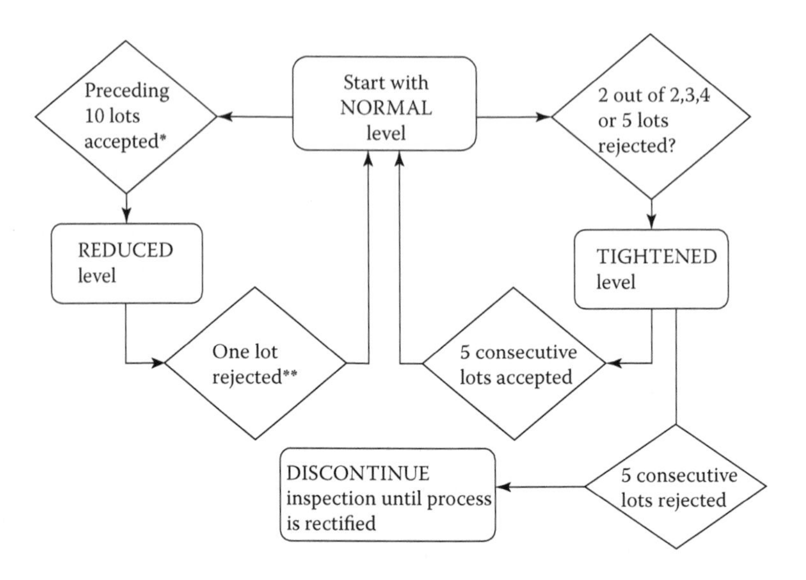

* In addition, the total number of defectives in the samples from the preceding 10 lots should not exceed the limiting number specified in a special table (Table VIII of the MIL STD.). Furthermore, production should be at steady rate, and reduced inspection is considered to be desirable by responsible authority

** In addition, normal level will be reinstated if a lot is accepted because the number of defectives from all samples is between the accept and reject numbers for double or multiple sampling plans. Also, normal level will be reinstated if production becomes irregular or the responsible authority considers the conditions warrant that normal inspection shall be instituted

FIGURE 7.10
Switching rules for military standard sampling plans (prepared from the "switching procedures" given in the MIL-STD-105E).

7.7.5.1 Selecting a Sampling Plan from MIL-STD-105E

A sampling plan from the military standard is selected through the following steps:

1. Enter Table 7.2 with lot size, and pick the sample size code letter. Use General Inspection Level II if no special conditions are specified.
2. Choose the appropriate table based on whether it is single or double sampling and whether a normal, tightened, or reduced level is required. The tables are titled according to level of inspection and whether the plans are single, double or multiple sampling plans. The SSP and DSP tables are reproduced in this text as Tables 7.3 through 7.8.

The military standard gives the OC curves for every plan listed in the tables. The OC curves are provided both as tables and as graphs. This helps in understanding how a chosen plan will perform against various lot quality.

TABLE 7.2

Sample Size Code Letters (MIL-STD-105E—Table 7.1)

	Special Inspection Levels				General Inspection Levels		
Lot or Batch Size	**S-1**	**S-2**	**S-3**	**S-4**	**I**	**II**	**III**
2–8	A	A	A	A	A	A	B
9–15	A	A	A	A	A	B	C
16–25	A	A	B	B	B	C	D
26–50	A	B	B	C	C	D	E
51–90	B	B	C	C	C	E	F
91–150	B	B	C	D	D	F	G
151–280	B	C	D	E	E	G	H
281–500	B	C	D	E	F	H	J
501–1200	C	C	E	F	G	J	K
1201–3200	C	D	E	G	H	K	L
3201–10,000	C	D	F	G	J	L	M
10,001–35,000	C	D	F	H	K	M	N
35,001–150,000	D	E	G	J	L	N	P
150,001–500,000	D	E	G	J	M	P	Q
500,001 and over	D	E	H	K	N	Q	R

An example page containing OC curves, as graphs and tables, is reproduced in Figure 7.11.

Example 7.7

Choose the SSPs for normal, tightened, and reduced inspection if the lot size is 200 and the *AQL* is 2.5%.

Solution

From Table 7.2, the code letter for the given lot size is G. The SSP for normal inspection from Table 7.3 is:

$$(n, c, r) : (32, 2, 3),$$

where *n*, *c*, and *r* indicate sample size, acceptance number, and rejection number, respectively. Similarly, the SSP for tightened inspection from Table 7.4 is (32, 1, 2), and the plan for reduced inspection from Table 7.5 is (13, 1, 3).

Note that for the normal and tightened SSPs, the rejection number equals the acceptance number + 1. For the reduced inspection, however, this is not so; there is a gap between the acceptance number and the rejection number. If the number of defectives in a chosen sample falls between the acceptance

TABLE 7.3

Single Sampling Plans for Normal Inspection (MIL-STD-105E—Table II-A)

Sample size code letter	Sample size	0.010	0.015	0.025	0.040	0.065	0.10	0.15	0.25	0.40	0.65	1.0	1.5	2.5	4.0	6.5	10	15	25	40	65	100	150	250	400	650	1000
		Acceptable quality levels (normal inspection)																									
		Ac Re	Ac Re	Ac Re	Ac Re	Ac Re	Ac Re	Ac Re	Ac Re	Ac Re	Ac Re	Ac Re	Ac Re	Ac Re	Ac Re	Ac Re	Ac Re	Ac Re	Ac Re	Ac Re	Ac Re	Ac Re	Ac Re	Ac Re	Ac Re	Ac Re	Ac Re
A	2	↓	↓	↓	↓	↓	↓	↓	↓	↓	↓	↓	↓	↓	↓	0 1	↓	↓	1 2	2 3	3 4	5 6	7 8	10 11	14 15	21 22	30 31
B	3	↓	↓	↓	↓	↓	↓	↓	↓	↓	↓	↓	↓	↓	0 1	↑	↓	1 2	2 3	3 4	5 6	7 8	10 11	14 15	21 22	30 31	44 45
C	5	↓	↓	↓	↓	↓	↓	↓	↓	↓	↓	↓	↓	0 1	↑	↓	1 2	2 3	3 4	5 6	7 8	10 11	14 15	21 22	30 31	44 45	↑
D	8	↓	↓	↓	↓	↓	↓	↓	↓	↓	↓	↓	0 1	↑	↓	1 2	2 3	3 4	5 6	7 8	10 11	14 15	21 22	30 31	44 45	↑	↑
E	13	↓	↓	↓	↓	↓	↓	↓	↓	↓	↓	0 1	↑	↓	1 2	2 3	3 4	5 6	7 8	10 11	14 15	21 22	30 31	44 45	↑	↑	↑
F	20	↓	↓	↓	↓	↓	↓	↓	↓	↓	0 1	↑	↓	1 2	2 3	3 4	5 6	7 8	10 11	14 15	21 22	↑	↑	↑	↑	↑	↑
G	32	↓	↓	↓	↓	↓	↓	↓	↓	0 1	↑	↓	1 2	2 3	3 4	5 6	7 8	10 11	14 15	21 22	↑	↑	↑	↑	↑	↑	↑
H	50	↓	↓	↓	↓	↓	↓	↓	0 1	↑	↓	1 2	2 3	3 4	5 6	7 8	10 11	14 15	21 22	↑	↑	↑	↑	↑	↑	↑	↑
J	80	↓	↓	↓	↓	↓	↓	0 1	↑	↓	1 2	2 3	3 4	5 6	7 8	10 11	14 15	21 22	↑	↑	↑	↑	↑	↑	↑	↑	↑
K	125	↓	↓	↓	↓	↓	0 1	↑	↓	1 2	2 3	3 4	5 6	7 8	10 11	14 15	21 22	↑	↑	↑	↑	↑	↑	↑	↑	↑	↑
L	200	↓	↓	↓	↓	0 1	↑	↓	1 2	2 3	3 4	5 6	7 8	10 11	14 15	21 22	↑	↑	↑	↑	↑	↑	↑	↑	↑	↑	↑
M	315	↓	↓	↓	0 1	↑	↓	1 2	2 3	3 4	5 6	7 8	10 11	14 15	21 22	↑	↑	↑	↑	↑	↑	↑	↑	↑	↑	↑	↑
N	500	↓	↓	0 1	↑	↓	1 2	2 3	3 4	5 6	7 8	10 11	14 15	21 22	↑	↑	↑	↑	↑	↑	↑	↑	↑	↑	↑	↑	↑
P	800	↓	0 1	↑	↓	1 2	2 3	3 4	5 6	7 8	10 11	14 15	21 22	↑	↑	↑	↑	↑	↑	↑	↑	↑	↑	↑	↑	↑	↑
Q	1250	0 1	↑	↓	1 2	2 3	3 4	5 6	7 8	10 11	14 15	21 22	↑	↑	↑	↑	↑	↑	↑	↑	↑	↑	↑	↑	↑	↑	↑
R	2000	↑	↑	1 2	2 3	3 4	5 6	7 8	10 11	14 15	21 22	↑	↑	↑	↑	↑	↑	↑	↑	↑	↑	↑	↑	↑	↑	↑	↑

↓ = Use first sampling plan below arrow. If sample size equals, or exceeds, lot or batch size, do 100% inspection

↑ = Use first sampling plan above arrow

Ac = Acceptance number

Re = Rejection number

TABLE 7.4

Single Sampling Plans for Tightened Inspection (MIL-STD-105E—Table II-B)

Sample size code letter	Sample size	Acceptable quality levels (tightened inspection)																									
		0.010	0.015	0.025	0.040	0.065	0.10	0.15	0.25	0.40	0.65	1.0	1.5	2.5	4.0	6.5	10	15	25	40	65	100	150	250	400	650	1000
		Ac Re	Ac Re	Ac Re	Ac Re	Ac Re	Ac Re	Ac Re	Ac Re	Ac Re	Ac Re	Ac Re	Ac Re	Ac Re	Ac Re	Ac Re	Ac Re	Ac Re	Ac Re	Ac Re	Ac Re	Ac Re	Ac Re	Ac Re	Ac Re	Ac Re	Ac Re
A	2	↓	↓	↓	↓	↓	↓	↓	↓	↓	↓	↓	↓	↓	↓	↓	↓	↓	↓	1 2	2 3	3 4	5 6	8 9	12 13	18 19	27 28
B	3	↓	↓	↓	↓	↓	↓	↓	↓	↓	↓	↓	↓	↓	↓	0 1	↓	↓	1 2	2 3	3 4	5 6	8 9	12 13	18 19	27 28	41 42
C	5	↓	↓	↓	↓	↓	↓	↓	↓	↓	↓	↓	↓	↓	0 1	↓	↓	1 2	2 3	3 4	5 6	8 9	12 13	18 19	27 28	41 42	↑
D	8	↓	↓	↓	↓	↓	↓	↓	↓	↓	↓	↓	↓	0 1	↓	↓	1 2	2 3	3 4	5 6	8 9	12 13	18 19	27 28	41 42	↑	↑
E	13	↓	↓	↓	↓	↓	↓	↓	↓	↓	↓	↓	0 1	↓	↓	1 2	2 3	3 4	5 6	8 9	12 13	18 19	27 28	41 42	↑	↑	↑
F	20	↓	↓	↓	↓	↓	↓	↓	↓	↓	↓	0 1	↓	↓	1 2	2 3	3 4	5 6	8 9	12 13	18 19	↑	↑	↑	↑	↑	↑
G	32	↓	↓	↓	↓	↓	↓	↓	↓	↓	0 1	↓	↓	1 2	2 3	3 4	5 6	8 9	12 13	18 19	↑	↑	↑	↑	↑	↑	↑
H	50	↓	↓	↓	↓	↓	↓	↓	↓	0 1	↓	↓	1 2	2 3	3 4	5 6	8 9	12 13	18 19	↑	↑	↑	↑	↑	↑	↑	↑
J	80	↓	↓	↓	↓	↓	↓	↓	0 1	↓	↓	1 2	2 3	3 4	5 6	8 9	12 13	18 19	↑	↑	↑	↑	↑	↑	↑	↑	↑
K	125	↓	↓	↓	↓	↓	↓	0 1	↓	↓	1 2	2 3	3 4	5 6	8 9	12 13	18 19	↑	↑	↑	↑	↑	↑	↑	↑	↑	↑
L	200	↓	↓	↓	↓	↓	0 1	↓	↓	1 2	2 3	3 4	5 6	8 9	12 13	18 19	↑	↑	↑	↑	↑	↑	↑	↑	↑	↑	↑
M	315	↓	↓	↓	↓	0 1	↓	↓	1 2	2 3	3 4	5 6	8 9	12 13	18 19	↑	↑	↑	↑	↑	↑	↑	↑	↑	↑	↑	↑
N	500	↓	↓	↓	0 1	↓	↓	1 2	2 3	3 4	5 6	8 9	12 13	18 19	↑	↑	↑	↑	↑	↑	↑	↑	↑	↑	↑	↑	↑
P	800	↓	↓	0 1	↓	↓	1 2	2 3	3 4	5 6	8 9	12 13	18 19	↑	↑	↑	↑	↑	↑	↑	↑	↑	↑	↑	↑	↑	↑
Q	1250	↓	0 1	↓	↓	1 2	2 3	3 4	5 6	8 9	12 13	18 19	↑	↑	↑	↑	↑	↑	↑	↑	↑	↑	↑	↑	↑	↑	↑
R	2000	0 1	↑	↓	1 2	2 3	3 4	5 6	8 9	12 13	18 19	↑	↑	↑	↑	↑	↑	↑	↑	↑	↑	↑	↑	↑	↑	↑	↑
S	3150			1 2																							

↓ = Use first sampling plan below arrow. If sample size equals, or exceeds, lot or batch size, do 100% inspection

↑ = Use first sampling plan above arrow

Ac = Acceptance number

Re = Rejection number

TABLE 7.5

Single Sampling Plans for Reduced Inspection (MIL-STD-105E—Table II-C)

Sample size code letter	Sample size	Acceptable quality levels (reduced inspection)†																									
		0.010	0.015	0.025	0.040	0.065	0.10	0.15	0.25	0.40	0.65	1.0	1.5	2.5	4.0	6.5	10	15	25	40	65	100	150	250	400	650	1000
		Ac Re	Ac Re	Ac Re	Ac Re	Ac Re	Ac Re	Ac Re	Ac Re	Ac Re	Ac Re	Ac Re	Ac Re	Ac Re	Ac Re	Ac Re	Ac Re	Ac Re	Ac Re	Ac Re	Ac Re	Ac Re	Ac Re	Ac Re	Ac Re	Ac Re	Ac Re
A	2	↓	↓	↓	↓	↓	↓	↓	↓	↓	↓	↓	↓	↓	↓	0 1	↓	↓	1 2	2 3	3 4	5 6	7 8	10 11	14 15	21 22	30 31
B	2	↓	↓	↓	↓	↓	↓	↓	↓	↓	↓	↓	↓	↓	0 1	↑	↓	0 2	1 3	2 4	3 5	5 6	7 8	10 11	14 15	21 22	30 31
C	2	↓	↓	↓	↓	↓	↓	↓	↓	↓	↓	↓	↓	0 1	↑	↓	0 2	1 3	1 4	2 5	3 6	5 8	7 10	10 13	14 17	21 24	↑
D	3	↓	↓	↓	↓	↓	↓	↓	↓	↓	↓	↓	0 1	↑	↓	0 2	1 3	1 4	2 5	3 6	5 8	7 10	10 13	14 17	21 24	↑	↑
E	5	↓	↓	↓	↓	↓	↓	↓	↓	↓	↓	0 1	↑	↓	0 2	1 3	1 4	2 5	3 6	5 8	7 10	10 13	14 17	21 24	↑	↑	↑
F	8	↓	↓	↓	↓	↓	↓	↓	↓	↓	0 1	↑	↓	0 2	1 3	1 4	2 5	3 6	5 8	7 10	10 13	↑	↑	↑	↑	↑	↑
G	13	↓	↓	↓	↓	↓	↓	↓	↓	0 1	↑	↓	0 2	1 3	1 4	2 5	3 6	5 8	7 10	10 13	↑	↑	↑	↑	↑	↑	↑
H	20	↓	↓	↓	↓	↓	↓	↓	0 1	↑	↓	0 2	1 3	1 4	2 5	3 6	5 8	7 10	10 13	↑	↑	↑	↑	↑	↑	↑	↑
J	32	↓	↓	↓	↓	↓	↓	0 1	↑	↓	0 2	1 3	1 4	2 5	3 6	5 8	7 10	10 13	↑	↑	↑	↑	↑	↑	↑	↑	↑
K	50	↓	↓	↓	↓	↓	0 1	↑	↓	0 2	1 3	1 4	2 5	3 6	5 8	7 10	10 13	↑	↑	↑	↑	↑	↑	↑	↑	↑	↑
L	80	↓	↓	↓	↓	0 1	↑	↓	0 2	1 3	1 4	2 5	3 6	5 8	7 10	10 13	↑	↑	↑	↑	↑	↑	↑	↑	↑	↑	↑
M	125	↓	↓	↓	0 1	↑	↓	0 2	1 3	1 4	2 5	3 6	5 8	7 10	10 13	↑	↑	↑	↑	↑	↑	↑	↑	↑	↑	↑	↑
N	200	↓	↓	0 1	↑	↓	0 2	1 3	1 4	2 5	3 6	5 8	7 10	10 13	↑	↑	↑	↑	↑	↑	↑	↑	↑	↑	↑	↑	↑
P	315	↓	0 1	↑	↓	0 2	1 3	1 4	2 3	3 6	5 8	7 10	10 13	↑	↑	↑	↑	↑	↑	↑	↑	↑	↑	↑	↑	↑	↑
Q	500	0 1	↑	↓	0 2	1 3	1 4	2 5	3 6	5 8	7 10	10 13	↑	↑	↑	↑	↑	↑	↑	↑	↑	↑	↑	↑	↑	↑	↑
R	800	↑	↑	0 2	1 3	1 4	2 5	3 6	5 8	7 10	10 13	↑	↑	↑	↑	↑	↑	↑	↑	↑	↑	↑	↑	↑	↑	↑	↑

↓ = Use first sampling plan below arrow. If sample size equals, or exceeds, lot or batch size, do 100% inspection

↑ = Use first sampling plan above arrow

Ac = Acceptance number

Re = Rejection number

† = If the acceptance number has been exceeded, but the rejection number has not been reached, accept the lot, but reinstate normal inspection

TABLE 7.6

Double Sampling Plans for Normal Inspection (MIL-STD-105E—Table III-A)

Sample size code letter	Sample	Sample size	Cumulative sample size	Acceptable quality levels (normal inspection)																									
				0.010	0.015	0.025	0.040	0.065	0.10	0.15	0.25	0.40	0.65	1.0	1.5	2.5	4.0	6.5	10	15	25	40	65	100	150	250	400	650	1000
				Ac Re	Ac Re	Ac Re	Ac Re	Ac Re	Ac Re	Ac Re	Ac Re	Ac Re	Ac Re	Ac Re	Ac Re	Ac Re	Ac Re	Ac Re	Ac Re	Ac Re	Ac Re	Ac Re	Ac Re	Ac Re	Ac Re	Ac Re	Ac Re	Ac Re	Ac Re
A				↓	↓	↓	↓	↓	↓	↓	↓	↓	↓	↓	↓	↓	↓	*	↓	↓	*	*	*	*	*	*	*	*	*
B	First	2	2	↓	↓	↓	↓	↓	↓	↓	↓	↓	↓	↓	↓	↓	*	↑	↓	0 2	0 3	1 4	2 5	3 7	5 9	7 11	11 16	17 22	25 31
	Second	2	4																	1 2	3 4	4 5	6 7	8 9	12 13	18 19	26 27	37 38	56 57
C	First	3	3	↓	↓	↓	↓	↓	↓	↓	↓	↓	↓	↓	↓	*	↑	↓	0 2	0 3	1 4	2 5	3 7	5 9	7 11	11 16	17 22	25 31	↑
	Second	3	6																1 2	3 4	4 5	6 7	8 9	12 13	18 19	26 27	37 38	56 57	
D	First	5	5	↓	↓	↓	↓	↓	↓	↓	↓	↓	↓	↓	*	↑	↓	0 2	0 3	1 4	2 5	3 7	5 9	7 11	11 16	17 22	25 31	↑	↑
	Second	5	10															1 2	3 4	4 5	6 7	8 9	12 13	18 19	26 27	37 38	56 57		
E	First	8	8	↓	↓	↓	↓	↓	↓	↓	↓	↓	↓	*	↑	↓	0 2	0 3	1 4	2 5	3 7	5 9	7 11	11 16	17 22	25 31	↑	↑	↑
	Second	8	16													1 2	3 4	4 5	6 7	8 9	12 13	18 19	26 27	37 38	56 57				
F	First	13	13	↓	↓	↓	↓	↓	↓	↓	↓	↓	*	↑	↓	0 2	0 3	1 4	2 5	3 7	5 9	7 11	11 16	↑	↑	↑	↑	↑	↑
	Second	13	26													1 2	3 4	4 5	6 7	8 9	12 13	18 19	26 27						
G	First	20	20	↓	↓	↓	↓	↓	↓	↓	↓	*	↑	↓	0 2	0 3	1 4	2 5	3 7	5 9	7 11	11 16	↑	↑	↑	↑	↑	↑	↑
	Second	20	40												1 2	3 4	4 5	6 7	8 9	12 13	18 19	26 27							
H	First	32	32	↓	↓	↓	↓	↓	↓	↓	*	↑	↓	0 2	0 3	1 4	2 5	3 7	5 9	7 11	11 16	↑	↑	↑	↑	↑	↑	↑	↑
	Second	32	64											1 2	3 4	4 5	6 7	8 9	12 13	18 19	26 27								
J	First	50	50	↓	↓	↓	↓	↓	↓	*	↑	↓	0 2	0 3	1 4	2 5	3 7	5 9	7 11	11 16	↑	↑	↑	↑	↑	↑	↑	↑	↑
	Second	50	100										1 2	3 4	4 5	6 7	8 9	12 13	18 19	26 27									
K	First	80	80	↓	↓	↓	↓	↓	*	↑	↓	0 2	0 3	1 4	2 5	3 7	5 9	7 11	11 16	↑	↑	↑	↑	↑	↑	↑	↑	↑	↑
	Second	80	160									1 2	3 4	4 5	6 7	8 9	12 13	18 19	26 27										
L	First	125	125	↓	↓	↓	↓	*	↑	↓	0 2	0 3	1 4	2 5	3 7	5 9	7 11	11 16	↑	↑	↑	↑	↑	↑	↑	↑	↑	↑	↑
	Second	125	250								1 2	3 4	4 5	6 7	8 9	12 13	18 19	26 27											
M	First	200	200	↓	↓	↓	*	↑	↓	0 2	0 3	1 4	2 5	3 7	5 9	7 11	11 16	↑	↑	↑	↑	↑	↑	↑	↑	↑	↑	↑	↑
	Second	200	400							1 2	3 4	4 5	6 7	8 9	12 13	18 19	26 27												
N	First	315	315	↓	↓	*	↑	↓	0 2	0 3	1 4	2 5	3 7	5 9	7 11	11 16	↑	↑	↑	↑	↑	↑	↑	↑	↑	↑	↑	↑	↑
	Second	315	630						1 2	3 4	4 5	6 7	8 9	12 13	18 19	26 27													
P	First	500	500	↓	*	↑	↓	0 2	0 3	1 4	2 5	3 7	5 9	7 11	11 16	↑	↑	↑	↑	↑	↑	↑	↑	↑	↑	↑	↑	↑	↑
	Second	500	1000					1 2	3 4	4 5	6 7	8 9	12 13	18 19	26 27														
Q	First	800	800	*	↑	↓	0 2	0 3	1 4	2 5	3 7	5 9	7 11	11 16	↑	↑	↑	↑	↑	↑	↑	↑	↑	↑	↑	↑	↑	↑	↑
	Second	800	1600				1 2	3 4	4 5	6 7	8 9	12 13	18 19	26 27															
R	First	1250	1250	↑	↑	0 2	0 3	1 4	2 5	3 7	5 9	7 11	11 16	↑	↑	↑	↑	↑	↑	↑	↑	↑	↑	↑	↑	↑	↑	↑	↑
	Second	1250	2500			1 2	3 4	4 5	6 7	8 9	12 13	18 19	26 27																

↓ = Use first sampling plan below arrow. if sample site equals, or exceeds lot or batch size, do 100% inspection

↑ = Use first sampling plan below arrow

Ac = Acceptance number

Re = Rejection number

* = Use corresponding single sampling plan (or alternatively, use double sampling plan below, where available)

TABLE 7.7

Double Sampling Plans for Tightened Inspection (MIL-STD-105E—Table III-B)

Sample size code letter	Sample	Sample size	Cumulative sample size	Acceptable quality levels (tightened inspection)																									
				0.010	0.015	0.025	0.040	0.065	0.10	0.15	0.25	0.40	0.65	1.0	1.5	2.5	4.0	6.5	10	15	25	40	65	100	150	250	400	650	1000
				Ac Re	Ac Re	Ac Re	Ac Re	Ac Re	Ac Re	Ac Re	Ac Re	Ac Re	Ac Re	Ac Re	Ac Re	Ac Re	Ac Re	Ac Re	Ac Re	Ac Re	Ac Re	Ac Re	Ac Re	Ac Re	Ac Re	Ac Re	Ac Re	Ac Re	Ac Re
A				⇩	⇩	⇩	⇩	⇩	⇩	⇩	⇩	⇩	⇩	⇩	⇩	⇩	⇩	⇩	⇩	⇩	⇩	*	*	*	*	*	*	*	*
B	First	2	2	⇩	⇩	⇩	⇩	⇩	⇩	⇩	⇩	⇩	⇩	⇩	⇩	⇩	⇩	*	⇩	⇩	0 2	0 3	1 4	2 5	3 7	6 10	9 14	15 20	23 29
	Second	2	4																		1 2	3 4	4 5	6 7	11 12	15 16	23 24	34 35	52 53
C	First	3	3	⇩	⇩	⇩	⇩	⇩	⇩	⇩	⇩	⇩	⇩	⇩	⇩	⇩	*	⇩	⇩	0 2	0 3	1 4	2 5	3 7	6 10	9 14	15 20	23 29	⇧
	Second	3	6																	1 2	3 4	4 5	6 7	11 12	15 16	23 24	34 35	52 53	
D	First	5	5	⇩	⇩	⇩	⇩	⇩	⇩	⇩	⇩	⇩	⇩	⇩	⇩	*	⇩	⇩	0 2	0 3	1 4	2 5	3 7	6 10	9 14	15 20	23 29	⇧	⇧
	Second	5	10																1 2	3 4	4 5	6 7	11 12	15 16	23 24	34 35	52 53		
E	First	8	8	⇩	⇩	⇩	⇩	⇩	⇩	⇩	⇩	⇩	⇩	⇩	*	⇩	⇩	0 2	0 3	1 4	2 5	3 7	6 10	9 14	15 20	23 29	⇧	⇧	⇧
	Second	8	16															1 2	3 4	4 5	6 7	11 12	15 16	23 24	34 35	52 53			
F	First	13	13	⇩	⇩	⇩	⇩	⇩	⇩	⇩	⇩	⇩	⇩	*	⇩	⇩	0 2	0 3	1 4	2 5	3 7	6 10	9 14	⇧	⇧	⇧	⇧	⇧	⇧
	Second	13	26														1 2	3 4	4 5	6 7	11 12	15 16	23 24						
G	First	20	20	⇩	⇩	⇩	⇩	⇩	⇩	⇩	⇩	⇩	*	⇩	⇩	0 2	0 3	1 4	2 5	3 7	6 10	9 14	⇧	⇧	⇧	⇧	⇧	⇧	⇧
	Second	20	40													1 2	3 4	4 5	6 7	11 12	15 16	23 24							
H	First	32	32	⇩	⇩	⇩	⇩	⇩	⇩	⇩	⇩	*	⇩	⇩	0 2	0 3	1 4	2 5	3 7	6 10	9 14	⇧	⇧	⇧	⇧	⇧	⇧	⇧	⇧
	Second	32	64												1 2	3 4	4 5	6 7	11 12	15 16	23 24								
J	First	50	50	⇩	⇩	⇩	⇩	⇩	⇩	⇩	*	⇩	⇩	0 2	0 3	1 4	2 5	3 7	6 10	9 14	⇧	⇧	⇧	⇧	⇧	⇧	⇧	⇧	⇧
	Second	50	100											1 2	3 4	4 5	6 7	11 12	15 16	23 24									
K	First	80	80	⇩	⇩	⇩	⇩	⇩	⇩	*	⇩	⇩	0 2	0 3	1 4	2 5	3 7	6 10	9 14	⇧	⇧	⇧	⇧	⇧	⇧	⇧	⇧	⇧	⇧
	Second	80	160										1 2	3 4	4 5	6 7	11 12	15 16	23 24										
L	First	125	125	⇩	⇩	⇩	⇩	⇩	*	⇩	⇩	0 2	0 3	1 4	2 5	3 7	6 10	9 14	⇧	⇧	⇧	⇧	⇧	⇧	⇧	⇧	⇧	⇧	⇧
	Second	125	250									1 2	3 4	4 5	6 7	11 12	15 16	23 24											
M	First	200	200	⇩	⇩	⇩	⇩	*	⇩	⇩	0 2	0 3	1 4	2 5	3 7	6 10	9 14	⇧	⇧	⇧	⇧	⇧	⇧	⇧	⇧	⇧	⇧	⇧	⇧
	Second	200	400								1 2	3 4	4 5	6 7	11 12	15 16	23 24												
N	First	315	315	⇩	⇩	⇩	*	⇩	⇩	0 2	0 3	1 4	2 5	3 7	6 10	9 14	⇧	⇧	⇧	⇧	⇧	⇧	⇧	⇧	⇧	⇧	⇧	⇧	⇧
	Second	315	630							1 2	3 4	4 5	6 7	11 12	15 16	23 24													
P	First	500	500	⇩	⇩	*	⇩	⇩	0 2	0 3	1 4	2 5	3 7	6 10	9 14	⇧	⇧	⇧	⇧	⇧	⇧	⇧	⇧	⇧	⇧	⇧	⇧	⇧	⇧
	Second	500	1000						1 2	3 4	4 5	6 7	11 12	15 16	23 24														
Q	First	800	800	⇩	*	⇩	⇩	0 2	0 3	1 4	2 5	3 7	6 10	9 14	⇧	⇧	⇧	⇧	⇧	⇧	⇧	⇧	⇧	⇧	⇧	⇧	⇧	⇧	⇧
	Second	800	1600					1 2	3 4	4 5	6 7	11 12	15 16	23 24															
R	First	1250	1250	*	⇧	⇩	0 2	0 3	1 4	2 5	3 7	6 10	9 14	⇧	⇧	⇧	⇧	⇧	⇧	⇧	⇧	⇧	⇧	⇧	⇧	⇧	⇧	⇧	⇧
	Second	1250	2500				1 2	3 4	4 5	6 7	11 12	15 16	23 24																
S	First	2000	2000			0 2																							
	Second	2000	4000			1 2																							

⇩ = Use first sampling plan below arrow. if sample site equals, or exceeds lot or batch size, do 100% inspection
⇧ = Use first sampling plan below arrow
Ac = Acceptance number
Re = Rejection number
* = Use corresponding single sampling plan (or alternatively, use double sampling plan below, where available)

TABLE 7.8

Double Sampling Plans for Reduced Inspection (MIL-STD-105E—Table III-C)

Sample size code letter	Sample	Sample size	Cumulative sample size	0.010	0.015	0.025	0.040	0.065	0.10	0.15	0.25	0.40	0.65	1.0	1.5	2.5	4.0	6.5	10	15	25	40	65	100	150	250	400	650	1000
				Acceptable quality levels (reduced inspection)†																									
				Ac Re	Ac Re	Ac Re	Ac Re	Ac Re	Ac Re	Ac Re	Ac Re	Ac Re	Ac Re	Ac Re	Ac Re	Ac Re	Ac Re	Ac Re	Ac Re	Ac Re	Ac Re	Ac Re	Ac Re	Ac Re	Ac Re	Ac Re	Ac Re	Ac Re	Ac Re
A				↓	↓	↓	↓	↓	↓	↓	↓	↓	↓	↓	↓	↓	↓	*	↓	↓	*	*	*	*	*	*	*	*	*
B				↓	↓	↓	↓	↓	↓	↓	↓	↓	↓	↓	↓	↓	*	↑	↓	*	*	*	*	*	*	*	*	*	*
C				↓	↓	↓	↓	↓	↓	↓	↓	↓	↓	↓	↓	*	↑	↓	*	*	*	*	*	*	*	*	*	*	↑
D	First	2	2	↓	↓	↓	↓	↓	↓	↓	↓	↓	↓	↓	*	↑	↓	0 2	0 3	0 4	0 4	1 5	2 7	3 8	5 10	7 12	11 17	↑	↑
	Second	2	4	↓	↓	↓	↓	↓	↓	↓	↓	↓	↓	↓		↑	↓	0 2	0 4	1 5	3 6	4 7	6 9	8 12	12 16	18 22	26 30	↑	↑
E	First	3	3	↓	↓	↓	↓	↓	↓	↓	↓	↓	↓	*	↑	↓	0 2	0 3	0 4	0 4	1 5	2 7	3 8	5 10	7 12	11 17	↑	↑	↑
	Second	3	6	↓	↓	↓	↓	↓	↓	↓	↓	↓	↓		↑	↓	0 2	0 4	1 5	3 6	4 7	6 9	8 12	12 16	18 22	26 30	↑	↑	↑
F	First	5	5	↓	↓	↓	↓	↓	↓	↓	↓	↓	*	↑	↓	0 2	0 3	0 4	0 4	1 5	2 7	3 8	5 10	↑	↑	↑	↑	↑	↑
	Second	5	10	↓	↓	↓	↓	↓	↓	↓	↓	↓		↑	↓	0 2	0 4	1 5	3 6	4 7	6 9	8 12	12 16	↑	↑	↑	↑	↑	↑
G	First	8	8	↓	↓	↓	↓	↓	↓	↓	↓	*	↑	↓	0 2	0 3	0 4	0 4	1 5	2 7	3 8	5 10	↑	↑	↑	↑	↑	↑	↑
	Second	8	16	↓	↓	↓	↓	↓	↓	↓	↓		↑	↓	0 2	0 4	1 5	3 6	4 7	6 9	8 12	12 16	↑	↑	↑	↑	↑	↑	↑
H	First	13	13	↓	↓	↓	↓	↓	↓	↓	*	↑	↓	0 2	0 3	0 4	0 4	1 5	2 7	3 8	5 10	↑	↑	↑	↑	↑	↑	↑	↑
	Second	13	26	↓	↓	↓	↓	↓	↓	↓		↑	↓	0 2	0 4	1 5	3 6	4 7	6 9	8 12	12 16	↑	↑	↑	↑	↑	↑	↑	↑
J	First	20	20	↓	↓	↓	↓	↓	↓	*	↑	↓	0 2	0 3	0 4	0 4	1 5	2 7	3 8	5 10	↑	↑	↑	↑	↑	↑	↑	↑	↑
	Second	20	40	↓	↓	↓	↓	↓	↓		↑	↓	0 2	0 4	1 5	3 6	4 7	6 9	8 12	12 16	↑	↑	↑	↑	↑	↑	↑	↑	↑
K	First	32	32	↓	↓	↓	↓	↓	*	↑	↓	0 2	0 3	0 4	0 4	1 5	2 7	3 8	5 10	↑	↑	↑	↑	↑	↑	↑	↑	↑	↑
	Second	32	64	↓	↓	↓	↓	↓		↑	↓	0 2	0 4	1 5	3 6	4 7	6 9	8 12	12 16	↑	↑	↑	↑	↑	↑	↑	↑	↑	↑
L	First	50	50	↓	↓	↓	↓	*	↑	↓	0 2	0 3	0 4	0 4	1 5	2 7	3 8	5 10	↑	↑	↑	↑	↑	↑	↑	↑	↑	↑	↑
	Second	50	100	↓	↓	↓	↓		↑	↓	0 2	0 4	1 5	3 6	4 7	6 9	8 12	12 16	↑	↑	↑	↑	↑	↑	↑	↑	↑	↑	↑
M	First	80	80	↓	↓	↓	*	↑	↓	0 2	0 3	0 4	0 4	1 5	2 7	3 8	5 10	↑	↑	↑	↑	↑	↑	↑	↑	↑	↑	↑	↑
	Second	80	160	↓	↓	↓		↑	↓	0 2	0 4	1 5	3 6	4 7	6 9	8 12	12 16	↑	↑	↑	↑	↑	↑	↑	↑	↑	↑	↑	↑
N	First	125	125	↓	↓	*	↑	↓	0 2	0 3	0 4	0 4	1 5	2 7	3 8	5 10	↑	↑	↑	↑	↑	↑	↑	↑	↑	↑	↑	↑	↑
	Second	125	250	↓	↓		↑	↓	0 2	0 4	1 5	3 6	4 7	6 9	8 12	12 16	↑	↑	↑	↑	↑	↑	↑	↑	↑	↑	↑	↑	↑
P	First	200	200	↓	*	↑	↓	0 2	0 3	0 4	0 4	1 5	2 7	3 8	5 10	↑	↑	↑	↑	↑	↑	↑	↑	↑	↑	↑	↑	↑	↑
	Second	200	400	↓		↑	↓	0 2	0 4	1 5	3 6	4 7	6 9	8 12	12 16	↑	↑	↑	↑	↑	↑	↑	↑	↑	↑	↑	↑	↑	↑
Q	First	315	315	*	↑	↓	0 2	0 3	0 4	0 4	1 5	2 7	3 8	5 10	↑	↑	↑	↑	↑	↑	↑	↑	↑	↑	↑	↑	↑	↑	↑
	Second	315	630		↑	↓	0 2	0 4	1 5	3 6	4 7	6 9	8 12	12 16	↑	↑	↑	↑	↑	↑	↑	↑	↑	↑	↑	↑	↑	↑	↑
R	First	500	500	↑	↑	0 2	0 3	0 4	0 4	1 5	2 7	3 8	5 10	↑	↑	↑	↑	↑	↑	↑	↑	↑	↑	↑	↑	↑	↑	↑	↑
	Second	500	1000	↑	↑	0 2	0 4	1 5	3 6	4 7	6 9	8 12	12 16	↑	↑	↑	↑	↑	↑	↑	↑	↑	↑	↑	↑	↑	↑	↑	↑

↑ = Use first sampling plan below arrow. if sample site equals, or exceeds lot or batch size, do 100% inspection
↓ = Use first sampling plan below arrow
Ac = Acceptance number
Re = Rejection number
* = Use corresponding single sampling plan (or alternatively, use double sampling plan below, where available)
† = If, after the second sample the acceptance number has been exceded, but the rejection number has not been reached,accept the lot reinstate normal inspection

Table X-A – Tables for sample size code letter: A individual plans

Chart A - Operating characteristic curves for single sampling plans

(Curves for double and multiple sampling are matched as closely as practicable)

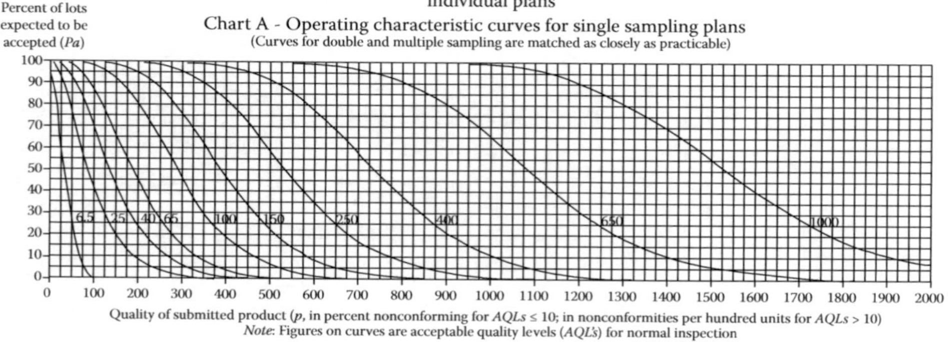

Table X-A-1–Tabulated values for operating characteristic curves for single sampling plans

Pa	Acceptable quality levels (normal inspection)														
	6.5	6.5	25	40	65	100	150	X	250	X	400	X	650	X	1000
	p (in percent nonconforming)	p (in nonconformities per hundred units)													
99.0	0.501	0.51	7.45	21.8	41.2	89.2	145	175	239	305	374	517	629	859	977
95.0	2.53	2.56	17.8	40.9	68.3	131	199	235	308	385	462	622	745	995	1122
90.0	5.13	5.25	26.6	55.1	87.3	158	233	272	351	432	515	684	812	1073	1206
75.0	13.4	14.4	48.1	86.8	127	211	298	342	431	521	612	795	934	1314	1354
50.0	29.3	34.7	83.9	134	184	284	383	433	533	633	733	933	1083	1383	1533
25.0	50.0	69.3	135	196	256	371	484	540	651	761	870	1087	1248	1568	1728
10.0	68.4	115	195	266	334	464	589	650	770	889	1006	1238	1409	1748	1916
5.0	77.6	150	237	315	388	526	657	722	848	972	1094	1334	1512	1862	2035
1.0	90.0	230	332	420	502	655	800	870	1007	1141	1272	1529	1718	2088	2270
	X	X	40	65	100	150	X	250	X	400	X	650	X	1000	X
	Acceptable quality levels (tightened inspection)														

Note: Binomial distribution used for percent nonconforming computations: Poisson for nonconformities per hundred units

FIGURE 7.11
Sample OC curves for MIL-STD-105E sampling plans (MIL-STD-105E, Table X-A and X-A-1).

and rejection numbers, the lot will be accepted, but normal inspection will be restored from the next lot. This happens only when using a reduced inspection plan.

Example 7.8

Choose DSPs for normal, tightened, and reduced inspection if the lot size is 200 and the *AQL* is 2.5%.

Solution

Again, the code letter is G. The DSP for normal inspection from Table 7.6 is:

$$(n_1, c_1, r_1, n_2, c_2, r_2) : (20, 0, 3, 20, 3, 4).$$

The DSP for tightened inspection from Table 7.7 is (20, 0, 2, 20, 1, 2), and that for reduced inspection from Table 7.8 is (8, 0, 3, 8, 0, 4).

The notations in the above examples must be clear. The gap between the acceptance number and the rejection number noticed in the case of reduced SSP exists in the reduced DSP as well. If the total number of defectives from both the first and second sample falls between c_2 and r_2 when using the reduced inspection plan, then the lot will be accepted, but normal inspection will be restored with the next lot inspected.

7.7.6 Average Outgoing Quality Limit

Average outgoing quality (*AOQ*) arises in the context of rectifying inspection, in which rejected lots from inspection by a sampling plan are subject to 100% inspection and all rejected units are replaced with acceptable units. This process of sorting rejected lots and making them 100% good quality before accepting them is known as "detailing." The scheme of a rectifying inspection is shown in Figure 7.12.

The output from this type of rectifying inspection consists of lots that are accepted in the first inspection with the original level of quality, and rejected lots that are detailed, coming in with 100% good units. The quality of the output from a rectifying inspection is indexed by "Average Outgoing Quality" (*AOQ*), which is the average quality of the lots (average proportion defective in each lot) leaving the rectifying inspection system. The *AOQ* is calculated using the formula:

$$AOQ = p \times Pa(p),$$

where p is the proportion defectives in the incoming lots and $Pa(p)$ is the probability of acceptance by the sampling plan, or the proportion of the original number of lots accepted by the sampling plan.

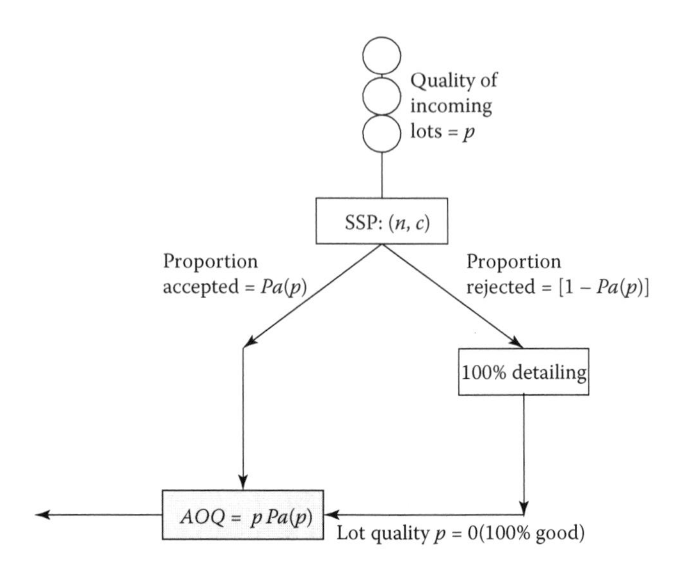

FIGURE 7.12
Rectifying inspection scheme and *AOQ*.

To see how this formula gives the *AOQ*, suppose that 1000 lots of quality p are submitted to the inspection station. Let the probability of acceptance by the sampling plan of lots of p quality be $Pa(p)$. Then, $1000Pa(p)$ of the lots will be accepted at the inspection station and will each have p fraction defectives. In addition, $1000(1 - Pa(p))$ of the lots will be rejected and detailed, and they will all have zero defectives in them. Overall, in the 1000 outgoing lots there will be $p \times 1000Pa(p)$ defectives. Thus, the average number of defectives per lot in the outgoing lots will be:

$$\left(p \times 1000 \times Pa\left(p\right)\right) / 1000 = pPa\left(p\right).$$

The example below shows how the *AOQ* is calculated.

Example 7.9

An SSP is used at an inspection station with $n = 20$ and $c = 1$. Rejected lots from inspection are inspected 100%, and defective units are replaced with good units. Find the *AOQ* for the various incoming lot quality (p values) shown in the table below.

Solution

The calculations are shown in the table below, and a graph of the *AOQ* verus p is shown in Figure 7.13. The $Pa(p)$ is the same as the OC function we calculated for the same SSP in an earlier example, Example 7.1.

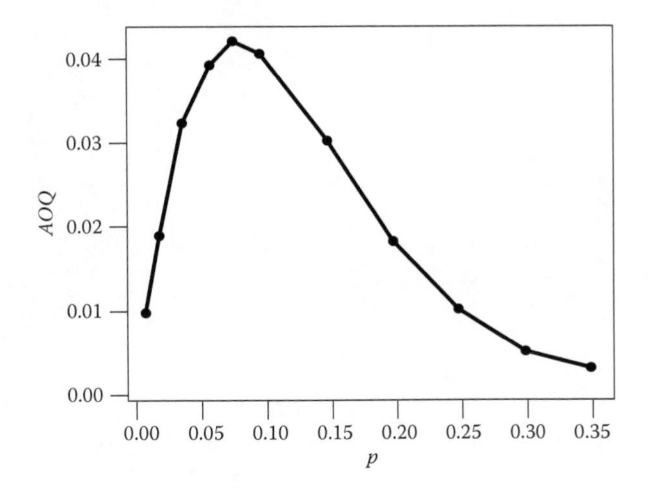

FIGURE 7.13
Graph of *AOQ* vs. incoming quality.

The graph of AOQ in Figure 7.13 shows the typical behavior of *AOQ*. The *AOQ* is small (good) with small (good) values for incoming quality *p*, and it is also small (good) with large (bad) values of *p*. This is because when the incoming quality is good, a large proportion of lots will be accepted in the first inspection, and the small proportion of rejected lots will be detailed. Thus, the *AOQ* will be good, because a large proportion of the accepted lots have only a small proportion of defectives. When the incoming quality is bad, many lots will be rejected on first inspection and rectified at detailing; as a result, a large proportion of lots will have 100% good quality. The *AOQ* will be good in this case as well. The *AOQ* hits a maximum value at a quality level that is in between the two extremes. This maximum value of *AOQ* is called the "average outgoing quality limit" (*AOQL*), and it is used as an index of the performance of sampling plans when a rectifying inspection is used. It represents the worst possible output quality when using a sampling plan with a rectifying inspection

p	**0.01**	**0.02**	**0.04**	**0.06**	**0.08**	**0.10**	**0.15**	**0.20**	**0.25**	**0.3**	**0.35**
$np = (20p)$	0.2	0.4	0.8	1.2	1.6	2.0	3.0	4.0	5.0	6.0	7.0
$Pa(p)$	0.982	0.938	0.808	0.662	0.524	0.406	0.199	0.092	0.04	0.017	0.007
$AOQ = pPa$	0.01	0.019	0.032	0.040	0.042	0.041	0.030	0.018	0.01	0.005	0.003

The MIL-STD-105E provides tables of factors for calculating the *AOQL* for sampling plans. Instructions are provided in the tables on how to calculate the *AOQL* for different sampling plans using the factors given in the tables. An example of an *AOQL* table for single sampling, normal, and tightened inspection plans is shown in Table 7.9. To illustrate how to use this table, suppose an SSP is used for lot size = 200 and *AQL* = 2.5%, then the code letter

TABLE 7.9

Factors for Calculating *AOQL* Values for Sampling Plans from MIL-STD-105E—an Example [MIL-STD-105E, Table V-A] for Single Sampling Normal Inspection

Code letter	Sample size	Acceptable quality levels																									
		0.010	0.015	0.025	0.040	0.065	0.10	0.15	0.25	0.40	0.65	1.0	1.5	2.5	4.0	6.5	10	15	25	40	65	100	150	250	400	650	1000
A	2															18			42	69	97	160	220	330	470	730	1100
B	3														12			28	45	65	110	150	220	310	490	720	1100
C	5													7.4			17	27	39	63	90	130	190	290	430	660	
D	8												4.6			11	17	24	40	56	82	120	180	270	410		
E	13											2.8			6.5	11	15	24	34	50	72	110	170	250			
F	20										1.8			4.2	6.9	9.7	16	22	33	47	73						
G	32									1.2			2.6	4.3	6.1	9.9	14	21	29	46							
H	50								0.74			1.7	2.7	3.9	6.3	9.0	13	19	29								
J	80							0.46			1.1	1.7	2.4	4.0	5.6	8.2	12	18									
K	125						0.29			0.67	1.1	1.6	2.5	3.6	5.2	7.5	12										
L	200					0.18			0.42	0.69	0.97	1.6	2.2	3.3	4.7	7.3											
M	315				0.12			0.27	0.44	0.62	1.00	1.4	2.1	3.0	4.7												
N	500			0.074			0.17	0.27	0.39	0.63	0.90	1.3	1.9	2.9													
P	800		0.046			0.11	0.17	0.24	0.40	0.56	0.82	1.2	1.8														
Q	1250	0.029			0.067	0.11	0.16	0.25	0.36	0.52	0.75	1.2															
R	2000			0.042	0.069	0.097	0.16	0.22	0.33	0.47	0.73																

Note: For the exect *AOQL*, the above values must be multiplied by $\left(1 - \frac{\text{Sample size}}{\text{Lot or batch size}}\right)$

will be G, and the plan will have $n = 32$ and $c = 2$. If this plan is used with a rectifying inspection, then the factor for *AOQL* from the Table 7.9 is 4.3%. Using the formula given at the bottom of the table, the exact value of *AOQL* would be 4.3(1 – (32/200)) = 3.612%.

In this chapter, we discussed some of the plans contained in the MIL-STD-105E. There are several other standard tables that give ready-to-use sampling plans, one of which is the Dodge-Romig inspection tables. These tables provide inspection plans indexed by *LTPD* and *AOQL*. The reader is referred to Duncan (1974) or Grant and Leavenworth (1996) for details.

7.7.7 Some Notes about Sampling Plans

7.7.7.1 What Is a Good AQL?

Most companies would settle for an *AQL* value for a product characteristic based on what has worked for them, both functionally and economically, for that product characteristic. Smaller *AQL* values will result in large sample sizes and, hence, in increased cost of inspection. Larger *AQL* values might result in lenient plans that could allow more defectives to pass through inspection, causing losses in the assembly or dissatisfaction to a final customer. The best *AQL* value is the trade-off between the cost of inspection and the cost of not inspecting enough, and mathematical calculations can be made to determine this optimal value. However, in practice, we may start using a value, such as 1.0%, and then adjust it up or down based on how the part or assembly performs at the place it is used, with the best value being selected based on what works satisfactorily.

Different *AQL* values can be used for different products, or characteristics of the same product, depending on the criticality of the characteristic. More critical characteristics or products should be inspected with smaller *AQL* values.

7.7.7.2 Available Choices for AQL Values in the MIL-STD-105E

First, we want to remember that the *AQL* values shown at the head of the columns of the MIL-STD tables are in percentages. Some of them ($AQL \leq 10\%$) are in percent defectives, or number of defectives per 100 units of a sample. Others ($AQL > 10\%$) are in number of defects per 100 units of a sample. The former is used when quality is measured by counting the number of defectives in a sample, and the latter is used when quality is checked by counting the number of defects in each unit of a sample. In the discussion on OC curves of a sampling plan, we assumed that the number of defectives in a sample follows the binomial distribution. However, the cumulative probabilities of the binomial variable were approximated using the Poisson cumulative probabilities. When the number of defects per unit is used as the quality measure, the defects per unit follow the Poison distribution, and the probability of acceptance is calculated using the cumulative Poisson probabilities from the Poisson table.

7.7.7.3 A Common Misconception about Sampling Plans

The *AQL* is used as an index for indexing sampling plans in the MIL-STD-105E. Suppose that a sampling plan chosen based on an *AQL* of 1.5% is used at an incoming inspection station. This does not mean that all the lots accepted at this inspection station will have 1.5% defectives in them. It only means that lots with less than 1.5% defectives will be readily accepted at this inspection station, and lots with more than 1.5% defects will not be readily accepted. The average quality of the lots accepted by a plan with $AQL = 1.5\%$ can be expected to be better than 1.5% (defectives <1.5%), especially if switching rules are applied. The switching rules put psychological pressure on the vendor to supply good quality. If a rectifying inspection is used, the average outgoing quality will certainly be smaller than the *AQL*.

7.7.7.4 Sampling Plans vs. Control Charts

As mentioned in Chapter 4, the control charts are preventive tools that are used during production by the producer. Sampling plans are acceptance tools that are used by the customer before accepting the product for use. If the results of a sampling inspection are fed back to the producer, however, and the producer uses the information to find the causes for the failures and then implements corrective action to repair the processes, then the sampling plans will also serve the same purpose as control charts.

Some people believe that sampling plans, because they are used after the product is produced, have no place in modern quality control systems, which should aim for the continuous reduction of product variability and zero, or near-zero, defective rates. No one can argue against such goals, or against the use of control charts to achieve those ends, but we must recognize that there will always be some suppliers that are unable to prove the quality of their products with evidence from control charts. Until all vendors can prove the quality of their supplies using control charts and capability measurements, customers may have to depend on sampling plans to verify quality at receiving inspection. It is then necessary to understand how the sampling plans work and how to use them correctly.

7.7.7.5 Variable Sampling Plans

The discussion in this chapter has been limited to attribute sampling plans only. Although they may not be the most efficient plans in terms of the number of units to be inspected for a given lot, they are simple to use. When the cost of inspection is high because of the destructive testing of sample units or an elaborate, time-consuming procedure for inspection, efficient sampling plans may be required. In such situations, variable sampling plans (sampling plans using measurements)—which are known to be efficient—may be appropriate. In variable sampling plans, some characteristic, such as length, strength, or percentage carbon, will be measured from the sample units. The average and range (or standard deviation) of the measurements will be calculated, and a measure that includes both the average and range (or standard deviation) will be computed to reflect the quality of the sample. This computed quality measure will be compared with critical values for these measures available in standard tables, which have been calculated to provide the α and β protection levels. One such set of standard tables is available in MIL-STD-414. Use of the variable plans is a bit involved, however, and they may require specially trained personnel to implement them. The reader is referred to the above military standard or books that discuss these plans, such as Wadsworth, Stephens, and Godfrey; (1986) and Grant and Leavenworth (1996).

7.8 Exercise

7.1 Answer the following questions:

- **a.** What are the basic tenets of a good supplier relationship?
- **b.** What are the advantages of having single versus multiple suppliers for purchased parts and materials?
- **c.** What are the criteria for selecting a supplier?
- **d.** What criteria are used by organizations for certifying suppliers?
- **e.** What are the objectives and the methodology of supply chain optimization?

7.2 Draw the OC curve of an SSP with $n = 30$ and $c = 2$. Choose $p = 0.01$, 0.02, 0.05, 0.1, 0.08, 0.15, 0.2, 0.25, and 0.3. (Use the Poisson table or binomial nomograph.)

7.3 Draw the OC curve of an SSP with $n = 60$ and $c = 1$. Choose $p = 0.01$, 0.02, 0.05, 0.08, 0.1, 0.15, and 0.2. (Use the Poisson table or binomial nomograph.)

7.4 An SSP is being used at an inspection station with $n = 90$ and $c = 3$.

- **a.** If the $AQL = 0.03$, what is the value of the producer's risk α?
- **b.** If the $LTPD = 0.08$, what is the value of the consumer's risk β?

7.5 An SSP has $n = 40$ and $c = 2$. If $\alpha = 0.05$ and $\beta = 0.05$, what are the values of the *AQL* and *LTPD*? Use the binomial nomograph.

7.6 Select an SSP for $AQL = 0.015$, $\alpha = 0.01$, $LTPD = 0.1$, and $\beta = 0.02$. Use the binomial nomograph.

7.7 Select an SSP for $AQL = 0.02$, $\alpha = 0.01$, $LTPD = 0.2$, and $\beta = 0.02$. Use the binomial nomograph.

7.8 Prepare the OC curve of a DSP with $n_1 = 50$, $c_1 = 1$, $r_1 = 3$, $n_2 = 50$, and $c_2 = 3$. Choose $p = 0.01, 0.03, 0.05, 0.07$, and 0.1.

7.9 Prepare the OC curve of a DSP with $n_1 = 20$, $c_1 = 0$, $r_1 = 3$, $n_2 = 20$, and $c_2 = 3$. Choose $p = 0.01, 0.05, 0.1$, and 0.15.

7.10 Draw the *ASN* curve of the DSP in Exercise 7.8. Choose an equivalent SSP, and compare the *ASN* of both the DSP and the SSP.

7.11 Draw the *ASN* curve of the DSP in Exercise 7.9. Choose an equivalent SSP, and compare the *ASN* of both the DSP and the SSP.

7.12 Select SSPs for normal, reduced, and tightened inspection from MIL-STD-105E for the following data: lot size = 200, and $AQL = 1.5\%$.

7.13 Select SSPs for normal, reduced, and tightened inspection from MIL-STD-105E for the following data: lot size = 2000, and $AQL = 2.5\%$.

7.14 Select DSPs for normal, reduced, and tightened inspection from MIL-STD-105E for the following data: lot size = 200, and $AQL = 1.5\%$.

7.15 Select DSPs for normal, reduced, and tightened inspection from MIL-STD-105E for the following data: lot size = 2000, and $AQL = 2.5\%$.

7.16 Calculate the *AOQ* for the normal inspection plan from Exercise 7.12 at $p = 0.02$.

7.17 Calculate the *AOQ* for the normal inspection plan from Exercise 7.13 at $p = 0.02$.

References

ANSI/ASQ Z1.5-2008: Sampling Procedures and Tables for Inspection by Attributes, Milwaukee, WI-ASQ.

ASQ, Customer-Supplier Division. 2004. *The Supplier Management Handbook*. 6th ed. Milwaukee, WI: American Society for Quality Control—Quality Press.

Bowker, A. H., and G. J. Lieberman. 1972. *Engineering Statistics*. 2nd ed. Englewood Cliffs, NJ: Prentice Hall.

Deming, W. E. 1986. *Out of the Crisis*. Cambridge, MA: MIT—Center for Advanced Engineering Study.

Duncan, A. J. 1974. *Quality Control and Industrial Statistics*. 4th ed. Homewood, IL: Irwin.

Donovan, J. B., and F. P. Maresca. 1999. "Supplier Relations," In *Juran's Quality Handbook*. 5th ed. Co-edited by J. M. Juran and A. B. Godfrey. New York, NY: McGraw-Hill.

Grant, E. L., and R. S. Leavenworth. 1996. *Statistical Quality Control*. 7th ed. New York, NY: McGraw-Hill.

Juran, J. M., and F. M Gryna. 1993. *Quality Planning and Analysis*. 3rd ed. New York, NY: McGraw-Hill.

Larson, H. R. 1966. "A Nomograph of Cumulative Binomial Distribution." *Industrial Quality Control* 23 (6): 270–278.

Montgomery, D. C. 2001. *Introduction to Statistical Quality Control*. 4th ed. New York, NY: John Wiley.

Schilling, E. G. 1982. *Acceptance Sampling in Quality Control*. New York, NY: Marcel Dekker.

U.S. Department of Defense. 1989. *Military Standard, Sampling Procedures and Tables for Inspection by Attributes (MIL-STD-105E)*. Washington, DC.

Wadsworth, H. M., K. S. Stephens, and A. B. Godfrey. 1986. *Modern Methods for Quality Control and Improvement*. New York, NY: John Wiley and Sons.

8

Continuous Improvement of Quality

8.1 The Need for Continuous Improvement

As has already been pointed out in previous chapters, improving product and service quality on a continuous basis is the key to improving customer satisfaction, increasing productivity, and remaining competitive in the marketplace. Opportunities always exist in productive enterprises for making improvements to the quality of products and services by using the resources already available. Many corporate managers believe that problems relating to quality can only be resolved by investing more capital to buy newer equipment, with all possible automation and the latest bells and whistles. Dr. Deming used to decry this tendency toward blaming lack of quality on lack of new and expensive equipment and then launching into extensive capital investments in the name of quality improvement. He cited several examples in his writings where improvements in quality and productivity were accomplished using the same machinery and resources that had been condemned earlier as being incapable of meeting quality needs. This author has seen this happen and can provide several examples to support this.

Poor quality, waste, and customer dissatisfaction all result largely from a lack of understanding of the processes and a lack of knowledge regarding how process variables and their interactions contribute to the quality of the final product or service. If proper tools are employed to understand the interrelationships among the variables governing a process—and if appropriate levels for these variables are chosen to provide the desired levels of the product characteristics—then quality and productivity can often be improved without much additional investment in new machinery. It does not, however, mean that new machinery will never be needed, just that the capability of existing machinery is more than likely adequate to achieve needed quality improvements.

In this chapter, we will discuss the tools that are used in discovering opportunities for improvement, making the improvements, and maintaining the improved positions. Some of the tools are simple, some involve a certain level of analysis, and some require a bit of advanced mathematics. We will discuss here some of the commonly used tools that are valuable in terms of the impact they make on the improvement process. Before discussing the improvement tools, however, we will discuss the general problem-solving methodology, or the framework for the process of completing improvement projects.

An opportunity for an improvement project exists wherever there is a problem. A problem exists wherever there are undesirable symptoms, such as an unusual number of customer complaints, excessive warranty charges, excessive internal failures, large number of accidents, or excessive worker absenteeism. Wherever the opportunity for a project is identified, the project should be completed by following a systematic process. The sequence of steps needed for solving problems successfully has been described under different names by different authors. We will discuss two such descriptions: one by Dr. Deming, and the other by Dr. Juran.

8.2 The Problem-Solving Methodology

8.2.1 Deming's PDCA Cycle

The PDCA cycle recommended by Dr. Deming has four steps—plan, do, check, and act—to be performed in that sequence. The PDCA cycle is a road map for making continuous improvements. According to Dr. Deming (1986), this sequence of steps was originally recommended by his mentor, Dr. Walter A. Shewhart. When he was teaching quality methods in Japan, Dr. Deming used to refer to this cycle as the "Shewhart Cycle." The Japanese started referring to the cycle as the "Deming Cycle," and that name stuck. The method is now known as the Deming Cycle among modern-day quality professionals. The activities needed in each step are described below:

1. *Plan*: Document and analyze the current process, gather relevant data to understand the causes and their effects, and propose theories on the root causes. Plan for testing the theories.
2. *Do*: Test the theories using experiments on a limited scale in order to understand the relationship among the important variables in the process.
3. *Check*: Check or study the data from the experiment to see if a good understanding of the process and its variables has been obtained. Check if the theories and the solutions proposed from them will produce the desired results, which is the removal of the undesirable symptoms to the desired extent. If not, modify the theories and solutions. This step, according to Dr. Deming, produces the knowledge (of the process)—a prerequisite for making any improvement.
4. *Act*: Implement the modified solutions by specifying the changes, documenting the revised instructions, and standardizing the new process.

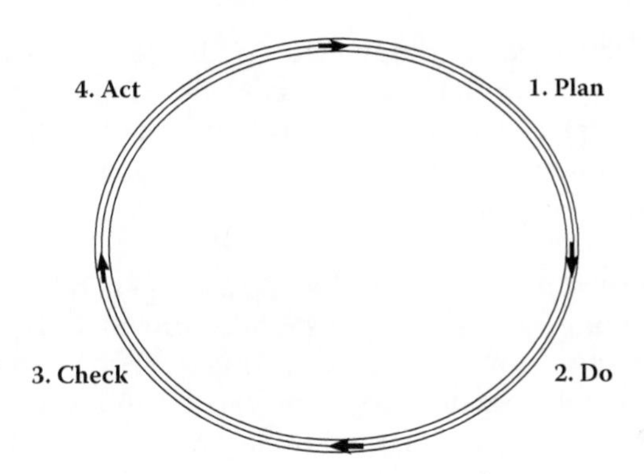

FIGURE 8.1
Deming's PDCA cycle.

The PDCA steps are presented in Figure 8.1, which shows the fourth step leading into the first step. This emphasizes that the PDCA cycle is an iterative process, with the end of the first iteration being the beginning of the next, and the iterations continuing forever. Such continuous iteration of the improvement cycle is necessary for improving quality, increasing productivity, reducing waste, and enhancing customer satisfaction, all leading to overall excellence in business performance.

The PDCA cycle describes, in simple terms, the process of solving a problem. It makes the important point that problem solving is not a one-shot attempt. Although the procedure does not describe the individual steps of problem solving in any great detail, this method was the forerunner of many problem-solving sequences created later by many others. This was also the process that the Japanese used with tremendous success (they referred it by the name "policy deployment") to solve many problems, from product rejection on a production line to product design for satisfying a customer need to strategic plans to transform organizations into quality organizations.

One important point Dr. Deming makes in this context is that "any step in the Shewhart (Deming) Cycle may need guidance of statistical methodology for economy, speed, and protection from faulty conclusions from failure to test and measure the effects of interactions" (Deming 1986, 89). He believed that learning and using statistical methods is critical to understanding the variations in populations, and in drawing proper conclusions about populations from sample observations in the presence of this variation.

8.2.2 Juran's Breakthrough Sequence

Dr. Juran's "breakthrough" approach is discussed in detail in Chapter 9 while discussing his quality trilogy as a model for a quality system. He recommends that improvements to processes and systems be accomplished through a "project-by-project" approach, meaning that quality improvements must be accomplished in discrete steps, with projects completed one by one, on a continuous basis. These projects should be completed following a sequence of steps that he called the "breakthrough sequence." The following steps of the breakthrough sequence are summarized from Juran and Gryna (1993):

1. *Establish proof of need*: Convince the top managers of the necessity for completing improvement projects. The approval and support of top managers is necessary before proceeding further, because help from many parts of the organization is needed for the successful completion of quality projects. Use preliminary data to highlight the extent of current problems and the projected savings to be made from solving them. Use examples of success stories from within or outside the organization.
2. *Identify the project*: There may be many project candidates, and these must be prioritized using their projected economic impact as the criterion. The Pareto analysis (explained later in this chapter) will help in prioritizing the projects. A mission statement, or a project statement, should be made that declares the major scope and boundaries of the project and the goals to be accomplished.
3. *Organize a project team*: Most chronic problems, or the problems that remain embedded in processes and cause unacceptable levels of rejects (see Chapter 9 for Dr. Juran's definition of a chronic problem), need to be addressed by interdepartmental or cross-functional teams of six to eight members. The team should have a leader, elected by the team or appointed by the quality council (i.e., the upper-level, company-wide team of senior managers). The team can use a facilitator who is a person trained and experienced in problem solving and team dynamics. The project team has the responsibility of diagnosing the root causes of problems and recommending solutions to them.
4. *Make the diagnostic journey:* Starting from the symptoms of the problem, analyze the causes for the symptoms by using diagnostic tools, and propose possible solutions. Test the solutions by experimenting on a pilot scale, and then recommend the best based on data gathered.
5. *Make the remedial journey:* Obtain approval from the quality council, and implement the chosen solution by specifying the new process along with its new parameters. There will be resistance to change from workers and managers. Good communication with people; that is, explaining the need for the change and possible benefits to the organization, will help in overcoming such resistance. Treat staff concerns and anxieties relating to change with sensitivity and respect.

6. *Institute controls to hold the gains:* Create new work instructions, and train people in the new methods. Make changes irreversible, and create control points downstream to check if the new method is being properly used.

Several variations of the problem-solving sequence can be found in the literature under different names. All, however, stress the point that problem solving, or project completion, must be done by a cross-functional team in a systematic, organized manner in order to reach a certain level of success. When not done in an organized manner—when people follow the "fire-ready-aim" (phrase borrowed from Carl Saunders of Caterpillar University) sequence in making changes—they end up solving the wrong problems or achieving inadequate results, or simply fail to make any impact on the problem.

The problem-solving methodology can be summarized, based on a study of the sequences suggested by several authors in the following generic steps. The tools used in each of the steps are listed under each one. Several of these tools have already been covered in earlier chapters. Those that have not will be discussed in the remaining part of this chapter. Figure 8.2 captures the essence of the continuous improvement process.

8.2.3 The Generic Problem-Solving Methodology

Step 1: Select a Problem, and Make a Problem Statement

Problems exist wherever there are dissatisfied customers, excessive waste, or low productivity in processes. Stating problems in measurable terms, such as "rejects account for 12% of sales," "customer returns accounted for $1.8 million dollars in the past year," or "market share decreased by 8% in the past two years" has more precision and impact than making vague, general statements. Take note of the constraints that may make some solutions infeasible. The constraints could be of a physical, time-based, or financial nature. These should be recognized at the time that the problem statement is formulated. The statement should include the goals and objectives to be accomplished by the time the project is completed.

Project selection can be initiated by calling for nominations from managers and workers to generate a pool of possible projects. If several project opportunities exist, then the one with the most economic impact, or one that will address an urgent need of a customer, should be chosen first.

The tools used in this step include:

Quality costs—to identify areas where deficiencies exist

Benchmarking—to compare the process performance with the best in class, and to select suitable goals

Pareto analysis—to prioritize projects based on their impact

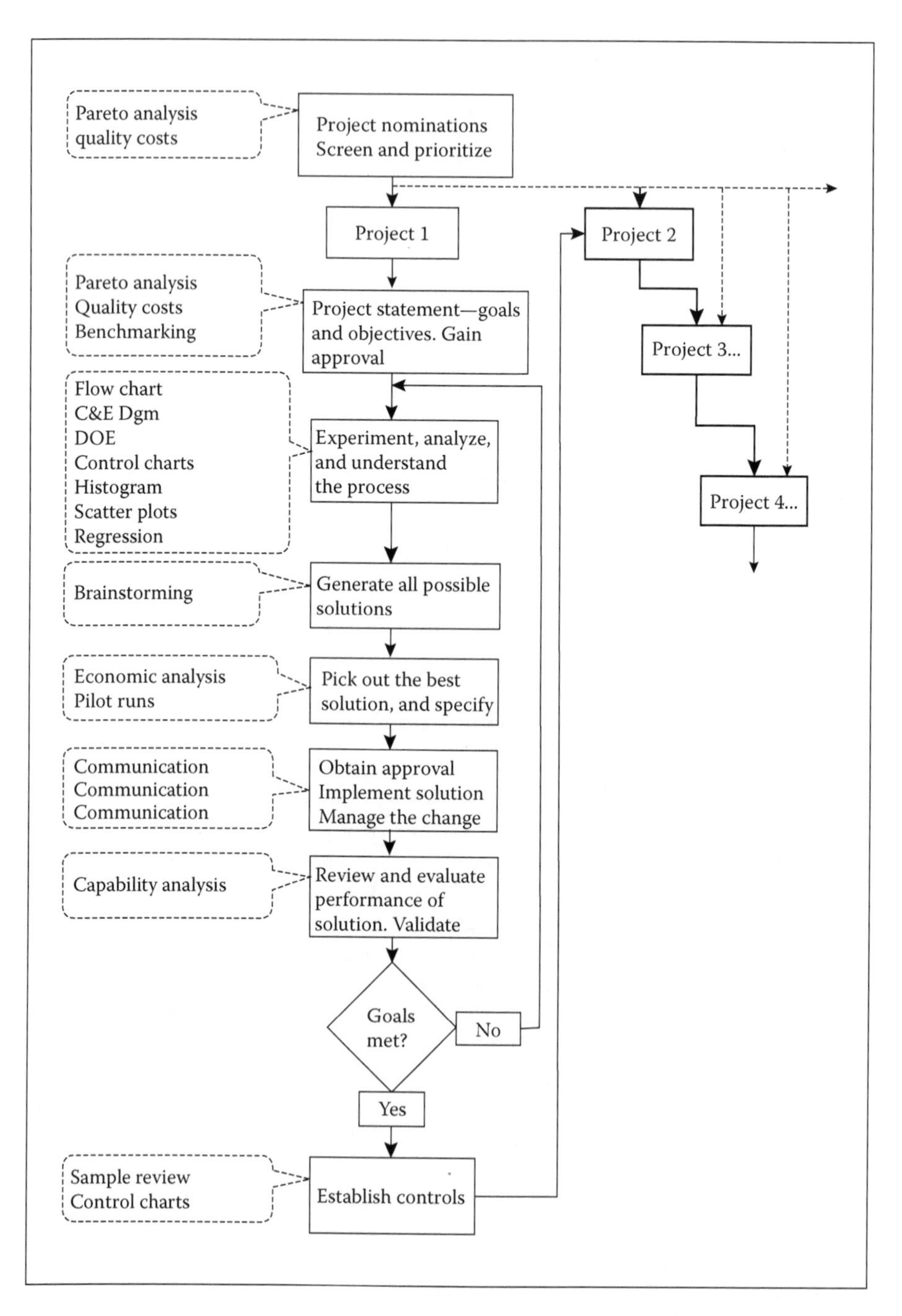

FIGURE 8.2
The process of continuous improvement.

Step 2: Analyze the Problem

Document the current process using process flow charts. Look for all possible causes of the problem, and understand the relationship between cause and effect variables. If the relationships are not apparent, conduct experiments to reveal those relationships. Only when the process is fully understood should solutions be attempted.

The tools used in this step include:

Process flow charts—to understand the relationships among the operations of a process

Cause and effect diagram—to understand everything responsible for causing the problem

Scatter-plot, regression analysis—to understand the relationship between variables

Designed experiments—to understand the relationship between process and product variables

Histogram—to understand the extent of variability in process and product variables

Control charts—to accomplish stability of a process over time

Step 3: Generate Alternative Solutions

There is always more than one way to solve any given problem. For the first attempt, every possible solution should be considered; no solution should be rejected. Brainstorming among the people who have an intimate knowledge of the process, including customers and suppliers, is a good way to obtain the possible solutions.

The tools used in this step include:

Brainstorming, creative thinking—to generate all possible solutions for solving the problem

Step 4: Choose the Best Solution(s)

Some of the solutions generated in Step 3 may not be feasible in the sense that they may not meet the constraints. The infeasible solutions must be eliminated from further consideration. Chose the best solution—or combination of solutions—from among the feasible solutions, using the criteria derived from the objectives laid down for the project. If the objective is to reduce the amount of rejects, the solution that produces the greatest reduction in the amount of rejects would be best. In many situations, the objectives can be translated into economic criteria, such as the amount of dollars saved or additional sales generated. When necessary, a short pilot run would help in verifying the projected performance of a solution or in comparing the performance of different solutions.

The tools used in this step include:

Economic analysis methods (e.g., rate of return, or return on investment calculations)—to compare solutions on economic basis.

Pilot runs—to verify projected benefits from solutions.

Step 5: Implement the Solution and Validate

The chosen solution(s) should be specified fully by making drawings, preparing bid documents for equipment, and writing out instructions for the new methods. The new method should receive approval from the managers who own the process, and the people who operate them. Good communication is very important, and the drawings, sketches, and visual aids can be used for communication. Explaining how the solutions were developed and how they would help to solve the problem at hand will go a long way toward generating acceptance. Implementing the solution may involve some of the team members working with the operating people and showing them how the new method is supposed to work. This would also give an opportunity for team members to see if the solution they generated works as intended. More often than not, solutions do not produce all the intended results during the first implementation. Modifications will then have to be made to the solutions to obtain the desired results.

Validation involves making a full-day or full-shift run using production workers at production capacity and under production conditions. The capability of the processes to hold the parameters at the specified levels, and to produce the product characteristics within specifications, must be established.

The tools used in this step include:

Engineering drawings, process charts, work instructions—to specify the new solution

Capability studies—to verify if the process is capable of meeting customer's specifications

Step 6: Hold the Gains

Processes, including people and hardware, have a tendency to revert back to the old ways if they are not monitored and controlled. Periodic reviews of results from the process, and the inspection and control methods installed downstream of the process are necessary to maintain the integrity of the new process and new methods.

The tools used in this step include:

Control charts—to monitor the behavior of the process

8.3 Quality Improvement Tools

As noted above, several tools are needed during the problem-solving process. Those that have not been covered elsewhere in this book are discussed below.

8.3.1 Cause-and-Effect Diagram

The cause-and-effect (C&E) diagram is the first level of dissecting the process to discover the root causes. This is also known as "Ishikawa's fishbone diagram," as it is named after the Japanese professor Kaoru Ishikawa (1915–1989), who first used it to investigate the causes of quality problems. This is one of those methods that helps a team to think together on paper.

The method consists in systematically identifying all of the sources that might contribute to the undesirable symptom(s) under investigation. This is generally done in a brainstorming session among a team of people who are knowledgeable in the problem area. The C&E diagram not only helps in the investigation of the causes, but also serves as a means of recording the problem-solving process, including the brainstorming sessions. To follow a uniform format, it is suggested (Ishikawa 1985) that the causes be investigated and recorded under the major categories shown as major stems, as shown in the diagram in Figure 8.3.

The C&E diagram is a communication tool, and to avoid confusion, a uniform format should be used in making the diagram. There are situations in which this format cannot be strictly followed. For example, this format may not be suitable when investigating a problem related to a service. Then, the format would have to be suitably modified to fit the problem situation without drastically changing the main structure. The case study that follows illustrates how the C&E diagram is made.

Case Study 8.1

The problem relating to the failure of bonds after curing was investigated by a team of students (Ramachandran and Xiaolan 1999) in a shop that assembles cable harnesses. The bonding takes place between rubber and plastic surfaces, with bonding material applied in between. Pressure is applied, and the bonded surfaces are cured in temperature- and humidity-controlled chamber. A bond is considered to be a failure when it does not meet the strength requirements. The C&E diagram that is shown in Figure 8.4 was prepared in a brainstorming session.

After creating the C&E diagram, the team votes to select the most important causes (usually three or four) that need to be investigated further with experimentation and analysis. Suppose the four top-ranking causes are to be picked. In this case, a simple voting procedure would be for each team member to pick the top-four causes according to his or her understanding of the problem situation, and rank them from one to four. For each cause, the rank given by the members will be totaled. The cause with the smallest total will be ranked first; the cause with the next smallest total will be ranked second, and so on. The causes that are ranked at the top will then be studied in further detail to discover solutions and implement remedies.

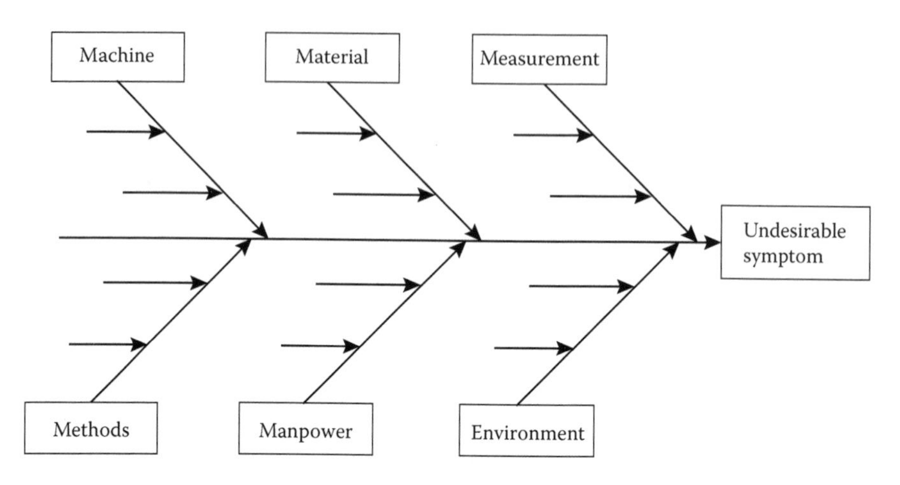

FIGURE 8.3
The general format of a C&E diagram.

8.3.2 Brainstorming

The term "brainstorming" has already been used earlier in this chapter, and there seemed no need to explain its meaning. However, a few remarks about brainstorming are appropriate.

Brainstorming involves exercising the brains of the members of a team engaged in improving a process who possess an intimate knowledge of the process under study. The objective is, with their help, to list all of the causes

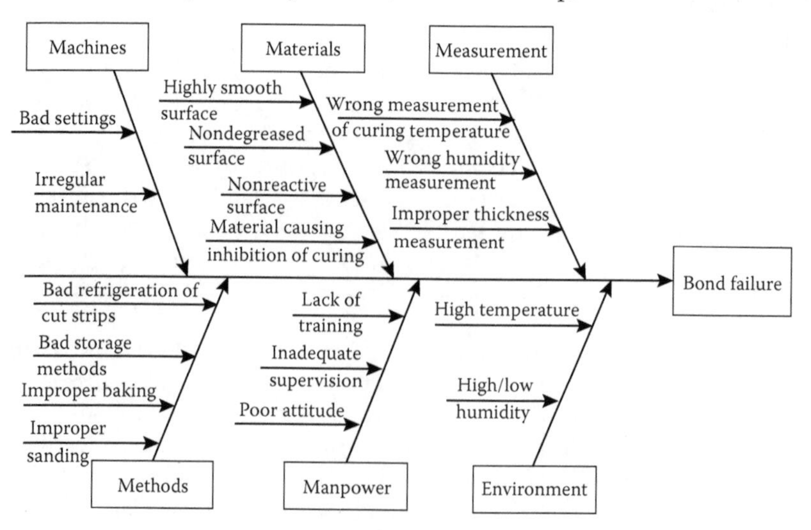

FIGURE 8.4
Example of a C&E diagram for bond failure. (From Ramachandran, B., and P. Xiaolan, "A Quality Control Study of a Bond Curing Process." Unpublished project report, IME 522, IMET Department, Bradley University, Peoria, IL, 1999.)

that contribute to the problem on hand and find possible ways of solving it. A certain set of conventions is usually adopted in those sessions to maximize the information generated:

- A facilitator, who is not part of the team, acts as a moderator and helps the team keep their focus on the goals.
- All ideas generated are recorded on a C&E diagram. No idea is rejected at the initial stages of idea generation, because negative reaction to suggestions might stymie creativity.
- The facilitator goes around the table for ideas, giving each member an opportunity to make his or her contributions. This avoids domination of the sessions by a few aggressive individuals, and helps the backbenchers to come forward with their ideas.
- At the end of a brainstorming session, the ideas are pruned to exclude those that are not relevant or may not be feasible under the given constraints. The resulting diagram is presented to the team for voting and a certain number (three or four) of top candidate causes are chosen for further investigation.

8.3.3 Benchmarking

The principle behind benchmarking is very simple. Suppose a business is looking to improve upon one of their critical performances. They may want, for example, to minimize the number of days needed to resolve a customer complaint or minimize the number of failures in the first 1000 hours of operation of a product. The benchmarking method recommends that the business first find out who has the best performance in those areas within their industry, among the competitors, or any industry where similar functions are performed. Then, they should find out how the best performers, called the best-of-class, accomplish those performances. Using the records of accomplishment of the best performers as goals, and using the procedures they employ, the business should implement an improvement process and continue it until the goals are achieved or exceeded. At the end of this process, the business would have performances in the selected areas that would be better than the best-in-class. Such goal setting and emulation of the methods of the best performers, or benchmarks, is referred to as the "benchmarking" process.

This benchmarking approach has several advantages. When the goals are set using those of the benchmark organization, they have the appeal of being realistic, because they have already been achieved by someone. This is a better alternative to the common practice of goal setting by choosing future goals based on the past history of a business. Goal setting using historical data suffers from the fact that the historical records might have been influenced by causes that had not been apparent to the people close to the processes. Benchmarking will lead people to think differently and to look for ways to make improvements that they had not thought of before. Moreover, customers' expectations are driven by the best-in-class performance. Therefore, basing one's goals on the performances of the best among one's competitors amounts to proactively anticipating customer expectations and taking steps to meet them.

The principle of benchmarking, although simple, needs to have a formal structure to obtain uniformity in approach when used by different people at different parts of a large organization. This formal structure was provided by the pioneering efforts of managers at Xerox (Camp 1989). The structured approach is contained in the 10-step process, summarized below from Camp and DeToro (1999). The process of benchmarking follows the problem-solving sequence in which the objective is to find and follow the best-in-class. The benchmarking process is implemented by a team.

Step 1: Select a Performance Measure to be Benchmarked

The measure to be benchmarked may be chosen as a result of a serious problem arising from customer complaints, loss of market share, or poor financial performance of the company. It could be the result of a new vision for the business and strategic goals established to accomplish that vision. It could also be from an opportunity for improvement that has been identified by a manager or an improvement team. Some examples of measures to be benchmarked are: number of inventory turnovers, time from concept to market, per-capita training dollars, and mean time between failures of production lines.

Step 2: Select the Benchmark Partner(s)

The institution(s) selected as benchmark(s) should be chosen after a search from within the team's organization (internal benchmarking), among the competitors in the same trade (external benchmarking), or in other trades where similar functions are performed (functional benchmarking). The search must be extensive. One rule of thumb is to identify at least 100 prospective benchmarking partners before selecting three or four for detailed study. There could be more than one benchmark for a performance measure, but there should not be too many.

Step 3: Study the Benchmark Partner(s)

This study includes the collection of data both indirectly from public documents, company financial statements, and corporate information to stockholders; and directly from phone calls, questionnaires, or even visits to partner locations. Obtaining data on a particular process becomes easier if the partner can be offered information of interest on other processes within the team's business. This two-way exchange of information for the benefit of both partners facilitates the benchmarking process. The site visit takes place only when there is *prima facie* evidence that the visit would be mutually beneficial to both parties.

A good preparation, including a planned agenda and a prepared set of questions to be asked during the meetings, should precede the visit. The questions should elicit information on the choice of goals and how they were accomplished, the lessons learned, and the roadblocks to be avoided. The questions should be worded discretely and not ask for information that the partner would rather not release. Some legal and ethical issues regarding intellectual property rights may arise while studying external benchmarking partners. The International Benchmarking Clearing House (Camp 1989) has developed a code of conduct to follow while studying a partner for benchmarking. The enquiries should not be too intrusive and should be sensitive to the confidentiality needs of the partners (ask not for anything that thou shall not want to be asked for). A report is written on the findings of the study, which forms the basis for further analysis and recommendations.

Step 4: Analyze the Benchmark Data

The benchmark data obtained from the partner(s) are analyzed and compared with those of the team's process, and the differences (gaps) must be identified. Often, the data may not be directly comparable, because the measures computed for a benchmark partner may be based on a set of assumptions that differ from those of the team's process. There may also be differences between the organizations in terms of size, geographical location, or workforce culture. Measures may have to be recalculated with new and appropriate assumptions in order to make them comparable.

The team should compare each of the steps in their process with those of the benchmarked process to discover avoidable operations and possible improvements. The team should prepare the "best-of-best-practices" flow chart as a result of comparing their process with that of the benchmarked partner. They should identify changes to be made to their current process to accomplish the benchmarked goals. These changes should be specified fully by writing new specifications and new work instructions. If new tools or equipment are needed, their specifications must be drawn up. The outcome of this analysis is the basis for any recommendations that will be presented to senior executives.

Step 5: Evaluate the Economic Impact of Changes to be Made

The impact of making changes to a process derived from the best-in-class practices should be evaluated not only for the current state of the team's process, but also for the projected changes to be made in the process in the near term (e.g., the next five years). The result of making all of the agreed upon changes should be evaluated using a summary statistic—preferably an economic statistic, such as annual savings in warranty costs or additional sales generated. It may well happen that when all the best practices are adopted, the resulting aggregate performance of the team's business will exceed the performance of the individual benchmarked businesses.

Step 6: Present Recommendations
The team should make a presentation of their findings to the executives, including a summary of their findings from the benchmark study, the changes they would like to make, and the expected improvements in results. The presentation should explain the procedures followed, data gathered, analysis, recommendations, and the impact on the business results. All these should be presented in a written report as well as in an oral presentation. If the team keeps executives updated on the progress of their work and receives feedback from them, then there will be no big surprises at the presentation. This makes it easier for recommendations to be accepted.

Step 7: Revise Operational Goals
When the recommendations are accepted, the team should seek revisions to the short- and long-term goals of the overall organization (assuming that the changes are implemented). The team should also study the effect of the recommended changes on the various agencies of the business, such as employees, support services, and suppliers. Any concerns regarding adverse effects on them should be adequately addressed as a way of preparing them for accepting the changes. If any of the changes has a significant effect on the work—or on the circumstances of work—of any individual person, then such a change must be approved by senior managers.

Step 8: Develop Action Plans
Not all changes recommended by the team will be equal. Some may produce a larger improvement in overall performance of the business compared to others, and some may take a longer time to implement than others. So, the changes must be prioritized based on some criterion, selected by the team that is appropriate for the given situation. For some teams, improvement in performance may be the important criterion; for others, time needed to implement changes may be the important criterion. For some others, however, the cost of making changes may be important. Based on the criterion selected by the team, the changes should be prioritized, and a chronological list of changes to be implemented should be prepared. For each change, the tasks to be completed, time frame for completion, resources needed as well as their allocation, and individuals responsible for accomplishing the tasks should be specified. A system for monitoring the progress in implementing the changes should be created.

Step 9: Implement Plans, and Monitor Results
When the plans are approved, the changes must be implemented one by one. In some situations, a pilot change (i.e., on a small scale) would make sense before implementing the entire change. Data must be collected, and the performance measures, such as percentage yield, scrap percentage, or cycle time, must be measured to evaluate the effect of implementing the best practices. If the results are not as expected, modifications to the plans may be warranted.

Step 10: Recalibrate Benchmarks

The benchmarks have a life, too, and they may become outdated after that time. So, the benchmarks, which may be the best-in-class this year, may not be so the next year. Therefore, the benchmarks must be reviewed and recalibrated periodically. Yearly review for most benchmarks would be reasonable. Annual reviews would keep the recalibration effort manageable. If measures are subject to changes because of rapid changes in technology or the market, however, more frequent review of the measures may be necessary. By the same token, some measures may remain robust for a long time and may not call for frequent reviews. The review frequency has to be decided case by case.

8.3.4 Pareto Analysis

The Pareto analysis is used when there are several opportunities to choose one from. For example, one project may have to be chosen as the first one from several project opportunities, or one defect category may have to be addressed first from among many defect categories. The Pareto analysis is used in these situations as it separates the *vital few from trivial many* opportunities.

The method is based on a distribution proposed by the Italian researcher Wilfredo Pareto (1848–1923) to describe how a small proportion of people in the free, Western societies controlled a large proportion of the wealth. Dr. Joseph Juran saw similar phenomenon in the quality area, where a small number of causes are responsible for a large proportion of losses. He adopted the Pareto distribution to describe the "mal-distribution" of quality losses among causes, and he formalized the method to separate the "vital few" causes from the "trivial many." (He used the terms *vital few* and *useful many* when referring to cause variables, such as customers or suppliers.) Such separation helps in prioritizing improvement opportunities and in concentrating on the vital few causes rather than spreading effort among the many trivial causes. A quality engineer will find this method to be useful in many situations when he or she should decide which project to select or product-defect to address first.

The Pareto analysis consists of obtaining data on the frequency at which the different causes have been occurring, in recent history, creating the problem. The causes are then rank-ordered based on the frequency, with the cause having the highest frequency being ranked first. Next, a diagram is made with percentage frequency on the y-axis and the causes on the x-axis arranged in ascending order of rank (descending order of frequency). The following example illustrates the method of making a Pareto analysis.

Case Study 8.2

Studying the causes of bond failures in the wire-harness assembly shop, referred to in Case Study 8.1, the student team decided, by voting among workers who knew the process, that data should be collected on seven causes for further study. The data available in documents for a period of

12 months were reviewed for the number of defectives caused by the seven causes. The reasons had been recorded as what an inspector suspected was the cause for a defective bond. The data are shown tabulated in Table 8.1. A Pareto analysis was prepared using the Minitab software, and the Pareto diagram is shown in Figure 8.5.

The Minitab program calculates the percentage frequency of occurrence for each of the causes and then rank orders them based on the percentage frequency. It also computes the cumulative percentage frequency of the causes, and graphs the cumulative frequency. The percentage frequency and cumulative percentage frequencies are shown in the table below the graphs. The cumulative frequency table and graph help in reading off the graph, the most important causes. For example, in the above case study, "curing humidity" and "curing temperature" can be read from the graph as being the two dominant causes, accounting for more than 60% of all bond failures.

The Pareto analysis thus helps in prioritizing project opportunities so that projects can be chosen based on their order of importance.

8.3.5 Histogram

The histogram provides the means of observing the variability and centering of a population. The method of drawing the histogram and examples of its use were discussed in Chapter 2. Many quality problems are known to result from excessive variability and/or poor centering (EV/PC) in process parameters or quality characteristics. The EV/PC in mating parts results in assemblies with inadequate clearance, or with excessive clearance, and cause early wear of the parts and early failure of machines. The EV/PC of in-process dimensions

TABLE 8.1

Data on the Causes of Bond Failures

Defect Category	May	June	July	Aug	Sept	Oct	Nov	Dec	Jan	Feb	Mar	Apr	Total
Degrease	3	3	4	2	1	1	2	2	1	7	1	—	**27**
Chemical treatments	—	—	—	—	—	—	—	—	—	12	—	—	**12**
Incompatibility	—	—	—	—	—	2	—	1	—	—	—	—	**3**
Bad storage	—	—	—	1	1	—	—	1	—	9	—	—	**12**
Bad refrigeration	—	—	3	—	6	4	3	—	—	1	—	—	**17**
Bad rolling	1	1	—	6	2	7	1	—	—	2	1	—	**21**
Curing temp.	4	3	4	—	—	6	—	6	5	3	16	12	**59**
Curing humidity	6	—	3	4	—	4	—	13	13	14	8	22	**87**
Total	14	7	14	13	10	24	6	23	19	48	26	34	**238**

Source: From Ramachandran, B., and P. Xiaolan, "A Quality Control Study of a Bond Curing Process." Unpublished project report, IME 522, IMET Department, Bradley University, Peoria, IL, 1999.

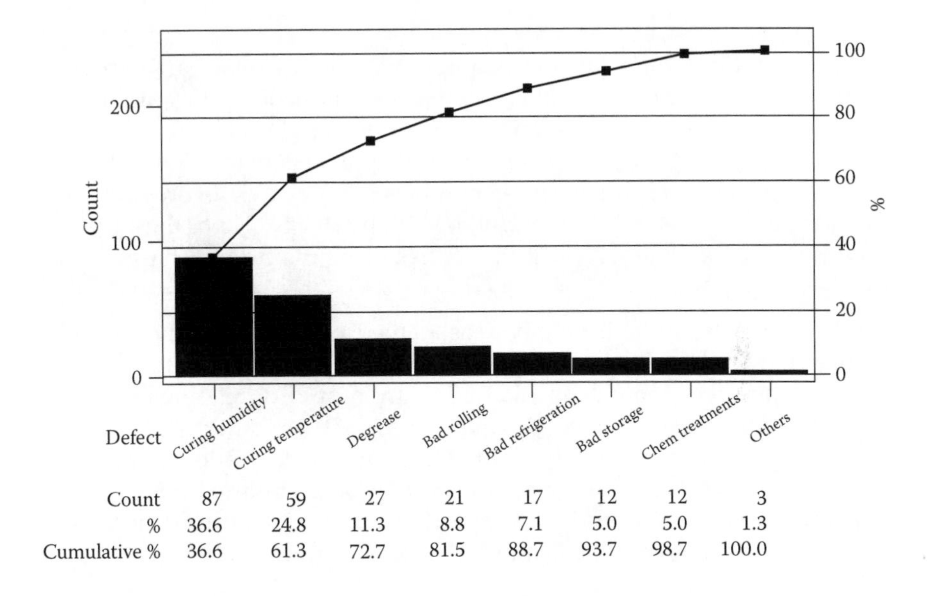

Count	87	59	27	21	17	12	12	3
%	36.6	24.8	11.3	8.8	7.1	5.0	5.0	1.3
Cumulative %	36.6	61.3	72.7	81.5	88.7	93.7	98.7	100.0

FIGURE 8.5
Pareto diagram for bond failure causes. (From Ramachandran, B., and P. Xiaolan, "A Quality Control Study of a Bond Curing Process." Unpublished project report, IME 522, IMET Department, Bradley University, Peoria, IL, 1999.)

causes assembly difficulties in subsequent stations. The EV/PC in the weight of packaged material results in an inability to meet specifications or in giving away too much material for free. In general, EV/PC in process parameters are the most common causes for poor quality in finished products. All these problems simply require plotting the histogram of the variable in question to understand the nature of the variability and the location of the process center, which usually facilitates discovering and eliminating the root causes of the problem.

Case Study 8.3

In the bond-failure problem referred to in Case Studies 8.1 and 8.2 the Pareto analysis revealed that curing chamber humidity was the most important reason for causing bond failures. The data on humidity were collected for a period of one month and a histogram of the humidity readings was made. The histogram and the specifications for humidity chosen by process engineers are shown in the Figure 8.6.

The histogram showed that almost all the humidity readings missed the specification and were below the lower spec limit. At least one clue to solving the problem in bond failures was apparent. The histogram confirmed the suspicion that the humidity was not maintained at the required level. This is not an uncommon situation where the process designers had chosen the specifications for an important process parameter, but the production people were not paying enough attention to that parameter. Possibly, the production people did not know that humidity was such an important variable and that it was so off-target. Now that the data were available, however, it was easy for them to see the facts of the case. It was not hard to convince the production people about the need for controlling the humidity. Then, they were advised on how to accomplish the control using the control charts.

In the above example, it may be reasonable to assume that the choice of specification for humidity by the process designers was correct. It should be more than mere coincidence that so many humidity readings were out of specification when there were so many cases of bond failures. However, a caution may be in order: We need not assume that specifications for process variables are always chosen correctly especially when failures are occurring. Sometimes we have to question the specifications and conduct experiments to determine the correct specifications that would give defect-free products.

8.3.6 Control Charts

Control charts were described in detail in chapters 4 and 5 as the method for keeping track of a process variable or a product characteristic at a consistent level. These charts can be used not only to monitor a process for consistency but also for troubleshooting and problem solving in quality improvement projects. In Dr. Juran's language, control charts can be used not only to watch for "sporadic deviations" but also to discover causes for some "chronic deviations" in processes. Thus, they can be employed as problem solving tools for process improvement.

When control charts are used on a process—say, to monitor a process variable—it is customary to maintain an event-log as a record of all those

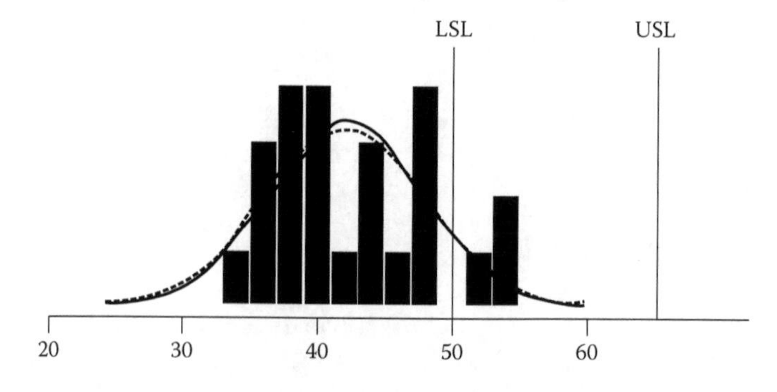

FIGURE 8.6
Histogram of curing humidity in bond failures. (From Ramachandran, B., and P. Xiaolan, "A Quality Control Study of a Bond Curing Process." Unpublished project report, IME 522, IMET Department, Bradley University, Peoria, IL, 1999.)

TABLE 8.2

Curing Chamber Humidity in Bond Failure Problem

Subgroup/ Week 1	Humidity %	Subgroup/ Week 2	Humidity %	Subgroup/ Week 3	Humidity %	Subgroup/ Week 3	Humidity %
1	37	8	28	15	50	22	54
	36		27		66		38
	37		28		61		53
2	35	9	26	16	40	23	35
	38		26		47		48
	39		28		47		47
3	40	10	27	17	68		
	43		28		74		
	43		29		67		
4	44	11	26	18	50		
	41		27		67		
	39		28		67		
5	34	12	27	19	64		
	32		28		64		
	33		70		53		
6	30	13	66	20	33		
	30		47		51		
	31		43		46		
7	29	14	53	21	43		
	29		76		50		
	30		81		60		

Source: From Ramachandran, B., and P. Xiaolan, "A Quality Control Study of a Bond Curing Process." Unpublished project report, IME 522, IMET Department, Bradley University, Peoria, IL, 1999.

things happening in the process or its environment, such as a change of tool, material, humidity, or operator. This is done either by writing remarks on the chart, if it is maintained manually, or by writing it in the event journal, which most computer software packages provide. So, when the control chart shows something has occurred making the process go out of control, the operator, or an improvement team, tries to relate the signal on the control chart with the happenings both in and around the process. More than likely, this will lead to the discovery of relationships between causes and effects and lead to resolution of problems. The example below illustrates the idea.

Case Study 8.4

The behavior of humidity in the curing chamber in the case of the bond-failure problem was studied using control charts. Chamber humidity was recorded on a daily basis for a period of 3 weeks. Three readings were taken each day, and the data are shown in Table 8.2. $\overline{X}$ & R-charts were drawn to study how the humidity changed in relation to wha-thappened around the curing chamber. The charts made from the data are shown in Figure 8.7.

The control charts showed that the process was not in control, indicating the humidity did not remain consistent over the period of study. The average humidity remained uniformly very low for almost 12 days, and it fluctuated widely up and down, then moved to a higher average level in the subsequent 11 days. The study of what happened in and around the curing chamber when the data were collected revealed some information.

The controller of the humidity generator that was supposed to maintain the humidity at the set point in the chamber was acting erratically. It was supposed to be supplied with dry gas that picked up the right amount of humidity while exiting from the generator. The gas that was supplied, however, was not dry. Indeed, it was very wet, causing swings in the behavior of the humidity generator. Also, the generator was supposed to be provided with distilled water per the operating instructions, which was not being done. The water supply had been connected to regular service water. The controller was replaced, and the operating personnel were advised to follow the manufacturer's instructions for maintaining the generator. They also were advised to continue using the control charts to watch the humidity of the chamber.

The other important variable—the temperature of the curing chamber—was studied as well. Data on temperature had also been collected while gathering data on humidity. When the chamber temperature was analyzed using $\overline{X}$ & R-charts, the temperature was found to be in-control at the prescribed level. This helped in deciding that the lack of humidity control was possibly a reason for the bond failures. The student team was not able to report the final results from the client after the recommendations were implemented, but it was not difficult to see how such monitoring and control of humidity would make a difference in the performance of the process.

8.3.7 Scatter Plots

Often, when a cause variable is suspected to be responsible for an undesirable effect on a quality characteristic, we want to verify this suspicion and establish a model for their relationship. Then, data are collected on the two variables, the cause variable X and the quality characteristic Y, usually as paired observations on the twin variable (X, Y). The data will be graphed with each value of Y plotted against the corresponding value of X. The resulting graph is called a *scatter plot* and shows the relationship of the two variables in a meaningful way. Figure 8.8 shows the types of scatter plots that may result from graphing the data as described above. Figure 8.8a shows that the variable Y has a linear relationship with X, in which increases in X cause increases in Y. Figure 8.8b also shows a linear relationship, but one in which increases in X cause decreases in Y. Figure 8.8c shows a relationship between X and Y, but one that is curvilinear rather than linear. Figure 8.8d

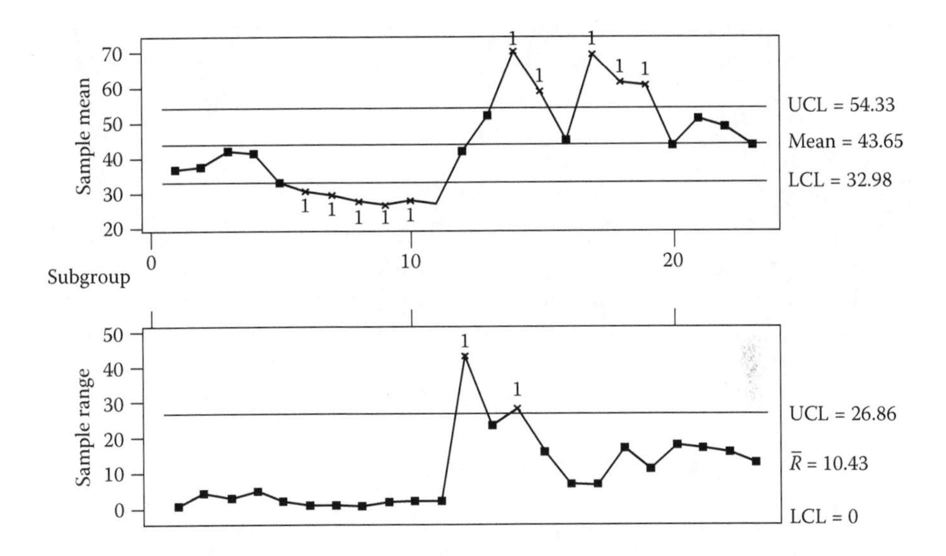

FIGURE 8.7
Control charts for humidity of the curing chamber in the bond-failure problem. (From Ramachandran, B., and P. Xiaolan, "A Quality Control Study of a Bond Curing Process." Unpublished project report, IME 522, IMET Department, Bradley University, Peoria, IL, 1999.)

shows a plot in which there seems to be no relationship between X and Y. Values of Y do not seem to be dependent on the values of X.

Case Study 8.5

Cores made of sand are dipped in a wash and dried in drying ovens in a foundry. These cores should be dried enough to expel the moisture, which strengthens the cores but over-drying will drive out the resin, the bonding chemical, and make the cores brittle. Therefore, the drying must be very carefully controlled. This was accomplished by controlling the dry- ability index of the ovens, which is an index that represents the amount of drying an oven is able to produce. A study was undertaken to understand how the dry-ability of the oven was influenced by several factors, one of them being atmospheric humidity. Data were collected in pairs of atmospheric humidity X and oven dry-ability Y and the data were plotted as a scatter plot as shown in Figure 8.9.

According to the scatter plot in Figure 8.9, dry-ability seems to be related to humidity, with larger values of humidity causing smaller values of dry-ability, and vice versa. In this case, humidity of the atmospheric air was not a factor that could be controlled, because air-conditioning the entire building was not feasible. The knowledge of this relationship, however, helped in understanding what to expect on a rainy day, when the air is humid. There were other variables in the equation, such as drying time and hot-air temperature, which could be adjusted to compensate for the loss in dry-ability due to increase in humidity. We needed to know the relationship among these variables so that when the humidity increased we would know how much adjustment should be made in the other variables in order to obtain consistent dry-ability.

In the above case study, all we could see from the scatter plot is that X and Y have a relationship. We possibly could say that an increase in X causes a decrease in Y. If we want a quantitative relationship to predict values of Y corresponding to values of X, then we need to use regression analysis, which provides quantitative measures to express the relationship between variables.

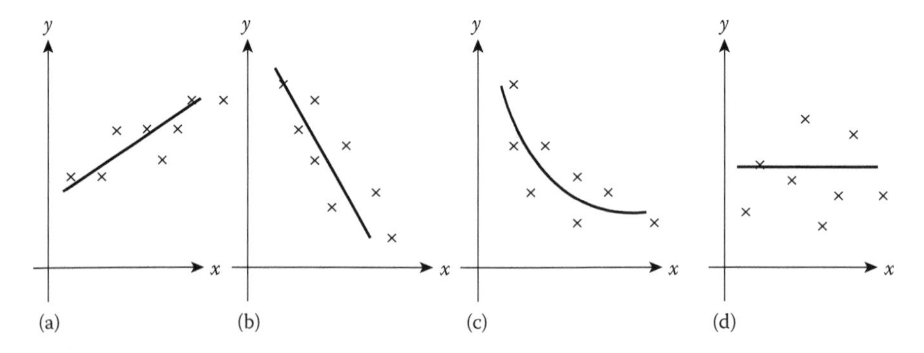

FIGURE 8.8
Examples of scatter plots.

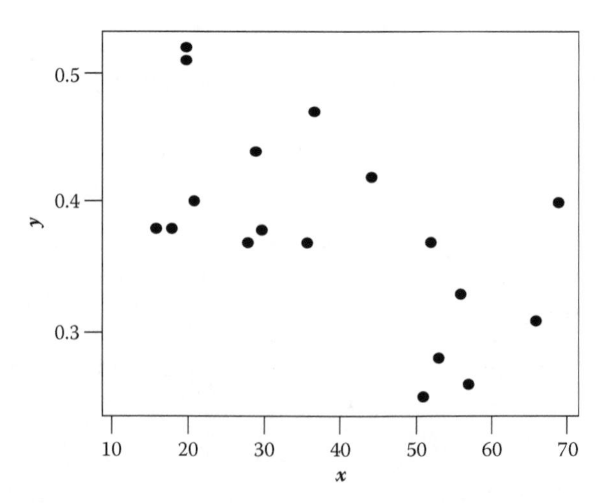

FIGURE 8.9
Example of a scatter plot.

8.3.8 Regression Analysis

Regression analysis is the formal, mathematical approach to studying the relationship between a cause-variable X and a resultant or effect-variable Y. When we want to study the linear relationship between a variable Y and just one other variable X, then we use the *simple linear regression*. When we suspect (these suspicions could arise from a study of their scatter plots) that the relationship is curvilinear, we use *nonlinear regression analysis*. When a study has more than two variables, for example, we want to study how a variable Y changes with changes in three cause variables X_1, X_2, and X_3, we then use *multiple-regression analysis*. In regression language, the resultant variable Y is called a *dependent variable*, or a *response*, and the cause variables X_1, X_2, and X_3 are called *independent variables*, or *regressors*.

8.3.8.1 Simple Linear Regression

In simple linear regression, we try to establish the relationship between the response Y and an independent variable X by fitting a straight line to define that relationship. We hypothesize that the relationship is defined by the line:

$$y = \alpha + \beta x$$

where α and β are the parameters of the line. Then, we collect certain number, say, n observations on the pair (x, y) and fit a line to the data. That is, we estimate the coefficients of a line:

$$y = a + bx$$

such that the line passes as close to "all" the observed (x, y) values as possible. This is done by choosing the values of a and b such that the sum of the squared deviations of the observed values of y from the values of y predicted by the line, at the different values of x, is minimized. Figure 8.10 shows an example of a line fitted to (x_i, y_i), $i = 1, \ldots, 5$.

For any value x_i, the line provides an estimate for y_i as $\hat{y}_i = a + bx_i$. The actual observation y_i, corresponding to x_i, may not be the same as this estimate, and the difference between y_i and $\hat{y}_i$, or $e_i = y_i - \hat{y}_i$, is called the residual at x_i. The sum of the squares of these residuals, or SS(residual) $= \Sigma_{i=1}^{n} (y_i - \hat{y}_i)^2$, represents how much the fitted line is off from the observed values. If the fitted line passed through all the observed (x_i, y_i) values, this quantity will be equal to zero. Fitting the line to minimize the SS(residual) is one way of making the line pass as close to "all" the observed values (x_i, y_i) as possible. Such a line $y = a + bx$, drawn to minimize the sum of squared deviations, is called the *least square line* for the given data and is an "estimate" for the hypothesized line $y = \alpha + \beta x$, where a and b are estimates for α and β, respectively.

The SS(residual) can be written as $\text{SS}(\text{residual}) = \Sigma_{i=1}^{n} (y_i - a - bx_i)^2$. Taking the partial derivatives of the SS(residual) with respect to a and b and equating

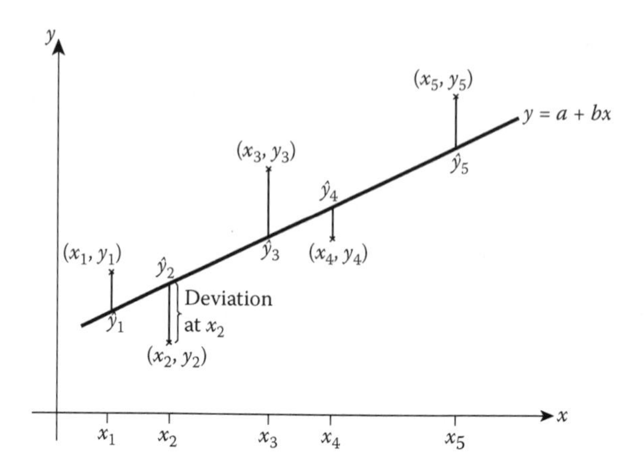

FIGURE 8.10
Fitting a straight line to data.

them to zero, and then solving the resulting set of equations for a and b, the values of a and b that minimize SS(residual) can be found. It can be shown (see Walpole et al. 2002) that such values of a and b that minimize SS(residual) are given by:

$$b = \frac{n\sum_{i=1}^{n} x_i y_i - \left(\sum_{i=1}^{n} x_i\right)\left(\sum_{i=1}^{n} y_i\right)}{n\sum_{i=1}^{n} x_i^2 - \left(\sum_{i-1}^{n} x_i\right)^2}, \quad a = \bar{y} - b\bar{x}$$

These are known as the *coefficients of regression*. Thus, we can find a line that provides a "good fit" to a set of data by calculating the values of these regression coefficients that define the least-squared line.

For any given value x_i, the value y_i will not be the same each time it is observed, because y_i is an observation of a random variable. The value y_i, we assume, comes from a normal distribution with mean = $E(Y \mid x_i)$ and variance = σ^2, where $E(Y \mid x_i)$ is the average of y_i at the given x_i and σ^2 is the variance, which is assumed to be constant for all values of x_i. So, the hypothesis we made above is equivalent to saying that the value of y_i corresponding to an x_i comes from a normal distribution, the variance of which is σ^2 and the mean of which, $E(Y \mid x_i)$, which is located on a straight line:

$$E(Y \mid x_i) = \alpha + \beta x_i$$

Equivalently, we can write $y_i = \alpha + \beta x_i + \varepsilon_i$, where $\varepsilon_i \sim N(0,\sigma^2)$, and the Figure 8.11 shows this graphically. The equation $y_i = \alpha + \beta x_i + \varepsilon_i$ is called the model for this regression, where α and β are the parameters of the model.

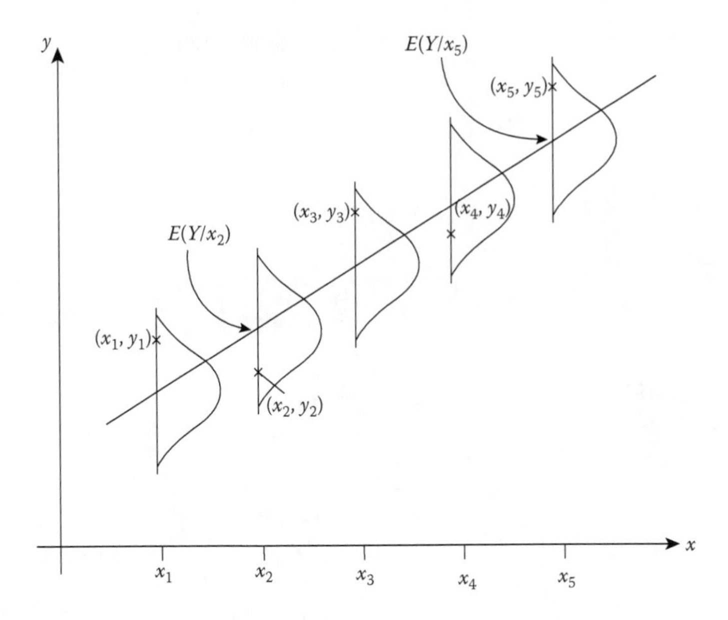

FIGURE 8.11
The regression line.

The least square line $y = a + bx$ with values of a and b calculated as above is an "estimate" for the regression model.

In practice, we use the estimated line $y = a + bx$ to represent the relationship between the two variables X and Y provided that estimated line can adequately represent the relationship. Now the question arises: When do we say the above straight line is an adequate representation of the relationship? This brings us to the issue of *model adequacy.*

8.3.8.2 Model Adequacy

Model adequacy is measured using a quantity called the *coefficient of determination,* which is denoted by R^2. The quantity R^2 represents the proportion of total variability in the observations of Y that is explained by the regression line.

The total variability in the observations is given by

$$\text{SS}(\text{total}) = \sum_{i=1}^{n} (y_i - \bar{y})^2.$$

The variability explained by the regression line is

$$\text{SS}(\text{regression}) = \sum_{i=1}^{n} (\hat{y}_i - \bar{y})^2.$$

The variability not explained by the regression line is

$$\text{SS}(\text{residual}) = \sum_{i=1}^{n} (y_i - \hat{y}_i)^2.$$

It can be shown (Hogg and Ledolter 1987) that

$$\sum_{i=1}^{n} (y_i - \bar{y})^2 = \sum_{i=1}^{n} (y_i - \hat{y}_i)^2 + \sum_{i=1}^{n} (\hat{y}_i - \bar{y})^2.$$

That is, SS(total) = SS(residual) + SS(regression).

The term R^2 is defined as the ratio of the variability explained by regression to the total variability. Therefore,

$$R^2 = \frac{\text{SS}(\text{regression})}{\text{SS}(\text{total})} = \frac{\text{SS(total) - SS(residual)}}{\text{SS}(\text{total})} = 1 - \frac{\text{SS}(\text{residual})}{\text{SS}(\text{total})}$$

If the value of R^2 for a model is large, close to 1.0, the regression line is able to explain a major portion of the variability in the values of y, so the fitted line is a good representation of the relationship between X and Y. If the value of R^2 is small, it may mean that the straight-line model does not fully represent the relationship between the variables. Figure 8.12a shows examples of data sets and corresponding fitted lines with varying values of R^2.

If the linear model does not adequately represent the relationship, other curvilinear models may have to be explored. Or, we may have to conclude that the values of Y are dependent on more than the single independent variable X. We then will seek the other variables that may help in explaining the behavior of Y.

8.3.8.3 Test of Significance

The coefficients a and b as obtained from the formulas above, are the intercept and slope, respectively, of the fitted straight line representing the relationship between X and Y. A question to be answered now: Does a straight line really explain the relationship? This question can be rephrased: Is β *significantly* different from zero? (see Figure 8.12b).

The question has to be answered using a test of significance with the following hypotheses:

$$H_0: \beta = 0$$
$$H_1: \beta \neq 0$$

An estimate for σ, which is the error or unexplained variability in the observations, is first necessary. An estimate for σ^2 is obtained from the residual sum of squares as

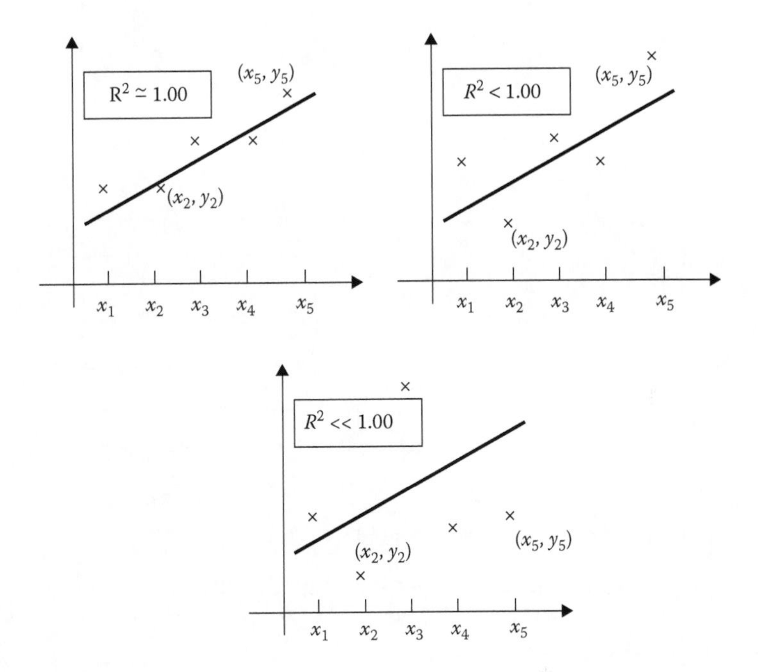

FIGURE 8.12a
Data Sets and Fitted Lines With Different R^2 Values.

$$S^2 = \frac{\sum_{i=1}^{n}(y_i - \hat{y}_i)^2}{n-2} = \frac{\sum_{i=1}^{n}(y_i - a - bx_i)^2}{n-2} = \frac{\sum_{i=1}^{n} y_i^2 - a\sum_{i=1}^{n} y_i - b\sum_{i=1}^{n} x_i y_i}{n-2}$$

The equivalence of the numerators in the above expressions for S^2 can be proved using algebra. The denominator in the above expressions, called the *degrees of freedom (df),* is obtained as the sample size n minus the number of parameters estimated from the sample data. For the simple regression, two parameters, α and β, are estimated from the sample data and, therefore, $df = (n - 2)$. The term S^2 is called the *mean square error.*

Using the estimate for σ^2, an estimate for the standard error (s.e.) of b can be shown as

$$\text{s.e.}(b) = \frac{S}{\sqrt{\sum_{i=1}^{n}(x_i - \bar{x})^2}}$$

Then, it can be shown that the statistic

$$\frac{b}{\text{s.e.}(b)} \sim t_{n-2} \text{(See Mendenhall and Sincich 1988)}.$$

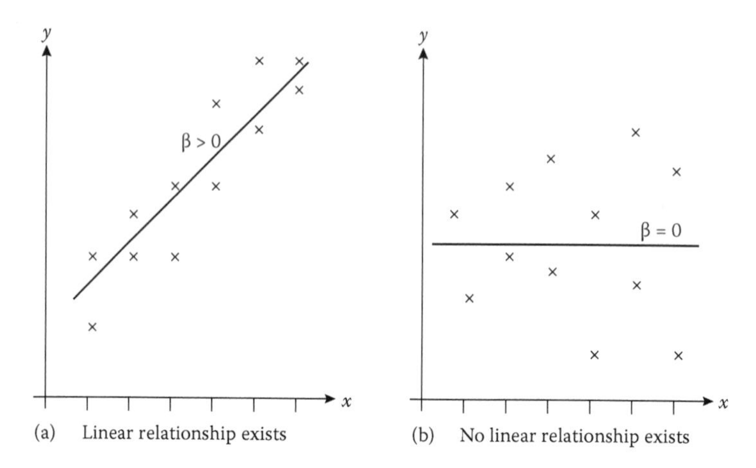

FIGURE 8.12b
Existence of a linear relationship between X and Y.

Using the above statistic as the test statistic, the critical region is chosen as defined below to reject the null hypothesis (H_0: $\beta = 0$) against the alternate hypothesis (H_1: $\beta \neq 0$).

Critical Region: If the absolute value of the observed value of the test statistic $|t_{obs}| > t_{\alpha/2,n-2}$, then reject H_0. Otherwise, do not reject H_0.

If H_0 is rejected, it means there *is* a straight-line relationship between X and Y. If not, there is no such relationship.

Example 8.1

A study of a drying oven was made to find the effect of atmospheric humidity and the drying capability of the oven as expressed by the dry-ability index. Data collected on dry-ability (Y) and humidity (X) are shown in the table below which includes some calculations to facilitate computation of the regression coefficients. Find out if a linear relationship exists between humidity and dry-ability.

Humidity X_i	Dryability Y_i	X_iY_i	X_i^2	Y_i^2	Humidity X_i	Dryability Y_i	X_iY_i	X_i^2	Y_i^2
36	0.37	13.32	1296	0.1369	**52**	0.37	19.24	2704	0.1369
51	0.25	12.75	2601	0.0625	**21**	0.4	8.4	441	0.16
56	0.33	18.48	3136	0.1089	**37**	0.47	17.39	1369	0.2209
66	0.31	20.46	4356	0.0961	**57**	0.26	14.82	3249	0.0676
30	0.38	11.4	900	0.1444	**20**	0.51	10.2	400	0.2601
16	0.38	6.08	256	0.1444	**53**	0.28	14.84	2809	0.0784
18	0.38	6.84	324	0.1444	**52**	0.37	19.24	2704	0.1369
44	0.42	18.48	1936	0.1764	**20**	0.52	10.4	400	0.2704
69	0.4	27.6	4761	0.16	**30**	0.38	11.4	900	0.1444
29	0.44	12.76	841	0.1936	**28**	0.37	10.36	784	0.1369
				Sum	**785**	7.59	284.46	36167	2.9801

From the data in the table, $n = 20$, $\sum x_i = 785$, $\sum y_i = 7.59$, $\sum x_i y_i = 284.46$, $\sum x_i^2 = 36{,}167$, and $\sum y_i^2 = 2.9801$.

Solution

$$b = \frac{20(284.46)-(785)(7.59)}{20(36167)-(785)^2} = \frac{-268.95}{107115} = -0.0025$$

$$a = \frac{7.59}{20} - (-0.0025)\left(\frac{785}{20}\right) = 0.478$$

$$s^2 = \frac{\text{SS(residual)}}{n-2} = \frac{\sum y_i^2 - a\sum y_i - b\sum x_i y_i}{n-2}$$

$$= \frac{2.9801-(0.478)(7.59)-(-0.0025)(284.46)}{18} = \frac{0.068}{18} = 0.00378$$

$$s = 0.0615$$

$$R^2 = 1 - \frac{\text{SS(residual)}}{\text{SS(total)}}$$

$$\text{SS(total)} = \sum y_i^2 - \frac{\left(\sum y_i\right)^2}{n} = 2.9801 - \frac{(7.59)^2}{20} = 0.0997$$

$$\text{SS(residual)} = 0.068$$

$$R^2 = 1 - \frac{0.068}{0.0997} = 0.318$$

$$\text{s.e.}(b) = \frac{0.0615}{\sqrt{\sum x_i^2 - \left(\sum x_i\right)^2/n}} = \frac{0.0615}{\sqrt{36167-(785)^2/20}} = \frac{0.0615}{73.18} = 0.00084$$

$$H_0: \beta = 0$$

$$H_1: \beta \neq 0$$

$$\text{Test statistic: } \frac{b}{\text{s.e.}(b)} \sim t_{n-2}$$

Critical region: $|t_{obs}| > t_{18,\,0.025}$ (see Figure 8.13a).

Observed value of the t statistic: $t_{18} = \dfrac{b}{\text{s.e}(b)} = \dfrac{-0.0025}{0.00084} = -2.976$

$$t_{crit} = t_{18,0.025} = 2.101 \text{ (from Table A.2 in the Appendix)}$$

Because $|t_{obs}| = 2.976 > t_{18,0.025}$, reject H_0. There is a significant linear relationship between dry-ability and humidity.

Figure 8.13b is the output of regression from Minitab for the data in the above example, wherein we see the values for a, b, R^2, S, s.e., and t_{obs} for X, all

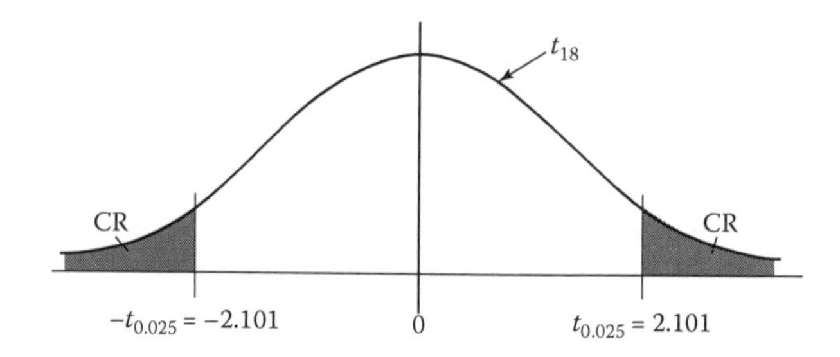

FIGURE 8.13a
Critical region (CR) for the test of significance of β.

checking out with calculated values within the round-off errors. Figure 8.14 shows a plot of the y vs. x given by Minitab. The Minitab output shows a quantity "R-sq(adj)," which stands for Adjusted R^2. This quantity conveys the same meaning as the R^2, except it has been adjusted for the number of independent variables in the regression, which is mainly useful in multiple regression. The adjustment is done as follows (Minitab 1995).

$$R^2 = 1 - \frac{\text{SS(residual)}}{\text{SS(total)}}$$

$$\text{Adjusted } R^2 = 1 - \frac{\text{SS(residual)}}{\text{SS(total)}} \times \frac{n-1}{n-p}$$

where p is the number of parameters estimated, which equals the number of independent variables in the regression plus one (for the intercept). For a simple linear regression, $p = 2$. Therefore, for the above example,

$$\text{Adjusted } R^2 = 1 - \frac{0.068}{0.0997} \times \frac{19}{18} = 0.28$$

The regression output also gives the P value corresponding to the observed t-values. In this case, since the P value corresponding to X is less than 0.05,

Regression Analysis: *Y*-Dryblty versus *X*-Humidity

The regression equation is
Y-Dryblty = 0.478 – 0.00251 *X*-Humidity

Predictor	Coef	SE Coef	*T*	*P*
Constant	0.47805	0.03517	13.59	0.000
X-Humid	–0.0025109	0.0008270	–3.04	0.007

S = 0.06052 *R*-Sq = 33.9% *R*-Sq(adj) =30.2%

FIGURE 8.13b
Minitab output from regression.

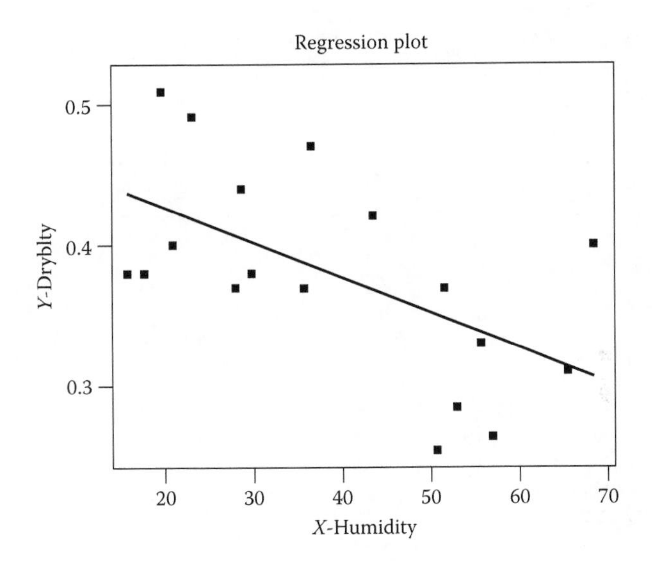

FIGURE 8.14
Minitab plot of the fitted line.

the chosen α, we reject H_0: $\beta = 0$ and conclude that X and Y have a significant linear relationship.

8.3.8.4 Multiple Linear Regression

When there is reason to believe that a dependent variable Y is influenced by several independent variables, say, X_1, X_2, and X_3, then we use the multiple linear regression. The regression model will be

$$y = \alpha + \beta_1 x_1 + \beta_2 x_2 + \beta_3 x_3 + \varepsilon$$

The relationship will be estimated by

$$y = a + b_1 x_1 + b_2 x_2 + b_3 x_3$$

Formulas for a, b_1, b_2, and b_3 can be developed the same way as for the simple regression to minimize the sum of squared deviations. The coefficient of determination R^2, which represents the proportion of the total variation in Y explained by the regression, is used for determining how adequately a linear model fits the data. The mathematics of deriving the formulas and the arithmetic for calculating the coefficients become complex; however, the availability of computer programs makes the work easier. We will use an example to show how the multiple linear regression is used to find the relationship of a response to several independent variables.

TABLE 8.3

Data from the Oven Study

X_1 (Rel. Humidity)	X_2 (Hot-air Temperature)	X_3 (Conveyor Speed)	Y (Dry-ability)	X_1 (Rel. Humidity)	X_2 (Hot-air Temperature)	X_3 (Conveyor Speed)	Y (Dry-ability)
36	393	1.8	0.37	52	345	1.4	0.37
51	347	1.8	0.25	21	399	1.4	0.4
56	404	1.8	0.33	37	397	1.4	0.47
66	411	1.8	0.31	57	350	1.8	0.26
30	395	1.8	0.38	20	370	1.4	0.51
16	395	1.8	0.38	53	395	1.8	0.28
18	401	1.8	0.38	52	345	1.4	0.37
44	401	1.4	0.42	20	372	1.4	0.52
69	397	1.4	0.4	30	395	1.8	0.38
29	370	1.4	0.44	28	401	1.8	0.37

Example 8.2

A study of a drying oven was made to find the effect of the variables atmospheric humidity, hot-air temperature, and conveyor speed on drying capability of the oven as represented by the dry-ability index. The data gathered on the variables and the dry-ability index are given in Table 8.3. Determine if there is a linear relationship between dry-ability and the three variables. Use $\alpha = 0.05$. Knowledge of the relationship will help in adjusting the parameters of the oven to obtain optimal drying performance.

Solution

Figure 8.15 shows the output from Minitab of the regression performed on the three variables. The computer output shows the estimates of the coefficients a, b_1, b_2, and b_3 along with the standard error in these estimates. For each variable, the output shows the value of the observed t statistic along with the P value for the

Regression Analysis: Dryblty versus *R*-Humidity, Air-Temperature, Conv. Spd.
The regression equation is

Dry-ability = 0.487 – 0.00119 RH + 0.000874 Hot-air – 0.260 conv spd

Predictor	Coef	SE Coef	*T*	*P*
Constant	0.4867	0.1585	3.07	0.005
Humidity	−0.0011853	0.0005206	−2.28	0.031
Air Temp	0.0008740	0.0004090	2.14	0.042
conv spd	−0.25963	0.04280	−6.06	0.000

S = 0.04482 *R*-Sq = 65.5% *R*-Sq(adj) =61.5%

FIGURE 8.15
Minitab output for the example in multiple regression.

observed value of t. If the P value for any variable is less than $\alpha = 0.05$, then that variable has significant impact on the response.

Each of the t values have $(n - p)$ degrees of freedom (*df*), where p is the number of parameters estimated, which is equal to the number of independent variables plus one. For the example, *df* for each of the observed t values = $[20 - (3 + 1)] = 16$. The critical value of t is $t_{0.025,16} = 2.12$. If the $|t_{\text{obs}}| > 2.12$ for any factor, reject H_0: $\beta_i = 0$ for that factor. In this example, all three factors are significant, which means they all have a significant impact on the dry-ability of the oven.

This information can be used to adjust the oven parameter to obtain optimal oven performance.

8.3.8.5 Nonlinear Regression

When curvature is suspected in the relationship between two variables as disclosed in the scatter plot, curvilinear models can be fitted to the data. For example, the following model may be appropriate for a simple nonlinear regression:

$$y = \alpha + \beta_1 x + \beta_2 x^2$$

When there are several independent variables, it is advisable to make scatter plots of the response variable with the independent variables individually and get a preliminary understanding of the nature of the relationships. This will help in deciding what relationships should be tried. When there are, for example, two independent variables, models with an interaction term of the form

$$y = \alpha + \beta_1 x_1 + \beta_2 x_2 + \beta_3 x_1 x_2$$

can be tried. The following example shows the use of nonlinear regression to investigate the relationship of variables.

Example 8.3

For the oven study in Example 8.2, try to fit a model that includes one interaction term.

Solution

We propose that there is interaction between air temperature and humidity based on the belief that the joint effect of changed air temperature and humidity is much more profound on dry-ability than the sum of the individual effects. We propose the model

$$y = \alpha + \beta_1 x_1 + \beta_2 x_2 + \beta_3 x_3 + \beta_4 x_1 x_2$$

The results of the nonlinear regression made using Minitab are reproduced in Figure 8.16. We see that the interaction term is also significant. We also see that

Regression Analysis: Dryblty versus Humidity (H), Air-temp (AT), conv-spd, $H \times AT$

The regression equation is

Dryblty = 1.98 – 0.0314 Humid – 0.00294 Air-temp – 0.221 conv-spd + 0.000075 $H \times AT$

Predictor	Coef	SE Coef	T	P
Constant	1.9762	0.4783	4.130	0.001
Humidity	–0.031406	0.009674	–3.25	0.005
Air Temp	–0.002942	0.001279	–2.30	0.036
conv-spd	–0.22115	0.03406	–6.49	0.000
$H \times AT$	0.000074840	0.00002475	3.02	0.009

$S = 0.02674$ R-Sq = 89.2% R-Sq(adj) = 86.4%

FIGURE 8.16
Minitab output of nonlinear regression analysis.

the value of R^2 has increased compared to that in the previous example showing that inclusion of the interaction term has improved the fit of the model to the data.

Regression analysis is a very powerful tool for understanding the relationships among variables in a process, which can be put to use for great advantage in improving process and product quality.

8.3.9 Correlation Analysis

Correlation analysis is another method to study the linear relationship between random variables. The correlation coefficient of two random variables X and Y is defined as

$$\rho_{XY} = \frac{\text{cov}(X,Y)}{\sqrt{V(X)V(Y)}} \equiv \frac{E\left[(X-\mu_X)(Y-\mu_Y)\right]}{\sqrt{V(X)V(Y)}} \equiv \frac{\sigma_{XY}}{\sigma_X\sigma_Y}$$

The empirical analog of this theoretical quantity, obtained from n sample observations of (x, y), is given by *Pearson product moment coefficient of correlation*, defined as

$$r_{XY} = \frac{\sum_{i=1}^{n}(x_i-\bar{x})(y_i-\bar{y})}{\sqrt{\left[\sum_{i=1}^{n}(x_i-\bar{x})^2\right]\left[\sum_{i=1}^{n}(y_i-\bar{y})^2\right]}}$$

It can be shown that ρ, and r, lie in the interval [–1, 1] (see Hines and Montgomery 1990).

When the value of r is near –1, we say the two variables have a strong *negative correlation*, which means that when the value of one variable increases, the value of the other decreases linearly, and vice versa. When the value of

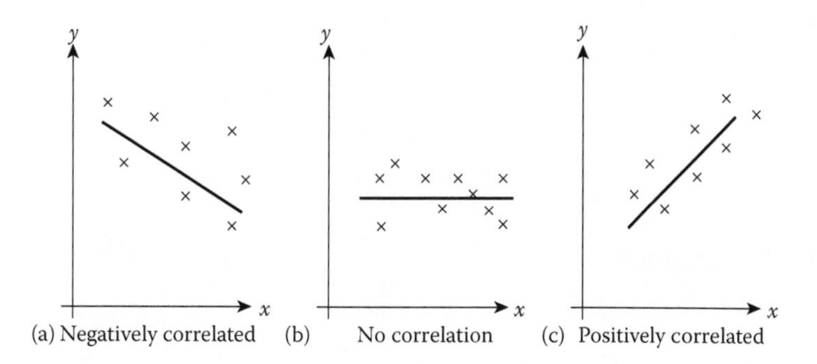

(a) Negatively correlated (b) No correlation (c) Positively correlated

FIGURE 8.17
Correlation between two variables.

r is near +1, we say the two variables have a strong *positive correlation*, which means that increasing/decreasing values of one variable produce increasing/decreasing values of the other in a linear fashion. When the value of r is near zero, we say the variables have no correlation, which means they have no linear relationship (see Figure 8.17).

8.3.9.1 Significance in Correlation

For normally distributed variables, Table 8.4 gives the 95% critical values of r for selected sample sizes. If the absolute value of the calculated r from sample observations exceeds the quantity in the table for a given sample size, then we conclude there is a significant correlation between the variables. Otherwise, we conclude there is no correlation.

The formula for the correlation coefficient can be rewritten in a form that is convenient for computation. The numerator and denominator of the expression can be shown to equal the following two expressions, which are computationally simpler.

TABLE 8.4
Critical Values of Correlation Coefficient ($\alpha = 0.05$)

Sample Size	Critical Value for r ($\alpha = 0.05$)
5	0.75
10	0.58
15	0.48
20	0.42
25	0.38
30	0.35
50	0.27
100	0.20

Source: From Chatfield, C., *Statistics for Technology*, John Wiley & Sons, New York, NY, 1978.

For the numerator

$$\sum_{i=1}^{n}(x_i-\bar{x})(y_i-\bar{y})=\sum_{i=1}^{n}x_iy_i-\frac{\left(\sum_{i=1}^{n}x_i\right)\left(\sum_{i=1}^{n}y_i\right)}{n}$$

For the denominator

$$\sqrt{\left[\sum_{i=1}^{n}(x_i-\bar{x})^2\right]\left[\sum_{i=1}^{n}(y_i-\bar{y})^2\right]}=\left(\sqrt{\sum_{i=1}^{n}x_i^2-\frac{\left(\sum_{i=1}^{n}x_i\right)^2}{n}}\right)\times\left(\sqrt{\sum_{i=1}^{n}y_i^2-\frac{\left(\sum_{i=1}^{n}y_i\right)^2}{n}}\right)$$

Example 8.4

Calculate the correlation coefficient between dry-ability and humidity from the data given in Table 8.3, and check if the correlation is significant.

TABLE 8.5

Calculation of the Correlation Coefficient for Example 8.4

X_1 (Humidity)	X_2 (Hot-air Temperature)	X_3 (Conveyor Speed)	Y (Dry-ability)	X_1Y	X_1^2	Y^2
36	393	1.8	0.37	13.32	1296	0.1369
51	347	1.8	0.25	12.75	2601	0.0625
56	404	1.8	0.33	18.48	3136	0.1089
66	411	1.8	0.31	20.46	4356	0.0961
30	395	1.8	0.38	11.4	900	0.1444
16	395	1.8	0.38	6.08	256	0.1444
18	401	1.8	0.38	6.84	324	0.1444
44	401	1.4	0.42	18.48	1936	0.1764
69	397	1.4	0.4	27.6	4761	0.16
29	370	1.4	0.44	12.76	841	0.1936
52	345	1.4	0.37	19.24	2704	0.1369
21	399	1.4	0.4	8.4	441	0.16
37	397	1.4	0.47	17.39	1369	0.2209
57	350	1.8	0.26	14.82	3249	0.0676
20	370	1.4	0.51	10.2	400	0.2601
53	395	1.8	0.28	14.84	2809	0.0784
52	345	1.4	0.37	19.24	2704	0.1369
20	372	1.4	0.52	10.4	400	0.2704
30	395	1.8	0.38	11.4	900	0.1444
28	401	1.8	0.37	10.36	784	0.1369
Sum = 785	7683	32.4	7.59	284.46	36167	2.9801

Correlations: X_1-Humidity, Y-Dryblty
Pearson correlation of X_1-Humidity and Y-Dryblty = −0.582
P-value = 0.007

FIGURE 8.18
Output from Minitab on correlation analysis.

Solution

Table 8.5 shows the calculation of sums and sums of squares of the variables that facilitate computation of the correlation coefficient.
To calculate the $r_{X,Y}$

$$\text{Numerator} = 284.46 - \frac{(785)(7.59)}{20} = -13.4475$$

$$\text{Denominator} = \sqrt{36{,}167 - \frac{(785)^2}{20}} \times \sqrt{2.9801 - \frac{(7.59)^2}{20}} = (73.183)(0.3157)$$

$$r_{X,Y} = \frac{-13.4475}{(73.183)(0.3157)} = -0.582$$

Referring to Table 8.4, we see that the critical value for r when $n = 20$ is 0.42. Because the absolute value of the observed value of r is greater than r-critical, we conclude there *is* a significant (negative) correlation between X_1 and Y at α = 0.05. In physical terms, the dry-ability of the oven does increase with a decrease in humidity in the atmosphere.

The results from Minitab on the correlation coefficient are shown in Figure 8.18 and the results are seen to agree with the calculated figures. The conclusion using the P values would also lead to the same conclusion as arrived at using Table 8.4.

8.4 Lean Manufacturing

The term "Lean Manufacturing" refers to the production system created by the Toyota Motor Corporation to deliver products of right quality, in right quantity, at the right price, to meet the needs of the customer. Taiichi Ohno (1912–1990), the Toyota engineer and a creative genius, who became V.P. of manufacturing, is mainly credited with creating the system, which is also known as the Toyota Production System. He had received assistance from Dr. Shigeo Shingo (1909–1990), an industrial engineer and author of the Single Minute Exchange of Die (SMED) procedure, who helped in strengthening parts of the system. Eiji Toyoda (1913–), the man who transformed the Toyota Motor Company that his uncle had founded into a global powerhouse, was searching for a model production system to adopt in order to improve the

Toyota production facilities that he was given charge to manage. At that time, in late 1940s, the Toyota Motor Company was producing cars at the rate of about 200 per year when the Ford Motor Co. at the Rouge Plant near Detroit was making 7,000 cars per day. The system he witnessed at the Rouge plant during a visit in 1950 offered a model he could adopt but he saw several drawbacks in the system. He saw those drawbacks as opportunities for improvement and built a new system with help from Ohno, which sought to eliminate all possible wastes in production and maximize the value-adding functions.

Seven types of wastes (called the *mudas*) were identified for elimination: waste in defective units produced, units overproduced, products or in-process material sitting in inventory, transportation of raw material or product moved unnecessarily over long distances, wasted motions of workers, waiting time of workers, and unnecessary operations performed. What resulted from continuously improving the system over two decades in the 1950s and 60s is now referred to as the Toyota Production System (TPS) or Lean manufacturing system. A brief discussion of the TPS is undertaken here to show its relationship to the traditional quality engineering methodology. Many of its components are indeed continuous improvement approaches to improve quality, eliminate waste, streamline operations and reduce overall costs, thus enabling quality products delivered to the customer at minimum price.

The major components of the TPS System and their interrelations are depicted in Figure 8.19 and are further elaborated below. The figure indicates only the dominant relationships among the component functions and there exist many subtle relationships that the reader will come to recognize, which cannot be expressed in any broad-brush portrayal of the system. Any reader with a background in industrial engineering would notice that many of the components of the TPS are tools industrial engineers have used traditionally as part of *methods engineering*, to make workplaces more efficient and more productive. However, the ideas of Just-in-time production was not part of the traditional IE discipline and, moreover, the tools of methods engineering have been refined, simplified, and packaged into one whole coherent system in the TPS, which has proven it's capability for achieving, quality, productivity and cost reduction.

Incidentally, methods engineering, methods analysis, or motion analysis, as it is variously called, is the process of analyzing how a work is accomplished in a workplace by breaking down the work into fundamental motions, or therbligs as they are called, and creating an optimal sequence of motions to accomplish the same work after eliminating wasteful motions. This same approach can be used to identify wasteful movements in a larger context of a factory, for improving flow of material and products among workplaces. The concept of methods analysis was introduced by Frank and Lillian Gilbreths in the 1910s and are well documented in books such as *Niebel's Methods, Standards, & Work Design* (Freivalds and

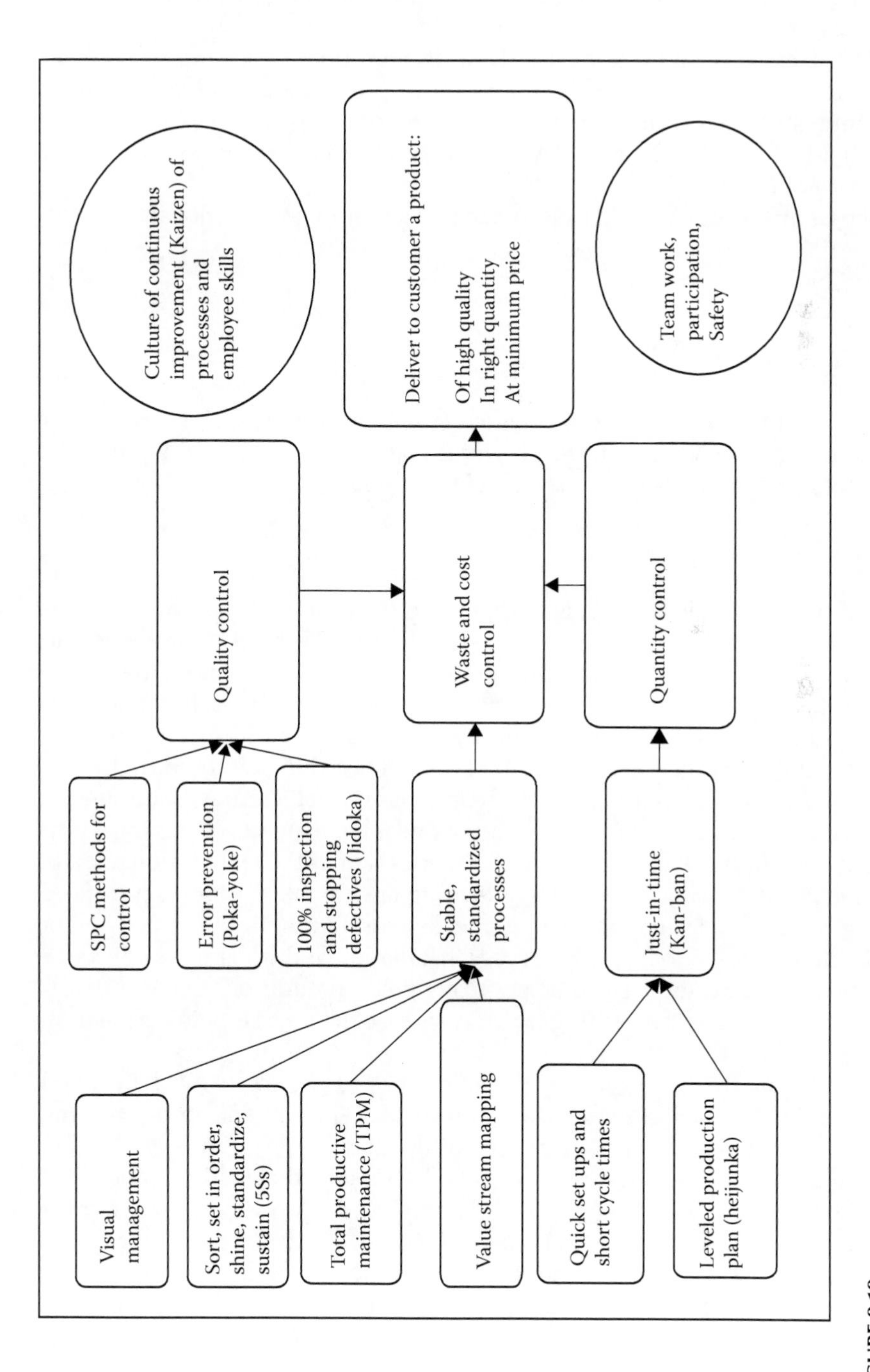

FIGURE 8.19
Lean Manufacturing System at a glance.

Niebel 2009) and *Motion and Time Study: Design and Measurement of Work* (Barnes 1980).

The objective of presenting the Lean System here is only to provide an overview of the system and is not meant to provide exhaustive guidelines on implementing a lean system in a production shop. The books by Pascal (2002), MacInnes (2002), and Black and Hunter (2003) can be considered for this latter purpose.

The Lean System can be considered to have three major functional modules:

- Quality Control
- Quantity Control
- Waste and Cost Control

There are a few minor modules which either contribute to the three major ones listed above or directly to the final objective as shown in Figure 8.19. Each of these modules is discussed below in some detail.

8.4.1 Quality Control

The objective of this module is to deliver to the customer a product (or service) that is designed and produced such that it will meet their needs and delight them in its use. If the product produced does not meet the needs, either due to poor design or due to poor production methods, it would have to be discarded and so is a waste causing losses to the producer. The methods described in other chapters of this book, starting from how to find what the customer wants, translating those needs into product features, choosing the targets and limits of variability for the key product features, designing a process to accomplish the product characteristics within the chosen limits, making the product using control procedures to ensure the product is produced according to the design, packaging and delivering the product to customer and helping the customer in its proper installation and use are the appropriate methodologies needed to achieve this goal. In addition, the TPS places special emphasis in preventing, at all cost, a defective unit being passed on to the customer.

Discovering the causes for the defectives or process errors (we had called these assignable causes or special causes in Chapter 4) and implementing error-prevention methods is employed religiously for this purpose. Mistake-proofing or fool proofing methods, called *Poke-yoke* that will not allow the errors to creep back into the process are employed to prevent defectives ever being produced. Further, the TPS uses 100% final inspection of the product so that not a single defective unit will be passed on to the customer. This is true whether the customer is the end user outside the producing organization or an internal customer who uses the "product" in the next stage of processing or assembly. The defect-free production is further facilitated by a practice, call it a culture, where any member of the production team would stop a production process if he/she sees the process producing defective units. This practice, called *Jidoka,* ensures delivery of 100% quality products to the customer and is considered (Wilson 2001) one of the pillars of the Toyota Production System.

8.4.2 Quantity Control

Quantity control refers to producing only the amount of product the customer has requested, not more, not less. This type of "lean" production is in contrast to producing large quantities of products and stocking them in inventory in anticipation of customer demand. Such storage of products in inventory results in losses which include cost of space and shelves, cost of handling in stocking and retrieving, cost of security and insurance, and the interest cost of the capital invested in the products being stored. Such "mass" production had been in vogue primarily because of large costs involved in "setting up" of production lines. When an assembly line has to be set up, for example, for changing over from one model to another model of a product, or when a tooling in a machine has to be changed from one set-up to another to make a different part number, it involves labor time and, more importantly, lost production of the production machinery. The traditional approach had been to produce products in batches for each set up so as to distribute the set-up cost over the larger number of units or a batch. The size of the batch is decided to minimize the total cost of the inventory operation and is called the *economic lot size.* Such production in large batches resulted in large inventories because products that are not immediately used by customers need to be stocked. Recognizing that the long set-up times were the reason for large batch-production, the TPS invented methods to reduce set up times, again using steps of methods engineering. They succeeded in reducing set-up times to a fraction of what used to be before improvement, which enabled production of small batches, or as much as needed by a customer's order. Even as small a batch size as *one* became possible and was economical. The Single Minute Exchange of Die (SMED), a procedure perfected Shigeo Shingo to minimize set up times in changing dies, one of the major set up operations in the production of automobile body parts, was a major driver in making the one-piece production (per set up) economical (Black and Hunter 2003, Chapter 6).

Next the TPS addressed the problem of overproduction resulting from lack of correlation between the number produced and the number demanded by the customer. The conventional wisdom was to anticipate or project the demand based on historical demand and "push" the projected quantities through the production system. This led to large inventories waiting for the customer to make the request for the product. The authors of the TPS invented the "pull" method where the product is made in quantity that is only pulled by the customer, leaving nothing (almost nothing) for storage. The pull system makes use of the *kanban*, which literally means a visual card.

A kanban is a card that is attached to a part with the part's ID number written on it. It may contain additional information such as name of supplier if it is a supplied part, location of stores if it is an item from storage, where it is needed, and so on. Every part that is made or received from a supplier gets a kanban and resides in a small storage location. When a part is withdrawn by a customer, the kanban is detached from the part and stacked in a kanban retrieval pouch located in a very visible spot on the assembly floor. Accumulation of the kanbans in this pouch indicates to processes upstream that more of this part is to be produced or procured in the quantity indicated by the number of kanbans that have accumulated. The kanbans could be transmitted to the upstream production processes or procurement sources at set time intervals or at predetermined number of units of product. When the kanbans are sent to the producer, it is a production-kanban and when it is detached at the time of withdrawal by the customer it is a withdrawal-kanban. For the kanban system to work well, some rules are followed:

- No new part will be made or procured unless there is a kanban for it.
- No part can be withdrawn unless a kanban is created.
- Customer withdraws only the quantity needed.
- Producer makes only the quantity indicated by the withdrawal kanbans.
- Never ship defective items.
- Level the production plan since kanban system cannot handle widely varying production plan.

Leveling implies that the demand placed on the system does not vary widely from day to day both in numbers of products as well in the product mix (model variation). This, though a bit of a compromise on the purely pull system, is considered necessary for the smooth functioning of the system. Leveling is accomplished by taking the demand for the product over a longer period of time, say, a week, or a month, and dividing evenly among the days.

The fundamental objective of the system is to make sure that no production is made beyond what the customer needs. It can be seen from the above discussion how the kanban system, also referred to (by Americans) as *take-one-make-one-system*, helps in accomplishing it.

8.4.3 Waste and Cost Control

We have already mentioned about efforts in reducing waste by avoiding production of defective units. We have also talked about reducing waste by preventing overproduction through the discipline enforced by the pull production. Yet another approach to waste reduction is by *value stream mapping*.

A value stream map is a flow process chart that enables identifying those operations that add value to the product and those that do not add value. A value adding operation is one that contributes to converting the raw material into the product in the shape and form that the customer wants. There may be elements of a process such as transportation, loading and unloading, and some delays that do not add to the value of the product but are still essential to the conversion process. Such unavoidable activities must be identified and the effort and resources expended in them must be minimized. The TPS uses specialized symbols for creating a value stream map, which adds more information to the conventional flow process charting discussed in Chapter 3. This additional information helps in deciding whether the operation is value-adding or not. The final objective for value stream mapping is to identify wasteful functions and eliminate them or minimize them.

Besides the above three main modules, viz., defect free production, just-in-time production with no inventory, and value stream analysis, the TPS employs a few other tools, which either contribute towards the objectives of the above three, or contribute directly to the final objective of producing quality product in the right quantity at the minimum cost. These tools include Total Productive Maintenance, stabilizing and standardizing processes and visual management. These are discussed below.

8.4.4 Total Productive Maintenance

Total Productive Maintenance (TPM) refers to keeping the production machinery and tools in perfect working condition through preventive maintenance, with total participation from all connected with planning, operating and maintaining the machinery. TPM is based on the premise that well maintained machinery and tooling is a requirement for maintaining the capability of processes to produce products of quality. Further, poorly maintained machines cause breakdowns resulting in unscheduled stoppage of production resulting in loss of production time of workers as well as machinery. The TPM process was created and perfected by the Japan Institute of Plant Maintenance (Suzuki 1992) and has been adopted as part of the Toyota Production System.

Preventive maintenance mainly involves oiling and greasing of seals and bearings at the optimally scheduled intervals, and keeping the machinery clean. When the machines and their surroundings are kept clean, any leaks, which, if left unobserved and unattended might lead to major damage to equipment, will be visible as soon as they occur and can be repaired. The TPM approach recommends *autonomous maintenance* in the sense that the production workers who are in charge of operating the machinery take responsibility for the regular oiling, greasing and cleaning of production machinery, without the assistance of the maintenance crew. Thus, a few minutes spent by workers in a day go a long way in preventing machine breakdowns, and spare the maintenance crew valuable time which they can use in other more productive functions in the upkeep of machinery for improving their reliability and longevity.

TPM relies for its success on educating and motivating the production and maintenance workers on the necessity for maintaining the equipment and training them in the tools for TPM. These tools are mainly problem solving tools that were identified in this chapter earlier, such as Pareto analysis, C&E diagram, histogram, etc. Problems relating to production line failures are identified, analyzed, their root causes discovered, and solutions implemented by teams made up of production workers and maintenance technicians. Results from successful implementation of TPM have been enormous. One Japanese company reported reduction in the number of equipment breakdown from 5,000 per month to 50 per month and the profits realized through use of TPM have been recorded as 10 times the cost for implementing the program (Suzuki 1992). The website TPMonline.com reports several successful stories. For example, the National Steel and Shipbuilding Co. (NASSCO), San Diego, CA, a full service shipyard that designs and builds ships, implemented TPM in their production shops in 1997 and increased equipment uptime in some instances from 74% to 99%.

8.4.5 Stable, Standardized Processes

Stabilizing a process means to make the process behave consistently the same. Standardization refers to providing standard operating procedures for each of the operations, preferably using clear graphics, so that the process can be repeated the same way each time it is performed. Standardization of operations helps in making a process stable. Standardization, the TPS cautions, is never meant to leave it in the same condition for ever (Dennis 2002). Standardization is the step needed for making it ready for further improvement. Standardization also helps in passing on the current expertise to next generation of workers. Before a process is standardized, the process must be improved in several dimensions. Of course, it should be improved to make defect-free products, made only in quantities that customer needs. The process must be subjected to value-stream analysis to eliminate non-value-adding functions before it is standardized. There are also a few other tools used to enhance the stability of a process. Visual management using the "5Ss," and proper layout of machinery to minimize unwanted motions are the important ones.

8.4.6 Visual Management

Visual management involves making the visual appearances of the workplace clear, symmetric, and uncluttered, so that any deviations will be detected easily just by looking at the process. The 5Ss recommended are: Sort, Set-in-order, Shine, Standardize, and Sustain.

Sorting simply means: keep in the workplace only the items that are needed by removing from the workplace those that are not needed in the daily operation. The unwanted items such as redundant parts, tools, jigs, old machinery, tables, chairs, shelves, obsolete computer screens, and so on, accumulate over time and take up shelf and floor space, and present safety hazards. Periodic house cleaning will get rid of the clutter and make the workplace clean. How often the clearing should be done, by whom, and how to decide which is an unwanted item, can be determined by a workplace team depending on what suits their workplace. However, these should be determined by the team, documented and strictly followed.

Set-in-order means: organize the machines, tools, staging areas, and parts-shelves in a layout that will minimize the movement of the job and the worker. This can be done by making a diagram of the movement of incoming material going through the process operations to the final finished part location. Such a diagram, known as the *flow diagram* in industrial engineering, will enable identification of zig-zagging and backtracking within the workplace. By rearranging the machines with respect to each other and other features of the workplace, these unwanted motions can be avoided leading to a more rational layout. The flow diagram for a workplace can be made on a grid paper where the items of machinery are shown located at their current locations with the distances between them drawn to scale. The path followed by the material as it goes through the conversion process is drawn on the paper with arrowheads indicating direction of travel. Such a diagram of a workplace, if it has not been improved already, will usually present a picture of unorganized loops of travel paths (thus earning the name "Spagetti diagram"). The total distance of travel is used as a metric of how well the work place is organized or disorganized. A critical eye will see in this diagram ways to untangle the loops and streamline the paths by rearranging the machinery so as to minimize the total distance travelled by the product. These diagrams and their analysis are best made by a team of people working in the workplace. At the end of such analysis the team will come up with the machinery laid out in such a way the product moves smoothly in the direction toward the final end point of the product. The U-shaped cell, one of the configurations of machinery layout, has been found to be the most suitable layout for workplaces, not only to facilitate smooth flow of the product but also to minimize the number of workers needed to provide loading, unloading and other services the machinery would need. A caution may be in order: the cells are capital intensive and should not be employed unless warranted by production volume.

The same approach used to smooth out and minimize the travel path of parts and subassemblies within a workplace can be used to analyze flow of material between workplaces in the larger context of the factory. A team of people working in the various workplaces will be able to identify unnecessary travel and create a layout to minimize travel distances, thus yielding an efficient layout for the entire production facility. U-shaped cells producing component parts, linked in the sequence needed in the assembly of the final product are known to produce the best layout for the total production facility (see Black and Hunter 2003, Chapter 5). Where many products are flowing through a workplace or a plant involving complex travel sequences, special computer tools and simulations programs available to an industrial engineer may have to be used.

Shine means: to clean the place of dirt, oil and unwanted grit that may make the workplace look dull. This will also disclose in a timely manner any leak that may spring in a machine leading to fixes that may save the machine from an unscheduled breakdown. A clean workplace also makes for a safe workplace, environmentally friendly and contributing to the health of workers. Proper lighting can also be included under this function to enhance the visibility and safety of the workplace. Cleaning must be done on a regularly scheduled time interval with responsibility for cleaning clearly identified among the team members working in the workplace.

Standardize means: to make the improved workplace the standard practice by documenting the new layout, schedules and responsibilities for sorting and cleaning. The entire team working in the workplace should be made aware of the new process as documented as the new standard.

Sustain means: to make sure that the sorted, cleaned and organized workplace remains in the improved condition in the future. This can be accomplished by implementing rules that everyone will follow regarding cleaning schedule and discarding unwanted material so that the team's objective of a clean and organized workplace remains accomplished.

8.4.7 Leveling and Balancing

Leveling was described earlier as the way of obtaining even schedule for the production system from day to day, both in product numbers and product mix. Balancing is the process of distributing the work within a workplace evenly among the operations to be performed in the workplace. The cycle time is the time needed to make once piece of a product in a workplace. If all the work needed to make a product is performed in one location, doing one operation after another, the cycle time will equal the sum of all operation times. If the operations are distributed to be performed at several locations, the cycle time will equal the sum of the operation times at the location where the largest number of operations is performed. (Of course, when the operations are distributed over multiple locations, more workers will be needed.) If, however, the operations are more evenly divided among the locations, the less will be the cycle time. Please see in Figure 8.20 the three configurations of a process to fabricate a welded part. The total time for welding the fabrication is 44 minutes, which, if performed all at one place as in Configuration 1, will result in a cycle time of 44 minutes. If the work is divided among two

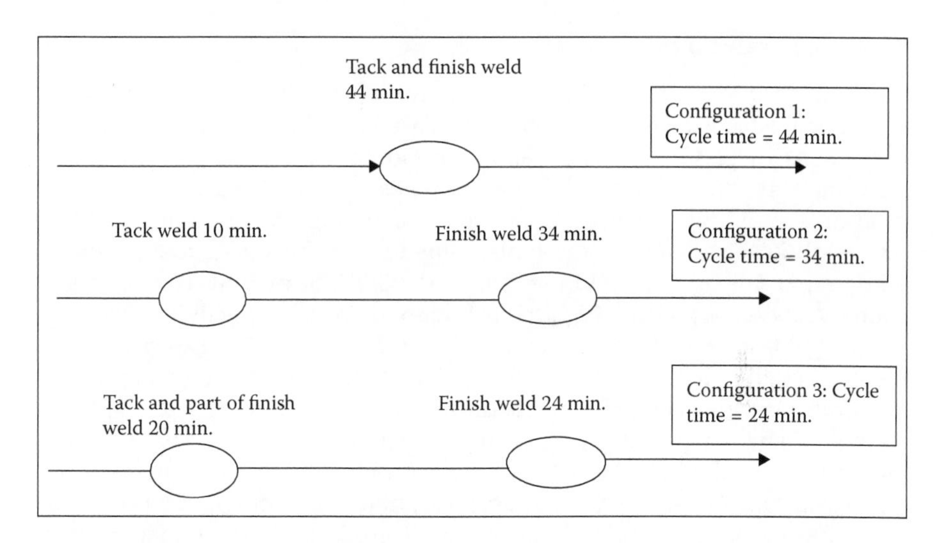

FIGURE 8.20
Distributing work among operations.

locations using two operators as shown in Configuration 2, cycle time will be 34 minutes. If however, the work is more evenly distributed between the two operations, a finished fabrication will be made every 24 minutes giving a cycle time of 24 minutes. This is the basic principle of *assembly line balancing*, covered in industrial engineering literature.

In the real world, the problems won't be this simple, and there will be many restrictions imposed by precedence requirements (which operation needs to be done first and which next, etc). There will also be opportunities for performing operations in parallel. This problem of assembly line balancing has been studied in great detail and tools, algorithms and computer programs, have been developed to handle complex situations. One has to refer to a book in production and operations management such as Buffa (1983) or Evans (1993) for more detailed information on this topic.

So, by properly configuring the operations, by splitting or combining and distributing the total work needed to accomplish a job among work-locations, the cycle time to produce a product can be adjusted and optimized. The TPS uses a metric called *takt-time*, which represents the maximum time within which a unit of product must be made at a workplace in order to meet customer's quantity demand. It is obtained by dividing total number of minutes available in a day by the number of units required to be made in the day:

Takt time = (Number of minutes in a day/Number of units required per day)

Thus, the takt-time is the upper bound for the cycle time. Therefore the design of the workplace should be such as to yield a cycle time which is less than or equal to the takt-time. These are the basic principles involved in balancing the assembly line and meeting the demand from the customer.

8.4.8 The Lean Culture

The Lean Production System depends heavily on a culture where trained workers willingly participate in continuous improvement of every aspect of the system, for its successful implementation. Providing continuous training to the team members in problem solving approaches (e.g., the PDCA method) and problem solving tools such as value stream mapping, making improved layouts and tools for quality control; planning and line balancing tools; coupled with project management and presentations skills improves the capabilities of team members for productive participation in improvement activities. Further, encouragement to the team should be provided by recognizing their achievements through publicity in company magazines and notice boards, and presenting awards through competitions conducted within plants and between plants.

Only when the workforce participates in lean implementation with knowledge and passion will there be suggestions for improving cycle times, reducing wastes, improved visual management, and many other functions that contribute to the success of the lean system (Dennis 2002). The Jidoka concept where a team member will stop a production process when a defective is seen being produced very much depends on the knowledge, involvement and willingness of a team member to take the action to ensure no defective is passed on to the next stage in production. Further, willing participation by employees and hence success of the lean manufacturing system can be realized only when the management practices openness in communication, trust, fairness and a sincere interest in the welfare of the workforce.

We have attempted to describe the Lean Manufacturing or the Toyota Production System in simple language so as to communicate the basic principles involved. We see from the above that the TPS is a collection of modules, which, though somewhat interdependent, can be progressively implemented one at a time. Although a thorough reengineering of a process starting from re-layout of the workplaces in cells that are linked following the assembly sequence will yield the best results through the use of "lean" concept, the individual modules of quality control, quantity control and cost control can be implemented in an existing facility in an incremental fashion. Such incremental implementation will yield improved results in terms of better quality and cost reduction and would help the system evolve into a lean or near-lean system progressively.

8.5 Exercise

8.5.1 Practice Problems

8.1 Prepare a Pareto analysis, first with paper and pencil and then using computer software, for the data given below. The data relate to the frequency of occurrence of short-picks in a book warehouse because of various reasons. A short-pick occurs when not enough books are in the shelves to fill an order. The data were collected over a period of 3 months to investigate the reasons for short-picks and to reduce/eliminate the short-picks in the warehouse.

Reason for Short-pick	No Display	Clean Up	Failed in Error	No Inventory	Key-in Error	Out of Stock	Up-price	Up-price-box	Unexplained
Frequency	7	87	9	120	4	23	42	41	3

Source: Data from Kamienski, K., and A. Murphy, "Line Shortage Analysis of a Publisher's Warehouse," unpublished project report, IME 522, IMET Department, Bradley University, Peoria, IL, 2000.

8.2 Draw a Pareto diagram for the data given below on the number of warranty claims made in a month for a car model at a dealership.

Cause	Engine Mechanical	Engine Fuel System	Engine Controls	Transmission and Axles	Brakes	Sound System	Lights and Signals	Doors, windows, and Locks	Heating and Cooling	Others
Frequency	48	22	25	76	14	11	10	40	12	8

8.3 The data in the table below come from a process that produces large iron castings for automobile engines. The castings showed occasional porosity in a particular location, which is a defect according to customer's specifications. The table shows data on several process variables, along with the measure on the porosity relative to 22 castings. The porosity measure was assigned by an inspector in the scale of 0 to 10, where 0 means no porosity and 10 means the casting has to be scrapped. A number of 1 to 3 for porosity requires no salvage work, and a number from 4 to 9 requires the castings be salvaged by cleaning the holes and filling them with welding.

Casting No.	Dip Baume	Dip Viscosity	Pour Temp	Tap Temp	Pour Time	Porosity
1	43.0	540	2417	2424	29	3
2	43.0	540	2444	2457	28	0
3	43.0	530	2441	2471	27	0
4	43.0	530	2447	2464	29	0
5	43.0	530	2449	2459	29	0
6	43.0	530	2450	2463	28	0
7	43.0	540	2449	2460	28	0
8	43.0	540	2430	2440	27	1
9	43.0	510	2449	2460	28	0
10	43.0	510	2422	2448	30	3
11	42.5	490	2449	2456	28	0
12	42.5	490	2428	2461	28	3
13	43.0	580	2440	2465	27	0

(continued)

Casting No.	Dip Baume	Dip Viscosity	Pour Temp	Tap Temp	Pour Time	Porosity
14	43.0	580	2435	2440	29	0
15	43.5	530	2434	2441	28	0
16	43.0	530	2447	2465	28	1
17	43.0	510	2449	2458	27	0
18	43.0	510	2447	2459	28	1
19	43.0	510	2428	2433	28	2
20	43.0	520	2428	2454	29	0
21	43.0	540	2426	2465	29	2
22	44.5	540	2433	2457	27	2

a. Make a simple regression analysis between porosity and pour temperature. Do the analysis both by hand and using computer software. Compare the results.

b. Using further regression analysis, including the use of multiple and curvilinear models, find out if there is any relationship between the porosity measure and the process variables of pour temperature, dip baume, and pour time.

8.4 Using the data from Problem 8.3:

a. Make a simple regression analysis between porosity and tap temperature. Do the analysis both by hand and using computer software. Compare the results.

b. Using further regression analysis, including the use of multiple and curvilinear models, find out if there is any relationship between the porosity measure and the process variables of tap temperature, dip viscosity, and pour time.

8.5 Calculate the correlation coefficient between pour temperature and tap temperature in the data of Problem 8.3, and compare the results with those from computer calculation.

8.6 Calculate the correlation coefficient between dip viscosity and dip baume in the data of Problem 8.3, and compare the results with those from computer calculation.

8.5.2 Term Project 8.1

There are quality problems all around us—in the cafeterias, residence halls, registration system, student retention programs, maintenance department, the laundromat, or any production facility with which you may be familiar. All these facilities produce a product or a service to meet certain needs of their customers, and if customers are not satisfied by what they pay for, there is a quality problem. Identify one such problem situation, and make a project statement using quantified measures of the symptoms to indicate existence of the problem. (In some situations, finding how well the customers are satisfied could itself be a project.)

Collect whatever data necessary, and use whatever tool necessary to analyze the data. Find the root causes of the problem, and understand why these causes exist. Identify solution alternatives to eliminate the root causes, and make recommendations for improving the quality of the product or service and, thus, the customer's satisfaction. If the improvements can be implemented, provide data on the performance measure to indicate the improvement has been accomplished. If changes cannot be implemented for want of time, explain, using some projections, how your solution will solve the problem.

A written report should be submitted, which should include the project statement, data collected, analysis, results, solutions, and projected outcomes from implementation of the solutions. The report should have no more than 15 typewritten pages, not including tables, graphs, diagrams, and photographs.

References

Barnes, R. M. 1980. *Motion and Time Study: Design and Measurement of Work*. 7th edition. New York, NY: John Wiley.

Black, J. T., and S. L. Hunter. 2003. *Lean Manufacturing Systems and Cell Design*. Dearborn, MI: Society of Manufacturing Engineers.

Buffa, E. S. 1983. *Modern Production/Operations Management*. 7th ed. New York, NY: John Wiley.

Camp, R. C., and I. J. DeToro. 1999. "Benchmarking," Section 12. In *Juran's Quality Handbook*, 5th ed., J. M. Juran and A. B. Godfrey, Co-editors-in-Chief. New York, NY: McGraw-Hill.

Camp, R. C. 1989. *Benchmarking: The Search for Industry Best Practices That Lead to Superior Performance*. Milwaukee, WI: ASQC Quality Press.

Chatfield, C. 1978. *Statistics for Technology*. New York, NY: John Wiley & Sons.

Deming, W. E. 1986. *Out of the Crisis*. Cambridge, MA: MIT-Center for Advanced Engineering Study.

Dennis, P. 2002. *Lean Production Simplified*. New York, NY: Productivity Press.

Evans, J. R. 1993. *Production/Operations Management*. 5th ed. West Publishing.

Freivalds, A., and B. Niebel. 2009. *Niebel's Methods, Standards, & Work Design*. 12th edition New York, NY: McGraw Hill.

Hines, W. W., and D. C. Montgomery. 1990. *Probability and Statistics in Engineering and Management Science*. 3rd ed. New York, NY: John Wiley and Sons.

Hogg, R. V., and J. Ledolter. 1987. *Engineering Statistics*. New York, NY: Macmillan.

Ishikawa, K. 1985. *What is Quality Control?—The Japanese Way*, translated by D. J. Lu, Prentice Hall, Englewood Cliffs, NJ.

Juran, J. M., and F. M. Gryna. 1993. *Quality Planning and Analysis*. 3rd ed. New York, NY: McGraw Hill.

Kamienski, K., and A. Murphy. 2000. "Line Shortage Analysis of a Publisher's Warehouse," unpublished project report, IME 522, IMET Department, Peoria, IL: Bradley University.

MacInnes, R. 2002. *The Lean Enterprise Memory Jogger*. Salem, NH: Goal/QPC.

Mendenhall, W., and T. Sincich, 1988. *Statistics for the Engineering and Computer Sciences*. San Francisco, CA: Dellen Publishing Co.

Minitab. 1995. *Reference Manual*, Release 10 Xtra, Minitab, Inc., www.minitab.com.

Ramachandran, B., and P. Xiaolan, 1999. "A Quality Control Study of a Bond Curing Process." Unpublished project report, IME 522, IMET Department, Bradley University, Peoria, IL.

Suzuki, T. 1992. *New Directions for TPM*, translated by John Loftus. Cambridge, MA: Productivity Press.

9

A System for Quality

9.1 The Systems Approach

The modern approach to producing quality products and services to satisfy customers' needs calls for creating a quality system wherein the responsibilities for various aspects of meeting customer needs are identified and assigned to the various agencies in the system. The different agencies of the system then perform their functions in a coherent manner, with a view to achieving the system's common goal of meeting customer needs while utilizing the system's resources efficiently. If such a system is organized and maintained well, that system *will* produce quality products while using resources efficiently. That was the premise upon which leaders in the quality field proposed and advanced the systems approach to quality.

This chapter explores some of the models proposed for creating a quality system. Some that were proposed in the early stages of systems thinking may not represent a complete model for a quality system, but they are worth reviewing for their historical value. More recent models have taken the best of the features of the earlier models and represent the state-of-the-art templates for quality management systems. The models reviewed are listed below, of which the first three represent the former kind and the next three represent the latter.

- Dr. Deming's system
- Dr. Juran's system
- Dr. Feigenbaum's system
- ISO 9000:2008 standards
- The Malcolm Baldrige National Quality Award criteria
- The Six Sigma system

A few other models have also been proposed by other authors, of which those by Philip B. Crosby, Dr. Kaoru Ishikawa, and Dr. Genichi Taghuchi have won acceptance among many users. We will restrict our discussions, however, to the six listed above; the reader is referred to books such as Evans and Lindsay (2005) for details of the other models. No single model will be the perfect fit for a given organization; a review of several models will provide a perspective on the strengths of the different models, and would help in tailoring a system that best suits the needs of the organization.

9.2 Dr. Deming's System

Dr. W. Edwards Deming was the guru who taught the Japanese how to organize and manage a system for quality. Born in Sioux City, Iowa, on October 14, 1900, Dr. Deming lived most of his early life in Wyoming, where his parents moved when he was seven years old. He earned a BS in electrical engineering from the University of Wyoming at Laramie in 1921, and later a master's degree in mathematics and physics from the University of Colorado at Boulder. He also earned a PhD in physics from Yale in 1928. He was employed after graduate school by one of the laboratories of the Department of Agriculture in Washington, D.C., and as part of his activities there, he organized lectures in statistics at the Graduate School of the Department of Agriculture. The trainees from these seminars included statisticians of the U.S. Census Bureau, who used statistical sampling surveys for the first time in determining the U.S. unemployment rate during the Great Depression. Dr. Deming also participated in the 1940 Census and used sampling techniques to evaluate and improve the accuracy of entering and tallying data at the Census Bureau.

Dr. Deming came in contact with Dr. Shewhart while working in Washington, and became one of the admirers of the author of the control chart methods. He organized seminars by Dr. Shewhart at the Graduate School of the Department of Agriculture, which offered the latter opportunities to expound the control chart method to audiences outside the AT&T telephone companies where he was employed. During World War II, Dr. Deming was called on to assist the Statistical Research Group at Columbia University in spreading statistical methods among the manufacturers of goods and ammunition for the war effort. He wrote the curriculum and personally taught classes in which thousands of engineers were trained in statistical process control techniques. Dr. Deming also spent a year studying statistical theory in London with Sir Ronald A. Fisher, the famous statistician who invented the methods of experimental design.

After the war, in 1947, Dr. Deming went to Japan. He was invited by General McArthur's administration, the occupying forces, to help the Japanese in their census work to evaluate the extent of rehabilitation and reconstruction work needed there. He went to Japan again in 1950, at the invitation of the Union of Japanese Scientists and Engineers (JUSE), to assist the organization in spreading knowledge of statistical quality control within Japanese industry. "With his simple explanations and adequate demonstrations, Dr. Deming's lectures were so effective and persuasive that they left an unforgettable impression upon our minds," wrote an official of the JUSE (Gabor 1990, 80). He taught them how to implement statistical quality control methods following the plan-do-check-act (PDCA) cycle. He also taught them his new management philosophy, which was contained in his 14 points. The Japanese learned the methods, adapted them to their culture, and institutionalized the continuous improvement process, which culminated in the enormous success of Japanese goods in world markets. In recognition of his contribution to the growth of a quality culture in their land, the Japanese instituted a prize, called the Deming Prize, to be awarded to corporations that achieve excellence in product quality, or to individuals who make an outstanding contribution to statistical theory or its application. In 1960, Emperor Hirohito awarded Dr. Deming the Order of the Sacred Treasure, Second Class—the highest honor bestowed by the emperor on a non-Japanese person. During the 1980s, Dr. Deming brought home to American industry the lessons of quality he had helped the Japanese to learn, and participated in the quality revolution that was to unfold within U.S. industry. He died in 1993 at the age of 93.

The system Dr. Deming recommended is contained in the 14 points he advocated to the Japanese to form a cogent set of guidelines for creating a management system that will enable an organization to develop, design, and produce products that satisfy customers. Dr. Deming later offered the same set of guidelines, with some minor modifications, in his book *Out of the Crisis* (Deming 1986) as the recipe for American managers to confront the enormous competition posed by foreign manufacturers in the 1980s. His recommendations include a vision for long-term growth of quality, productivity, and business in general, as well as recommendations on how the vision can be accomplished through process improvements to reduce variability with the willing participation of a well-trained workforce. The 14 points recommended by Dr. Deming are discussed in detail below. These "points" are reproduced verbatim from his book; the explanation he provided under each point has been summarized. Subheadings have been added to facilitate comparison with other models. [Note: All quotes in the discussion of Deming's system come from the book *Out of the Crisis* (Deming 1986), a veritable source of wisdom for quality engineers.]

9.2.1 Long-Term Planning

Point 1: Create Constancy of Purpose for Improvement of Product and Service
This point relates to having a strategic, long-term vision for an organization regarding growth in quality and productivity, to become competitive, to stay in business, and to provide jobs. Organizations should have a long-term plan for creating new products, investing in research for new technology, and investing in people through education and training, to be able to satisfy the

needs of their customers. Toward this end, they should first find out, through customer surveys, what the customers' needs are, and then develop products to meet those needs. When the products and services are in the hands of the customer, organizations should ascertain, again through enquiry, if the products and services meet the needs, as intended. If they do not meet the needs, the products should be redesigned to satisfy the unmet needs. This should be an ongoing activity for productive organizations.

Dr. Deming said, "The customer is the most important part of the production line, and providing product and service through research and innovation to satisfy them is the best way to stay in business and to provide jobs. This should be the vision and it should be made clear to all in the organization and to those that are related to it."

9.2.2 Cultural Change

Point 2: Adopt the New Philosophy

According to Dr. Deming, the old management philosophy practiced by U.S. industry (during the 1960s and 1970s) allowed workers on jobs that they did not know how to perform, employed supervisors who neither knew the jobs they were supervising nor the skills of supervision, and employed managers who had no loyalty to the organization and were job-hopping. Those management practices also considered a certain level of mistakes (i.e., defects) to be acceptable. Those practices, according to Dr. Deming would not work in the new competitive environment that was emerging in the global market, which had competitors such as the Japanese, who had adopted a new management philosophy for increasing quality and making continuous improvement. The old philosophy caused too much waste, increased the cost of production, and resulted in a noncompetitive product.

For Dr. Deming, doing everything "right the first time" should be the new culture. Whether taking down a customer order, choosing specifications for a dimension, writing work instructions, making the product, preparing invoices, or answering a service call, everything should be done correctly the first time around.

9.2.3 Prevention Orientation

Point 3: Cease Dependence on Mass Inspection

Quality cannot be achieved through inspection. It must be built into the product through the use of the right material and the right processes by trained operators. Dr. Deming claimed that: "Quality comes not from inspection, but from improvement of the production process." Furthermore, routine inspection becomes unreliable through mistakes caused by boredom and fatigue.

Use of control charts, which need small samples taken at regular intervals, will help in achieving and maintaining statistical control of processes. In turn, this will assure the production of products with a minimum of variation.

9.2.4 Quality in Procurement

Point 4: End the Practice of Awarding Business on the Basis of Price Tag Alone
Dr. Deming placed great importance in buying quality material in order to produce quality products. Buying from the lowest bidder, with no regard for quality, is detrimental to producing quality and satisfying customers. "He that has a rule to give his business to the lowest bidder deserves to be rooked."

Purchasing managers must be educated to understand the need for quality material through exposure to how that material is used in production processes. They should learn how to specify quality in purchasing contracts. They should know that sometimes, even if the material meets the specifications that are written down, it might not meet the needs of the production process adequately, because all the requirements of a material cannot be fully written into specifications and contracts. Suppliers should be made to understand where and how the materials are used so that they know the requirements for the supplied materials.

Dr. Deming emphasized the need to have a single source for each material or part, and to develop long-term relationships with such single suppliers. The reasons he gave for having a single supplier for each of the supplies are elaborated in Chapter 7 under the section dealing with customer relationships. Briefly, the reasons include: a long-term relationship with a single supplier enables research and innovation; the variability in supplies from a single supplier will be smaller compared to multiple suppliers; each supplier will try to improve their performance through their motivation to become the single supplier; accounting and administrative expenses will be smaller, as will the total inventory for the supplies. The single supplier can also participate in the design activities of the customer.

9.2.5 Continuous Improvement

Point 5: Cotinuously Improve the System of Production and Service
Quality starts at the design stage, with a good "understanding of the customer's needs and of the way he uses and misuses a product." This understanding must be continuously updated, and any newly discovered needs should be met through changes to the product design.

There should be continuous improvement in every activity, such as procurement, transportation, production methods, equipment maintenance, layout of the work area, handling, worker training, supervisor training, sales, distribution, accounting, payroll, and customer service. The processes must be improved to reduce variability in key characteristics such that their distribution becomes "so narrow that specifications are lost beyond the horizon." (The spread in process variability must be reduced to a size such that the spread in the specification will appear huge in comparison. In terms of setting goals for process capability, this goes far beyond Motorola's 6-sigma capability.)

To understand process variation and make process improvements, continuous learning of new methods is necessary. "There is no substitute for knowledge." People must be given the opportunity to educate themselves and to learn new skills in experimentation, process control, and improvement methods to maintain processes in statistical control.

9.2.6 Training, Education, Empowerment, and Teamwork

Point 6: Institute Training

Everyone in an organization needs training on how to do his or her job. Managers should learn about the jobs on the production floor, in distribution, in process maintenance, in accounting, and so on. Supervisors should learn the jobs their workers or subordinates perform. Workers must have formal training on how to do the job they are expected to do. Inspectors must be trained on what is acceptable and what is not. Lack of such training leads to inconsistencies in how a particular job is performed and in the quality of the products shipped to the customer, which ultimately results in poor customer satisfaction.

Any training should be preceded by removing inhibitors of good work, such as lack of proper tools and handling methods, work standards for daily performance (piece rate), and so on, that "rob" workers of their pride in workmanship. Training should include learning the customer's ultimate needs for the product. Also, everyone in the organization should understand the concept of variation, how it is measured, and how it is to be controlled.

Point 7: Adopt and Institute Leadership

Managers should become leaders or coaches who facilitate and help their teams in performing their jobs. Management by objective (MBO), which is based solely on results or outcomes, should be replaced with *leadership*, with a focus on quality of product and service. Managers should be able to recognize opportunities for improvement and have the know-how to devise and implement the improvements. These leaders should remove the barriers that make it impossible for workers to do their jobs with pride. They should focus on quality rather than quantity of products produced. They must know the work they supervise; otherwise, they cannot help or train their workers. They should understand that errors, or defectives, when they occur, are produced by the system and not by the people. They should work to improve the system to prevent errors rather than blame the workers for them. They should also know how to recognize when a system is stable (i.e., in-control) and when it is not, so that they can apply remedies to prevent defectives.

The new leaders should understand the laws of nature on how worker performance will vary from worker to worker and is generally distributed as a normal distribution, with half the people performing below the average and half above. There will be some outliers, or performances outside the system—that is, outside of the 3σ-limits. The outliers above the upper limits should be examples to be emulated by others, and those below the lower limits should be helped to improve their performance. Such an understanding of the statistical behavior of populations will help to eliminate practices, such as celebrating someone as an excellent performer based on a single day's or month's performance, or denigrating another as a poor performer based on a single instance.

Point 8: Drive Out Fear

Fear among employees prevents them from reporting problems in product design, or problems arising from process deterioration that can cause poor quality products, which then get shipped out to customers. Fear of losing one's job prevents workers from suggesting innovations or improved methods. Fear prevents workers from acting in the best interest of the company because some short-term, narrowly specified goals will be violated.

Some managers cause fear in their workers because they believe in managing by fear. Fear is caused by: production quotas, which are used in evaluating performance; by annual merit ratings, which are used for determining salary increases; by decisions made by managers through erratic judgments based on one's feelings rather than on real data. Unless fear is removed, workers will not come forward with suggestions for improvement, will not accept new knowledge and new methods for quality improvement, and will not stop poorly made products from being shipped to customers.

Point 9: Break Down Barriers between Staff

The work involved in achieving product quality is done at many places in an organization, and by many different people. They all have information that needs to be shared. For example, marketing people have information on consumer preferences, which the product design people need, and the manufacturing people generate information on process capabilities, which the designers need. Dr. Deming gives an example of a service person fixing a problem in each of the new machines built by a company, which he never shared with the designers. If he had informed the designers of the problem in the machines when he first noticed it, the problem could have been fixed at the source and all the service calls and expensive repairs needed later at the customer locations could have been avoided.

Many of the points made by Dr. Deming can be identified as the basis for several new methodologies later developed by others for quality improvement and waste reduction. Point 9, perhaps, gave the impetus for growth of concurrent engineering (CE), which has been adopted by many design organizations to create products using cross-functional teams. Under the CE model, the team members exchange information on market research, product design, manufacturing capability, maintainability, and so on—even at the design stage. Design issues are decided with the team members working concurrently rather than exchanging information "over the wall." Such concurrent designs have reduced the concept-to-market time, improved quality, and avoided redesigns. The whole concept of concurrent engineering is based on breaking down barriers among functional areas and exchanging information at the early stages of product and process design, which is the thrust of Point 9.

Point 10: Eliminate Slogans, Exhortations, and Targets for the Workforce
Posters and slogans on walls, mainly addressed to workers, do not have any positive results. In fact, they have negative effects, such as creating mistrust, frustration, and demoralization among workers. These posters create the perception that the defects and errors are caused by workers, and suggest that if workers would pay proper attention, the errors could be avoided. The truth, however, is that the management has not addressed the causes of the failures and defectives, such as poor material, bad machinery, and inadequate training. "No amount of entreating or exhortation of the workers can solve problems of the system, which only the management can fix."

Slogans such as "be a quality worker," "take pride in your work," "do it right the first time," and so on, may have "some fleeting temporary effect on removing some of obvious problems in processes, but will be eventually recognized as a hoax." The management should start working on purchasing better-quality material from fewer suppliers, maintaining machinery in better working conditions, providing better training, using statistical tools for stabilizing processes, and publicizing these activities among workers. These will create faith in the words of management among workers, improve their morale, and encourage their further cooperation.

Point 11(a): Eliminate Numerical Quotas for the Workforce
According to Dr Deming, a numerical quota or work standard—that is, requiring so many pieces to be produced per day—"is a fortress against improvement of quality and productivity." Standards are often set for the average worker, which means that half the workers will exceed them and the other half will not. The first half will not make any more than this standard (they stop work before the end of the day). The second half are demoralized, because they cannot do any better than what they are currently doing under the given conditions. Once the standard is set, no one makes any effort to improve upon it, which is counter to the concept of continuous improvement.

Work standards are used by many organizations to create production schedules and prepare budgets. According to Dr. Deming, a better way to obtain information for budget and schedules is to collect data on the production time, find out the distribution, find the special causes for the outliers, and eliminate the causes of these outliers by avoiding the reasons that generate the special causes. Better training of workers would also improve performance. Such actions will improve productivity and provide data for budgets and schedules.

Point 11(b): Eliminate Numerical Goals for People in Management
Numerical goals set for managers, such as *decrease cost of warranty by 50% next year*, with no plans as to how to accomplish it "is just a farce." Those

goals will never be accomplished. A manager should learn the job that he is expected to supervise, and seek methods to improve the processes by identifying and eliminating wasteful steps. Mere goal setting without the knowledge and ability to make improvements will not help.

Point 12: Remove Barriers that Rob People of Pride of Workmanship
According to Dr. Deming, every worker wants to do a good job and be proud of it. If defectives are produced and waste generated, it is because the management does not provide the opportunity for workers to do their jobs well. Workers have many obstacles to doing their jobs with pride. Being treated as a commodity—such as being hired one week when needed and then fired the next week when not—is one such obstacle. Lack of work instructions on how to do a given job, and lack of standards for what is acceptable work and what is not are others. Poor quality of raw material (bought at a cheap price), poorly maintained equipment, inadequate tools, out-of-order instruments, and foremen pushing to meet the daily production quota are still more causes that "rob the hourly worker of his birthright, the right to be proud of his work, the right to do a good job."

When workers are denied the chance to do a good job and be proud of it, they are no longer eager to come to work and widespread absenteeism results. "He that feels important to do the job will make every effort to be on the job." According to Dr. Deming, "Barriers against realization of pride of workmanship may in fact be one of the most important obstacles in reduction of cost and improvement of quality in the United States."

Point 13: Encourage Education and Self-Improvement for Everyone
Dr. Deming claimed, "There is no shortage of good people, shortage exists at the high levels of knowledge, and this is true in every field. People must be continually educated, not for the short-term benefit of being able to do their current job better, but to improve themselves in acquiring new knowledge. Advances in competitive position will have their roots in knowledge." For, Dr. Deming, everyone in an organization should learn continually and become knowledgeable not just in their own area of work, but overall.

Point 14: Take Action to Accomplish the Transformation
Dr. Deming laid down an action plan for initiating and accomplishing quality in an organization:

1. Management must first understand and agree to the 13 points enunciated above.
2. Management must be prepared to change its philosophy and explain to the rest of the organization, through seminars and other means, the need for this change.

3. When a critical mass of people—especially in middle management—understand and agree with the changes needed, then these change will happen.
4. Every job in an organization can be divided into stages, or operations, and can be analyzed using a flow diagram. Each stage has a customer and a supplier, with the last stage having the end-user as the customer. Such analysis helps in continuously improving the methods and procedures to better satisfy the customer who receives the output from each of the stages in the system. The Deming cycle of plan-do-check-act (PDCA) explained in Chapter 8 will help in the continuous improvement.
5. An organization should be created to guide and monitor the continuous improvement in the process stages using qualified and trained statisticians, who will help in conducting experiments and in completing process improvement projects.

We can see in the above discussion that Dr. Deming intended a systemic change for an organization to become a quality organization. His 14-point recipe contains a model that, when followed, will create a quality management system to enable the production of quality products and services.

9.3 Dr. Juran's System

Dr. Joseph M. Juran was another guru who enormously influenced the direction of the quality movement, both in the United States and around the world. He was born in Romania on December 2, 1904, and came to the United States in 1912 when his family migrated to Minneapolis. He graduated from the University of Minnesota with a degree in electrical engineering in 1924. After graduation, he joined the Western Electric Co.—the manufacturing arm of the former Bell Telephone Co. (now known as AT&T)—in the inspection department of their Hawthorne plant. He was promoted as manager of the department and, when he was only 24 years old, became chief of the inspection division. He earned a JD degree in 1936 from Loyola University Law School in Chicago, and then moved to New York to become the corporate industrial engineer at the AT&T headquarters. After a tour of duty with the Lend-Lease Administration in procuring and leasing arms, equipment, and supplies during World War II, he started teaching industrial engineering as a professor and department chair at New York University. After working for a consulting firm as a quality management consultant, he then started his own consulting practice in 1949. This practice grew into the Juran Institute, which today provides education, training, and consulting to a wide range of industries worldwide.

Dr. Juran's main contribution to the quality field was in the management area. He emphasized that quality professionals should become "bilingual"—meaning they should learn to speak the language of finance, which the executives understand; and also the technical language that the workers and engineers understand. His adoption of the Pareto principle to differentiate between the vital few and the trivial many causes while investigating causes responsible for quality losses is one of his major contributions. The differentiation between chronic and sporadic problems to facilitate the application of appropriate methods to resolve them is another. The systematic "breakthrough" approach to solving chronic problems that Dr. Juran postulated is yet another important contribution to the quality field.

He did not see the value of statistics in quality improvement work as much as, for example, Dr. Deming did. He even claimed that the field of statistics was being overdone. In his view, it was an important element for achieving quality but should not be treated as the "be-all and end-all." That was a contentious statement to make even in the early 1950s (Donaldson 2004), and it may be even more contentious today in view of the revival of interest in statistical methods among quality professionals, as evidenced by the widespread use of the Six Sigma methodology, which makes extensive use of statistics.

Dr. Juran has written many books and many articles, of which the *Quality Control Handbook*, now in its sixth edition (Juran and Defeo 2010), stands out today as the standard reference on quality topics. In the early 1950s, invited by the JUSE, Dr. Juran went to Japan and provided training in the management aspects of quality. In 1981, he was, like Dr. Deming, awarded the Order of the Sacred Treasure, Second Class, in recognition of his contribution to the Japanese quality movement. Dr. Juran died in February 2008 at the age of 104.

The system recommended by Dr. Juran is contained in the "quality trilogy," also called the "Juran trilogy," which he proposed for producing quality products and satisfying customers. The three components of the trilogy (Juran 1988) are:

1. Quality planning
2. Quality control
3. Quality improvement

9.3.1 Quality Planning

Quality planning consists of the activities carried out during the product development and design stages—as well as during process engineering—before the product is put into production. These activities include:

- Determining who the customers are for the product
- Determining what their needs are

- Developing product features that respond to these needs
- Developing processes that are able to produce those product features
- Transferring the product and process plans to the operating or production function

The road map for quality planning according to Dr. Juran is shown in Figure 9.1. In this figure, the boxes indicate the action taken, and the lines between the boxes represent the result from the previous action. The first step is to identify the customer. The customer is someone who is impacted by the product. A customer can be external or internal. The external customer is someone outside the organization who buys and uses the product, and

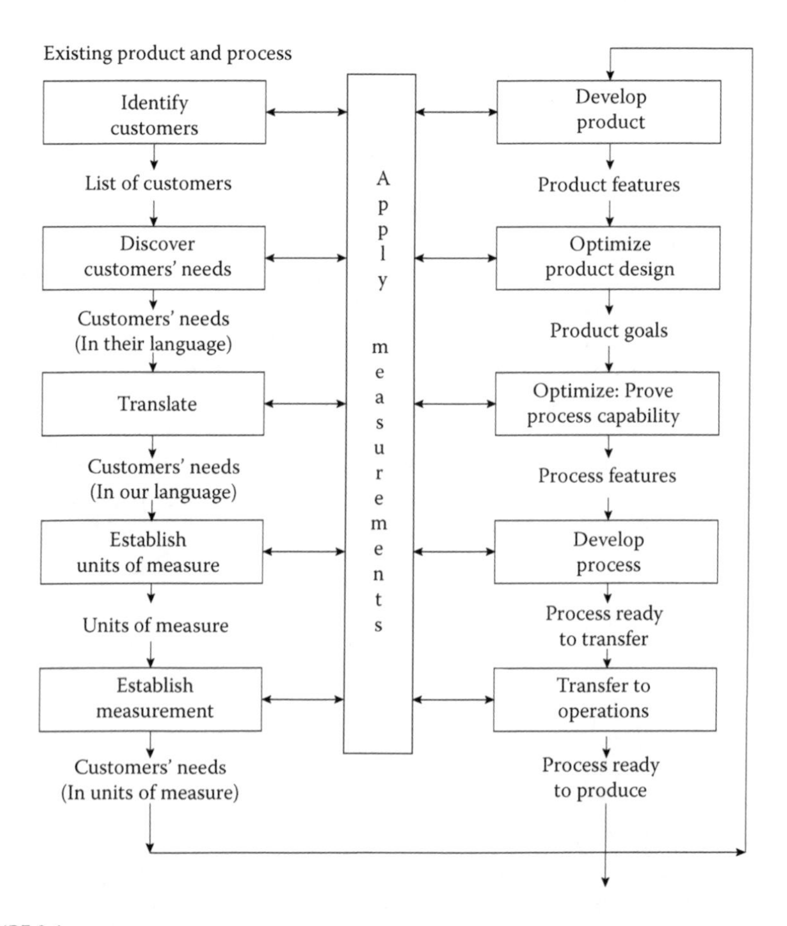

FIGURE 9.1
Road map for quality planning. (From Juran, J. M., and F. M. Gryna, eds., *Juran's Quality Control Handbook*. 4th ed., McGraw-Hill, New York, NY, 1988. With permission.)

the internal customer is someone inside the organization who further processes the product to be delivered to the next internal or external customer. There are multiple customers for many products. For example, with a new medicine, in addition to the ultimate patient who receives the treatment, the doctor, the pharmacist, the hospital, and the government regulatory agencies are all customers too. When there are numerous customers, it is necessary to classify them into the vital few and the useful many categories so that planning resources may be directed to meeting the needs of the most important segment of customers.

The major approaches for discovering the customer's needs are:

1. By being the customer (i.e., by putting the planner in the shoes of the customer)
2. Simulating the customer's needs in the laboratory
3. Communicating with the customer

Communicating with the customer (i.e., market survey) is the most widely used method. Some of the communications are customer initiated, such as customer complaints and warranty claims. Translation of the customer needs, which are in the language of the customer, into the language of the planner is done through spreadsheets, which are akin to the matrices of the quality function deployment procedure discussed in Chapter 3.

Product development must be done in order to meet the needs of the customers and the needs of the suppliers, and to optimize the costs to both. The final cost to the end-user must also be competitive. Product development should take into account the vital few (rather than the useful many) needs of the customers and consider how, and to what extent, competitors attempt to meet those needs. The opinions and perceptions of the customers that guide their buying habits, the value that the product will provide for the price that customers pay, and the failure-free operation of the product in customer hands should all influence the determination of product features.

Monopoly of product development vested in the product development or engineering functions can cause difficulties to other functions, such as manufacturing and marketing. It can also result in products that elicit complaints from end-users. To avoid this, participation from those involved with other functions must be sought by the product developers through design reviews.

The concept of dominance is a useful tool for process planners. Manufacturing processes can be identified as set-up dominant, time dominant, component dominant, worker dominant, or information dominant, based on what is important to the process so that it can function error-free. Identification of the dominant variable in the process helps the planner in establishing suitable control for the variable. For example, for set-up-dominant processes, in which the quality of subsequent batches depends on how well the initial set-up is done, "first-piece inspection" and approval would serve well.

Information on the capability of various production processes in meeting tolerances is vital for process planners. Knowledge regarding the available capabilities of production machinery helps the planner to choose the right process, or right machinery, for producing a given product characteristic. They are also used to specify to the production people the level of capability that must be maintained in the process machinery.

Process control is the means of maintaining a process in its planned state. This involves selecting the critical operations of a process, based on the potential for serious danger to human lives or the environment, or for serious waste and monetary loss at subsequent stages of processing. For these critical operations, process control procedures must be specified, including the variable to be checked, by whom it is to be checked, and the qualification of the people making the checks. How critical the incidents of deviation must be before triggering an investigation and elimination of root causes must also be stipulated.

The final step in process planning is process validation. This is done during a pilot run when the process is run using production equipment and production workers, at production capacity, and its capability is evaluated under normal operating conditions. The capability must be within acceptable limits before the process is handed over to the operating function. The process planning function ends with a transfer of knowledge on how the product should be produced, including documents on process parameters, procedures, cautions, and lessons learned during the pilot run.

9.3.2 Quality Control

Quality control is defined by Dr. Juran as "the regulatory process through which we measure actual quality performance, compare it with quality goals, and act on the difference" (Juran 1988, 6.31). To exercise this control, a limited number of centers—or control stations—must be established. These control stations are chosen using the following guiding principles:

1. At changes of jurisdiction to protect the recipients (e.g., between major departments or between a supplier and a customer)
2. Before embarking on an irreversible path (e.g., set-up approval before production)
3. After creation of a critical quality characteristic
4. At dominant process variables
5. At natural "windows" for economical control (e.g., chemistry of molten iron provides a window on strength of iron when it solidifies)

The choice of control stations is made on process flow diagrams. For each control station, the process variable or product characteristic to be measured, instruments to be used, interval between taking measurements, tolerances to be allowed, and decisions to be taken based on deviation from tolerance, should be specified. These are all chosen by planners and are included in the process planning documents. The designated operating personnel should take those measurements and make the decisions as laid out.

Tools exist for analyzing the measurements made at the control stations, which would indicate any significant deviations in the measurements from the expected goal. The control chart is one such tool for seeing if a statistically significant deviation exists—that is, whether the deviation results from a chance cause or a real cause. Significant deviations are acted on for eliminating the root cause. Dr. Juran identifies a difference between statistically significant and economically significant deviations. According to him, not all statistically significant deviations can be economically significant calling for immediate corrective action. When numerous deviations are found, priorities must be established based on economic significance, and when "economic significance of some nonconformance is at a very low level, corrective action may not be taken for a long time" (Juran 1988).

One of the very significant contributions made by Dr. Juran relative to process management is in the differentiation he made between two possible sources of problems when measurements indicate deviations from expected goals. The two sources, according to him, are sporadic sources of deviation and chronic sources of deviation. Sporadic sources cause occasional deviations and can be detected by control mechanisms, leading to corrective action. Chronic sources, on the other hand, cause deviations in the process that are likely to be ignored by control mechanisms and thus remain embedded in the system—even accepted as a fact of life. Their root causes are usually difficult to detect and to eliminate. The chronic sources of variations should be addressed through quality improvement projects using a certain sequence of steps, which Dr. Juran called the "breakthrough" sequence. The difference between sporadic sources and chronic sources of variation are illustrated in Figure 9.2.

9.3.3 Quality Improvement

Dr. Juran advocated making quality improvement "project-by-project" using the breakthrough sequence of steps. He claimed that the quality improvements and reduction of chronic wastes are not capital intensive, and they produce high levels of return on investment. The improvement process must be "institutionalized," meaning that it should become part of the culture of the organization. The breakthrough approach includes the following sequence of steps.

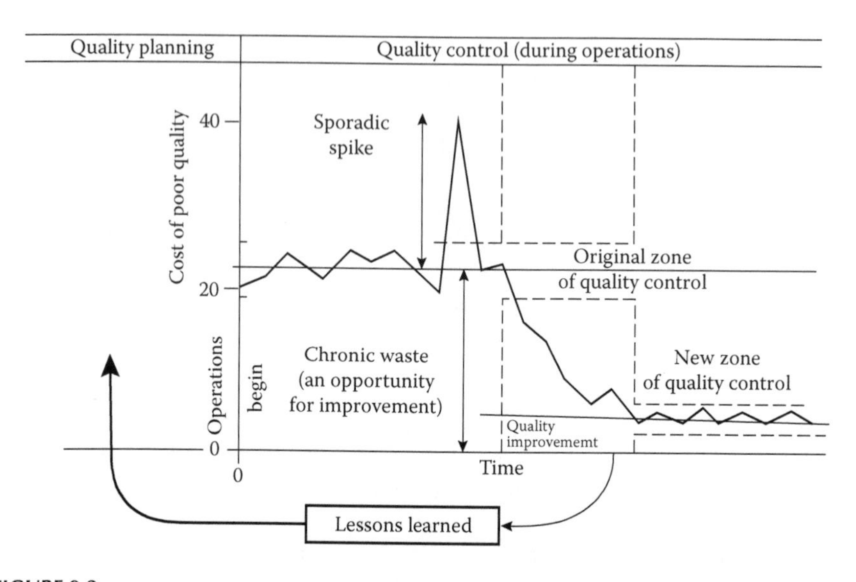

FIGURE 9.2
Sporadic vs. chronic deviations in a process. (From Juran, J. M., and A. B. Godfrey, eds., *Juran's Quality Handbook.* 5th ed. New York, NY: McGraw-Hill, 1999. With permission.)

Step 1: Establishing Proof of Need for the Project

The upper management must be convinced of the need to invest in the improvement project. Data must be gathered—for example, on the losses incurred by the company because of poor quality, the number of customer complaints in the recent past, and the market share lost because of unsatisfied customers—and then presented to the upper management, along with the estimated potential savings from the proposed project. The estimated expenditure should be compared with the potential savings to show the possible return, which is usually high. Dr. Juran (1999) gives examples in which an investment of $15,000 for quality improvement produced an average benefit of $100,000. Use of successful examples would help in selling the project to the upper management and winning their approval.

Step 2: Mobilizing for Quality Improvement

There must be some formal structure in the organization to initiate, approve, and encourage quality improvement, and to then execute the projects that have been approved. Organization is needed at two levels:

1. A *quality council* consisting of upper-level management, whose responsibilities will be to select projects, write mission statements, assign teams, provide resources, review progress, and provide rewards and recognition.
2. *Improvement teams* made up of people with intimate knowledge of the processes involved and the problem-solving know-how. The

improvement teams will gather data, analyze root causes, propose theories and possible solutions, test solutions, and make recommendations for remedies.

Step 3: Project Nominations and Selection

A list of candidate projects is generated by calling for nominations from both management and workers. The quality council members could nominate projects or obtain them through brainstorming sessions organized for this purpose. Joint projects with customers and suppliers are also good candidates for projects. The screening of nominations and selection of projects is done by the quality council.

Selecting the first project is more critical than selecting the subsequent ones. "The first project should be a winner" (Juran and Godfrey 1999). It should deal with a chronic problem and be a feasible project in the sense that it can be completed with reasonable success, and in a reasonable time frame. The project results must be measurable—for example, in dollars or in terms such as the number of customer complaints resolved or the number of breakdowns avoided. The results must be significant. The performance of the first project will greatly influence how well quality improvement will do in an organization.

Subsequent projects must be selected based on the estimated potential return on investment, urgency based on customer needs, safety, ease of finding a solution, and the anticipated ease with which the project will be approved by those associated with it.

The Pareto principle (see Chapter 8) helps in prioritizing the candidate projects into the vital few and the useful many based on their potential benefits. The vital few may generate huge savings, but the useful many will create participation from a wide segment of workers and management, in addition to contributing to savings. According to Dr. Juran, both types of projects must be tackled in order for the organization to reach both economic and noneconomic goals. He further recommends that "elephant-sized" projects be reduced into "bite-sized" projects so that they can be addressed by several improvement project teams, thus creating a wider participation and, possibly, a shorter completion time.

Every selected project must be accompanied by a mission statement, which should delineate the results expected in numerical terms and specify the time frame for completion. The results expected must be reasonable and practical—"perfection is not a good goal" (Juran and Godfrey 1999)—and the mission statement must be made public in order to make it official.

Projects are best assigned to teams rather than to individuals, because most quality projects require the input of people from different functional areas and with different experience, knowledge, and expertise. Generally, a sponsor or champion, who is a member of the quality council, is assigned to the team to act as an intermediary between the team and the quality council. The team will have a leader, appointed by the council or elected by the team;

a secretary, appointed by the leader; and a facilitator, who is not a member of the team. The facilitator who is trained and experienced in team dynamics and problem solving, helps in team building and training activities, and generally assists the team leader in setting agendas, running meetings, keeping the team focused on the goals, and writing final reports.

Step 4: Making the Improvement

Making the improvement involves two "journeys:"

1. The *diagnostic journey* for discovering the root causes of the problem, starting from the symptoms.
2. The *remedial journey* for devising solutions to the problem, testing and implementing them, and establishing control to ensure their continued effectiveness.

The diagnostic and remedial journeys require the use of investigative tools, such as the cause-and-effect (C&E) diagram, Pareto analysis, designed experiments, and regression analysis.

Implementation of the solutions may encounter resistance from the manager who owns the process, workers who are unwilling to accept new methods, or the union that wishes to protect workers' jobs. This is a human relations issue, and it should be approached with respect and sensitivity. Some of the approaches to addressing the resistance are communicating the reasons for the changes and the effects they will have on the parties involved; using parties as participants in the problem-solving process; making adjustments to the solutions when reasonable objections are raised; and allowing enough time for people to accept the changes. Once the solution has been implemented, controls must be established at key locations in the process to make sure the new methods are followed and people do not go back to the old methods either by design or by default.

When the expected goals are reached and there is a fair degree of assurance that the solution will continue to perform as intended, then the project can be considered as being completed. A final report must then be written that contains the record of the journeys made, from the symptoms to the final implementation. The data collected, methods used, analytical results, new configuration of the process (depicted in flow diagrams), specifications of the final solution, and the final results on process performance should all be included in the report.

Dr. Juran's system is more pragmatic than Dr. Deming's, and it can be implemented within the existing framework of a business organization. Dr. Deming's system is, of course, more comprehensive, calling for a complete philosophical and cultural change in the organization and revamping the entire system—from the way in which customer needs are ascertained to the way in which those needs are met.

9.4 Dr. Feigenbaum's System

Dr. Armand V. Feigenbaum was born on April 6, 1920, in New York City. He received a bachelor's degree from Union College, Schenectady, New York, and his MS and PhD from the Massachusetts Institute of Technology.

Dr. Feigenbaum was the one who first proposed a formal method for studying the costs associated with producing quality products and those arising from not producing quality products, which we have come to know as a "quality cost study." He was also the originator of the concept of *total quality control*—the approach to quality and profitability that has profoundly influenced management strategy in the competition for world markets. He postulated that contributions from many parts of an organization are needed to make a quality product that will satisfy the customer; and that these activities must be coordinated to optimize the output of the entire organization. This concept was embraced by the Japanese, who used it as the basis for their *company wide quality control* philosophy. This was also the genesis for the systems approach to quality that was later adopted by quality professionals worldwide.

Dr. Feigenbaum served as the founding chairman of the board of the International Academy for Quality, the worldwide quality body. He served two terms as president of the American Society for Quality. He was awarded the Edwards Medal, the Lancaster Award, and Honorary Membership by the American Society for Quality. The Union College awarded him the Founders Medal for his distinguished career in management and engineering. In December 1988, Dr. Feigenbaum was awarded the Medaille G. Borel by France—the first American to be so honored—in recognition of his international leadership in quality as well as his contributions to France. In 1993, he was named a Fellow of the World Academy of Productivity Science, and was awarded the Distinguished Leadership Award by the Quality and Productivity Management Association. In 1996, he was the first recipient of the Ishikawa/Harrington Medal for outstanding leadership in management excellence for the Asia-Pacific region. Dr. Feigenbaum is the president and CEO of General Systems Company, which is the leading international company in the design and implementation of management operating systems in major manufacturing and services companies throughout the world.

As mentioned, Dr. Feigenbaum was one of the early thinkers who recognized the need for a systems approach for achieving quality in products and satisfying customers. Although Dr. Deming had been propagating the value of a systems approach to optimize the output of any organization since his early days in Japan (Deming 1993), the first clear enunciation of the need for a systems approach to produce quality products, and how the system should be organized and managed, came from Dr. Feigenbaum. He identified four "jobs" (Feigenbaum 1983, 64) that were necessary for assuring quality in products:

1. New design control

   This includes designing product features to satisfy customer needs; establishing quality standards; verifying manufacturability; analyzing tolerance, failure mode, and effects; establishing reliability, maintainability, and serviceability standards; and conducting pilot runs when needed.

2. Incoming material control

   This includes receiving and stocking material and parts at the most economic levels of quality. Only those parts and material that conform to the specification requirements should be received and stocked. The vendor will have the "fullest practical responsibility" for the quality of the supplies.

3. Product control

   This includes control of production, assembly, and packaging processes to prevent the production or delivery of defective products to the customer. The customer will be provided with proper field service "to assure full provision of the intended customer quality."

4. Special process studies

   This includes investigating and eliminating the causes for product failures, thus improving product quality. The effort from special studies should "ensure that improvement and corrective action are permanent and complete" in order for there to be product quality in the future.

On the need for a quality system, Dr. Feigenbaum (1961, 109) said:

> Since the work is most generally part of an over-all "team" effort, it must be related to that of the other members of the team. Hence, not only does the division of labor become a consideration, but also the integration of labor becomes an equally important consideration. The whole purpose of organization is to get division of labor but with integrated effort leading to singleness of purpose. If the individuals are working at cross-purposes or are interfering with each other's efforts, either the people or the system are not working as they should be. The quality system provides the network of procedures that the different positions in a company must follow in working closely together to get the four jobs of total quality control done.

Dr. Feigenbaum showed in a matrix (Figure 9.3) how the responsibilities for various quality activities should be shared by different agencies in a quality system. One can see the genesis of the basic structure for a quality system in this matrix and observe that many of his ideas for creating and managing a quality system, some proposed as early as 1951 (Feigenbaum 1951), being reflected in later models for quality management systems, such as the ISO 9000 standards and the Baldrige Award criteria.

Relationship chart
(Applied to product quality)

Code: (R) = Responsible
C = Must contribute
M = May contribute
I = Is informed

Areas of responsibility	General manager	Finance	Marketing	Engineering	Manager manufacturing	Manufacturing engineering	Quality control	Materials	Shop operations
Determine needs of customer			(R)						
Establish quality level for business	(R)		C	C	C				
Establish product design specs				(R)					
Establish manufacturing process design				C	M	(R)	M	M	C
Produce product design specs			M	C	C	C	C	C	(R)
Determine process capabilities					I	C	(R)	M	C
Qualify suppliers on quality							C	(R)	
Plan the quality system	(R)		C	C	C	C	(R)	C	C
Plan inspection and test procedures						C	(R)	C	C
Design test and inspection equipment						C	(R)		M
Feed back quality information			C	C	I	M	(R)	C	C
Gather complaint data			(R)						
Analyze complaint data			M	M			(R)		
Obtain corrective action			M	C	C	C	(R)	C	C
Compile quality costs		(R)	C	C	C				
Analyze quality costs		M					(R)		
In-process quality measurements							(R)		C
In-process quality audit				C		C	(R)		
Final product inspection			C	C	M	C	(R)		

FIGURE 9.3
Interrelationships in a quality system. (From Feigenbaum, A. V., *Total Quality Control*. 3rd ed. New York, NY: McGraw-Hill, 1983. With permission.)

9.5 Baldrige Award Criteria

The Malcolm Baldrige National Quality Award (MBNQA) was established in 1987, during the presidency of Ronald Reagan, by an act of the U.S. Congress, for recognizing U.S. corporations who achieve excellence in organizational performance using quality and productivity improvement methods. This was a recognition that the U.S. corporations would not be able to compete in the global market unless they successfully adapted to a world where competitors were excelling in quality and productivity. The award was named after Malcolm Baldrige, President Reagan's secretary of commerce who died in an accident while in office. The award is administered by the Baldrige National Quality Program (BNQP) of the National Institute of Standards and Technology (NIST), assisted by the American Society for Quality (ASQ). The award administrators call for applications for the award each year through publicity documents, wherein the criteria for the award and the process of selection of the winners are explained. The awards are made under the six categories of: manufacturing, service, small business, education, healthcare, and nonprofit.

The MBNQA program publishes three sets of criteria: one for businesses/nonprofit, one for educational institutions, and one for health organizations. We discuss here the criteria for businesses/nonprofit.

The criteria are created and updated by a board of overseers, who are appointed by the U.S. secretary of commerce, and consist of distinguished leaders from all sectors of the U.S. economy. The awards categories and criteria are currently revised and published every two years. The following account is based on the 2009–2010 documents.

According to the BNQP, the criteria serve two main purposes:

- Identify Baldrige Award recipients to serve as role models for other organizations
- Help organizations assess their improvement efforts, diagnose the overall performance of the management systems, and identify their strengths and opportunities for improvement

There are seven criteria:

- Leadership
- Strategic planning
- Customer focus
- Measurement, analysis, and knowledge management
- Workforce focus
- Process management
- Results

Each criterion is divided into a certain number of "items," with the total number of items being 18. The award uses a point system, with a certain number of points being assigned to each item, and the total for all items being equal to 1000 (see Figure 9.4). The points assigned to the items, and thus to the criteria, can be interpreted as the level of importance that the board of overseers attribute to the items, and to the criteria, for achieving

	Categories and Items		Point Values
P	**Preface: Organizational Profile**		
	P.1 Organizational Description		
	P.2 Organizational Situation		
1	**Leadership**		**120**
	1.1 Senior Leadership	70	
	1.2 Governance and Societal Responsibilities	50	
2	**Strategic Planning**		**85**
	2.1 Strategy Development	40	
	2.2 Strategy Deployment	45	
3	**Customer Focus**		**85**
	3.1 Customer Engagement	40	
	3.2 Voice of the Customer	45	
4	**Measurement, Analysis, and Knowledge Management**		**90**
	4.1 Measurement, Analysis, and Improvement of Organizational Performance	45	
	4.2 Management of Information, Knowledge, and Information Technology	45	
5	**Workforce Focus**		**85**
	5.1 Workforce Engagement	45	
	5.2 Workforce Environment	40	
6	**Process Management**		**85**
	6.1 Work Systems	35	
	6.2 Work Processes	50	
7	**Results**		**450**
	7.1 Product Outcomes	100	
	7.2 Customer-Focused Outcomes	70	
	7.3 Financial and Market Outcomes	70	
	7.4 Workforce-Focused Outcomes	70	
	7.5 Process Effectiveness Outcomes	70	
	7.6 Leadership Outcomes	70	
	TOTAL POINTS		**1,000**

FIGURE 9.4
Malcolm Baldrige National Quality Award criteria and items. (From Baldrige National Quality Program, 2009–2010. *Criteria for Performance Excellence.* NIT, Gaithersburg, MD.)

performance excellence. Performance excellence here implies excellence in product or service quality, customer satisfaction, operational and financial results, relationship with employees and business partners, and discharge of public responsibility as a corporate citizen.

The interrelationship among the criteria, showing how together they form an integrated process for managing an organization and contribute to excellence in performance, is shown in Figure 9.5, which is reproduced from a BNQP document (Baldrige National Quality Program 2009–2010).

The criteria, the items, and the requirements under them are explained through a series of questions in the MBNQA brochure. These questions have been recast below into positive statements under each criterion. These statements can serve as guidelines for setting up a quality management system modeled after the MBNQA criteria. One important point is to be made here: the BNQP does not specify or prescribe how an organization's quality system should be set up to qualify for the award. The recommendations made here are an interpretation by the author of the questions laid down in the documents. A particularly good book on how to implement a quality management system in line with the MBNQA criteria is Brown (2008).

Among the criteria, the leadership, strategic planning, and customer focus categories are together called "leadership triad," as it emanates from the organizational leadership. The workforce focus, process management and results categories are together called the "results triad." The category

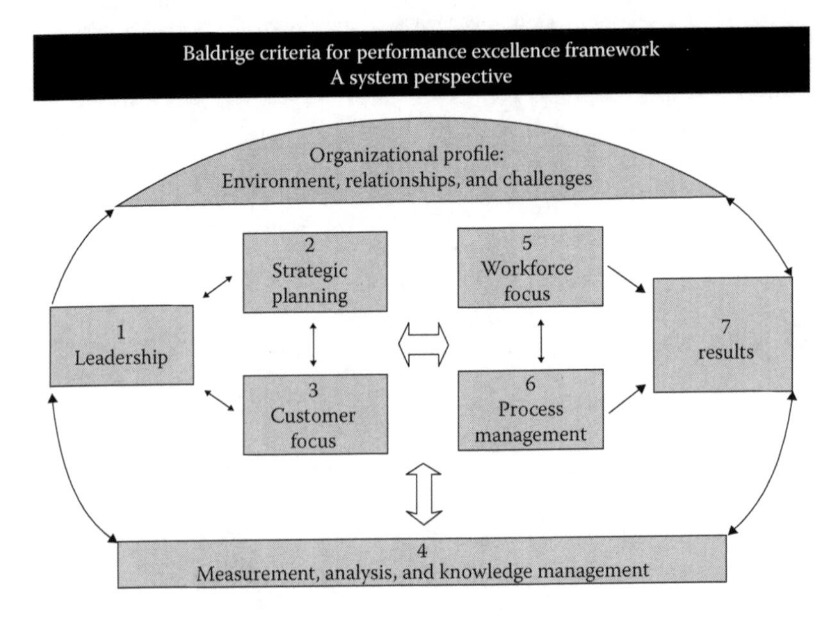

FIGURE 9.5
Malcolm Baldrige National Quality Award criteria—a systems perspective. (From Baldrige National Quality Program, 2009–2010. *Criteria for Performance Excellence*. NIT, Gaithersburg, MD.)

measurement, analysis, and knowledge management is called the "system foundation," as it is central and critical to the fact-based, knowledge-driven system for improving performance and competitiveness.

9.5.1 Criterion 1: Leadership

Senior Leadership

a. Vision, Values, and Mission

The senior leaders of an organization should set organizational vision and values and deploy them through the leadership system to the workforce, to key suppliers and partners, and to customers and other stakeholders, as appropriate. Their personal actions should reflect a commitment to the chosen values.

Senior leaders should personally promote an environment that fosters, requires, and results in legal and ethical behavior by people in the organization.

Senior leaders should create an environment for the improvement of organizational performance, accomplishment of mission and strategic objectives, innovation, competitive leadership, and organizational agility. They should foster an environment for organizational and workforce learning, in which they should develop and enhance their own leadership skills and participate in the learning process themselves. They should also be involved in succession planning for the development of future organizational leaders, thus creating a sustainable organization.

b. Communication and Organizational Performance

Senior leaders should communicate with and engage the entire workforce, and encourage frank, two-way communication throughout the organization. They should take an active role in reward and recognition programs to reinforce and encourage high performance.

Senior leaders should create measures to reflect the achievement of the organizational vision, objectives, and performance improvement; and should monitor them continuously and take corrective action when warranted. They should also focus on creating and balancing value for customers and other stakeholders in their expectations for the performance of the organization.

Governance and Societal Responsibilities

a. Organizational Governance

The organization should have procedures for reviewing and achieving the following key aspects of their governance system: accountability for the management's action, fiscal accountability, transparency in operations, selection and disclosure policies for board members, independence in internal and external audits, and protection of stakeholder and stockholder interests.

The organization should have procedures for evaluating the senior leaders, including the chief executive and the governance board. There should be evidence that senior leaders and the governance board use their performance evaluations to improve their personal leadership effectiveness as well as that of the board.

b. Legal and Ethical Behavior

The organization should have plans for addressing any adverse impacts on society from their operations and products. They should prepare for these concerns in a proactive manner. They should have plans for conserving natural resources and measures to evaluate and monitor compliance to regulatory and legal requirements. They should have processes, measures, and goals for addressing the risks associated with their products and operations.

The organization should promote and assure ethical behavior in all their interactions, and they should have measures and indicators to reflect ethical behavior in their governance structure and interactions with customers, suppliers, partners, and other stakeholders. There should be processes to monitor and respond to breaches of ethical behavior.

c. Societal Responsibilities and Support of Key Communities

Organizations should function as good citizens in their communities by participating in key activities, including the environment, social, and economic systems of the communities. They should identify their key communities and the areas where they can participate based on their core competencies. The senior leaders and workforce should participate in improving their communities.

Measures should be defined to evaluate how the organization identifies its key communities and meets its responsibilities to those communities. The measures should be monitored and improved.

9.5.2 Criterion 2: Strategic Planning

Strategy Development

a. Strategy Development Process

Organizations should have plans for the short and long term, with the time horizons for each term defined. They should identify the steps in making the plans, along with the key participants in the planning process. They should identify their core competencies, strategic advantages and challenges, and potential blind spots.

The planning process should establish—through collecting data—their strengths, weaknesses, opportunities, and threats; early indications of major shifts in technology, markets, products, customer preferences, competition, or the regulatory environment. They should be able to assess the long-term sustainability of their organization, and be able to execute the strategic plan.

b. Strategic Objectives

The organization should define their key strategic objectives and their timetable. They should choose the strategic objectives aligned with their strategic advantages and challenges, balanced between short- and long-term plans, and balanced with respect to the needs of all key stakeholders.

Strategy Deployment

a. Action Plan Development and Deployment

Organizations should use short- and long-term action plans to accomplish the objectives set out in the strategic plans, and should use measures to assess the progress and effectiveness of implementation of these plans. The strategies for deploying the action plans throughout the organization must be defined along with plans for securing the resources, including financial resources.

The impact of the action plans on human resources, with respect to their effect on the workplace, should be identified. The action plans should include measures for appraising their performance in terms of the achievement of objectives.

b. Performance Projection

Projections must be made with respect to the above performance measures and compared with those of the competitors, benchmarks, or past performance. Should there be gaps between the projected performance of the organization and those of the comparables, modifications to the action plans to address those gaps should be outlined.

9.5.3 Criterion 3: Customer Focus

Customer Engagement

a. Product Offerings and Customer Support

Organizations should have processes in place to obtain knowledge of current and future customers and markets, and to offer products and services to meet their requirements and exceed their expectations. The processes should also include efforts to innovate in order to attract new customers and expand relationships with existing customers.

Key mechanisms should be in place to provide support to the customers in their use of the products supplied by the organization, and to facilitate their conducting business with the organization. Key communication channels—based on the needs of the various customer segments—should be identified for customer contact and customer support. These provisions of customer support should be evaluated periodically, monitored, and updated to ensure that those supports are in line with the current business needs and directions.

b. Building a Customer Culture

Organizations should create a culture that will ensure consistently positive experiences for the customer. The workforce performance evaluation system, and the leader development system, should reinforce this culture.

Organizations should build and manage relationships in such a manner in order to acquire new customers, meet their requirements, and exceed their expectations.

Organizations should also keep their approaches for creating customer-focused culture current with business needs and directions.

Voice of the Customer

a. Customer Listening

Organizations should have identified the processes by which they will gather feedback from customers on the products and services provided. The methods should be tailored according to the type of customers the feedback is collected from.

Organizations should also have plans to listen to former customers, potential customers, and the customers of competitors.

Organizations should have a process through which customers' complaints can be lodged and then promptly resolved, so as to recover their confidence, and enhance their satisfaction and engagement. The process should also aggregate and analyze the complaints so as to obtain information to improve the operations of the organization, as well as those of its partners.

b. Determination of Customer Satisfaction and Engagement

The feedback received from customers should be such that action can be initiated based on it to exceed customer expectations, and analysis of the feedback could lead to process improvements throughout the organization and at its partners.

The data collected should be such that it can be used to find the organization's customers' satisfaction with a competitor's product, which can then be compared with their satisfaction of the organization's product.

c. Analysis and Use of Customer Data

Organizations should gather data on customers, the market, and product offerings in order to identify potential customer groups and market segments. They should gather information on competitors' customers and use this information to determine which customers to pursue for current and future products.

Organizations should use the data on customers, the market, and product offerings to identify and anticipate key customer requirements and their changing expectations. The information on how these changing expectations differ across customers, across market segments, and across customer life cycle should also be gathered.

All these data should then be used in improving the marketing of products, building more customer-focused culture, and in identifying opportunities for further innovation.

All the above approaches for listening to the customer should be kept current with business needs and trends.

9.5.4 Criterion 4: Measurement, Analysis, and Knowledge Management

Measurement, Analysis, and Improvement of Organizational Performance

a. Performance Measurement

Organizations should have processes for collecting data for tracking daily operations and overall organizational performance, including progress relative to strategic objectives and action plans. They should have defined their key organizational performance measures, as well as key short- and long-term financial measures. They should have determined the frequency for collecting data on these measures and defined how they will use the information for decision making and for making improvements and innovations.

The performance measurement system should be kept current to meet the current business needs of the organization, and should be sensitive to rapid or unexpected changes in the organization or in its environment.

b. Performance Analysis and Review

Organizations should have well-defined procedures for analyzing performance data, and should use the results to assess their success, competitive performance, and progress toward meeting strategic objectives. They should use reviews to assess their ability to respond to changing organizational needs and challenges in the environment.

c. Performance Improvement

Organizations should have procedures to translate the performance reviews into action plans for continuous and breakthrough improvements and innovations. The review information should be passed on to workgroups and functional-level operations to support their decision making. Where appropriate, the information from these reviews should be shared with suppliers for their actions to support the organization's goals.

Management of Information, Knowledge, and Information Technology

a. Data, Information, and Knowledge Management

There must be assurance that the data gathered are accurate and the analysis reliable, timely, and available to employees, suppliers, partners, and customers, as appropriate. Appropriate levels of security and confidentiality must be established.

The process to manage knowledge in the organization should include methods of collection and transfer of knowledge of the workforce; transfer of knowledge among customers, suppliers, partners, and collaborators; rapid identification, sharing, and implementation of best practices; and the transfer of relevant knowledge for use in strategic planning process.

b. Management of Information Resources and Technology

Adequate investment in computer hardware and software must be made according to requirements, so that the system remains reliable, secure, and user friendly. There must be provisions so that the hardware and software are available in the event of an emergency. The hardware and software for the collection of data and analysis should be kept current according to business needs in the context of technological changes in the operating environment.

9.5.5 Criterion 5: Workforce Focus

Workforce Management

a. Workforce Enrichment

Organizations should have plans to identify the key factors that affect workforce engagement and workforce satisfaction, and these factors should be determined as they vary among different segments of the workforce.

They should foster an organizational culture with open communication and a high-performing workforce. They should have policies in force for taking advantage of the diverse ideas, cultures, and thinking of the workforce.

The organization should be supported by a workforce performance evaluation system that ensures a well-engaged, high-performance workforce through proper compensation, reward, recognition, and incentive practices. The performance management system should also reinforce a customer and business focus, and ensure alignment of the workforce for achieving objectives of their operational and strategic action plans.

b. Workforce and Leader Development

The organization's system for learning and the development of personnel should ensure that the workforce and leaders are prepared to meet the needs of the organization's core competencies, strategic challenges, and the accomplishment of short- and long-term action plans. The learning system should also ensure ethics and ethical business practice goals through education, training, coaching, and mentoring.

The system for learning and development should also address the needs of the workforce with regards to: their own personal learning needs, as well as those identified by supervisors; the transfer of knowledge of retiring workers; and training for new knowledge and skills on the job.

The organization should have procedures for evaluating the effectiveness and efficiency of their learning and development system.

The organization should have an effective career progression program for the entire workforce and, in addition, they should have succession planning for management and leadership positions.

c. Assessment of Workforce Engagement

Organizations should have defined procedures for measuring the level of workforce engagement and workforce satisfaction; they may be formal and/or informal and may be different for separate workforce groups and segments. They should also take into account other indicators, such as retention rates, absenteeism, grievances, safety and productivity in assessing workforce engagement, and satisfaction.

Workforce Environment

a. Workforce Capability and Capacity

Organizations should have procedures to assess their needs in respect of workforce capability and capacity. They should have plans for recruiting, hiring, placing, and retaining new members of their workforce in line with their needs.

They should ensure that their workforce represents the diverse ideas, cultures, and thinking of their customer communities. They should manage and organize the workforce so as to accomplish the organization's goals using core competencies. They should also reinforce the need for customer and business focus, address strategic challenges, and achieve agility to face changing business needs.

They should have plans to prepare the workforce for facing the changes in requirements of capability and capacity. They should have plans for facing situations when workforce reduction becomes a necessity, although they should try to prevent such reductions. They should try to minimize the impact of reduction on the workforce and ensure continuity of operations at the organization.

b. Workforce Climate

Organizations should have procedures to ensure and improve workforce health, safety, and security. They should have performance measures to assess these factors in the work environment, and set goals for these measures and strive to achieve them. If there is such a need, different measures and goals must be created for different workplace environments.

They should have policies, services, and benefits to enhance the climate for the workforce. These may have to be different for the different groups to suit their needs.

9.5.6 Criterion 6: Process Management

Work Systems

a. Work Systems Design

Organizations should have plans for designing and innovating overall work systems, differentiating between those that are internal to the organization—their key work processes—and those that are external to them. The work systems should relate to and capitalize on the core competencies of the organization.

b. Key Work Processes

The organization should identify the key work processes (value-adding processes, such as product design, production and delivery, customer support, supply chain management, and business and support processes) and their ability to contribute to delivering customer value, profitability, and organizational success. They should determine the requirements for these processes based on customer inputs, and on those of suppliers and partners.

c. Emergency Readiness

The organization should have plans for emergency preparedness to ensure that, in case of emergencies and natural disasters, they are prepared for the continuity of operations and recovery with minimal disruption.

Work Processes

a. Work Process Design

The organization should design their work processes to meet all the key requirements, taking into account any new technology that is available, organizational knowledge, and the potential need for agility. The design should incorporate cycle time and cost control, productivity, and other efficiency and effectiveness factors in the design.

b. Work Process Management

The organization should implement and manage the processes to ensure that they meet the design and process requirements in the day-to-day operations. Input from the workforce, customers, suppliers, partners, and collaborators should be used in managing these processes. The key performance measures and in-process measures used for the control and improvement of these processes must be identified.

The organization should have procedures for controlling the overall costs of the work processes. Procedures must be in place to prevent defects, service errors, rework, and minimize warranty costs. The procedures should also minimize the costs of inspection, tests, and performance audits.

c. Work Process Improvement

The organization should have practices that seek to improve the work processes (continuously) in order to achieve better performance, reduce variability, improve products, and keep the processes current with business needs and directions. They should incorporate results from organizational performance reviews in the systematic evaluation and improvement of work processes. The lessons learned from process improvements from any one process should be shared with other units of the organization in order to enhance organizational learning and innovation.

9.5.7 Criterion 7: Results

Product Outcomes

a. Product Results

Organizations should keep track of their products' performance that are important to their customers by following the key measures of performance. The performance measures should be compared with those of competitors that are offering similar products.

Customer-Focused Outcomes

a. Customer-Focused Results

Organizations should be able to present the levels and trends in key measures of customer satisfaction and dissatisfaction, and present a comparison with the customer satisfaction levels of their competitors. They should also be able to present data on key measures on building customer relations and engagement, and provide the comparison with similar measures of their competitors.

Financial and Market Outcomes

a. Financial and Market Results

Organizations should use key measures to evaluate the current levels and trends in their financial performance, including aggregate measures of financial performance, financial viability, and budgetary performance. They should also have similar measures to evaluate their marketplace performance, including market share, market share growth, and new markets entered.

Workforce-Focused Outcomes

a. Workforce Results

Organizations should use key measures to evaluate the current levels and trends in workforce engagement and satisfaction. They should use key measures to evaluate the current levels and trends in workforce and leader development. They should use key measures to evaluate the current levels and trends in workforce capability and capacity, including staffing levels and appropriate skills. They should use key measures to evaluate the current levels and trends in workforce climate, including the health, safety, security, services, and benefits that they receive.

Process Effectiveness Outcomes

a. Process Effectiveness Results

Organizations should use key measures to evaluate the current levels and trends in the operational performance of their work systems, including workplace preparedness for disasters and emergencies.

They should have key measures to evaluate the current levels and trends in the operational performance of their key work processes, including productivity, cycle time, and other appropriate measures of process effectiveness, efficiency, and innovation.

9.6 ISO 9000 Quality Management Systems

9.6.1 The ISO 9000 Standards

This is a set of international standards that contains the requirements for building a basic quality management system. The standards were initially issued in 1987 by the International Organization for Standardization (ISO), which is a worldwide federation of national standards bodies. A minor revision of the standards was issued in 1994, and the third edition, issued in 2000, incorporated many changes to the original version to make it a modern template for building a quality management system. The standards were revised again as described below.

The original edition of the ISO 9000 standards consisted of five documents—ISO 9000, ISO 9001, ISO 9002, ISO 9003, and ISO 9004. Of these, ISO 9000 had the preamble, definition of terms, and instructions on how to use the rest of the documents; and ISO 9004 contained guidelines on how to establish a quality management system that is specified, through the "requirements," in ISO 9001, ISO 9002, and ISO 9003. A quality management system could be certified by independent auditors, upon verification, that the system satisfies the requirements. The three standards—ISO 9001, ISO 9002, and ISO 9003—differed in the scope of the system to be certified. ISO 9001 was to be used with a system of the largest scope, in which the organization's responsibilities included finding the needs of the market, designing the product to meet those needs, and making, delivering, and helping in the product's installation and use. The ISO 9002 and ISO 9003 were to be used by organizations with more limited responsibilities, ISO 9002 being applicable when products are made according to a customer's design, and ISO 9003 being applicable when the function was only distributing products designed and made by another organization. In a way, ISO 9002 and ISO 9003 were subsets of ISO 9001.

In the 2000 version of the standards, modifications made to ISO 9001 rendered ISO 9002 and ISO 9003 redundant, and so these were eliminated, resulting in the new set consisting only of ISO 9000, ISO 9001, and ISO 9004. The numbers 9002 and 9003 had been retired while keeping the numbers 9001 and 9004, presumably to maintain an alignment between the older and newer versions of ISO 9004. This should explain why the newer sets, ISO 9000:2000, and later versions contain only ISO 9001 and ISO 9004, and are missing the numbers 9002 and 9003.

ISO 9000 was revised in 2005 and describes, as the earlier edition did, the fundamentals of quality management systems and defines the terminology relative to them.

ISO 9001 was revised in 2008 and the discussion that follows is based on this 4th edition of the standard. ISO 9001 specifies the requirements of a quality management system, which, when adopted by an organization, will help the organization provide products that will satisfy customer requirements while making efficient use of the organization's resources. As mentioned earlier, an organization's quality system can be certified against this standard when there is a need to demonstrate its ability to meet customer requirements.

ISO 9004 was revised in 2009, and is also a standard that provides guidelines for organizing a quality management system but it has a wider scope than ISO 9001. The guidelines are meant to improve the performance of the organization to enhance the satisfaction of all stakeholders—customers as well as other interested parties, such as suppliers, employees, management, stockholders, and the community. However, it is not intended to be used for certification purposes.

So, counting all the latest revisions, we have the new standards: ISO 9000:2005, ISO 9001:2008, and ISO 9004:2009. Although ISO 9000 is just one of the documents in the set of standards, the entire set is often referred to as "ISO 9000 standards."

The standards recognize eight principles of quality management that can help the top management to lead an organization toward improved performance. These eight principles form the basis of formulating the requirements in the standards.

9.6.2 The Eight Quality Management Principles

Customer Focus: Organizations should understand the current and future needs of their customers and strive to meet and exceed the customers' needs and expectations.

Leadership: Leaders should establish the goals and create—as well as maintain—an environment within the organization so that people will become committed, and will participate in achieving the goals. They should periodically review the quality management system to verify its effectiveness.

Involvement of People: People at all levels should be involved in the pursuit of the chosen quality goals so that all their abilities are fully utilized for the benefit of the organization.

Process Approach: A productive organization is a collection of interrelated and interacting processes, with each process transforming some inputs into outputs through the use of some resources. Figure 9.6 has been drawn to show that an enterprise is a process made up of several subprocesses, which are interrelated. If the organization and its activities are analyzed as processes, it helps in gaining a good understanding of the individual processes and their interdependencies. This improves the chances of discovering opportunities for making improvements to the processes, and making them both effective and efficient. (*Effectiveness* refers to choosing the right objectives and achieving them fully; *efficiency* refers to achieving the objectives making economic use of resources.)

System Approach to Management: The entire organization should be managed as a system by identifying and understanding interrelated and interacting

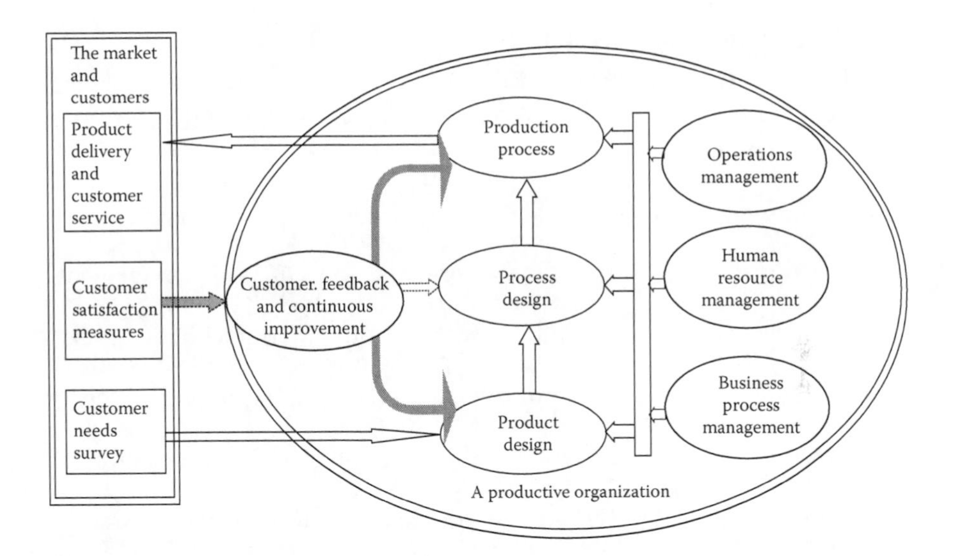

FIGURE 9.6
Model of a quality management system (the system is a process made up of several interconnected subprocesses).

components so as to optimize the overall system's goals, rather than optimizing subsystem goals, in a manner that enhances the effectiveness and efficiency of the entire organization.

Continual Improvement: Organizations should continually look for opportunities to improve their processes in order to improve customer satisfaction and improve the efficiency of internal processes.

Factual Approach to Decision Making: Organizations should encourage decision making based on evidence from data, or information gathered from processes, rather than on the feelings and beliefs of people.

Mutually Beneficial Supplier Relationship: Organizations should enter into interdependent, mutually beneficial relationships with their suppliers in order to enhance the abilities of both to create value in the supply chain.

9.6.3 Documentation in ISO 9000

The ISO 9000 standards require documentation at several stages for the following reasons.

Documentation enables communication of intent and consistency of action, which contribute to:

1. Achievement of conformity to customer requirements and quality improvement
2. Provision of appropriate training

3. Repeatability and traceability
4. Provision of objective evidence
5. Evaluation of effectiveness and continuing suitability of the quality management system

Several different types of document are required:

Quality manual	Describes an organization's quality management system
Quality plans	Describe how the quality management system is applied to a specific product, project, or contract
Specifications	Describe the requirements of a product per the needs of the customer
Guidelines	Describe recommendations or suggestions for managing a process
Procedures, work instructions, or drawings	Provide information on how to perform activities and manage processes in a consistent manner
Records	Provide objective evidence of activities performed and results achieved

9.7 ISO 9001:2008 Requirements

The ISO 9001:2008 requirements for establishing and maintaining a quality management system that would enable an organization to satisfy customer requirements as stipulated in the standard are discussed below. The numbers given in parentheses after each heading or subheading refer to the numbers given to the headings and subheadings in ISO 9001. The requirements discussed here start with Subheading (4) as the subheadings numbered (1) to (3) relate to preliminaries, such as definition of terms. The details of all the definitions are not reproduced here except for the following important definition:

> Throughout the text of this standard, the term "product" can also mean "service."

The reader will notice that the requirements of the standard are not reproduced in full in this text. For each requirement, only a small part of it is reproduced from the standard (ANSI/ISO/ASQ 2008) just to convey the intent of the requirement. Interested readers should refer to the standards themselves for complete details of the requirements. However, at the end of each subheading, or wherever appropriate, a brief summary of the requirements is provided in simple language by the author to help first-time readers understand the requirements better. These summary passages, shown in square parentheses are meant to convey the essentials of the requirements. They are not exhaustive, and are not meant to supplant the text of the original standard.

9.7.1 Quality Management System (4)

General Requirements (4.1)

The organization shall establish, document, implement, and maintain a quality management system and continually improve its effectiveness in accordance with the requirements of this International Standard.

The organization shall

a. determine the processes needed for the quality management system and their application throughout the organization, … [b–f follow]

[The standard requires that a documented quality management system be implemented, adequately supported with resources, and continually improved. The processes that underlie the productive (and business) activities should be identified, charted, their interactions determined, and their effectiveness measured by appropriate criteria. The measurements should be analyzed to discover opportunities for improvement and the improvements should be implemented.]

Documentation Requirements (4.2)

General (4.2.1)

The quality management system documentation shall include

a. documented statements of a quality policy and quality objectives, … [b–e follow]

[The standard requires documentation that lays down the policies and objectives of the quality management system; procedures to be followed for various activities; planning, operating, and control instructions considered necessary by the organization for the effective performance of the activities; a quality manual; and records of data as evidence for the output of the activities.]

Quality Manual (4.2.2)

The organization shall establish and maintain a quality manual that includes

a. the scope of the quality management system … [b–c follow]

[The standard requires a quality manual that describes the scope of the system, documented procedures or reference to them, and description of the interactions among the processes within the system.]

Control of Documents (4.2.3)

Documents required by the quality management system shall be controlled according to the requirements given below. A documented procedure shall be established to define the controls needed

a. to approve documents for adequacy prior to issue, ... [b–g follow]

> [Written procedures are needed to specify how the documents are controlled. The document control should ensure how documents are prepared and approved; how they are reviewed and updated; how they are identified by proper numbering, including revision status; how the latest revisions are made available at the point of use; how they are maintained to be legible and readily identifiable; and how obsolete documents are removed from workplaces and suitably marked if they are retained for any purpose. If any documents from external agencies are considered necessary in the system, they must be suitably identified and their distribution controlled.]

Control of Records (4.2.4)

Records are special type of documents and shall be created and maintained to provide evidence of conformity to requirements and of the effective operation of the quality management system. Records shall remain legible, readily identifiable, and retrievable. A documented procedure shall be established to define the controls needed for the identification, storage, protection, retrieval, retention time, and disposition of records.

> [Records provide important evidence of how the system functions, and they should be properly identified and stored according to a documented procedure. The procedure should specify how records should be maintained to be legible, readily identifiable, and easily retrievable]

9.7.2 Management Responsibility (5)

Management Commitment (5.1)

Top management shall provide evidence of its commitment to the development and implementation of the quality management system and to continually improving its effectiveness by

a. communicating to the organization the importance of meeting customer as well as statutory and regulatory requirements. ... [b–e follow]

> [Top management should establish quality policy and objectives for the organization and communicate to all the importance of meeting the needs of the customer as well as meeting statutory and regulatory requirement. They should conduct periodic management reviews to verify proper functioning of the quality system and make certain that adequate resources are available for implementing and maintaining the quality management system.]

Customer Focus (5.2)

Top management shall ensure that customer requirements are determined and are met with the aim of enhancing customer satisfaction.

> [Top management is responsible for creating a culture in which customer satisfaction is the focus at all levels of the organization.]

Quality Policy (5.3)

Top management shall ensure that the quality policy

a. is appropriate to the purpose of the organization, ... [b–e follow]

> [Top management should formulate quality policy appropriate to the organization in order to provide the basis for creating and implementing quality objectives at lower levels. The quality policy should show a commitment to comply with the requirements of the quality management system, and to continually review and revise it. The policy must be communicated to all within the organization.]

Planning (5.4)

Quality Objectives (5.4.1)

Top management shall ensure that quality objectives, including those needed to meet requirements for product, are established at relevant functions and levels within the organization. The quality objectives shall be measurable and consistent with the quality policy....

Quality Management System Planning (5.4.2)

Top management shall ensure that

a. the planning of the quality management system is carried out in order to meet the requirements given in 4.1 as well as the quality objectives, and ... [b follows]

> [Top management is responsible for ensuring that the quality management system is planned well to meet the needs of the organization. They must also make sure that planning for the quality of products and services is made at suitable functional levels within the organization. They should determine quality objectives in measurable terms in accordance with the quality policy. Any changes in the policies should be suitably planned and approved so as to maintain the integrity of the system.]

Responsibility, Authority, and Communication (5.5)

Responsibility and Authority (5.5.1)

Top management shall ensure that responsibilities and authorities are defined and communicated within the organization.

Management Representative (5.5.2)

Top management shall appoint a member of the organization's management who, irrespective of other responsibilities, shall have responsibility and authority that includes … [a–c follow]

> [Top management should assign various responsibilities for quality (e.g., design approval, conflict resolution, and responding to customer complaints) to appropriate people. They will also appoint a "management representative" who will be the focal point for creating an awareness of customer needs within the organization and for organizing, maintaining, and improving the quality system. The management representative will be responsible for reporting to top management on the performance of the system and be the liaison between external agencies and the organization in matters relating to the quality system.]

Internal Communication (5.5.3)

Top management shall ensure that appropriate communication processes are established within the organization and that communication takes place regarding the effectiveness of the quality management system.

> [Top management should also ensure that organization-wide communication exists regarding how well the system is performing.]

Management Review (5.6)

General (5.6.1)

Top management shall review the organization's quality management system, at planned intervals, to ensure its continuing suitability, adequacy, and effectiveness. This review shall include assessing opportunities for improvement and the need for changes to the quality management system, including the quality policy and quality objectives. Records from management reviews shall be maintained….

Review Input (5.6.2)

The input to management review shall include information on

a. results of audits … [b–g follow]

Review Output (5.6.3)

The output from the management review shall include any decisions and actions related to

a. improvement of the effectiveness of the quality management system and its processes … [b–c follow]

> [Top management should review the performance of the quality management system at planned intervals through internal audits done by a

group of trained personnel. The internal auditors verify and report to the top management on the conformance of the various aspects of the system to this standard. The report should include audit results, customer feedback, process performance and product conformity, status of preventive and corrective actions, follow-up actions from prior reviews, changes that could affect the quality management system, and recommendations for improvement. Top management will respond to the report by making decisions that will result in improvement in the effectiveness of the system, improvement in product-related customer requirements, and any resource requirements for the system.]

9.7.3 Resource Management (6)

Provision of Resources (6.1)

The organization shall determine and provide the resources needed

a. to implement and maintain the quality management system and continually improve its effectiveness, and
b. to enhance customer satisfaction by meeting customer requirements.

Human Resources (6.2)

General (6.2.1)

Personnel performing work affecting product quality shall be competent on the basis of appropriate education, training, skills, and experience.

Competence, Training, and Awareness (6.2.2)

The organization shall

a. determine the necessary competence for personnel performing work affecting product quality, … [b–e follow]

[There should be an adequate number of people to implement and improve the quality management system as well as to address customer concerns. Personnel engaged in activities relating to product quality and customer satisfaction must have adequate qualification through education and training. Records must be maintained on the education, training, skills, and experience of such personnel.]

Infrastructure (6.3)

The organization shall determine, provide, and maintain the infrastructure needed to achieve conformity to product requirements. Infrastructure includes, as applicable:

a. buildings, workspace, and associated utilities, … [b–c follow]

Work Environment (6.4)

The organization shall determine and manage the work environment needed to achieve conformity to product requirements.

> [Production and service facilities in terms of equipment, hardware and software, buildings, handling facilities, communication, and transport infrastructure must be adequate. The work environment (clean rooms or environmentally controlled spaces) that is needed to assure product quality must be available. The work environment including, noise, temperature, humidity, and lighting should be conducive to achieving conformity in product requirements.]

9.7.4 Product Realization (7)

Planning of Product Realization (7.1)

The organization shall plan and develop the processes needed for product realization. Planning of product realization shall be consistent with the requirements of the other processes of the quality management system. In planning product realization, the organization shall determine the following, as appropriate:

a. quality objectives and requirements for the product, ... [b–d follow]

> [There must be planning for quality at the product and process design stages so as to be able to meet the product quality. Quality objectives and requirements for the product should be established, process and product documents should be prepared, and resources needed for the product should be arranged. The control of processes should be planned through the use of control procedures along with measurements, inspection, and testing. Product acceptance criteria should be established along with procedures for verification and validation in the production stage. The plan should call for records to be kept of processes and products to show they met the established requirements.]

Customer-Related Processes (7.2)

Determination of Requirements Related to the Product (7.2.1)

The organization shall determine

a. requirements specified by the customer, including the requirements for delivery and post-delivery activities, ... [b–d follow]

> [The organization should determine the requirements as specified by the customer, those that are needed for the product's performance even if not specified by the customer, those needed to satisfy legal requirements, and any others the organization may decide as necessary.]

Review of Requirements Related to the Product (7.2.2)

The organization shall review the requirements related to the product. This review shall be conducted prior to the organization's commitment to supply a product to the customer (e.g., submission of tenders, acceptance of contracts or orders, acceptance of changes to contracts or orders), and shall ensure that

a. product requirements are defined, ... [b–c follow]

[The organization should make sure, prior to agreeing to supply the product, that the product requirements are defined, that all previously expressed differences are resolved, and that the organization has the ability to meet the requirements. Records must be kept of all results from reviews made as above.]

Customer Communication (7.2.3)

The organization shall determine and implement effective arrangements for communicating with customers in relation to

a. product information, ... [b–c follow]

[The customer should have easy access to the organization for obtaining product information, making changes to the contracts, or making complaints.]

Design and Development (7.3)

Design and Development Planning (7.3.1)

The organization shall plan and control the design and development of product. During the design and development planning, the organization shall determine

a. the design and development stages, ... [b–c follow]

[Product design and development should be preceded by planning, with stages for reviews, details of reviews, verification and validation needed, and responsibilities and authorities established.]

Design and Development Inputs (7.3.2)

Inputs relating to product requirements shall be determined and records maintained. These inputs shall include

a. functional and performance requirements, ... [b–d follow]

[The input to the product design and development include functional and performance requirements, legal requirements, information from previous design, and any other requirement essential for the product. The inputs should be reviewed for adequacy and completeness.]

Design and Development Outputs (7.3.3)

The outputs of design and development shall be provided in a form that enables verification against the design and development input, and shall be approved prior to release.

Design and development outputs shall

a. meet the input requirements for design and development, ... [b–d follow]

[The design and development of products should be made by authorized personnel, with planned reviews at appropriate stages. The design outputs, such as drawings, bill of materials, and inspection and test requirements should adequately respond to inputs, such as customer needs and government requirements. They should provide adequate information for purchase or production decisions. Characteristics of the product that reflect on safety in use should be specified. Records must be kept of the inputs and outputs of the design process.]

Design and Development Review (7.3.4)

At suitable stages, systematic reviews of design and development shall be performed in accordance with planned arrangements

a. to evaluate the ability of the results of design and development to meet requirements, and
b. to identify any problems and propose necessary actions....

[The design should be subject to reviews and checks at appropriate stages of product development. The design outputs, the drawings and specifications, should be subjected to design reviews by cross-functional teams for their adequacy in meeting established customer and government requirements, and for making suggestions for improvements, where necessary.]

Design and Development Verification (7.3.5)

Verification shall be performed in accordance with planned arrangements....

[The design must be "verified" for its ability to meet established input requirements. Records of verification and any necessary action suggested should be maintained.]

Design and Development Validation (7.3.6)

Design and development validation shall be performed in accordance with planned arrangements...

[Design validation is done to ensure that the resulting product is capable of meeting the intended use. Validation should be completed, wherever possible, before the design is handed off to production. Records should be kept of validation results and any actions taken pursuant to validation study.]

Control of Design and Development Changes (7.3.7)

Design and development changes shall be identified and records maintained...

[Changes to designs should be subjected to the same review, verification, validation, and approval process, as the original design. A review of changes to the original design should include evaluation of the effects of changes on parts and products already delivered. Records of changes, reviews, and subsequent actions should be maintained.]

Purchasing (7.4)

Purchasing Process (7.4.1)

The organization shall ensure that purchased product conforms to specified purchase requirements. The type and extent of control applied to the supplier and the purchased product shall be dependent upon the effect of the purchased product on subsequent product realization or the final product....

[The organization should select suppliers based on established criteria. The criteria should assure that the suppliers are able to supply goods that meet the needs of the processes. Records of supplier selection and evaluation should be maintained.]

Purchasing Information (7.4.2)

Purchasing information shall describe the product to be purchased, including, where appropriate:

a. requirements for approval of products, procedures, processes, and equipment, ... [b–c follow]

[The purchasing contract, or the P.O., should clearly specify the product requirements and how products are checked before acceptance. It should also describe, where appropriate, how the supplies are to be produced, including the qualifications of production personnel and the quality management system at the supplier, to assure quality of supplies.]

Verification of Purchased Product (7.4.3)

The organization shall establish and implement the inspection or other activities necessary for ensuring that purchased product meets specified purchase requirements....

[If the organization intends to perform inspection at the supplier's facilities, it must be stated in the P.O.]

Production and Service Provision (7.5)

Control of Production and Service Provision (7.5.1)

The organization shall plan and carry out production and service provision under controlled conditions. Controlled conditions shall include, as applicable:

a. the availability of information that describes the characteristics of the product.... [b–f follow]

[Production of products and creation of services should be done under controlled conditions using suitable equipment and measuring devices, with product/service characteristics being well defined and guided by proper work instructions. Clear instructions must be available on how the product/service will be handed off to the next stage for processing.]

Validation of Processes for Production and Service Provision (7.5.2)

The organization shall validate any processes for production and service provision where the resulting output cannot be verified by subsequent monitoring or measurement. This includes any processes where deficiencies become apparent only after the product is in use or the service has been delivered.

Validation shall demonstrate the ability of these processes to achieve planned results. The organization shall establish arrangements for these processes, including, as applicable:

a. defined criteria for review and approval of the processes, ... [b–e follow]

[Process validation through the approval of equipment, personnel, and procedures shall be done, especially when the performance of the product can be verified only after being used by the customer. Requirements that the operations adhere to specified procedures, and periodic revalidation of the processes, will assure the quality of products in such instances.]

Identification and Traceability (7.5.3)

Where appropriate, the organization shall identify the product by suitable means throughout product realization. The organization shall identify the product status with respect to monitoring and measurement requirements. Where traceability is a requirement, the organization shall control and record the unique identification of the product....

[When traceability is required—that is, when information is needed on where, when, using what batch of material, and by whom a product is produced—the organization should uniquely identify each unit of product produced, gather the relevant information, and keep a record of the data.]

Customer Property (7.5.4)

The organization shall exercise care with customer property while it is under the organization's control or being used by the organization. The organization shall identify, verify, protect, and safeguard customer property provided for use or incorporation into the product. If any customer property is lost, damaged or otherwise found to be not suitable for use, this shall be reported to the customer and records maintained.

[In situations where the customers provide material or machinery to be used in the production of the product, the organization has the responsibility to protect and preserve such customer's property. Customer property may also include intellectual.]

Preservation of Product (7.5.5)

The organization shall preserve the conformity of product during internal processing and delivery to the intended destination. This preservation shall include identification, handling, packaging, storage, and protection. Preservation shall also apply to the constituent parts of a product.

[The organization has the responsibility to preserve and protect the product until delivered to the customer through proper packaging, handling, and delivery.]

Control of Monitoring and Measuring Devices (7.6)

The organization shall determine the monitoring and measurement to be undertaken and the monitoring and measuring devices needed to provide evidence of conformity of product to determined requirements....

Where necessary to ensure valid results, measuring equipment shall

a. be calibrated or verified at specified intervals, or prior to use, against measurement standards traceable to international or national measurement standards; where no such standards exist, the basis used for calibration or verification shall be recorded;... [b–e follow]

In addition, the organization shall assess and record the validity of the previous measuring results when the equipment is found not to conform to requirements. The organization shall take appropriate action on the equipment and any product affected. Records of the results of calibration and verification shall be maintained....

[The organization has the responsibility for the selection and provision of appropriate measuring devices to verify product conformity. The instruments, gages, and monitoring equipment should be periodically calibrated and adjusted so that their traceability to national or international standards can be maintained. They must be protected from damage, deterioration, and unauthorized adjustment. The calibrated instruments and gages must be suitably labeled to indicate the status of their calibration. If an instrument is not in calibration, the organization should take appropriate action to assess any possible damage to the quality of products that might have been approved using the instrument that was not in calibration. Suitable action is necessary for recall and repair if product quality had been affected by use of a faulty instrument. A record of such action must be maintained. When using computer software in the measurement processes, the computer software should be tested for conformance to requirement before use. They must be periodically retested as necessary.]

9.7.5 Measurement, Analysis, and Improvement (8)

General (8.1)

The organization shall plan and implement the monitoring, measurement, analysis and improvement processes needed

a. to demonstrate conformity of the product, ... [b–c follow]

[Measurements are needed on how products meet specifications and how the quality management system conforms to this standard. Based on these measurements, deficiencies should be identified and improvement projects implemented. This should be done on a continuing basis.]

Monitoring and Measurement (8.2)

Customer Satisfaction (8.2.1)

As one of the measurements of the performance of the quality management system, the organization shall monitor information relating to customer perception as to whether the organization has met customer requirements. The methods for obtaining and using this information shall be determined.

Internal Audit (8.2.2)

The organization shall conduct internal audits at planned intervals to determine whether the quality management system

a. conforms to the planned arrangements, to the requirements of this International Standard and to the quality management system requirements established by the organization, and
b. is effectively implemented and maintained....

[Measurements must be taken on customer satisfaction with the organization's product. Measurements are needed on the conformance of the quality management system to this standard. An internal audit program by trained, impartial auditors should be instituted to verify conformance of the quality system to this standard. Measurements must be taken to verify if the processes of the quality management system perform per the plans of the system. Improvements must be implemented by process owners in a timely fashion when deficiencies are reported by the auditors. Records must be kept of audit results. The internal audit should be guided by a written procedure.]

Monitoring and Measurement of Processes (8.2.3)

The organization shall apply suitable methods of monitoring and, where applicable, measurement of the quality management system processes ...

[Measurement of process quality, including that of the quality management system process, should be made to verify if they are capable of producing planned outputs to meet requirements. When planned outputs are not meeting requirements, action should be taken for improving the processes.]

Monitoring and Measurement of Product (8.2.4)

The organization shall monitor and measure the characteristics of the product to verify that product requirements have been met. This shall be carried out at appropriate stages of the product realization process in accordance with the planned arrangements. Evidence of conformity with the acceptance criteria shall be maintained. Records shall indicate the person(s) authorizing release of product....

[Measurement of product characteristics must be made to verify if they meet the requirements specified for them. There must be control to assure that products are not released to the customer until their conditions are verified.]

Control of Nonconforming Product (8.3)

The organization shall ensure that product which does not conform to product requirements is identified and controlled to prevent its unintended use or delivery. The controls and related responsibilities and authorities for dealing with nonconforming product shall be defined in a documented procedure.

The organization shall deal with nonconforming product by one or more of the following ways:

a. by taking action to eliminate the detected nonconformity; ... [b–c follow]

[The handling of nonconforming products must be guided by a written procedure. Nonconforming products, when detected, should be identified and isolated so there is no chance that they will get mixed in with the other, conforming products. They must be kept quarantined until their disposition is decided by the proper authority, possibly in consultation with the customer. When a nonconforming product is detected and corrected, it should be subjected to a retest to demonstrate conformity. Records must be kept of the nonconforming units and on how they were disposed.]

Analysis of Data (8.4)

The organization shall determine, collect, and analyze appropriate data to demonstrate the suitability and effectiveness of the quality management system and to evaluate where continual improvement of the effectiveness of the quality management system can be made. This shall include data generated as a result of monitoring and measurement, and from other relevant sources.

The analysis of data shall provide information relating to

a. customer satisfaction, … [b–d follow]

[Data must be gathered and analyzed on system conformance, customer satisfaction, product conformance, and supplier performance, to identify opportunities for preventive action and for improvement.]

Improvement (8.5)

Continual Improvement (8.5.1)

The organization shall continually improve the effectiveness of the quality management system through the use of the quality policy, quality objectives, audit results, analysis of data, corrective and preventive actions, and management review.

Corrective Action (8.5.2)

The organization shall take action to eliminate the cause of nonconformities in order to prevent recurrence. Corrective actions shall be appropriate to the effects of the nonconformities encountered.

A documented procedure shall be established to define requirements for

a. reviewing nonconformities (including customer complaints), … [b–f follow]

Preventive Action (8.5.3)

The organization shall determine action to eliminate the causes of potential nonconformities in order to prevent their occurrence. Preventive actions shall be appropriate to the effects of the potential problems and include:

a. determining potential nonconformities and their causes, … [b–e follow]

[The organization should continually make improvements to the quality management system by identifying opportunities for improvement and implementing improvement projects. When nonconformities are found, the root causes of the problem must be identified and remedial action implemented. The remedial action should prevent recurrence of the nonconformities. The continuous improvement process must be guided by a documented procedure.]

9.8 The Six Sigma System

"The Six Sigma system is a comprehensive and flexible system for achieving, sustaining and maximizing business success" (Pande, Neuman, and Cavanagh 2000). The Six Sigma process strives to achieve this by careful understanding of customer needs, use of facts through data collection and statistical analysis, and managing, improving, and re-engineering productive as well as business processes to increase customer satisfaction and business excellence. Business excellence here includes cost reduction, productivity improvement, growth of market share, customer retention, cycle time reduction, defect reduction, change to a quality culture, and new product/service development.

The Six Sigma methodology was born in 1987 in Motorola's communication sector as an approach to track and compare performance against customer requirements, and to achieve an ambitious target of near-perfect, 6σ (or 6-sigma) quality, in the products produced. The 6σ quality means that the defect rate in the production of each component of an assembly (e.g., a cellular phone) will not be more than 3.4 parts per million (ppm) opportunities. The Six Sigma process, or the systematic approach to process improvement to attain 6σ quality, later spread throughout the company with strong backing from the then-chairman Robert Galvin. In the 1980s, this process helped the company to achieve enormous improvements in quality: 10-fold improvement every two years, or 100-fold improvement in four years. Two years after they set out on the Six Sigma journey, they received the prestigious national award: MBNQA. In the 10 years between 1987 and 1997, the company increased its sales fivefold, saved $14 billion from Six Sigma projects, and saw its stock prices increase at an annual rate of 21.3% (Pande et al., 2000). All this happened to Motorola when it was facing tough competition from Japanese competitors.

Several other organizations have followed the Six Sigma process model and reported enormous success in reducing waste and winning customer satisfaction. Two important examples are Allied Signal (Honeywell) and the General Electric Co.

[Note: the term "6σ" is used here to refer to the quality of a process, and the term "Six Sigma" is used to refer to the systematic approach used to achieve the 6σ quality.]

9.8.1 Six Themes of Six Sigma

Theme 1: Focus on the Customer
The Six Sigma process begins with the measurement of customer satisfaction on a dynamic basis, and the Six Sigma improvements are evaluated based on how they impact the customer. Customer requirements are assessed first, and the performance of the organization's product is then evaluated against those requirements. Next, the unmet needs are addressed through product or process change. Customers get the highest priority in a Six Sigma organization, which is any organization that adopts the Six Sigma process to improve "all" its processes.

Theme 2: Data and Fact-Driven Management
The Six Sigma philosophy emphasizes the need for taking measurements on process performance, product performance, customer satisfaction, and so on. The Six Sigma process has no room for decisions based on opinions, assumptions, and gut feelings. Process managers and problem solvers should decide what information is needed and arrange to gather it. The data should then be analyzed using the appropriate tools, and the information generated must be used to make decisions on process and product improvements. This "closed-loop" system, in which measurements from output are used to take corrective action to improve a process, is an important hallmark of the Six Sigma system.

Theme 3: Process Focus
Every product or service is created or produced by a process; whether it is designing a product, preparing an invoice, answering a customer complaint, or solving a problem, the activities can be mapped as a process involving a sequence of steps. When processes are mapped using charts on paper, they can be better understood, analyzed, improved, and managed. Understanding a process fully in order to be able to control and improve it is the key precept leading to the success of Six Sigma systems.

Theme 4: Proactive Management
Proactive management involves setting ambitious goals and clear priorities, reviewing them frequently, and implementing changes in the system to prevent errors and defects from reaching the customer. Waiting for customer complaints puts the organization in a "firefighting" mode. Preventing problems from occurring is a lot cheaper and less time consuming than trying to solve them after they have occurred.

Theme 5: Boundaryless Collaboration
Lack of good communication among the functions in the design or production stage results in delays, redesigns, and rework, all of which cause wasted

resources. Teamwork across the functions, toward the common goal of customer satisfaction, has enormous potential for creating savings and customer confidence. The Six Sigma approach emphasizes process improvement and process redesign through cross-functional teams, and thus creates an environment in which people work together and support one another.

Theme 6: Drive for Perfection (with Tolerance for Failure)

The most important theme of the Six Sigma process is driving toward near-perfection—that is, not more than 3.4 defects per million opportunities (DPMO) in every process. This is achieved through diligent efforts in process analysis and process improvement, in repeated iterations, on a never-ending basis. This is the way to satisfy and then delight the customer, whose standards usually keep changing—and increasing. There may be occasional failures when driving toward perfection, but a Six Sigma organization will not be deterred by these. A Six Sigma organization learns from failures and makes progress toward perfection (Pande et al., 2000).

9.8.2 The 6σ Measure

The 6σ measure used to signify the capability of processes comes from the representation of a process variation using the normal distribution. Assuming that a process follows a normal distribution, if the measure of variability of this process, the standard deviation σ, is equal to one-sixth of the distance of a specification limit from the specification center, then the process is said to have 6σ quality or 6σ capability. If the variability is larger—that is, if the value of σ is larger, such that it is only, say, one-fourth of the distance between the specification center and a specification limit, then the process is said to have 4σ quality, and so on. When a process has 6σ quality, there will be no more than 3.4 ppm outside specification limits, even if the process is off-center from the target by a 1.5σ distance. (A little bit of reflection would show that a 6σ process is the same as a process with $Cp = 2.0$.) The proportions outside specifications under other process quality levels (sigma conditions), in defects per million (DPM), and the corresponding yield or acceptable proportions produced by such processes are shown in Table 9.1. Please see Chapter 5 for an explanation of how the figures in Table 9.1 are computed. Table 9.1 is a repeat of Table 5.10.

The above explanation of a 6σ process is valid if the quality characteristic is a measurement, such as height, weight, and so on, and the proportion falling in any region can be calculated using the normal distribution. If the characteristic is a countable characteristic, such as number of pin holes per square-foot of glass, number of dirty bottles in a skid, or number of mails delivered to the wrong address, then these measures cannot be assumed to be normally distributed and so the process quality in number of sigmas cannot be computed using the above approach; therefore, a different approach is used.

TABLE 9.1

Quality of Different Sigma Processes

Process Quality in No. of Sigmas	ppm Outside Spec (DPMO)	% Yield
1	697,700	30.23
1.5	501,300	49.87
2	308,700	69.13
2.5	158,700	84.13
3	66,800	93.32
3.5	22,800	97.72
4	6200	99.38
4.5	1300	99.87
5	230	99.977
5.5	30	99.997
6	3.4	99.99966

A ("large") sample is taken from the process and the proportion defectives in the sample are calculated. The sigma level of the process is equated to the sigma level of a normal process producing the same level of defectives as in the sample data. Table 9.1 facilitates reading the sigma level of a process given the proportion defectives in the sample. For example, if the sample data for an attribute showed that there were 6200 defects per million (0.62%), then the quality level of the process is 4σ, because, a normal process with 4σ capability will produce 0.62% outside specification in the worst condition of its center. The defects per million can be read as the number of defective units out of every million units produced, or as number of defects per million opportunities (DPMO). The latter reading is appropriate for evaluating service processes such as postal deliveries, data entry, or inventory verifications.

The people at Motorola, who created the sigma measure to designate quality levels of processes, were motivated by the fact that this provides a simple, uniform manner of designating the quality levels of processes, regardless of whether the quality is evaluated by measuring a characteristic or by counting units with a certain attribute. Because of the availability of tables such as Table 9.1, the people at Motorola claimed that one does not have to understand any statistical principles to be able to compute process quality levels.

The following are the advantages claimed by the Six Sigma advocates for the 6σ measure. According to them, the measure:

1. Helps in defining a customer's requirement clearly
2. Provides a metric to define the quality of processes using different types of quality characteristics from both manufacturing and service areas
3. Provides a link to an ambitious goal of reaching 6σ level

This 6σ terminology has become widely accepted among quality professionals for evaluating process quality (or capability) levels; although some honest statisticians contend that the 6σ terminology has been a cause of confusion among quality workers. For example, we hear people say "the larger the value of σ for a process, the better the quality," which may be correct in the Six Sigma language but is incorrect in the language of statistics. What the Six Sigma people really mean is that "the larger the number of sigmas that can fit within a customer's specification and process center, the smaller the value of σ, hence the better the process quality."

9.8.3 The Three Strategies

The Six Sigma system recognizes three major strategies for enhancing process quality. These strategies are used in three different situations:

1. Process improvement
2. Process design/redesign
3. Process management

These three strategies are not mutually exclusive. They can be used together in any given situation as appropriate. The process management strategy will usually follow after the improvement or design/redesign strategies have been implemented.

Process Improvement

This strategy will be appropriate when a problem results in customer discontent, which calls for an analysis to find the root causes and the implementation of solutions to eliminate them. The goal of the process improvement is limited to seeking solutions to a specific problem of concern while letting the overall structure of associated processes remain the same.

Process Design/Redesign

While the process improvement strategy will produce incremental changes, it sometimes may be necessary to revamp an entire process because incremental improvements will not produce enough of a desired result. Then, the process design/redesign strategy must be used.

This involves a whole rethinking on the process, based on inputs from customers. This is similar to the concepts of re-engineering used by some, to denote redoing an entire process when it has become ineffective due to changes in technology or customer expectations.

Process Management

This is the strategy equivalent to what Dr. Juran called "holding the gains." The difference here, however, is in maintaining a watch as to whether the process is responding to changes in the needs of the customer. In this strategy, processes are documented, and customer needs are defined and updated on a regular basis. Meaningful measures of process performance are defined and compared with the measures of customer needs, in real time. Quick, responsive action is taken when the process performance falls short of customer needs and expectations.

9.8.4 The Two Improvement Processes

The Six Sigma system recommends a five-step problem-solving methodology comparable to Dr. Deming's plan-do-check-act (PDCA) cycle, or Dr. Juran's "breakthrough" sequence of steps. The five steps differ when they are applied to "the process improvement" strategy from when they are used for "the process redesign/redesign" strategy.

When they are used for process improvement, the methodology recommends use of the sequence: define, measure, analyze, improve, and control—which is abbreviated as DMAIC (pronounced "deh-may-ihk").

When they are used in the context of process design/redesign, the sequence of steps will be: define, measure, explore, develop, and implement—which is abbreviated as DMEDI (pronounced "deh-may-di").

The steps of the improvement processes are shown in Figure 9.7, which is self-explanatory.

9.8.5 The Five-Step Road Map

A five-step Six Sigma road map is recommended for organizations that want to start implementing a Six Sigma system throughout the organization and become a Six Sigma organization. The road map, shown in Figure 9.8, is explained further in the following steps.

Step 1: Identify Core Processes and Key Customers

The objective of this step is to get a perspective on the core processes in the organization, their interactions with one another, and a clear understanding of the products they produce and the customers they serve. When this step is completed, process charts, or "maps" of the core processes would have been generated. This is like taking stock of what is going on in the organization as preparation for the next step.

Step 2: Define Customer Requirements

The objective of this step is to establish standards for the products produced and the services provided based on input from the customer so that the process performance can be judged and its capabilities predicted. It is important not only to obtain the current customer's needs, but also to watch for changes expected in these requirements over time.

Six Sigma Improvement Processes			
Process Improvement (DMAIC)		**Process Design/Redesign (DMEDI)**	
1. Define	Identify the problem Identify requirements Set goals	**1. Define**	Identify specific or broad problems Define goals/change vision Clarify scope and customer requirements
2. Measure	Validate problem/process Refine problem/goal Measure key steps/inputs	**2. Measure**	Measure performance to requirements Gather process efficiency data
3. Analyze	Develop causal hypotheses Identify "vital few" root causes Validate hypotheses	**3. Explore**	Identify "best practices" Assess process design: • Value/non-value-adding • Bottlenecks/disconnects • Alternate paths Refine requirements
4. Improve	Develop ideas to remove root causes Test solutions Standardize solutions/measure results	**4. Develop**	Design new process: • Challenge assumptions • Apply creativity • Workflow principles Implement new process, structures, and systems
5. Control	Establish standard measures to maintain performance Correct problems as needed	**5. Implement**	Establish measures and reviews to maintain performance Correct problems as needed

FIGURE 9.7
The five-step improvement models for processes. (From Pande, P. S., R. P. Neuman, and R. L. Cavanagh, *The Six Sigma Way*, McGraw-Hill, New York, NY, 2000. With permission.)

Step 3: Measure Current Performance

In this step, the performance of each process is evaluated against the established needs of the customer for the products and services created. At the end of this step, baseline evaluations of performance and process capabilities will be known in a quantifiable manner. The process capabilities can be expressed in terms of the number of sigmas. When no sufficient or satisfactory measuring scheme exists to assess the process performance, newer measurements must be created so those performances can be compared with customer needs. The efficiencies of the process—such as productivity, cycle time, cost per output, rework, and reject rate—are also measured in this step.

Step 4: Prioritize, Analyze, and Implement Improvements

In this step, priorities are established among opportunities for improvement, and the improvement projects, or design/redesign projects, are completed by

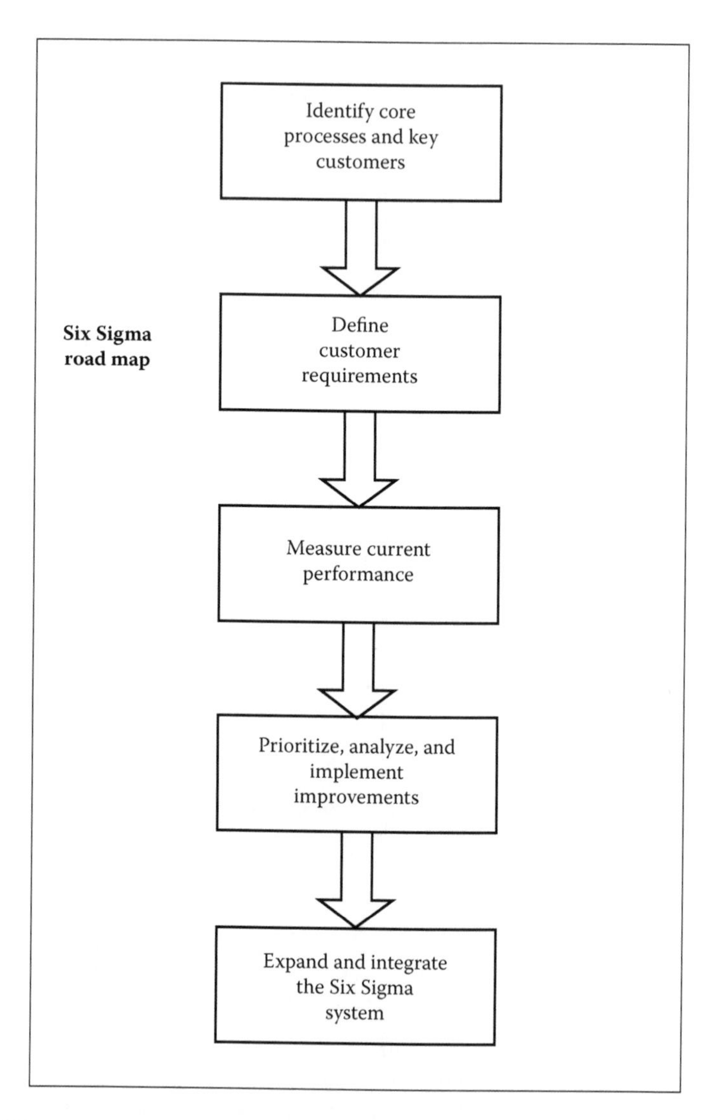

FIGURE 9.8
The Six Sigma road map for an organization. (From Pande, P. S., R. P. Neuman, and R. L. Cavanagh, *The Six Sigma Way*, McGraw-Hill, New York, NY, 2000, with permission of McGraw-Hill.)

teams, resulting in optimal solutions. The solutions are then implemented to obtain better process performance and customer satisfaction.

Step 5: Expand and Integrate the Six Sigma System

Achieving Six Sigma performance for an entire organization, as opposed to achieving Six Sigma performance for individual processes, is the objective of this step. This result has to come from a long-term commitment to the themes of the Six Sigma system, along with the identification of a series of process improvements and process redesigns to produce organization-wide excellence in terms of customer satisfaction and process efficiency.

Process controls must be established where improvements have been obtained, and these controls must be sustained. Plans must be available for responding to changes in processes or customer requirements. A cultural change in the organization toward a commitment to applying Six Sigma principles to all business processes must be accomplished.

9.8.6 The Organization for the Six Sigma System

The Six Sigma system calls for a special organizational structure as well as continuous and rigorous cooperation among all employees in an organization. The key players in a Six Sigma organization and their roles in the Six Sigma process are as follows:

Executive leaders: Show highly visible top-down commitment, assume ownership of the Six Sigma process, create vision and goals, identify opportunities, allocate resources, and provide inspired leadership.

Project sponsor (champion): A line manager or owner of a process who identifies and prioritizes project opportunities. Selects projects, provides resources, participates in project execution, and removes barriers.

Master black belt: Works full-time for Six Sigma implementation. Has responsibility for planning and providing technical support for the entire organization. Trains black belts, acts as a coach and mentor, and provides overall leadership in Six Sigma implementation.

Black belts: Experts on Six Sigma tools, work full-time on Six Sigma projects, trains green belts, lead teams, and provide assistance with Six Sigma tools (e.g., improvement methods, diagnostic tools, and statistical methods).

Green belts: Work part-time on projects with black belts, and integrate Six Sigma methodology in daily work. Can lead small projects.

Yellow belts: All employees. Trained in quality awareness and are part-time participants in teams. Contribute with process expertise.

Financial rep: Independent of project team. Determines the project costs and savings. Reports project benefits.

9.9 Summary of Quality Management Systems

The idea that a quality system should be created and maintained to produce quality products and services to meet customer needs was recognized by people such as Dr. Deming, Dr. Juran, and Dr. Feigenbaum in the early 1950s, and each proposed a model to create such a system. Although those models were only initial attempts to create a quality management system, they provided the basic building blocks, which were later used by others to create more complete models, such as the ISO 9000 standards, the MBNQA criteria, and the Six Sigma system. These three systems, which incorporate the best experiences with quality systems, are the best models currently available for creating a quality management system.

In the case of the ISO 9000 standards, certification can be obtained from registrars accredited by a government or quasi-governmental agency. (The Registrar Accreditation Board in the United States is a quasi-governmental agency.) The registrars, through their certified auditors, review the documents of an organization that is aspiring to become certified, make a site visit, and, if satisfied, certify that the organization meets the requirements of the standard in the production of specified products. Such certification by a third party (i.e., a party other than the producer and the customer) provides an objective evaluation of an organization's quality system and its ability to supply quality products and services.

In a similar manner, the MBNQA organization evaluates businesses that wish to receive an award, through volunteer groups of examiners and judges, and decides on the final recipients of the award. An applicant business will receive feedback from the examiners and judges on the status of their quality management system, irrespective of whether or not they receive the award. When an organization receives finalist status and a site visit by the judges, it is already recognition of the healthy status of their quality management system. The availability of such formal evaluation and feedback procedures by independent parties provides an added incentive for organizations to adopt one of these two models to create their quality system.

The Six Sigma system has also proved to be a powerful model for a quality management system, but it does not have the formalized structure of the MBNQA criteria or the ISO 9000, and the rigor of application depends largely on the level of commitment of the organization or the consultants helping to implement the system.

Between the ISO 9000 and the MBNQA criteria, the ISO 9000 requirements seek to build a basic foundation for a quality management system, whereas the MBNQA criteria seek to reach excellence in business results. ISO 9000 is very detailed and thorough in its requirements for the technical side of the system, such as product and process design, process management, instrument control, and document control. On the other hand, the MBNQA criteria devote much attention to the business and management side of an organization, such as strategic planning, human resource management, and supplier relationships. They also place greater emphasis on the results rather than on the processes for achieving those results.

Quality products and services are the output of processes that are both effective and efficient. Learning the needs of the customer correctly, developing the products that will meet those needs, designing the processes to make those products, producing the products according to the designs, and delivering the products to customers along with appropriate services will produce customer satisfaction—the objective of a quality management system. Because the ISO 9000 requirements pay meticulous attention to the details of designing and implementing such core processes, it is a good vehicle for creating a sound and solid infrastructure for a quality system—especially in the manufacturing sector.

We should recognize that a collection of excellent technical processes alone does not guarantee success for an organization in terms of business results, such as market share and profitability. The MBNQA model accordingly emphasizes the need for managerial components such as leadership, strategic planning, employee relationships, information systems, and supplier partnerships. It seeks to integrate the technical system with the management philosophy in an optimal way to achieve customer satisfaction while producing successful business results for the organization.

Thus, the MBNQA criteria and ISO 9000 place different degrees of emphasis on the two different components of a quality management system: the technical system, and the management system. They could even be used to complement each other. The Baldrige National Quality Program, for example, recommends the use of both models, along with the Six Sigma system, to organize a quality management system to achieve overall organizational success. In one of their communications (Baldrige National Quality Program 2002) the program administrators say, in relation to choice of system: "Your needs drive the choice. It shouldn't be 'either/or.' It can be 'one, two, and/or three.'" They cite examples where organizations have used both models, or all three models, and have achieved excellent business results. In those cases, ISO 9000 is used first to build the foundation for the system, and then the Baldrige model is used to strengthen and enhance the system.

Thus, an organization would do well to exploit the individual strengths of the models and implement a quality management system that best suits its own needs and circumstances.

9.10 Exercise

9.10.1 Practice Problems

Deming System

9.1 What did Dr. Deming mean when he said, "Adopt a new philosophy"?

9.2 Why is 100% inspection undesirable? Why will 100% inspection not result in 100% good products?

9.3 Why is a single source for supplies better than multiple sources?

9.4 According to Dr. Deming, what were the most important inhibitors for workers doing their work in American industry?

9.5 What is the role of a supervisor in Deming's System?

9.6 According to Dr. Deming, why do slogans and numerical quotas not help in achieving quality?

Juran System

9.7 What are the three components of the "Juran Trilogy"?

9.8 How does the concept of dominance help in process planning?

9.9 How does one select the locations where control must be exercised in a process?

9.10 What is the difference between sporadic and chronic deviations in processes?

9.11 Why is the first project chosen for quality improvement important?

9.12 What are the reasons for resistance to change while making improvements to a process? How should this resistance be handled?

Baldrige System

9.13 What were the reasons for creating the MBNQA in the United States?

9.14 Why is the criterion: Measurement, analysis, and knowledge management, fourth among seven, central to the MBNQA criteria?

9.15 What are the business results on which the MBNQA focuses?

9.16 Why are business processes important to the excellence of a business?

9.17 The MBNQA focuses not only on product quality, but also on the business results. Explain.

9.18 Figure 9.5 is commonly referred to as the "Baldrige burger." Where is the meat?

ISO 9000 System

9.19 What is the systems approach, and what is the process approach, in the ISO 9000 standards?

9.20 What is the difference between a procedure, a specification, and a record?

9.21 What is document control, and why is it necessary?

9.22 What are the major responsibilities of the top management in an ISO 9000 system?

9.23 What is included in the infrastructure of an ISO 9000 system?

9.24 What is the purpose of the design and development review in the ISO 9000 system?

Six Sigma System

9.25 What is meant by boundaryless collaboration as one of the themes of the Six Sigma process?

9.26 How does one determine the capability of a process in number of sigmas:

a. When the characteristic is a measurement?

b. When the characteristic is evaluated in percentage of defectives?

9.27 Show mathematically that a 6σ process will not have more than 3.4 DPMO. (Assume the process is producing a measurement characteristic that has a normal distribution with the maximum possible shift in the process mean being 1.5σ).

9.28 What is the difference between the process improvement strategy and the process design/redesign strategy?

9.29 What do the acronyms DMAIC and DMEDI stand for? What are the steps in each?

9.30 What is the difference between a 6-sigma process and a Six Sigma organization?

9.10.2 Mini-Projects

Mini-Project 9.1

The above set of 30 questions has been created to help students understand the various systems in good detail. However, it is only one of several possible sets. Generate another set of 30 questions, six from each system, similar to but different from the above set.

Mini-Project 9.2

Compare the three modern systems—MBNQA, ISO 9000, and Six Sigma—and identify their differences.

References

ANSI/ISO/ASQ Q9000:2005. 2005. Milwaukee, WI: ASQ Quality Press.
ANSI/ISO/ASQ Q9001:2008. 2008. Milwaukee, WI: ASQ Quality Press.

ANSI/ISO/ASQ Q9004:2009. 2009. Milwaukee, WI: ASQ Quality Press.
Baldrige National Quality Program. Summer 2002. "Baldrige, Six Sigma, & ISO: Understanding Your Options." *CEO Issue Sheet*.
Baldrige National Quality Program. 2009–2010. *Criteria for Performance Excellence*. NIST, Gaithersburg, MD.
Brown, M. G. 2008. *Baldrige Award Winning Quality*. 17th ed. Boca Raton: CRC Press.
Deming, W. E. 1986. *Out of the Crisis*. Cambridge, MA: MIT—Center for Advanced Engineering Study.
Deming, W. E. 1993. *The New Economics for Industry, Government, Education*. Cambridge, MA: MIT—Center for Advanced Engineering Study.
Donoldson, D. P. 2004. "100 Years of Juran." *Quality Progress* 37 (5): 25–39.
Evans, J. R., and W. M. Lindsay. 2005. *The Management and Control of Quality*. 6th ed. Mason, OH: Thompson/Southwestern.
Feigenbaum, A. V. 1951. *Quality Control: Principles, Practice, and Administration*. New York, NY: McGraw-Hill.
Feigenbaum, A. V. 1961. *Total Quality Control—Engineering and Management*. New York, NY: McGraw-Hill.
Feigenbaum, A. V. 1983. *Total Quality Control*. 3rd ed. New York, NY: McGraw-Hill.
Gabor, A. 1990. *The Man Who Discovered Quality*. Time Books. Republished, New York, NY: Penguin.
Juran, J. M., and J. Defeo, eds. 2010. *Juran's Quality Handbook*. 6th ed. New York, NY: McGraw-Hill.
Juran, J. M., and A. B. Godfrey, eds. 1999. *Juran's Quality Handbook*. 5th ed. New York, NY: McGraw-Hill.
Juran, J. M., and F. M. Gryna, eds. 1988. *Juran's Quality Control Handbook*. 4th ed. New York, NY: McGraw-Hill.
Pande, P. S., R. P. Neuman, and R. L. Cavanagh. 2000. *The Six Sigma Way*. New York, NY: McGraw-Hill.

Appendix 1

Statistical Tables

Table A.1 Cumulative Probabilities of the Standard Normal Distribution
Table A.2 Percentiles of the t-Distribution
Table A.3 Percentiles of the χ^2-Distribution
Table A.4 Factors for Calculating Limits for Variable Control Charts
Table A.5 Cumulative Poisson Probabilities
Table A.6 Percentiles of the F-Distribution

TABLE A.1

Cumulative Probabilities of the Standard Normal Distribution

Normal Tables: gives $F(z) = P(Z \leq z)$, where $Z \sim N(0, 1)$

z	0.00	0.01	0.02	0.03	0.04	0.05	0.06	0.07	0.08	0.09
−3.4	0.00034	0.00033	0.00031	0.00030	0.00029	0.00028	0.00027	0.00026	0.00025	0.00024
−3.3	0.00048	0.00047	0.00045	0.00043	0.00042	0.00040	0.00039	0.00038	0.00036	0.00035
−3.2	0.00069	0.00066	0.00064	0.00062	0.00060	0.00058	0.00056	0.00054	0.00052	0.00050
−3.1	0.00097	0.00094	0.00090	0.00087	0.00085	0.00082	0.00079	0.00076	0.00074	0.00071
−3.0	0.00135	0.00131	0.00126	0.00122	0.00118	0.00114	0.00111	0.00107	0.00104	0.00100
−2.9	0.0019	0.0018	0.0018	0.0017	0.0016	0.0016	0.0015	0.0015	0.0014	0.0014
−2.8	0.0026	0.0025	0.0024	0.0023	0.0023	0.0022	0.0021	0.0021	0.0020	0.0019
−2.7	0.0035	0.0034	0.0033	0.0032	0.0031	0.0030	0.0029	0.0028	0.0027	0.0026
−2.6	0.0047	0.0045	0.0044	0.0043	0.0041	0.0040	0.0039	0.0038	0.0037	0.0036
−2.5	0.0062	0.0060	0.0059	0.0057	0.0055	0.0054	0.0052	0.0051	0.0049	0.0048
−2.4	0.0082	0.0080	0.0078	0.0075	0.0073	0.0071	0.0069	0.0068	0.0066	0.0064
−2.3	0.0107	0.0104	0.0102	0.0099	0.0096	0.0094	0.0091	0.0089	0.0087	0.0084
−2.2	0.0139	0.0136	0.0132	0.0129	0.0125	0.0122	0.0119	0.0116	0.0113	0.0110
−2.1	0.0179	0.0174	0.0170	0.0166	0.0162	0.0158	0.0154	0.0150	0.0146	0.0143
−2.0	0.0228	0.0222	0.0217	0.0212	0.0207	0.0202	0.0197	0.0192	0.0188	0.0183
−1.9	0.0287	0.0281	0.0274	0.0268	0.0262	0.0256	0.0250	0.0244	0.0239	0.0233
−1.8	0.0359	0.0352	0.0344	0.0336	0.0329	0.0322	0.0314	0.0307	0.0301	0.0294
−1.7	0.0446	0.0436	0.0427	0.0418	0.0409	0.0401	0.0392	0.0384	0.0375	0.0367
−1.6	0.0548	0.0537	0.0526	0.0516	0.0505	0.0495	0.0485	0.0475	0.0465	0.0455
−1.5	0.0668	0.0655	0.0643	0.0630	0.0618	0.0606	0.0594	0.0582	0.0571	0.0559
−1.4	0.0808	0.0793	0.0778	0.0764	0.0749	0.0735	0.0722	0.0708	0.0694	0.0681
−1.3	0.0968	0.0951	0.0934	0.0918	0.0901	0.0885	0.0869	0.0853	0.0838	0.0823
−1.2	0.1151	0.1131	0.1112	0.1093	0.1075	0.1056	0.1038	0.1020	0.1003	0.0985
−1.1	0.1357	0.1335	0.1314	0.1292	0.1271	0.1251	0.1230	0.1210	0.1190	0.1170
−1.0	0.1587	0.1562	0.1539	0.1515	0.1492	0.1469	0.1446	0.1423	0.1401	0.1379
−0.9	0.1841	0.1814	0.1788	0.1762	0.1736	0.1711	0.1685	0.1660	0.1635	0.1611
−0.8	0.2119	0.2090	0.2061	0.2033	0.2005	0.1977	0.1949	0.1922	0.1894	0.1867
−0.7	0.2420	0.2389	0.2358	0.2327	0.2296	0.2266	0.2236	0.2206	0.2177	0.2148
−0.6	0.2743	0.2709	0.2676	0.2643	0.2611	0.2578	0.2546	0.2514	0.2483	0.2451
−0.5	0.3085	0.3050	0.3015	0.2981	0.2946	0.2912	0.2877	0.2843	0.2810	0.2776
−0.4	0.3446	0.3409	0.3372	0.3336	0.3300	0.3264	0.3228	0.3192	0.3156	0.3121
−0.3	0.3821	0.3783	0.3745	0.3707	0.3669	0.3632	0.3594	0.3557	0.3520	0.3483
−0.2	0.4207	0.4168	0.4129	0.4090	0.4052	0.4013	0.3974	0.3936	0.3897	0.3859
−0.1	0.4602	0.4562	0.4522	0.4483	0.4443	0.4404	0.4364	0.4325	0.4286	0.4247
−0.0	0.5000	0.4960	0.4920	0.4880	0.4840	0.4801	0.4761	0.4721	0.4681	0.4641

TABLE A.1 (Continued)

Cumulative Probabilities of the Standard Normal Distribution

Normal Tables: Gives $F(z) = P(Z \leq z)$, Where $Z \sim N(0, 1)$										
z	**0.00**	**0.01**	**0.02**	**0.03**	**0.04**	**0.05**	**0.06**	**0.07**	**0.08**	**0.09**
0.0	0.5000	0.5040	0.5080	0.5120	0.5160	0.5199	0.5239	0.5279	0.5319	0.5359
0.1	0.5398	0.5438	0.5478	0.5517	0.5557	0.5596	0.5636	0.5675	0.5714	0.5753
0.2	0.5793	0.5832	0.5871	0.5910	0.5948	0.5987	0.6026	0.6064	0.6103	0.6141
0.3	0.6179	0.6217	0.6255	0.6293	0.6331	0.6368	0.6406	0.6443	0.6480	0.6517
0.4	0.6554	0.6591	0.6628	0.6664	0.6700	0.6736	0.6772	0.6808	0.6844	0.6879
0.5	0.6915	0.6950	0.6985	0.7019	0.7054	0.7088	0.7123	0.7157	0.7190	0.7224
0.6	0.7257	0.7291	0.7324	0.7357	0.7389	0.7422	0.7454	0.7486	0.7517	0.7549
0.7	0.7580	0.7611	0.7642	0.7673	0.7704	0.7734	0.7764	0.7794	0.7823	0.7852
0.8	0.7881	0.7910	0.7939	0.7967	0.7995	0.8023	0.8051	0.8078	0.8106	0.8133
0.9	0.8159	0.8186	0.8212	0.8238	0.8264	0.8289	0.8315	0.8340	0.8365	0.8389
1.0	0.8413	0.8438	0.8461	0.8485	0.8508	0.8531	0.8554	0.8577	0.8599	0.8621
1.1	0.8643	0.8665	0.8686	0.8708	0.8729	0.8749	0.8770	0.8790	0.8810	0.8830
1.2	0.8849	0.8869	0.8888	0.8907	0.8925	0.8944	0.8962	0.8980	0.8997	0.9015
1.3	0.9032	0.9049	0.9066	0.9082	0.9099	0.9115	0.9131	0.9147	0.9162	0.9177
1.4	0.9192	0.9207	0.9222	0.9236	0.9251	0.9265	0.9278	0.9292	0.9306	0.9319
1.5	0.9332	0.9345	0.9357	0.9370	0.9382	0.9394	0.9406	0.9418	0.9429	0.9441
1.6	0.9452	0.9463	0.9474	0.9484	0.9495	0.9505	0.9515	0.9525	0.9535	0.9545
1.7	0.9554	0.9564	0.9573	0.9582	0.9591	0.9599	0.9608	0.9616	0.9625	0.9633
1.8	0.9641	0.9649	0.9656	0.9664	0.9671	0.9678	0.9686	0.9693	0.9699	0.9706
1.9	0.9713	0.9719	0.9726	0.9732	0.9738	0.9744	0.9750	0.9756	0.9761	0.9767
2.0	0.9772	0.9778	0.9783	0.9788	0.9793	0.9798	0.9803	0.9808	0.9812	0.9817
2.1	0.9821	0.9826	0.9830	0.9834	0.9838	0.9842	0.9846	0.9850	0.9854	0.9857
2.2	0.9861	0.9864	0.9868	0.9871	0.9875	0.9878	0.9881	0.9884	0.9887	0.9890
2.3	0.9893	0.9896	0.9898	0.9901	0.9904	0.9906	0.9909	0.9911	0.9913	0.9916
2.4	0.9918	0.9920	0.9922	0.9925	0.9927	0.9929	0.9931	0.9932	0.9934	0.9936
2.5	0.9938	0.9940	0.9941	0.9943	0.9945	0.9946	0.9948	0.9949	0.9951	0.9952
2.6	0.9953	0.9955	0.9956	0.9957	0.9959	0.9960	0.9961	0.9962	0.9963	0.9964
2.7	0.9965	0.9966	0.9967	0.9968	0.9969	0.9970	0.9971	0.9972	0.9973	0.9974
2.8	0.9974	0.9975	0.9976	0.9977	0.9977	0.9978	0.9979	0.9979	0.9980	0.9981
2.9	0.9981	0.9982	0.9982	0.9983	0.9984	0.9984	0.9985	0.9985	0.9986	0.9986
3.0	0.99865	0.99869	0.99874	0.99878	0.99882	0.99886	0.99889	0.99893	0.99896	0.99900
3.1	0.99903	0.99906	0.99910	0.99913	0.99915	0.99918	0.99921	0.99924	0.99926	0.99929
3.2	0.99931	0.99934	0.99936	0.99938	0.99940	0.99942	0.99944	0.99946	0.99948	0.99950
3.3	0.99952	0.99953	0.99955	0.99957	0.99958	0.99960	0.99961	0.99962	0.99964	0.99965
3.4	0.99966	0.99967	0.99969	0.99970	0.99971	0.99972	0.99973	0.99974	0.99975	0.99976

TABLE A.2

Percentiles of the t-Distribution

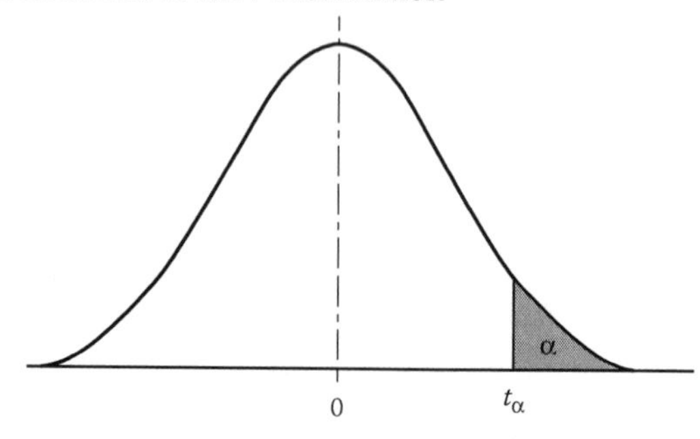

	α				
ν	**0.10**	**0.05**	**0.025**	**0.01**	**0.005**
1	3.078	6.314	12.706	31.821	63.657
2	1.886	2.920	4.303	6.965	9.925
3	1.638	2.353	3.182	4.541	5.841
4	1.533	2.132	2.776	3.747	4.604
5	1.476	2.015	2.571	3.365	4.032
6	1.440	1.943	2.447	3.143	3.707
7	1.415	1.895	2.365	2.998	3.499
8	1.397	1.860	2.306	2.896	3.355
9	1.383	1.833	2.262	2.821	3.250
10	1.372	1.812	2.228	2.764	3.169
11	1.363	1.796	2.201	2.718	3.106
12	1.356	1.782	2.179	2.681	3.055
13	1.350	1.771	2.160	2.650	3.012
14	1.345	1.761	2.145	2.624	2.977
15	1.341	1.753	2.131	2.602	2.947
16	1.337	1.746	2.120	2.583	2.921
17	1.333	1.740	2.110	2.567	2.898
18	1.330	1.734	2.101	2.552	2.878
19	1.328	1.729	2.093	2.539	2.861
20	1.325	1.725	2.086	2.528	2.845
21	1.323	1.721	2.080	2.518	2.831
22	1.321	1.717	2.074	2.508	2.819
23	1.319	1.714	2.069	2.500	2.807
24	1.318	1.711	2.064	2.492	2.797
25	1.316	1.708	2.060	2.485	2.787
26	1.315	1.706	2.056	2.479	2.779
27	1.314	1.703	2.052	2.473	2.771
28	1.313	1.701	2.048	2.467	2.763
29	1.311	1.699	2.045	2.462	2.756
∞	1.282	1.645	1.960	2.326	2.576

TABLE A.3

Percentiles of the χ^2-Distribution

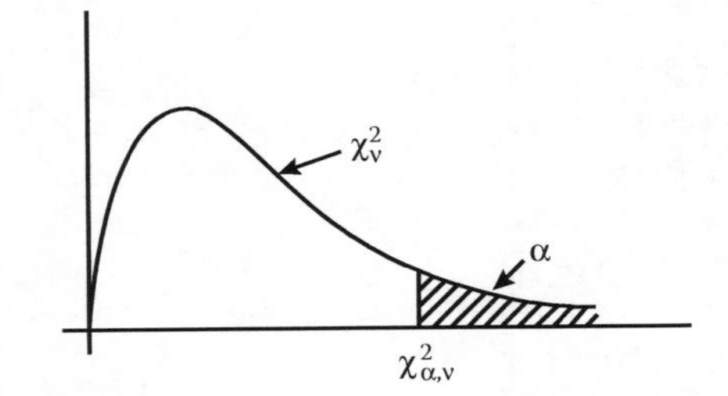

$\nu\backslash\alpha$	.995	.99	.975	.95	.90	.75	.5	.25	.10	.05	.025	.01	.005	.001
1					0.02	0.10	0.45	1.32	2.71	3.84	5.02	6.63	7.88	10.83
2	0.01	0.02	0.05	0.10	0.21	0.58	1.39	2.77	4.61	5.99	7.38	9.21	10.60	13.82
3	0.07	0.11	0.22	0.35	0.58	1.21	2.37	4.11	6.25	7.81	9.35	11.34	12.84	16.27
4	0.21	0.30	0.48	0.71	1.06	1.92	3.36	5.39	7.78	9.49	11.14	13.28	14.86	18.47
5	0.41	0.55	0.83	1.15	1.61	2.67	4.35	6.63	9.24	11.07	12.83	15.09	16.75	20.52
6	0.68	0.87	1.24	1.64	2.20	3.45	5.35	7.84	10.64	12.59	14.45	16.81	18.55	22.46
7	0.99	1.24	1.69	2.17	2.83	4.25	6.35	9.04	12.02	14.07	16.01	18.48	20.28	24.32
8	1.34	1.65	2.18	2.73	3.49	5.07	7.34	10.22	13.36	15.51	17.53	20.09	21.96	26.12
9	1.73	2.09	2.70	3.33	4.17	5.90	8.34	11.39	14.68	16.92	19.02	21.67	23.59	27.88
10	2.16	2.56	3.25	3.94	4.87	6.74	9.34	12.55	15.99	18.31	20.48	23.21	25.19	29.59
11	2.60	3.05	3.82	4.57	5.58	7.58	10.34	13.70	17.28	19.68	21.92	24.72	26.76	31.26
12	3.07	3.57	4.40	5.23	6.30	8.44	11.34	14.85	18.55	21.03	23.34	26.22	28.30	32.91

(continued)

TABLE A.3 (Continued)

Percentiles of the χ^2-Distribution

$\nu \backslash \alpha$	.995	.99	.975	.95	.90	.75	.5	.25	.10	.05	.025	.01	.005	.001
13	3.57	4.11	5.01	5.89	7.04	9.30	12.34	15.98	19.81	22.36	24.74	27.69	29.82	34.53
14	4.07	4.66	5.63	6.57	7.79	10.17	13.34	17.12	21.06	23.68	26.12	29.14	31.32	36.12
15	4.60	5.23	6.26	7.26	8.55	11.04	14.34	18.25	22.31	25.00	27.49	30.58	32.80	37.70
16	5.14	5.81	6.91	7.96	9.31	11.91	15.34	19.37	23.54	26.30	28.85	32.00	34.27	39.25
17	5.70	6.41	7.56	8.67	10.09	12.79	16.34	20.49	24.77	27.59	30.19	33.41	35.73	40.79
18	6.26	7.01	8.23	9.39	10.86	13.68	17.34	21.60	25.99	28.87	31.53	34.81	37.16	42.31
19	6.84	7.63	8.91	10.12	11.65	14.56	18.34	22.72	27.20	30.14	32.85	36.19	38.58	43.82
20	7.43	8.26	9.59	10.85	12.44	15.45	19.34	23.83	28.41	31.41	34.17	37.57	40.00	45.32
21	8.03	8.90	10.28	11.59	13.24	16.34	20.34	24.93	29.62	32.67	35.48	38.93	41.40	46.80
22	8.64	9.54	10.98	12.34	14.04	17.24	21.34	26.04	30.81	33.92	36.78	40.29	42.80	48.27
23	9.26	10.20	11.69	13.09	14.85	18.14	22.34	27.14	32.01	35.17	38.08	41.64	44.18	49.73
24	9.89	10.86	12.40	13.85	15.66	19.04	23.34	28.24	33.20	36.42	39.36	42.98	45.56	51.18
25	10.52	11.52	13.12	14.61	16.47	19.94	24.34	29.34	34.38	37.65	40.65	44.31	46.93	52.62
30	13.79	14.95	16.79	18.49	20.60	24.48	29.34	34.80	40.26	43.77	46.98	50.89	53.67	59.70
40	20.71	22.16	24.43	26.51	29.05	33.66	39.34	45.62	51.80	55.76	59.34	63.69	66.77	73.40
50	27.99	29.71	32.36	34.76	37.69	42.94	49.33	56.33	63.17	67.50	71.42	76.15	79.49	86.66
60	35.53	37.48	40.48	43.19	46.46	52.29	59.33	66.98	74.40	79.08	83.30	88.38	91.95	99.61
70	43.28	45.44	48.76	51.74	55.33	61.70	69.33	77.58	85.53	90.53	95.02	100.42	104.22	112.32
80	51.17	53.54	57.15	60.39	64.28	71.14	79.33	88.13	96.58	101.88	106.63	112.33	116.32	124.84
90	59.20	61.75	65.65	69.13	73.29	80.62	89.33	98.64	107.56	113.14	118.14	124.12	128.30	137.21
100	67.33	70.06	74.22	77.93	82.36	90.13	99.33	109.14	118.50	124.34	129.56	135.81	140.17	149.45

TABLE A.4

Factors for Calculating Limits for Variable Control Charts

n	A	A_2	A_3	B_3	B_4	c_4	D_1	D_2	D_3	D_4	d_2	d_3
2	2.121	1.880	2.659	0	3.267	0.798	0	3.686	0	3.267	1.128	0.853
3	1.732	1.023	1.954	0	2.568	0.886	0	4.358	0	2.574	1.693	0.888
4	1.500	0.729	1.628	0	2.266	0.921	0	4.698	0	2.282	2.059	0.880
5	1.342	0.577	1.427	0	2.089	0.940	0	4.918	0	2.114	2.326	0.864
6	1.225	0.483	1.287	0.030	1.970	0.952	0	5.078	0	2.004	2.534	0.848
7	1.134	0.419	1.182	0.118	1.882	0.959	0.205	5.204	0.076	1.924	2.704	0.833
8	1.061	0.373	1.099	0.185	1.815	0.965	0.387	5.306	0.136	1.864	2.847	0.820
9	1.000	0.337	1.032	0.239	1.761	0.969	0.546	5.393	0.184	1.816	2.970	0.808
10	0.949	0.308	0.975	0.284	1.716	0.973	0.687	5.469	0.223	1.777	3.078	0.797
11	0.905	0.285	0.927	0.321	1.679	0.975	0.812	5.535	0.256	1.744	3.173	0.787
12	0.866	0.266	0.886	0.354	1.646	0.978	0.924	5.594	0.283	1.717	3.258	0.778
13	0.832	0.249	0.850	0.382	1.618	0.979	1.026	5.647	0.307	1.693	3.336	0.770
14	0.802	0.235	0.817	0.406	1.594	0.981	1.118	5.696	0.328	1.672	3.407	0.763
15	0.775	0.223	0.789	0.428	1.572	0.982	1.204	5.741	0.347	1.653	3.472	0.756

Source: Abridged from Table M of Duncan, A. J., *Quality Control and Industrial Statistics*. 4th ed. Homewood, IL: Richard D. Irwin, 1974.

TABLE A.5

Cumulative Poisson Probabilities

c/λ	.01	.05	.10	.20	.30	.40	.50	.60	.70	.80	.90	1.00
0	.990	.951	.904	.818	.740	.670	.606	.548	.496	.449	.406	.367
1	.999	.998	.995	.982	.963	.938	.909	.878	.844	.808	.772	.735
2		.999	.999	.998	.996	.992	.985	.976	.965	.952	.937	.919
3				.999	.999	.999	.998	.996	.994	.990	.986	.981
4							.999	.999	.999	.998	.997	.996
5										.999	.999	.999

c/λ	1.10	1.20	1.30	1.40	1.50	1.60	1.70	1.80	1.90	2.00	2.10	2.20
0	.332	.301	.272	.246	.223	.201	.182	.165	.149	.135	.122	.110
1	.699	.662	.626	.591	.557	.524	.493	.462	.433	.406	.379	.354
2	.900	.879	.857	.833	.808	.783	.757	.730	.703	.676	.649	.622
3	.974	.966	.956	.946	.934	.921	.906	.891	.874	.857	.838	.819
4	.994	.992	.989	.985	.981	.976	.970	.963	.955	.947	.937	.927
5	.999	.998	.997	.996	.995	.993	.992	.989	.986	.983	.979	.975
6		.999	.999	.999	.999	.998	.998	.997	.996	.995	.994	.992
7						.999	.999	.999	.999	.998	.998	.998
8										.999	.999	.999

c/λ	2.30	2.40	2.50	2.60	2.70	2.80	2.90	3.00	3.50	4.00	4.50	5.00
0	.100	.090	.082	.074	.067	.060	.055	.049	.030	.018	.011	.006
1	.330	.308	.287	.267	.248	.231	.214	.199	.135	.091	.061	.040
2	.596	.569	.543	.518	.493	.469	.445	.423	.320	.238	.173	.124
3	.799	.778	.757	.736	.714	.691	.669	.647	.536	.433	.342	.265
4	.916	.904	.891	.877	.862	.847	.831	.815	.725	.628	.532	.440
5	.970	.964	.957	.950	.943	.934	.925	.916	.857	.785	.702	.615
6	.990	.988	.985	.982	.979	.975	.971	.966	.934	.889	.831	.762
7	.997	.996	.995	.994	.993	.991	.990	.988	.973	.948	.913	.866
8	.999	.999	.998	.998	.998	.997	.996	.996	.990	.978	.959	.931
9			.999	.999	.999	.999	.999	.998	.996	.991	.982	.968
10								.999	.998	.997	.993	.986
11									.999	.999	.997	.994
12											.999	.997
13												.999

c/λ	5.50	6.00	6.50	7.00	7.50	8.00	8.50	9.00	9.50	10.0	15.0	20.0
0	.004	.002	.001	.000	.000	.000	.000	.000	.000	.000	.000	.000
1	.026	.017	.011	.007	.004	.003	.001	.001	.000	.000	.000	.000
2	.088	.061	.043	.029	.020	.013	.009	.006	.004	.002	.000	.000
3	.201	.151	.111	.081	.059	.042	.030	.021	.014	.010	.000	.000
4	.357	.285	.223	.172	.132	.099	.074	.054	.040	.029	.000	.000
5	.528	.445	.369	.300	.241	.191	.149	.115	.088	.067	.002	.000

TABLE A.5 (Continued)

Cumulative Poisson Probabilities

c/λ	5.50	6.00	6.50	7.00	7.50	8.00	8.50	9.00	9.50	10.0	15.0	20.0
6	.686	.606	.526	.449	.378	.313	.256	.206	.164	.130	.007	.000
7	.809	.743	.672	.598	.524	.452	.385	.323	.268	.220	.018	.000
8	.894	.847	.791	.729	.661	.592	.523	.455	.391	.332	.037	.002
9	.946	.916	.877	.830	.776	.716	.652	.587	.521	.457	.069	.005
10	.974	.957	.933	.901	.862	.815	.763	.705	.645	.583	.118	.070
11	.989	.979	.966	.946	.920	.888	.848	.803	.751	.696	.184	.021
12	.995	.991	.983	.973	.957	.936	.909	.875	.836	.791	.267	.039
13	.998	.996	.992	.987	.978	.965	.948	.926	.898	.864	.363	.066
14	.999	.998	.997	.994	.989	.982	.972	.958	.940	.916	.465	.104
15		.999	.998	.997	.995	.991	.986	.997	.966	.951	.568	.156
16			.999	.999	.998	.996	.993	.988	.982	.974	.664	.221
17					.999	.998	.997	.994	.991	.985	.748	.297
18						.999	.998	.997	.995	.992	.819	.381
19							.999	.998	.998	.996	.875	.470
20								.999	.999	.998	.917	.559
21										.999	.946	.643
22											.967	.720
23											.980	.787
24											.988	.843
25											.993	.887
26											.996	.922
27											.998	.947
28											.999	.965
29												.978
30												.986

TABLE A.6

Percentiles of the F-Distribution

95th Percentiles of the $F(\nu_1, \nu_2)$

$\nu_2 \backslash \nu_1$	1	2	3	4	5	6	7	8	9	10	12	15	20	24	30	40	60	120	∞
1	161	200	216	225	230	234	237	239	241	242	244	246	248	249	250	251	252	253	254
2	18.51	19.0	19.16	19.25	19.30	19.33	19.35	19.37	19.38	19.40	19.41	19.43	19.45	19.45	19.46	19.47	19.48	19.49	19.5
3	10.13	9.55	9.28	9.12	9.01	8.94	8.89	8.85	8.81	8.79	8.74	8.70	8.66	8.64	8.62	8.59	8.57	8.55	8.53
4	7.71	6.94	6.59	6.39	6.26	6.16	6.09	6.04	6.00	5.96	5.91	5.86	5.80	5.77	5.75	5.72	5.69	5.66	5.63
5	6.61	5.79	5.41	5.19	5.05	4.95	4.88	4.82	4.77	4.74	4.68	4.62	4.56	4.53	4.50	4.46	4.43	4.40	4.36
6	5.99	5.14	4.76	4.53	4.39	4.28	4.21	4.15	4.10	4.06	4.00	3.94	3.87	3.84	3.81	3.77	3.74	3.70	3.64
7	5.59	4.74	4.35	4.12	3.97	3.87	3.79	3.73	3.68	3.64	3.57	3.51	3.44	3.41	3.38	3.34	3.30	3.27	3.23
8	5.32	4.46	4.07	3.84	3.69	3.58	3.50	3.44	3.39	3.35	3.28	3.22	3.15	3.12	3.08	3.04	3.01	2.97	2.93
9	5.12	4.26	3.86	3.63	3.48	3.37	3.29	3.23	3.18	3.14	3.07	3.01	2.94	2.90	2.86	2.83	2.79	2.75	2.71
10	4.96	4.10	3.71	3.48	3.33	3.22	3.14	3.07	3.02	2.98	2.91	2.85	2.77	2.74	2.70	2.66	2.62	2.58	2.54
12	4.75	3.89	3.49	3.26	3.11	3.00	2.91	2.85	2.80	2.75	2.69	2.62	2.54	2.51	2.47	2.43	2.38	2.34	2.30
15	4.54	3.68	3.29	3.06	2.90	2.79	2.71	2.64	2.59	2.54	2.48	2.40	2.33	2.29	2.25	2.20	2.16	2.11	2.07
20	4.35	3.49	3.10	2.87	2.71	2.60	2.51	2.45	2.39	2.35	2.28	2.20	2.12	2.08	2.04	1.99	1.95	1.90	1.84
24	4.26	3.40	3.01	2.78	2.62	2.51	2.42	2.36	2.30	2.25	2.18	2.11	2.03	1.98	1.94	1.89	1.84	1.79	1.73
30	4.17	3.32	2.92	2.69	2.53	2.42	2.33	2.27	2.21	2.16	2.09	2.01	1.93	1.89	1.84	1.79	1.74	1.68	1.62
40	4.08	3.2	2.84	2.61	2.45	2.34	2.25	2.18	2.12	2.08	2.00	1.92	1.84	1.79	1.74	1.69	1.64	1.58	1.51
60	4.00	3.15	2.76	2.53	2.37	2.25	2.17	2.10	2.04	1.99	1.92	1.84	1.75	1.70	1.65	1.59	1.53	1.47	1.39
120	3.92	3.07	2.68	2.45	2.29	2.17	2.09	2.02	1.96	1.91	1.83	1.75	1.66	1.61	1.55	1.50	1.43	1.35	1.25
∞	3.84	3.00	2.60	2.37	2.21	2.10	2.01	1.94	1.88	1.83	1.75	1.67	1.57	1.52	1.46	1.39	1.32	1.22	1.00

99th Percentiles of the $F(\nu_1, \nu_2)$

$\nu_2 \backslash \nu_1$	1	2	3	4	5	6	7	8	9	10	12	15	20	24	30	40	60	120	∞
1	4052	5000	5403	5625	5764	5859	5928	5981	6022	6056	6106	6157	6209	6235	6261	6287	6313	6339	6366
2	98.5	99.00	99.17	99.25	99.30	99.33	99.36	99.37	99.39	99.4	99.42	99.42	99.45	99.46	99.47	99.47	99.48	99.49	99.5
3	34.12	30.82	29.46	28.71	28.24	27.91	27.67	27.49	27.35	27.23	27.05	26.87	26.69	26.60	26.50	26.41	26.32	26.22	26.12
4	21.20	18.00	16.69	15.98	15.52	15.21	14.98	14.80	14.66	14.55	14.37	14.20	14.02	13.93	13.84	13.75	13.65	13.56	13.46
5	16.26	13.27	12.06	11.39	10.97	10.67	10.46	10.29	10.16	10.05	9.89	9.72	9.55	9.47	9.38	9.29	9.20	9.11	9.02
6	13.75	10.92	9.78	9.15	8.75	8.47	8.26	8.10	7.98	7.87	7.72	7.56	7.40	7.31	7.23	7.14	7.06	6.97	6.88
7	12.25	9.95	8.45	7.85	7.46	7.19	6.99	6.84	6.72	6.62	6.47	6.31	6.16	6.07	5.99	5.91	5.82	5.74	5.65
8	11.26	8.65	7.59	7.01	6.63	6.37	6.18	6.03	5.91	5.81	5.67	5.52	5.36	5.28	5.20	5.12	5.03	4.95	4.86
9	10.56	8.02	6.99	6.42	6.06	5.80	5.61	5.47	5.35	5.26	5.11	4.96	4.81	4.73	4.65	4.57	4.48	4.40	4.31
10	10.04	7.56	6.55	5.99	5.64	5.39	5.20	5.06	4.94	4.85	4.71	4.56	4.41	4.33	4.25	4.17	4.08	4.00	3.91
12	9.33	6.93	5.95	5.41	5.06	4.82	4.64	4.50	4.39	4.30	4.16	4.01	3.86	3.78	3.70	3.62	3.54	3.45	3.36
15	8.68	6.36	5.42	4.89	4.56	4.32	4.14	4.00	3.89	3.80	3.67	3.52	3.37	3.29	3.21	3.13	3.05	2.96	2.87
20	8.10	5.85	4.94	4.43	4.10	3.87	3.70	3.56	3.46	3.37	3.23	3.09	2.94	2.86	2.78	2.69	2.61	2.52	2.42
24	7.82	5.61	4.72	4.22	3.90	3.67	3.50	3.36	3.26	3.21	3.07	2.93	2.78	2.70	2.62	2.54	2.45	2.35	2.26
30	7.56	5.39	4.51	4.02	3.70	3.47	3.30	3.17	3.07	2.98	2.84	2.70	2.55	2.47	2.39	2.30	2.21	2.11	2.01
40	7.31	5.18	4.31	3.83	3.51	3.29	3.12	2.99	2.89	2.80	2.66	2.52	2.37	2.29	2.20	2.11	2.02	1.92	1.80
60	7.08	4.98	4.13	3.65	3.34	3.12	2.95	2.82	2.72	2.63	2.50	2.35	2.20	2.12	2.03	1.94	1.84	1.73	1.60
120	6.85	4.79	3.95	3.48	3.17	2.96	2.79	2.66	2.56	2.47	2.34	2.19	2.03	1.95	1.86	1.76	1.66	1.53	1.38
∞	6.63	4.61	3.78	3.32	3.02	2.8	2.64	2.51	2.41	2.32	2.18	2.04	1.88	1.79	1.70	1.59	1.47	1.32	1.00

Appendix 2

Answers to Selected Exercises

CHAPTER 1

1.1 (d)
1.3 (a)
1.5 (b)
1.7 (b)
1.9 (b)
1.11 (b)
1.13 (b)
1.15 (e)

CHAPTER 2

2.1 The chance the game will end in 180 minutes in American League: 80%.

The chance the game will end in 180 minutes in National League: 85%.

The time before which 95% of the games will end in American League: 200 minutes.

The time before which 95% of the games will end in National League: 200 minutes.

2.3 $\bar{X} = 1891$; $S = 4.06$; Median = 1891; Mode = 1891; $R = 25$; $IQR = 2$.

2.5 **a.** S = {ME1, ME00, ME01, ME02, EE1, EE00, EE01, EE02, CE1, CE00, CE01, CE02, IE1, IE00, IE01, IE02, MfE1, MfE00, MfE01, MfE02}.

b. A = {ME01, EE01, CE01, MfE01}.

2.7 S is shown with cross hatch in the figure below. Event A: $\{X > 24, Y < 120\}$ is shown double hatch.

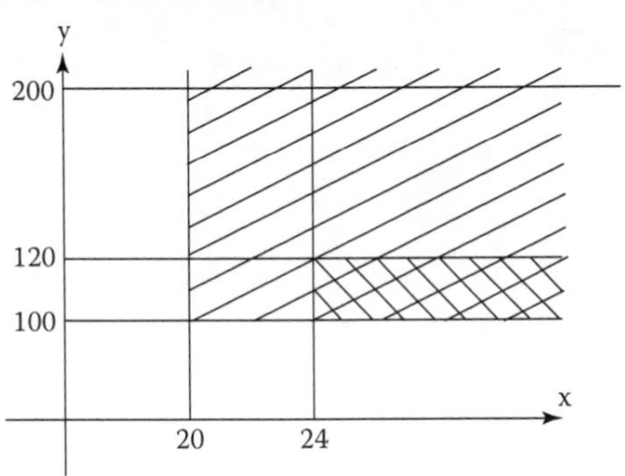

2.8 **a.**

$$S = \begin{Bmatrix} B_1B_1, B_1B_2, B_1B_3, B_1B_4, B_1W_1, B_1W_2, B_1W_3, B_1W_4 \\ B_1B_2, B_2B_2, \ldots\ldots B_2W_3, B_2W_4 \\ B_3B_1, B_3B_2, \ldots\ldots B_3W_3, B_3W_4 \\ B_4B_1, B_4B_2, \ldots\ldots B_4W_3, B_4W_4 \\ W_1B_1, W_1B_2, \ldots\ldots W_1W_3, W_1W_4 \\ W_2B_1, W_2B_2, \ldots\ldots W_2W_3, W_2W_4 \\ W_3B_1, W_3B_2, \ldots\ldots W_3W_3, W_3W_4 \\ W_4B_1, W_4B_2, \ldots\ldots W_4W_3, W_4W_4 \end{Bmatrix}$$

Number of the elements in sample space = 64

b. $A(\text{both black}) = \begin{Bmatrix} B_1B_1, B_1B_2, B_1B_3, B_1B_4 \\ B_2B_1, B_2B_2, B_2B_3, B_2B_4 \\ B_3B_1, B_3B_2, B_3B_3, B_3B_4 \\ B_4B_1, B_4B_2, B_4B_3, B_4B_4 \end{Bmatrix}$

Number of elements in Event $A = 16$.

2.9 **a.** S = {same as in Problem 2.8 except omit the eight elements B_1B_1, B_2B_2, B_3B_3, B_4B_4, W_1W_1, W_2W_2, W_3W_3, W_4W_4}. Number of elements in $S = 56$.

b. A = {same as above except omit the four elements B_1B_1, B_2B_2, B_3B_3, B_4B_4}. Number of elements in $E = 12$.

2.11 1/4

2.12 3/14

2.13 18/36

2.15 3/5

2.17 **a.** 0.1

b. 0.4

2.19 **a.** 0.32

b. 0.2048

2.21 144 ways

2.23 0.299

2.25 0.0043

2.27 0.4

2.29 pmf of X:

x	1	2
$p(x)$	.5	.5

$\mu_X = 1.5$

$\sigma_x^2 = 0.25$

2.31 $a = 4$

$F(x) = x^4, 0 \leq x \leq 1$

$F(1/2) = 1/16$

$F(3/4) = 81/256$

$P(1/2 \leq X \leq 3/4) = 65/256$

$\mu_x = 4/5$

$\sigma_x^2 = 4/150$

2.33 0.85

2.35 7

2.37 **a.** 0.0

b. 0.0228

c. 0.9772

2.39 LSL = 0.436

2.41 0.614

2.43 [168.85, 191.15]

2.45 [0.99, 1.31]

2.47 99% CI for μ: [6.215, 6.253]

99% CI for σ^2: [0.000273, 0.0025]

99% CI for σ: [0.0165, 0.0464]

2.49 There is no reason to believe the yield is less than 90% at $\alpha = 0.05$.

2.51 Both machines are filling equal volumes.

2.53 The data do not come from a normal population.

CHAPTER 3

3.1 Pearson correlation coefficient of 1st survey and 2nd survey results = 0.668.

P-value = 0.025. So, Reject H_o: Pearson coefficient = 0, the two sets of results are correlated. The questionnaire is reliable.

3.3 $n = 21$

3.5 The exponential seems a good fit for the data. The MTTF = 76.6 months.

3.7 MTTF = 10.625 years.

3.9 22.1% of the washing machines will need service during warranty.

3.11 Effects A = 2.25, B = 6.75, AB = 0.25.

Only Effect B is significant.

3.13 Effects: A = −17, B = 55, C = 165, AB = −20, AC = 20, BC = −8, ABC = 3. Only Effects B and C are significant.

3.15 25.8% of the assemblies will have less then the desired clearance of 0.01.

3.17 The tolerance on the total weight of 25 bags should be = ±20 lb.

When the truck operator allows a tolerance of ±100 lbs., instead of the ±20, we should expect many underweight bags to pass through this check.

CHAPTER 4

4.1 $UCL(\bar{X}) = 25.19$, $CL(\bar{X}) = 24.61$, $LCL(\bar{X}) = 24.03$.

$UCL(R) = 1.80$, $CL(R) = 0.79$, $LCL(R) = 0$.

Process not-in-control.

4.3 $UCL(\bar{X}) = 37.19$, $CL(\bar{X}) = 35.72$, $LCL(\bar{X}) = 34.18$.

$UCL(R) = 4.58$, $CL(R) = 2.01$, $LCL(R) = 0$.

The process is not-in-control. There is an upward drift in the process.

4.4 $UCL(\bar{X}) = 37.19$, $CL(\bar{X}) = 35.72$, $LCL(\bar{X}) = 34.25$.

$UCL(S) = 2.04$, $CL(S) = 0.9$, $LCL(S) = 0$.

We see the same phenomenon we saw in the $\bar{X}$- and R-charts for the same data. The average weight of the packages increases from the beginning to the end of the period.

4.5 $UCL(P) = 0.147$, $CL(P) = 0.04$, $LCL(P) = 0$.

The process is in-control, but there are on average 4% defective bottles. Steps must be taken to reduce the average level of defectives.

4.7 We will use a C-chart.

$UCL(C) = 13.3$, $CL(C) = 5.95$, $LCL(C) = 0$.

The process is not-in-control. Some days the errors are too many. There seems to be an opportunity for controlling the process at a consistent level.

4.9 A P-chart with varying sample size would be appropriate. $\bar{P} = 0.033$

The process is not-in-control. There are hours when the defective rate is low and hours when the defective rate is high. An investigation for the reasons is called for.

4.10 We will use a U-chart since we are tracking the number of occurrences of defects per house and the house size is changing. As an example, the limits for the 23rd value of u are:

$UCL(u_{23}) = 0.393$, $CL(u_{23}) = 0.108$, $LCL(u_{23}) = 0$.

There are two houses where the average number of defects per room were above the limits. Both houses were cleaned by Crew "C."

On the capability of the process: The cleaning business does not have the capability to deliver what they promised. They should either improve their capability or revise their guarantee.

4.11 The trial control limits:

$UCL(\bar{X}) = 25.19$, $CL(\bar{X}) = 24.61$, $LCL(\bar{X}) = 24.03$.

$UCL(R) = 1.80$, $CL(R) = 0.79$, $LCL(R) = 0$.

After going through a process of eliminating plots outside limits and recalculating limits with remaining data, we get to a stable process with:

$\bar{\bar{X}} = 24.44$, $\bar{R} = 0.6769$.

$C_p = 1.01$, $C_{pk} = 0.44$

The C_{pk} is much smaller than the C_p indicating that the process is quite off center.

4.13 Repeatability error: $\sigma_e = 0.0315$. Reproducibility error: $\sigma_0 = 0.0492$.

Gage error: $\sigma_g = 0.0584$. Overall standard deviation: $\sigma_{all} = 0.1837$.

The variability in the product: $\sigma_p = 0.1742$. $\sigma_g/\sigma_p = 0.33$. Not very good. This ratio should be less than 10%. The precision to tolerance ratio (PT Ratio): $6\sigma_g/(USL - LSL) = 0.7$.

This is also not good. The PT ratio should be less than 10%. The gage has too much variability both from the instrument as well as from the operators.

To check the resolution and variability of the instrument using control charts, we draw $\bar{X}$- and R-charts for the data from Operator 1 and Operator 2.

$\bar{X}$-charts from both operators show that the instrument variability is smaller than the variability in product, which is a different conclusion from the quantitative analysis performed above. The quantitative analysis is more dependable.

The R-charts from both operators show that the resolution of the instrument is adequate.

CHAPTER 5

5.1 **a.** $p(\bar{X} > 3.24) = 0.0082$

b. The answer will be approximately the same because, according to the central limit theorem, the $\bar{X}_9$ will have the same normal distribution used to calculate the probability in Part (a).

5.3

k	0.5	1.0	1.5	2.0
β	0.9332	0.5	0.0668	0.00135

5.5 $UCL(P) = 0.127, CL(P) = 0.05, LCL(P) = 0.0.$

5.7 $UCL(C) = 16.5, CL(C) = 9, LCL(C) = 1.5.$

5.9 **a.** Control chart for current control:

$UCL(\bar{X}) = 894.4, CL(\bar{X}) = 891, LCL(\bar{X}) = 887.5.$

$UCL(R) = 5.997, CL(R) = 1.835, LCL(R) = 0.$

A few $\bar{X}$ values and a few R values are outside limits.

b. $\bar{X}$- and R-charts for controlling at the given standard $\mu = 882$.

$UCL(\bar{X}) = 885.5, CL(\bar{X}) = 882, LCL(\bar{X}) = 878.5.$

$UCL(R) = 5.997, CL(R) = 1.835, LCL(R) = 0.$

Numerous $\bar{X}$ values are outside the limits because a standard for the process mean is being imposed.

c. $\bar{X}$- and R-charts for controlling at the given standard $\mu = 882$ and the given $\sigma = 3$.

$UCL(\bar{X}) = 888.4, CL(\bar{X}) = 882, LCL(\bar{X}) = 875.6.$

$UCL(R) = 11.06, CL(R) = 3.384, LCL(R) = 0.$

A large number of $\bar{X}$ values are outside the limits.

5.10 **a.** The control for individual and MR values.

$UCL(X) = 898.4, CL(X) = 890.7, LCL(X) = 882.9.$

$UCL(MR) = 9.489, CL(MR) = 2.904, LCL(R) = 0.$

The chart shows process not-in-control.

b. The MA- and MR-charts with $n = 3$. [use $n = 1$ and the length of $MA = 3$ in the Minitab for drawing the MR and MA chart]

$UCL(\bar{X}) = 895.1, CL(\bar{X}) = 890.7, LCL(\bar{X}) = 886.2.$

$UCL(MR) = 9.489, CL(MR) = 2.904, LCL(MR) = 0.$

The chart shows the process is not-in-control.

c. The $EWMA$ chart with $\lambda = 0.2$.

The steady-state control limits:

$UCL(EWMA) = 893.2, CL(EWMA) = 890.7, LCL(EWMA) = 888.1$

The $EWMA$ chart with $\lambda = 0.6$.

The steady-state control limits:

$UCL(EWMA) = 895.7, CL(EWMA) = 890.7, LCL(EWMA) = 885.6.$

It is interesting to see that the MA chart and both the $EWMA$ charts showed the same set of sample plots outside limits.

5.11 **a.** To make the $DNOM$ chart:

The pooled standard deviation $S = 1.61$.

The limits for $(\bar{X}_i - T_i)$: $UCL = 2.79, CL = 0, LCL = -2.79.$

None of the $(\bar{X}_i - T_i)$ values are outside the limits; the process is in-control.

b. To make the standardized *DNOM* chart:

The estimate for the pooled $CV = 0.041$.

The control limits for the standardized *DNOM* chart:

$UCL = 1.071$, $CL = 1.0$, $LCL = 0.929$.

The process is in-control.

5.13 $C_p = 0.146$

$C_{pk} = -1.28$

To calculate C_{pm} Use Target = (31.5 + 32.3)/2 = 31.9

$C_{pm} = 0.033$

The very small value of the C_{pm} indicates that the process center is very much off target.

5.15 The proportion outside spec when the specs are 2.2σ from the nominal (center of specification), and the process center is 1.5σ away from the nominal, is equal to 242,000 ppm.

5.16 The process has 3.38-sigma capability.

CHAPTER 7

7.3

p	0.01	0.02	0.05	0.08	0.10	0.15	0.20
$Pa(p)$	0.878	0.662	0.199	0.05	0.017	0.003	0.00

7.5 $(1 - \alpha) = 0.95$ $AQL = 0.022$

$\beta = 0.05$ $LTPD = 0.15$

7.7 $n = 45$, $c = 3$

7.9

p	0.01	0.03	0.05	0.07	0.10	0.15
$Pa(p)$	0.998	0.956	0.840	0.728	0.428	0.157

7.11

p	0.01	0.02	0.03	0.04	0.05	0.06	0.07	0.08	0.09	0.10	0.12
ASN	23.6	26.4	28.6	30.0	31.0	31.6	31.7	31.6	31.3	30.8	29.6

The equivalent SSP: (32, 2).

7.13 Code letter: K.

SSP for normal inspection: (125, 7). Tightened inspection: (125, 5). Reduced inspection: (50, 3, 6).

7.15 Code letter: K.

DSP for Normal inspection: $(n_1, c_1, r_1, n_2, c_2, r_2)$: (80, 3, 7, 80, 8, 9)

Tightened inspection: $(n_1, c_1, r_1, n_2, c_2, r_2)$: (80, 2, 5, 80, 6, 7)

Reduced inspection: $(n_1, c_1, r_1, n_2, c_2, r_2)$: (32, 1, 5, 32, 4, 7)

7.17

p	0.01	0.02	0.04	0.05	0.06	0.08	0.10	0.12	0.14	0.16
AOQ	0.01	0.02	0.035	0.035	0.031	0.018	0.011	0.002	0.001	0.000

CHAPTER 8

8.3 **a.** The results of regression show that the "pour temp" has a significant impact on "porosity."

b. Pour temp and dip viscosity have significant impact on porosity, but pour time does not.

8.5 Pearson correlation coefficient of tap-temp and pour-temp = 0.616 *P*-value = 0.002.

There is significant correlation between tap-temp and pour-temp.

Index

A

Acceptable quality level (AQL), 426
Addition theorem of probability, 72–74
American National Standards Institute (ANSI), 415
American Society for Quality (ASQ), 530
American Society for Testing Materials (ASTM), 415
Analysis of variance (ANOVA), 175–176
ARL. *See* Average run length (ARL)
ASQ. *See* American Society for Quality (ASQ)
Attribute control charts
 C-chart, 248–251
 LCL meaning on *P*-/*C*-chart, 258–259
 nP-chart, 254–255
 P-chart, 245–248, 259
  with varying sample sizes, 251–254
 percent defectives chart (*100P*-chart), 255
 rational subgrouping, 259–261
 runs use, 259
 U-chart, 255–258
Attribute data, 51
Automatic test equipment (ATE), 30
Average outgoing quality (AOQ), 447–448
 behavior of, 449
 MIL-STD-105E, sampling plans from, 449–450
Average outgoing quality limit (AOQL), 449–450
Average run length (ARL)
 average value of, 310
 control chart, 309
 probability mass function, 309–310
 3-sigma $\bar{X}$-chart
  calculations, 310
  curves for, 310–311

B

Baldrige National Quality Program (BNQP), 530
Benchmarking
 action plans, 470
 data analysis of, 469
 economic impact, 469
 goal setting, 467–468
 operational goals, 470
 partner(s) study, 468–469
 performance measure, 468
 plans and monitor results, 470
 process of, 468–471
 recalibration effort, 471
 recommendations, 470
Binomial distribution
 mean and variance, 98–99
 parameters, 96–97
 pmf, 96
 probability of success and failure, 97
 random variable, 96–97
 trials, 97–98
Boundary-less organizations, 377–378
Box-and-whisker (B&W) plot. *See also* Populations
 median, 63
 tap-out and pouring temperatures, 63–64
 values and dispersion, 63
Brainstorming, 466–467. *See also* Improvement tools for quality

C

Cause-and-effect (C&E) diagram. *See also* Improvement tools for quality
 case study, 465
 communication tool, 465
 format of, 466
 for problem-solving process, 465
C-chart
 example of, 249–250

limits, 248–249
defects, average number of, 303
results, 304
3-sigma limits, 303–304
operating characteristics
drawing of, 313–314
function, 314
probability of acceptance, 315
procedure, 248
scratches, 251
use, 248
χ^2-Distribution percentiles, 579–580
Central limit theorem (CLT), 110–111
Certified Quality Manager Handbook, 376
Combinations theorem on number, 83–85
Company wide quality control (CWQC), 8
Complement theorem of probability, 74–75
Concurrent engineering (CE)
approach results, 195
Lean Production Method, 196
manufacturability/assembly
suggestions, 196–197
systematic procedures, 197
market, reduced time to, 195–196
model, 195
reviews
stages for, 197
verification of, 197
and sequential engineering (SE), 195
Conditional probability, 75–76
Confidence interval (CI). *See also* Populations
interpretation of, 116–117
μ of normal population with σ known
estimator, 115
standard deviation, 116
for μ when σ unknown, 118
degrees of freedom, 117
t distribution, 117
for σ of normal population
χ^2 (chi-squared) distribution, 118–119
models, 119–120
sample variance, 118
Continuous improvement of quality
lean manufacturing, 493–495
culture, 504
leveling and balancing, 502–503
quality control, 496–497
quantity control, 497–498
stabilizing and standardization processes, 500
TPM, 499–500
visual management, 500–502
waste and cost control, 498–499
methodology for
Deming's PDCA cycle, 458–459
generic, 461–464
Juran's breakthrough sequence, 459–461
need for
problem-solving methodology, 458
quality of products, 457
process of, 462
tools for, 464
benchmarking, 467–471
brainstorming, 466–467
cause-and-effect (C&E) diagram, 465–466
control charts, 474–475
correlation analysis, 490–493
histogram, 472–474
Pareto analysis, 471–472
regression analysis, 479–490
scatter plots, 476–477
Continuous sampling plans, 421
Control charts
assignable causes, 219
basic charts, 262
case study, 474–476
chance causes, 218
common and special cause variability, 219
cumulative sum chart, 262–263
data types
attribute, 220
measurement, 220
EWMA chart, 262–263
for fraction defectives, 259–261
for individuals (*see* $\overline{X}$-chart)
measurement control charts (*see* Measurement control charts)
median chart, 262–263

product characteristic, variability in, 218–219
SPC implementation (*see* Statistical process control (SPC))
typical control chart
benefits, 220
centerline (CL), 219–220
examples, 220
in-control and not-in-control process, 219–220
SPC tools, 220
UCL and LCL, 219–220
Correlation analysis
Pearson product moment coefficient, 490
significance in, 491–493
between two variables, 490–491
Cost of quality
appraisal, components, 20
case study, 32–38
categories, 18
relationship among, 31–32
data, 31
external failure
complaint adjustment, 21
product returns, 21
warranty charges, 21
index by American Society for Quality, 30
internal failure
penalty for not meeting schedules, 21
retest, 21
rework/salvage, 20
scrap, 20
prevention
improvement projects, 19
information system, 19
planning, 18–19
process control, 19
system development, 19–20
training, 19
study
approval from upper management, 21–22
data analysis, 23–26
data collection, 22–23
not included in TQC, 29–31
program, 29
projects arising from, 26
scoreboard, 26–28
Critical region (CR), 121
Cumulative distribution function (CDF). *See also* Mathematical models for population
random variable, 92–93
Cumulative Poisson probabilities, 581–583
Cumulative probabilities of standard normal distribution, 576–577
Cumulative sum (CUSUM) control charts, 326
Customer survey
for book in quality engineering, 146–147
characteristics, list of, 146
instrument reliability, 146, 148
methods and advantages, 146
quality-related function, 150
results, 148
sample size, 149
satisfaction questionnaire, 146, 148
simple random sampling, 148–149
stratified random sampling, 148–149
tools used, 147
CUSUM. *See* Cumulative sum (CUSUM) control charts

D

Data types, 51
Defects per million (DPM), 346–347, 563
Defects per million opportunities (DPMO), 344–345, 563
Defects-per-unit chart. *See* C-chart
Define, measure, analyze, improve, and control (DMAIC), 566
Define, measure, explore, develop, and implement (DMEDI), 566
Degrees of freedom, 117
Deming's PDCA cycle, 458–459
2^3Design, 180
confidence intervals, 182
contrast coefficients, table, 181
design columns, 179, 181
experimental error, 181–182
factor and interaction effects, 181
graphical representation, 178–179

model building
factor-level space, 183
fractional factorial designs, 184
residual, 183–184
response surface methodology, 183
screening designs, 184
results interpretation
graphical representation, 182–183
standard order for, 179
Dr. Taguchi's
noise factors, 184–185
signal-to-noise ratio, 184–185
use of, 178
Design of experiments (DOE), 167, 349
analysis of variance (ANOVA)
a × b factorial experiment, 349–350
error, 351
2^2 experiment for lawnmower design, 354–355
experiments, analyzing data, 349
fixed-effect model, 350
interaction effects, 350–351
mean and factor effects, 350
MSE, 353
sum of squares (SS), 351–352, 355
2^3design, 178–181
Dr. Taguchi's, 184–185
model building, 183–184
results interpretation, 182–183
2^4 design, 356–357
2^k design
from machining process, 357–358
normal probability plot, 359
screening designs, 355–356
with single trial, 357–359
two-level designs, 356
2^2factorial, 168–170
effects calculation shortcut, 175
experimental results, 171–172
factor effects, 172–173
interaction effects, 174–175
main effects, 173–174
randomization, 170–171
significance determination, 175–178
fractional factorials, 359
effects calculation, 361–362
2^3 design, 360–361
generation of, 361
resolution
calculation of effects from 2^{4-1}, 368
case study, 366–367
chemical process yield improvement, 2^{4-1} design, 366–367
2^{4-1} design, example, 364
2^4 factorial design, 363–364, 366–367
2^5 factorial design, 365
fractional factorials examples, 365–366
normal probability plot, 367–368
number, 362
treatment combinations, 167–168
Design, quality in
product creation cycle
activity stages, 143
quality planning timing chart, 143
tools employed, 144
product planning
customer needs, 145–150
QFD, 150–155
reliability fundamentals (*see* Reliability)
Deviation from nominal (DNOM) chart
calculations for, 335–336
pooled standard deviation, 334–335
sample average deviation, 333–334
3-sigma limits, 334
standard deviation, 334
standardized
coefficient of variation, 335, 337
plot of, 339
3-sigma control limits, 338
χ^2 (chi-squared) Distribution, 118–119
t-Distribution percentiles, 578
DMAIC. *See* Define, measure, analyze, improve, and control (DMAIC)
DMEDI. *See* Define, measure, explore, develop, and implement (DMEDI)
DNOM. *See* Deviation from nominal (DNOM) chart
Double/multiple sampling plans, 422

Double sampling plan (DSP), 431
 for normal inspection, 443
 for reduced inspection, 445
 for tightened inspection, 444
DPM. *See* Defects per million (DPM)
DPMO. *See* Defects per million opportunities (DPMO)
Dr. Deming's system
 barriers between staff, 515–516
 continuous improvement
  production and service, 513–514
 cultural change
  management philosophy, 512
 education and self-improvement, 517
 goals, 516–517
 institute training, 514
 leadership, 514–515
 long-term planning
  constancy of purpose, 511–512
 numerical quotas, 516
 personal and professional details of Dr. Deming
  control chart method, 510
  JUSE invitation, 510–511
  Order of the Sacred Treasure, Second Class, 510–511
  plan-do-check-act (PDCA) cycle, 511
 prevention orientation
  control charts use, 512
 pride of workmanship, 517
 quality in procurement
  supplied materials, 513
 statistical process control techniques, 510
 transformation, 517–518
 worker's fear, 515
 workforce, 516
Dr. Feigenbaum's system
 jobs for quality in products, 527–528
 need for, 528
 professional and personal details, 527–528
 quality cost study, 527
 total quality control, 527
Dr. Juran's system
 break-through approach, 519, 523
 professional, personal details, 518–519
 quality control
  contributions and control station, 523
  guiding principles, 522–523
  regulatory process, 522
  sporadic and chronic deviations, 523–524
  tools, 523
 quality improvement
  breakthrough approach, 523–524
  diagnostic and remedial journeys, 526
  Dr. Deming's system, comparison with, 526
  formal structure, 524–525
  potential savings, 524
  project nominations and selection, 525–526
 quality planning
  capability of production machinery, 522
  concept of dominance, 521–522
  customer's needs, 521
  process engineering activities, 519–520
  process validation, 522
  road map, 520

E

Economics of quality, 15
Equipment quality costs, 29–30
Excessive variability and/or poor centering (EV/PC), 472–474
Exponentially weighted moving average (EWMA) chart, 262–263
 calculation, 326
 construction and use, 326–327
 CUSUM control charts, 326
 data and calculation, 327–328
 limits for, 331–332
 for 1.5σ change in process mean, 327, 330
 with 1σ change in process mean, 327, 329
 sensitivity improvement, 326
 Shewhart chart, 327, 331
 smoothing constant, 326

F

2^2Factorial design
effects calculation, 175
experimental results
difference in, 171–172
factorial, 168
graphical representation, 171–172
treatment combination codes, 171–172
factor effects, 172–173
graphical representation, 169
interaction effects, 174–175
main effects, 173–174
noise, 170
randomization
noise factors, 170–171
randomized incomplete block, 171
treatment combinations, 171
replicates needed, 169–170
significance determination
ANOVA method, 175–176
confidence interval, 177
data, layout of, 176
factor levels, response against, 177–178
response surface methodology, 177–178
running with winner method, 177–178
standard error, 175–177
treatment combinations responses, 177–178
Failure mode and effects analysis (FMEA)
advantages, 194
design, 191–192
example, 192–194
process, 191–192
RPN, 194
weakness prioritization, 191–192
F-Distribution percentiles, 584–585
Fraction defectives. *See P*-chart
Fraction nonconforming chart. *See P*-chart

G

Generic problem-solving methodology
steps for
alternative solutions, 463
best solution(s), selection, 463
gains, 464
solution implementation and validation, 464
stating problems and project selection, 461
tools used in, 463
Goodness-of-fit test. *See also* Tests for normality
χ^2 statistic, 133–134
parameter of hypothesized distribution, 133
total deviation, 133
Good supplier relationship
essentials of
integrity, 410
mutual understanding, 410
objectivity, 410
personal behavior, 410
product definition and quality, 410
proprietary information, 410
quality evaluation, 410
reputation safeguard, 410
rewards, 410
technical aid, 410

H

Hazard function, 156–157
Histogram
case study, 473–474
EV/PC process parameters, 472–473
House of quality (HOQ), 150–152, 154
Human-oriented TQM philosophy, 14–15
Human resources
customer focus, 381
dual responsibility, 383
external customer and product, 383
needs and expectations, 382
research and product quality, 382
suggestions and, 382–383
"three corners of quality," 382–383
education and training
basic skills, 387
benefits, 388
continuous change in technology, 387

effectiveness, 392
global competition, 387
improve processes, need of, 388
methodology, 390–391
planning, 388–390
resources finding, 391–392
workplace diversity, 388
empowerment
and commitment, relationship, 385
decision making, 385
directive command stage, 385
high-performance organizations, 386
self-managed teams, 386
sense of ownership, 385
traditional and empowered organizations, 386
importance of
Dr. Deming's view, 375–376
workforce contribution, 375
leadership
characteristics of, 380–381
motivation methods, 401–402
open communications
benefits, 384
exchange of information, 383–384
fear of employees, 384
open-book management, 384
quality organizations, 384
organizations
culture, 378–379
structures, 376–378
principles
controlling, 402
leading, 402
organizing, 402
strategic planning, 402
teamwork, 392–400
Hypothesis testing. *See also* Populations
critical region (CR), 121, 124
difference of two means
distribution-free/nonparametric tests, 129
ratio of variances, 129
for testing, 127–128
test model, 127–129
mean of normal population, 121
location, 126–127
sample standard deviation, 125
testing, 122–123
test statistic, 122
outcomes of statistical test, 120–121
possible alternate of
cases, 124–125
test statistic, 125
possible sets, 123–124
relationship of parameters, 120
steps, 121
supplier's claim, 124
types, 120

I

Improvement tools for quality, 464
benchmarking, 467–471
brainstorming, 466–467
cause-and-effect (C&E) diagram, 465–466
Independent events, 76–77
Indirect quality costs, 29–30
Instantaneous failure rate. *See* Hazard function
Instruments. *See* Measurement system analysis (MSA)
Intangible quality costs, 29–30
International Organization for Standardization (ISO 9000)
Standards for Quality Assurance Systems, 3
Ishikawa's fishbone diagram. *See* Cause-and-effect (C&E) diagram
ISO 9000 quality management systems
documentation in, 545–546
ISO 9001:2008 requirements
management responsibility, 548–551
measurement, analysis and improvement, 558–560
product realization, 552–558
quality management system, 547–551
resource management, 551–552
principles
continual improvement, 545
customer focus, 544
factual approach, 545
involvement of people, 544

leadership, 544
mutually beneficial supplier relationship, 545
process approach, 544
system approach, 544–545
standards
ISO 9000, 543–544
ISO 9001, 543–544
ISO 9002, 543–544
ISO 9003, 543–544
ISO 9004, 543–544

J

Joint occurrence of events, theorems on, 75
Juran's breakthrough sequence, 461
"project-by-project" approach, 459–460
steps of, 460–461
Juran's Quality Handbook, 376
"Juran trilogy." *See* "Quality trilogy"

L

Leadership quality
ambitious, caring and commitment, 381
democratic, intuition and learners, 380
values and vision, 380
Lean manufacturing
cost and waste control approach, 498–499
culture, 504
leveling and balancing
assembly line balancing principle, 502–503
takt-time, 503
process, standardization and stabilization, 500
quality control, 497
quantity control, 497–498
SMED procedure, 493
Toyota Production System, 493–494
TPM, 499–500
visual management, 500
set-in-order, 501–502
shine and sustain, 502
sorting, 501
standardize, 502
wastes types, 494
Liability costs, 29–30
Life-cycle quality costs, 30
Lot tolerance percent defective (LTPD), 426
Lower control limit (LCL), 219–220, 222–226, 242–250, 252–259

M

Malcolm Baldrige National Quality Award (MBNQA), 5
analysis and review, 537
ASQ and BNQP, 530
criteria and items, 530–532
customer focus
engagement, 535–536
voice of, 536–537
customer-focused outcomes, 542
data and knowledge management, 538
financial and market outcomes, 542
governance and societal responsibilities, 533–534
information resources and technology, 538
leadership triad, 532
performance
improvement, 538
measurement, 537
process effectiveness outcomes, 542
process management
emergency readiness, 540–541
work processes, 540–541
work systems design, 540
product outcomes, 541
program, 530
results triad, 532
senior leadership, 533
strategic planning
deployment, 535
development process, 534–535
systems perspective, 532
workforce
capability and capacity, 539–540
climate, 540
engagement and leader development, 539

management, 538–539
workforce-focused outcomes, 542
Management of quality
human resources
customer focus, 381–383
education and training, 387–392
empowerment, 385–387
importance of, 375–376
leadership, 380–381
motivation methods, 401–402
open communications, 383–384
organizations, 376–379
principles, 402
strategic planning for
deployment, 405–407
history of, 402–404
making of, 404–405
Mathematical models for population
histogram, 67
probability, 85–87
analysis method, 69–70
case, 70
definition of, 68
event, 68
relative frequency method, 71–72
sample point and space, 68, 80–85
theorems, 72–80
trial, 67–68
probability distributions, 111–113
cumulative distribution function (*See* Cumulative distribution function (CDF))
mean and variance of, 93–96
probability density function (*See* Probability density function (pdf))
Probability mass function (*See* Probability mass function (pmf))
random variable (*See* Random variable)
sample average $\overline{X}$, 109–110
Mean square error (MSE), 353
Mean time between failures (MTBF), 162–165
Mean time to failure (MTTF), 162–165
Measurement control charts, 221
short runs, control charts for, 333–339
slow processes, 319–326
$\overline{X}$- and *R*-charts (*see* $\overline{X}$- and *R*-charts)
$\overline{X}$- and *R*-charts, 315–319
$\overline{X}$ and S-charts (*see* $\overline{X}$ and S-charts)
Measurement data, 51
Measurement system analysis (MSA). *See also* Production, quality in
evaluating instrument
adequacy, quick check, 287–289
bias, 283–284
checking for resolution, 280–281
gage R&R study, 279–280, 282
methods, 279–280
repeatability and repeatability error, 285
resolution, 280–281, 283
variability (precision), 284–287
instruments properties
accuracy, 275–276
bias and precision definition, 275–276
error standard deviation, 276
linearity and stability, 276–277
resolution, 276
standards
calibration and traceability, 278
hierarchy, 277–278
National Institute of Standards and Technology (NIST), 277–278
primary and secondary, 278
Measurement System Analysis (MSA)-Reference Manual, 275, 279, 283, 288
Minimum variance unbiased (MVUB), 115
Mortality rate. *See* Hazard function
Motion and Time Study: Design and Measurement of Work, 494, 496
Moving average (MA) and moving range (MR) charts
data for, 323–324
limits for, 322–323
mathematical expression, 322
plots outside control limits and, 325–326
power advantage, 324
samples/subgroups formation, 322
subgroup sizes, 324–325

MSA. *See* Measurement system analysis (MSA)
MSE. *See* Mean square error (MSE)
MTBF. *See* Mean time between failures (MTBF)
MTTF. *See* Mean time to failure (MTTF)
Multiplication rule, 80
  tree diagram, 81
Multiplication theorems of probability, 77–79

N

National Institute of Standards and Technology (NIST), 277–278
Niebel's methods, standards, & work design, 494, 496
Nonconformities, control chart. *See* *C*-chart
Normal distribution, 106
  application of, 107–109
  area, 104–105
  CDF, 103
  cumulative probabilities, 103
  example of, 107–108
  measurements, 102
  normal curve, 101–102
  normal table, 103–104
  notations, 103
  parameters, 102
  pdf, 101
  proportion in normal population, 102–103
  random variable, 101–102
Normal probability paper (NPP), 129
nP-chart
  advantage of, 255
  defectives in, 254
  limits, formulas for, 254

O

Operating characteristic curve (OC curve), 422–423
Organizations
  culture
    characteristics of, 379
    customer's needs, 379
    Dr. Deming's view, 379
    leadership, 378–379
    and members, 378
  "open-book management," policy, 384
  structures
    boundary-less, 377–378
    customer-based, 377
    design of, 378
    functional and product, 377
    matrix and team, 377
    technical and social systems, 376
    unity of command, principle, 378
Out of the Crisis, 17, 511

P

Pareto analysis of continuous improvement of quality
  case study, 471–472
  Minitab program, 472
P-chart
  control limits formula, 245–246
  example of, 246–248
  fraction, monitor and control, 245
  limits derivation
    defective units, 302
    expected value and standard deviation, 302
    results, 303
    3-sigma limits, 303
  operating characteristics
    acceptance probability, 313
    calculations for, 312
    control process, designed to, 312–313
  requirements of, 245
  with varying sample sizes
    example, 247
    sample size and defective castings number, 252–254
    stair-step limits, 251–252
Percent defectives chart (*100P*-chart), 255
Permutations, 81
  theorem on number of, 82–83
PFC. *See* Process flow chart (PFC)
Poisson distribution
  mean and variance of, 100–101
  pmf, 99
  random variable, 99–100

Populations
definitions, 50–51
empirical methods
frequency distribution, 53–59
graphical methods
box-and-whisker plot (*see* Box-and-whisker (B&W) plot)
stem-and-leaf diagram (*see* Stem-and-leaf (S&L) diagram)
inference procedures of, 135–137
confidence interval (*see* Confidence interval (CI))
estimator, 114–115
hypothesis testing (*see* Hypothesis testing)
MVUB estimator, 115
parameters, 113–114
point estimate, 115
P-value, 134–135
random variables, 114
statistic, 114
tests for normality (*see* Tests for normality)
mathematical models (*see* Mathematical models for population)
numerical measures
dispersion, 65
location, 64
numerical methods
average and standard deviation, 61
distributions types, 59–60
variability in, 49–50
Probability density function (pdf). *See also* Mathematical models for population
random variable, 91
Probability mass function (pmf). *See also* Mathematical models for population
elements of range space, 90
nonnegative function, 89
probability histogram, 90
random variable, 89, 91
Problem-solving tools, 26
Process capability, 339
with attribute output, 274–275
calculation difference, 340
capability indices, confidence interval for, 342–344
conditions and C_p values, 269–270
$c_p/c_{pk}/c_{pm}$, comparison of, 341–342
C_{pk} index measurement, 272
C_{pm} index, 340–341
good C_p but poor C_{pk}, process with, 272–273
indices C_p and C_{pk}, 269–274
with measurable output, 268–269
Motorola's 6σ capability
advantage, 349
DPM, 346–347
DPMO, 344–345
normal distribution, 345
process conditions with, 346–348
process defects, 344–345
and proportion out of specification, 347
uniformity in, 347
qualitative evaluation, 268–269
with same C_p, 271
Process capability analysis, 268
Process control. *See also* Production, quality in
defective units, 217
environment effect, 217
parameters, 217–218
scheme of, 217–218
Process design. *See also* Design, quality in
capability results, 208
control plan, 205
feedback/assessment/corrective action, 208–209
floor plan layout, 204–205
FMEA, 205–206
MSA, 208
packaging standards, 207
parameter selection, 204
case study, 200–203
experimental units treatments, 202
treatment combinations, assigning experimental units, 202
variables, experiment to choose, 203

PFC, 198–200
preliminary capabilities, 207
process instructions, 205, 207
product/process approval, 208
quality-related outcomes, 197–198

Process flow chart (PFC)
checks and control activities, 199–200
communication tool, 198
for making foundry mold, 199–200
operations, schematic representation, 198
symbols, 198

"Process parameters," 217–218

Product creation cycle
activity stages, 143
quality planning timing chart, 143
tools employed, 144

Product design
concurrent engineering (CE), 195
manufacturability/assembly, 196–197
reviews, 197
DOE
2^3 design, 178–185
2^2 factorial, 168–170
treatment combinations, 167–168
FMEA, 191–194
parameter
characteristics selection, 166
DOE, 167
QFD exercise, 166–167
quality-related activities, 166
tolerance
assembly, 187
Dr. Taguchi, view, 186–187
traditional approaches, 185–186

Production, quality in
attribute control charts
C-chart, 248–251
LCL meaning on *P*-/*C*-chart, 258–259
nP-chart, 254–255
P-chart, 245–248, 259
P-chart with varying sample sizes, 251–254
percent defectives chart (*100P*-chart), 255
rational subgrouping, 259–261
runs use, 259
U-chart, 255–258
control charts, 218
data types, 220
measurement, 221 (*see also* $\overline{X}$- and *R*-charts; $\overline{X}$ and S-charts)
SPC implementation, 263–267
typical control chart, 219–220
control charts, operating characteristics
ARL, 309–311
C-chart, 313–315
P-chart, 312–313
R-chart, 307–309
$\overline{X}$-chart, 304–307
design of experiments (DOE), 349
ANOVA, 349–355
2^4 design, 356–357
2^k design, 355–359
fractional factorials, 359–362
resolution, 362–368
limits derivation
C-chart, 303–304
P-chart, 302–303
R-chart, 301–302
$\overline{X}$-chart, 298–301
measurement control charts
EWMA chart, 326–332
short runs, control charts for, 333–339
slow processes, 319–326
$\overline{X}$- and *R*-charts, 315–319
MSA (*see* Measurement system analysis (MSA))
process capability, 339 (*see also* Process capability)
capability indices, confidence interval for, 342–344
$c_p/c_{pk}/c_{pm}$, comparison of, 341–342
C_{pm} index, 340–341
Motorola's 6σ capability, 344–349
process control (*see* Process control)
types of, 297–298

Product planning
customer needs
approach determination, 145–146
customer satisfaction questionnaire, 146, 148
new book in quality engineering, 146–147

survey, 146–150
voice of, 145
features determination, 144–145
planning team, 145
QFD, 150–155
quality and reliability goals, 145
reliability fundamentals (*see* Reliability)

Q

QFD. *See* Quality function deployment (QFD)
Quality
costs, 2 (*see also* Cost of quality)
definitions of, 9–10
dimensions, 9
engineering, 7–8
in healthcare industry, 7
history, 1
management approach, 3
movement in United States, 4
events related, 6
as parameters for business planning, 5
process variables and product characteristics, 3
product and service, 11
products, old and new models of making, 16–17
quality control (QC) circles
birth of, 3
movement, 4
revolution, 5–6
Six Sigma process, 4–5
in software industry, 7
total quality system, 2
Quality Control for the Foreman, 3
Quality Control Handbook, 519
Quality function deployment (QFD)
competitor as benchmark, 154
customer requirements and design features
importance-weight, 153
preference numbers, 152–153
strength levels, 153
design features prioritizing
normalization of, 154
relationship matrix, 153
strength relationships, numerical equivalents, 153
HOQ, 150–152
major component, 150
preference numbers, 150–151
product, design features, 150
targets
benchmark, comparison with, 154
design features and customer requirements, 154–155
quality and reliability goals, 155
triangular matrix, 152
Quality improvement tools, 464
cause-and-effect diagram (C&E)
case study, 465–466
Quality management system, model, 545
Quality planning activities, 143–144
"Quality trilogy," 519

R

Random variable. *See also* Mathematical models for population
continuous, 88–89
discrete, 88–89
in notations, 87
range space, 87–88
Rational subgrouping, 233
assignable causes discovering, 259
case study, 234–235, 259–261
fraction defectives, control chart for, 259–261
RBD. *See* Reliability block diagram (RBD)
R-chart
limits derivation
relative range, 301
results from, 302
rule for, 301
operating characteristics
acceptance/rejection, probability of, 309
new to old process standard deviation, 308
process standard deviation, change, 307
Reference Manual of Advanced Product Quality Planning and Control Plan, 143

Regression analysis
model adequacy, 481–482
multiple linear, 487–489
nonlinear, 489–490
simple linear, 481
coefficients, 479–480
straight line to data fitting, 479–480
test of significance
data sets and fitted lines, 482–483
degrees of freedom, 483
linear relationship, existence of, 482, 484
Minitab output and plot, 485–487
MSE, 483
standard error, 483
Rejectable quality level (RQL), 426
Reliability. *See also* Design, quality in
bathtub curve
chance failures/useful life period, 161
early failures, 159–160
failure rate curves, 158–159
infant mortality region, 159–160
wear-out period, 161
definition, 156
engineering
defined, 165
field failure and laboratory test data, 165–166
goals, 165
improvement, 165–166
RBD, 165
exponential distribution
failure rate, 162
function form, 161
mean and variance, 162
shape and standard deviation, 162
failure rate/life distribution calculations, 157–159
hazard function, 156–157
life, frequency distribution, 155–156
MTTF
exponential distribution, 164–165
life distribution, 163–164
units, average life, 162–163
product life distribution, 155
characteristics, 161
exponential/Weibull/log-normal/gamma, 161
Reliability block diagram (RBD), 165
Risk priority number (RPN), 194, 205
Root sum of squares (RSS), 188–190

S

Sample definitions, 50–51
Scatter plots
case study, 477–478
paired observations, 476–477
Shewhart chart, 327
EMWA chart, comparison with, 331
for individuals on
1σ change in process mean, 327, 329
1.5σ change in process mean, 327, 330
Single Minute Exchange of Die (SMED), 493–494, 497
Single sampling plan (SSP), 422
for normal inspection, 440
for reduced inspection, 442
for tightened inspection, 441
Six Sigma (6σ) system, 4–5
five-step improvement models, 567
five-step road map
current performance, 567
customer requirements, 566
expansion and integration, 568–569
improvements, 567–568
processes and customers, 566
measure
advantages, 564
DPM, 563
quality of, 563–565
organization for
black belts, 569
executive leaders, 569
financial rep, 569
green belts, 569
master black belt, 569
project sponsor (champion), 569
yellow belts, 569
origin of, 561
strategies
design/redesign, 565
management, 565–566
process improvement, 565

themes of, 562–563
two improvement processes, 566
SMED. *See* Single Minute Exchange of Die (SMED)
SPC. *See* Statistical process control (SPC)
Standard Handbook of Machine Design, 185
Statistical process control (SPC), 8, 220, 228, 242, 245, 263, 266–267
implementation
assignable causes elimination, 264–265
attribute charts, use, 265–266
capabilities of processes, 266
cause-and-effect diagram, 264
in-control and not in-control process, 264
manual plotting, 263–264
organization-wide survey, 266
prioritization criterion, 263
process variables, 265
scheme for, 266–267
team organization, 263
type determination, 264
Statistical Quality Control Handbook, 240
Stem-and-leaf (S&L) diagram. *See also* Populations
example of
cells of histogram, 61–62
cumulative counts, 62
distribution percentiles, 62–63
Strategic planning
deployment
core process, 406–407
developing measurement tools, 407
ISO 9000 standards and Balridge Award criteria, 407
performance measures, 406
history of
goals, quality, 403–404
monitoring process, 403
quality methods, growth, 402–403
making of
external parameters, 405
gap analysis, 404–405
internal parameters, 404–405
mission statement, 404
set of values, 404
strategies, 405, 406
vision statement, 404
Supplier Management Handbook, 410
Suppliers
auditing, 415
ISO 9000 standards, 416
certifying, 413
criteria list, 414
choosing, 412–413
Dr. Deming's advice, 412
maximum failure-rate requirements, 415
minimum mean-time-to-failure (MTTF), 415
single and multiple, 411–412
specifications by
ANSI and ASTM, 415
supplied items, 414
Supplies, 409
average outgoing quality limit, 447–448
average sample number of sampling plan
function of lot quality, 434
probability, 434
double sampling plans for attributes
OC curve, 432
scheme of, 431
MIl-STD-105E ANSI Z1.5, 437
sample OC curves for, 446
sample size code letters, 439
sampling plan, 438
switching rules, 438
misconception about, 452
sampling plans and control charts, 452
single sampling plans for attributes
choosing, 426–428
designing, 426
nomograph, 429–430
operating characteristic curve, 422–424
probabilities, 428
statistical sampling for acceptance
categories, 420–421
characteristics, 421
continuous sampling plans, 421

double/multiple sampling plans, 422
inspection, 420–421
variable sampling plans, 452–453
Supply chain optimization
components of, 416
control, 418
improvement
cost reduction and customer needs, 419
information sharing, 419
resource sharing, 419–420
value enhancement, 419
planning phase, 417
traditional and strategic view of purchasing process, 418
trilogy of supplier relationship components, 417
Systems for quality
interrelationships in, 529
management systems, 569–571
models, 509
Baldrige Award criteria, 530–542
Dr. Deming's system, 510–518
Dr. Feigenbaum's system, 527–529
Dr. Juran's system, 518–526
ISO 9000 quality management systems, 543–546
Six Sigma system, 561–569

T

The Team Handbook, 397
Teamwork. *See also* Human resources
basic training
basic tools, quality improvement, 395
breakthrough sequence, 395
magnificent seven tools, 395
plando-check-act (PDCA) cycle, 395
problem-solving process, 395
characteristics, 393
enthusiasm and initiative, 396
perseverance and tolerance, 397
resourcefulness, 396–397
responsibility and selflessness, 396
trust, 396
defining team mission, 393
making work, 398–399
meeting, ground rules for, 397–398
motivation methods
achievement and affiliation, 401
Dr. Deming's theory, 401
equity and expectancy theory, 401
Maslow hierarchy of needs, 401
quality effort, 401
reinforcement theory, 401
objectives in organization, 392
quality control circle (QCCs), 400
reason for, 397
selecting team members, 393
team building, 393
weakness, 394–395
The Team Handbook, 392
team's strength
accountability, 394
adequacy of expertise and resources, 394
direction of, 394
personal characteristics, 394
types of team
cross-functional, 399
process improvement, 399
self-managed, 400
Tests for normality. *See also* Populations
normal probability plot on computer
goodness-of-fit test (*see* Goodness-of-fit test)
mean rank for data, 130, 132
by Minitab software, 130, 132
normal probability plot use
cumulative frequency distribution, 129–131
frequency distribution, 129–130
mean and standard deviation, 130
NPP, 129
Tolerance design. *See* Tolerancing
Tolerancing
assembly, 187
Dr. Taguchi, according to, 186–187
RSS formula, 188–190
setting assembly tolerance, 190
specification limits, 191
traditional approaches, 185–186

Total probability theorem, 79–80
Total Productive Maintenance (TPM), 499–500
Total quality management (TQM), 8
- model, 13
- philosophy components, 14

Total quality system (TQS), 11
- components of, 12

Training. *See also* Human resources
- benefits, 388
- effectiveness
  - improvement tools, 392
  - projects results, 392
- finding resources
  - cost benefit, 391–392
  - qualified instructors, 391–392
  - statistical methods, 392
- methodology
  - design of experiments (DOE), 390
  - peer group, 390–391
- need for
  - basic skills, 387
  - continuous change, technology, 387
  - global competition, 387
  - improve process need, 388
  - workplace diversity, 388
- planning
  - executive, 389
  - management, 389–390
  - quality awareness, 389
  - technical, 390

U

U-chart
- *C*-chart, variation of, 255
- example of, 258
- limits calculation, 255–256
- salvaged holes in castings, data on, 256–257

Union of Japanese Scientists and Engineers (JUSE), 510–511, 519
Upper control limit (UCL), 219–220, 222–226, 242–250, 252–259

V

Variability
- methods of, 49–50
- products, 49
- and quality, 52–53
- sample of units, 49
- science of, 49

Variable control charts, factors for calculating limits, 581
Vendor quality costs, 29–30
Virtual corporations. *See* Boundary-less organizations

W

Western Electric rules, 240

X

$\overline{X}$- and *R*-charts
- assignable causes, 225
- check sheets preparation, 230–231
- control and capability
  - in-control process, 240, 242
  - not-in-control process, 242
  - study, 240, 242
- example of, 224–225
- false alarm in $\overline{X}$-charts
  - in-control process, 231
  - Type I and Type II errors, 231
- frequency of sampling, 233
- instruments preparing, 230
- limits, 225–224
  - calculation, 222–223
  - using "remaining" data, 226
- measurement control charts, 222, 315
  - μ and σ given, 316–319
  - μ given, σ not given, 316–318
- objective of, 226
- patterns, 240–241
- process observation, 227
- process standard deviation, 227–228
- rational subgrouping, 233
  - case study, 234–235
- remaining data, recalculations, 225
- with revised limits, 226
- runs use
  - below and above CL, 238–240
  - down, 238–240
  - Western Electric rules, 240

sample size
central limit theorem effect, 231–232
changes in, 235–236
data with varying, 237–238
determination, 231–232
increasing, 236, 238
sampling and testing, 227
sensitivity improvement
conventional charts, 236
insensitivity advantage, 236
3-sigma limits
cost benefits, 233
process interruption, 232
2-sigma distance and, 232–233
statistical methods, propagation of, 233
trial control limits, 226
uses
as acceptance tool, 229
control process, 228
maintain process, 228–229
process maintaining, 228
as troubleshooting tool, 229
variable selection, 229–230
warning limits, use, 238–239

$\overline{X}$ and S-charts
calculations of, 243–244
control limits, 242
sample size effect, 242
standard deviation, 242
$\overline{X}$-chart
limits derivation
central limit theorem, 298
in-control and not-in-control, 298–299
normal distribution, 298
process distribution effect, 299
result of, 300–301
3-sigma limits, 299–300
measurement control charts
data for, 320–321
individuals and successive differences, 321–322
limits for, 320
moving range (MR) chart, 320
operating characteristics (OC)
computing, 305–306
curves example, 304–305
process mean, change in, 305
with 3-sigma limits functions, 306–307